Innovative Pharmaceutical Excipients: Natural Sources

Jailani. S • Kuldeep Vinchurkar
Meghraj Suryawanshi
Sheetal Mane
Editors

Innovative Pharmaceutical Excipients: Natural Sources

Springer

Editors
Jailani. S
R&D Formulation Department
Alpha Pharma Industries
Rabigh, Saudi Arabia

Meghraj Suryawanshi
Department of Pharmaceutics
Sandip Institute of Pharmaceutical
Sciences (SIPS), Affiliated to Savitribai
Phule Pune University (SPPU, Pune)
Nashik, Maharashtra, India

Kuldeep Vinchurkar
Department of Pharmaceutics
Professor and HOD, Sandip Institute of
Pharmaceutical Sciences (SIPS), Affiliated
to Savitribai Phule Pune University
(SPPU, Pune)
Nashik, Maharashtra, India

Sheetal Mane
NMT Gujrati College of Pharmacy
Indore, Madhya Pradesh, India

ISBN 978-981-96-7958-4 ISBN 978-981-96-7959-1 (eBook)
https://doi.org/10.1007/978-981-96-7959-1

This Springer imprint is published by the registered company Springer Nature Singapore Pte Ltd.
The registered company address is: 152 Beach Road, #21-01/04 Gateway East, Singapore 189721, Singapore

Foreword

In the ever-evolving realm of pharmaceutical sciences, the search for safer, more effective, and sustainable drug delivery solutions has brought natural excipients to the forefront of innovation. *Innovative Pharmaceutical Excipients: Natural Sources* stands as a comprehensive exploration of this promising frontier, shedding light on the transformative potential of naturally derived excipients in modern pharmaceutical development.

This edited volume is a reflection of the growing global interest in harnessing the therapeutic and functional versatility of natural polymers, lipids, gums, proteins, and other biologically sourced materials. Through a carefully curated series of chapters authored by eminent scientists, researchers, and academicians, the book presents a holistic view of the sources, properties, applications, and regulatory perspectives surrounding natural pharmaceutical excipients.

Each chapter intricately details the journey from natural source to pharmaceutical utility—highlighting advancements in extraction techniques, characterization methods, and novel applications in oral and controlled release formulations. With topics spanning plant-derived polymers, marine-based materials, and polysaccharides to regulatory frameworks and future directions, this volume not only enriches academic discourse but also offers practical insights for formulation scientists and industry professionals.

As the pharmaceutical industry leans toward greener, patient-friendly, and biocompatible solutions, the role of natural excipients becomes increasingly crucial. This book serves not only as an academic resource but also as a catalyst for future research and innovation in excipient science, ensuring that sustainability and efficacy remain at the heart of pharmaceutical advancement.

I extend my sincere gratitude to the editors, contributors, and all collaborators involved in this endeavor. May *Frontiers in Pharmaceutical Excipients: Natural Sources and Biomaterials—Volume I* ignite new ideas, foster collaboration, and inspire continued exploration in the fascinating world of natural excipients.

Authors

Department of Pharmaceutics
Sandip Institute of Pharmaceutical Sciences (SIPS)
Affiliated to Savitribai Phule Pune University (SPPU, Pune)
Nashik, Maharashtra, India

Kuldeep Vinchurkar

Preface

Pharmaceutical excipients have long played a vital role in drug formulation, evolving from simple inert carriers to functional entities capable of enhancing solubility, stability, bioavailability, and patient compliance. In recent years, the spotlight has increasingly turned toward excipients derived from natural sources—driven by the global demand for safer, biocompatible, and eco-friendly alternatives in pharmaceutical development.

This edited volume, *Innovative Pharmaceutical Excipients: Natural Sources,* represents a comprehensive exploration of the vast potential of natural polymers, lipids, gums, and other biomaterials in pharmaceutical applications. Through the combined efforts of renowned researchers, academicians, and industry experts, this book offers critical insights into the sourcing, extraction, characterization, and application of natural excipients across a wide spectrum of drug delivery systems.

From plant-derived materials and marine-based polymers to functional polysaccharides and peptides, the chapters in this volume illuminate the diverse roles that natural biomaterials can play—not just as carriers, but as active participants in modulating drug release, enhancing therapeutic efficacy, and supporting novel formulation strategies. Regulatory perspectives and future trends further enrich this compilation, making it a well-rounded resource for contemporary pharmaceutical science.

This book is intended to serve as a valuable reference for formulation scientists, researchers, academicians, and students in the field of pharmaceutics and drug delivery. It also aims to foster a deeper understanding of the potential of natural excipients and promote their integration into next-generation pharmaceutical products.

As editors, we extend our sincere appreciation to all contributors who have shared their knowledge, research, and passion in developing this volume. We hope this book not only reflects the current advancements in natural excipient science but also inspires future exploration into sustainable, innovative, and patient-centric pharmaceutical solutions.

We invite you to explore the pages of *Innovative Pharmaceutical Excipients: Natural Sources* and join us in this exciting journey toward greener and smarter pharmaceutical formulation.

Rabigh, Saudi Arabia	Jailani. S
Nashik, Maharashtra, India	Kuldeep Vinchurkar
Nashik, Maharashtra, India	Meghraj Suryawanshi
Indore, Madhya Pradesh, India	Sheetal Mane

Acknowledgments

The realization of *Innovative Pharmaceutical Excipients: Natural Sources* is the result of collective effort, intellectual synergy, and unwavering support from numerous individuals and institutions. As editors, we wish to extend our deepest gratitude to everyone who has contributed to the successful completion of this volume.

First and foremost, we express our sincere thanks to the contributing authors, whose scholarly expertise, innovative research, and unwavering dedication have been pivotal in shaping the richness and depth of this book. Your contributions have brought forth a comprehensive and interdisciplinary understanding of natural excipients and their growing relevance in pharmaceutical sciences.

We are equally indebted to the reviewers who graciously devoted their time and expertise in critically evaluating the submitted chapters. Your constructive insights and suggestions have played a vital role in enhancing the academic quality and scientific rigor of this work.

Our heartfelt appreciation goes to Springer's editorial and publishing team for their professionalism, guidance, and consistent support throughout the publication process. Your efforts have been instrumental in turning this vision into a high-quality academic resource.

We are grateful to the academic institutions, research laboratories, and industry collaborators who supported the authors and editors in their research endeavors. Your encouragement and provision of a conducive environment for scholarly work have significantly contributed to the advancement of knowledge in this field.

Lastly, we express our profound thanks to our families, friends, and colleagues for their constant encouragement, patience, and support throughout this journey. Your belief in our vision and your understanding of our commitment to this project have been a source of strength and inspiration.

This book stands as a symbol of collaborative achievement, and we hope it will serve as a valuable reference and source of inspiration for researchers, academicians, and professionals in the field of pharmaceutical excipients and drug delivery science.

Thank you.

Kuldeep Vinchurkar
Sheetal Mane
Meghraj Surywanshi
Jailani. S

Contents

About the Editors and Contributors

About the Editors

Kuldeep Vinchurkar Department of Pharmaceutics, Professor and HOD, Sandip Institute of Pharmaceutical Sciences (SIPS), Affiliated to Savitribai Phule Pune University (SPPU, Pune), Nashik, Maharashtra, India

Sheetal Mane NMT Gujrati College of Pharmacy, Indore, Madhya Pradesh, India

Meghraj Surywanshi Department of Pharmaceutics, Sandip Institute of Pharmaceutical Sciences (SIPS), Affiliated to Savitribai Phule Pune University (SPPU, Pune), Nashik, Maharashtra, India

Jailani. S R&D Formulation Department, Alpha Pharma Industries, Rabigh, Saudi Arabia

Contributors

Sneha Agrawal Bharati Vidyapeeth's College of Pharmacy, Navi Mumbai, Maharashtra, India

L. Priyanka Dwarampudi JSS College of Pharmacy, JSS Academy of Higher Education & Research, Ooty, Tamil Nadu, India

Kevinkumar Garala School of Pharmaceutical Sciences, Atmiya University, Rajkot, Gujarat, India

Pranjal Gujarathi Department of Pharmacology, Vidhyadeep Institute of Pharmacy, Vidhyadeep University, Surat, Gujarat, India

Jatin Kumar Chitkara University School of Pharmacy, Chitkara University, Baddi, Himachal Pradesh, India

Neha Minocha School of Medical and Allied Sciences, K.R. Mangalam University, Gurugram, Haryana, India

Nikunjkumar Parekh Department of Pharmaceutical Chemistry and Analysis, Ramanbhai Patel College of Pharmacy, Charotar University of Science and Technology, CHARUSAT Campus, Anand, Gujarat, India

Anil Pawar Sennos Biotech Private Limited, Risod, Maharashtra, India

Prashant Pingale Department of Pharmaceutics, Gokhale Education Society's Sir Dr. M. S. Gosavi College of Pharmaceutical Education and Research, Nashik, Maharashtra, India

Satish Polshettiwar School of Health Sciences and Technology, Dr Vishwanath Karad, MIT World Peace University, Pune, Maharashtra, India

Hardik Rana Department of Pharmaceutics, Anand Pharmacy College, Anand, Gujarat, India

Pragya Sharma Amity Institute of Pharmacy, Amity University Rajasthan, Jaipur, Rajasthan, India

Meghraj Suryawanshi Department of Pharmaceutics, Sandip Institute of Pharmaceutical Sciences (SIPS), Affiliated to Savitribai Phule Pune University (SPPU, Pune), Nashik, Maharashtra, India

Poonam Taru School of Pharmacy, Vishwakarma University, Kondhwa Pune, Maharashtra, India

Kuldeep Vinchurkar Department of Pharmaceutics, Professor and HOD, Sandip Institute of Pharmaceutical Sciences (SIPS), Affiliated to Savitribai Phule Pune University (SPPU, Pune), Nashik, Maharashtra, India

Introduction to Pharmaceutical Excipients: History and Evolution

1

Anil Pawar, Alka Zade, and Harshada Khandelwal

1.1 Introduction to Pharmaceutical Excipients

Pharmaceutical excipients have been in use from time immemorial when natural products were used in making medical remedies. Practitioners of the earliest civilisations employed honey, waxes, plant-derived gel and resins as vehicles to improve the stability, taste and tractability of plant-based medicines. With the onset of more advanced forms of drug dosage formulation and the growth of the field of modern pharmaceutical science, the uses of excipients changed from simple solid fillers to active components of the drug formulation delivery systems. The vast improvements of the nineteenth and twentieth centuries can be largely attributed to industrialisation of the pharmaceutical manufacturing and synthetic chemistries which enhanced excipients' advancement. It was during this time that synthetic excipients like polyethene glycols were first used and semi-synthetic derivatives like cellulose ethers were incorporated into formulations based on their superior physical and chemical characteristics to the naturally occurring compounds. In addition, this period saw changes in the formulation processing techniques, which improved the reproducibility of the final dosage form. This enhanced the production of stable dosage forms, controlled release and patient acceptability of the products [1, 2].

Over the last few decades, shifts in the types of dosage forms, technologies and trends towards patients' preferences have contributed to the evolution of excipients. New vehicles have therefore been designed and optimised to meet particular tasks including improving the dissolution of biopharmaceutics classification system (BCS) class II drugs, targeting the drug delivery and controlling the release profile of the drug. Further, the availability of biopharmaceuticals has led to the need to incorporate their compatible excipients, such as peptides and monoclonal antibodies. The

A. Pawar (✉) · A. Zade · H. Khandelwal
Sennos Biotech Private Limited, Risod, Maharashtra, India
e-mail: Anil.Pawar@sennosbiotech.com

J. S et al. (eds.), *Innovative Pharmaceutical Excipients: Natural Sources*,
https://doi.org/10.1007/978-981-96-7959-1_1

evolution of pharmaceutical excipients represents not only important contributors to drug product performance, safety and stability but also as highly technical and engineered components in the development of modern drugs. Such evolution highlights their relevance in progressing the field of pharmaceutical science as well as countering current healthcare challenges [3, 4].

1.1.1 Definition and Significance of Excipients

Pharmaceutical excipients are a class of adjuvants that make up a proportion of pharmaceutical preparations and are inactive solids, liquids or gases that allow the formulation of active pharmaceutical ingredient (API) to be balanced for quality, efficacy, performance and acceptability. Some of them may be virtually inactive in the pharmacologic sense but are essential for the processing, stability and performance of the product to guarantee that the drug produces the requisite therapeutic effect. They have a myriad of uses including serving as excipients that bind all the components together to form tablets, others serving to disintegrate once the tablet is ingested and yet others that simply provide bulk where required or as lubricants to ease the manufacturing processes by having friction. Furthermore, excipients play the role of decorative and sensory qualities of the drug when used as colouring agents, flavouring agents and sweetening agents with the view to improving patients' compliance. Classification of pharmaceutical excipients is carried out based on the functionality of the excipients, their source and their chemical composition. From a functional perspective, they are classified as binders, fillers, disintegrants, preservatives, solubilisers, coating agents, etc. According to the source of derivation, excipients can be natural semi-synthetic or synthetic, which includes lactose and starch as natural sources, cellulose derivatives as semi-synthetic options and polyethene glycol as synthetic options. The selection of excipients in a formulation depends on certain key factors which involve compatibility with API and other formulation ingredients, toxicity profile and regulatory requirement as well as the demand put forward by the drug delivery system [5, 6].

Recent developments in excipients have involved new excipients with capabilities to respond to new drug delivery complications. Smart excipients, for example, are capable of changing their properties in response to stimuli like pH and temperature and therefore can control the rate and site of drug release. Co-processed excipients can be defined as two or more excipients whose functionalities are handled in one entity, increasing performance and decreasing formulation complexity. In addition, advances in the use of biodegradable and environmentally friendly excipients also demonstrate the industry's environmental consciousness. However, they are not without some drawbacks; for instance, they can be incompatible with the APIs, derived from natural resources with variations, and have to meet a variety of regulatory requirements. However, due to their presence in formulations for the production of pharmaceuticals, they remain quite important in determining the therapeutic value, safety of the patient and feasibility of production. The value of excipients, therefore, is rooted in utility in handling multifaceted issues in formulation especially where the API has issues with solubility, stability or bioavailability. Most of the advanced drug delivery systems like sustained-release tablets, transdermal patches or even orally dispersing systems are made possible by the excipients, which create the means for efficient and accurate control of the rate and place of the drug elution process. Similarly, they are crucial for the assurance of product quality, as well as for reviewing the appropriateness of the batch manufacturing processes and verification procedures, thus helping the organisation to comply with established regulations and to provide quality products in batches. Over the years, and as there is further research in pharmaceutical science, excipients are not only viewed as just carriers of the dosage form but are rather appreciated for their ability to contribute to the creation and development of new dosage forms that will help make personalised medicines possible together with other drug delivery systems [7, 8].

1.1.2 Historical Perspective on Excipients

The evolution of pharmaceutical excipients can arguably be linked with the growth of medical applications from the use of native resources to the present more advanced technological systems for drug delivery. In the early days, excipients referred to naturally derived materials such as honey, water, animal and plant fats, and extractions that were utilised to make raw remedies. Many of these substances had practical purposes including acting as a binder, sweetener or base material for the active components. For instance, the Egyptians and Greeks mentioned the use of beeswax and gums as carriers and stabilisers for various herbs used for medicine. Ayurvedic and traditional Chinese medicines also used a similar form of natural materials in their therapeutic preparations. The nineteenth-century Industrial Revolution played a major role in changing the role of the excipients as the whole pharmaceutical sector moved from craft-based production to a more industrialised process. These disclosures occurred in an era of excipients with more uniform formulations hailing from natural raw materials such as starch, lactose and gelatin. At the same time, the findings of the analytical chemistry work created the basis for the rationally controlled selection of excipients in drug formulations. Excipient advancement in the twentieth century was revolutionised by the findings of synthetic and semi-synthetic ones. The usage of cellulose derivatives (hydroxypropyl methyl cellulose [HPMC]) and the synthetic polymer (polyethene glycol) increased the variable properties of the drug delivery system because of the controlled release of drugs, solubility of the active pharmaceutical ingredient and stability of the API [9].

The middle of the twentieth century was characterised by increased specialisation whereby excipients were designed to fit specific functional standards such as disintegrating agents for fast-dissolving systems, and emulsifying agents for emulsified drugs. As the products became more complex during this period with the appearance of sustained-release and parenteral formulations, excipients had to be of high performance and with predictable attributes. At the same time, the rules of regulation became more concrete and required increased quality control and safety assessment for additives, which led to the establishment of the pharmacopoeia. Processes of the modern development of additives are associated with modern technologies such as biotechnology, nanotechnology and personalised medicine. New agents, including co-processed agents and stimuli-sensitive smart agents activated by stimuli such as pH or temperature, are being designed to overcome the issues closest to the delivery of advanced APIs which can be unstable, poorly soluble small molecules and biologics. Also, advancement in sustainable and green chemistry has forced researchers to go for biodegradable and environmentally friendly excipients. This evolutionary pathway of excipients proves the exceptional importance of these compounds for the development of pharmaceutical science and better treatment outcomes for patients [10].

1.1.3 Evolution of the Excipients Concept

The connotation concerning excipients has changed over hundreds of years from simple fillers of active substances to multifunctional components that are crucial to drug delivery, stability and patient acceptance. In the earliest times when the formal practice of medicine had not developed, excipients were natural products like honey, water, animal fats and vegetable-derived gums which were used in preparing the herbal preparations for administration. These initial excipients served mainly for dilution, coalescing or solvent purposes without bothering how their physicochemical characteristics or their impacts on API behaved. Their role was only to present, store and stabilise traditional medicine to a certain extent. The increasing growth of pharmaceutical science in the Industrial Revolution led to the expanded use of excipients. Since the quantity produced became a mass item, there was a need to have a uniform and repeatable formula, which resulted in refined and standardised natural excipients such as lactose, starch and gelatin. During this period, excipients were appreciated not only as non-active components and carriers but also as the factor that influences the manufacturability and physical stability of preparations. With the improving comprehension of material science and chemistry, skilled experts began to seek new excipients that can help achieve specific desired characteristics like solubility and stability duration. The emergence of cellulose derivatives—half synthetic category of excipients—was another significant step in the progress of the excipients' classification as they demonstrated controlled performance yet had eased compatibility with numerous APIs [11].

The centennial change in excipients came with the advent of synthetic polymers and the changing face of drug delivery systems in the twentieth century. This involves the synthesis of polyethylene glycols (PEGs), polyvinyl alcohols and other synthetic polymers to produce controlled-release, sustained-release and target delivery system formulations. With these advances, excipients were indicated to contribute to the optimisation of the pharmacokinetics and pharmacodynamics of APIs and turn from carriers to functional components of dosage forms. The discovery of biologics and biopharmaceuticals simply added to the repertoire of excipients. This further required materials capable of supporting the stability of proteins, peptides and other bioactive molecules while retaining their function.

In the past few decades, the progress of excipients has been forged by creativity to some of the existing problems within today's advancing field of medicine. The new line of thinking that has developed is known as 'smart excipients', which are materials that can respond to signals such as pH, temperature and enzyme activity to release drugs at the desired site and rate. To enhance functionality and minimise formulation complexity, the concept of co-processed excipients, which possess properties of more than one individual excipient, has evolved. Additionally, the highly valued paradigms of patient- and sustainability-oriented pharmaceutical developments have fuelled the search for more appropriate excipients for improving the sensorial characteristics, functionality as well as ecological footprint of dosage forms. Nowadays, excipients are not considered complementary to the active substance but as critical determinants of the drug's effectiveness, stability and compliance, proving their essential part in the modern development of pharmaceutical technology [12].

1.2 Evolution of Excipients Over Time

Since the emergence of pharmaceutical science, excipients have evolved from being basic naturally occurring substances to being the most advanced multifunctional coated systems. This development can though be classified under broad categories, which are determined by the progress of medicine, chemistry and technology.

1.2.1 Early Excipients: Ancient and Pre-modern Usage

The use of pharmaceutical excipients dates back to early Egyptian civilisation when natural materials were used in preparing medicines. More than 1500 years ago, honey, waxes, animal fats and water were utilised as vehicles for herbal and mineral preparations. For instance, honey had other uses in addition to sweetening the product and in the process dissolved the bitter hits of most herbal preparations; it also preserved the product by acting as an antimicrobial agent. Similarly, beeswax was used to create consistency and organisation for ointments, whereas plant-derived gummy substances such as gum arabic and tragacanth were used as binders and emulsifiers. These materials though crude were useful in the manipulation of early drugs and their storage. Early medicines in the context of the Egyptian civilisation are as follows. Medicines in the ancient Egyptian civilisation involved some kind of preparation of plant materials and sometimes animal products. The ancient Egyptians used simple solvents such as water or wine for the preparation of decoctions based on herbs. Fats and oils were used to prepare topical solutions. Greeks and Romans also developed other uses of excipients by including clay and chalk to hold powders, and by creating simple pills, with applications made from plant extracts. Hippocrates and Dioscorides, Ayurvedic and traditional Chinese scripts contain copious evidence of natural excipients such as sugars and starchy materials for their adhesive or protective or suspending capabilities. Throughout this period, the incorporation of excipients was arbitrary, with little or no scientific knowledge informing decisions. The history of the compounding of medicines in apothecary shops saw the elaboration of pre-modern excipients. Anyone who uses the history of materia medica, today's apothecaries or pharmacy, has come to realise the importance of how those functional materials are involved in their work such as using sugar syrup for stabilisation and sweetening of liquid preparations in addition to incorporating milk-derived lactose as a diluent in powders. New techniques for purifying and standardising excipients in medieval alchemy and pharmacology were developed and used in the preparation of dosage forms; these new techniques included the production of purified gelatin coming from animal sources and the extraction of plant-based resins. Natural acids such as citric and tartaric acids were used as additional acidifying agents in formulations for the stabilisation of the products.

While it would be generations before the specific functions of various excipients would be understood, their purpose in aiding in the preparation, storage and administration of medications seemed self-evident. These early materials provided the groundwork for the more deliberate search and study of excipients, and the future innovations that rose out of the excipient focus. The historical usage of excipients in pre-modern and ancient systems affirms the gravity of the excipient's place in the formulation history, signifying the move to enhanced and complex systems of medication delivery.

1.2.2 Functional Evolution in Modern Times

Changes observed in the functional roles of excipients in recent times relate to the advancement in the field of pharmaceuticals and technology (Table 1.1). With the advent of industrialised drug production in the nineteenth and beginning of the twentieth century, excipients evolved from crude 'bulking agents' to components with functionality designed to respond to issues arising from the development and manufacture of drugs. This came with purified natural materials of production including starch, lactose and gelatin and was associated with more enhanced consistencies in the manufacturing processes. With the advent of semi-synthetic excipients like cellulose derivatives, like carboxymethyl cellulose and hydroxypropyl methyl cellulose, a new era was ushered in with regard to functionality, controlled-release profiles and poor solubility of water marker. These excipients also made possible the formation of sustained-release tablets, which was a milestone in the oral dosage forms. The mid of this twentieth century marked the entry of synthetic excipients; the polymers are polyethene glycols (PEGs), polyvinylpyrrolidone (PVP) and polymethacrylates. They supplied the means to prepare formulations possessing delayed- or sustained-release characteristics of the drug. Polysorbates and sorbitan esters are synthetic surfactants and emulsifiers used in stabilising emulsions, enhancing the solubility of lipophilic active ingredients and for injectables and ophthalmic products. In this period, it was not sufficient and reasonable for excipients to be considered mere vehicles or carriers but components that played

Table 1.1 Timeline of key developments in the use of excipients

Time	Key development	Details
Ancient era	Use of natural substances	Honey, beeswax and gums are used for stability, binding and delivery in herbal medicines
2000 BCE	Egyptian formulations	Use of fats, waxes and water as excipients in ointments and decoctions
400 BCE	Greek and Roman advancements	Gums, chalk and plant extracts are used for binding and coating pills
1000 CE	Introduction of sugar-based syrups	Sugar is used as a stabiliser and sweetener in liquid formulations
Sixteenth–seventeenth century	Refinement of natural excipients	Apothecaries began standardising excipients like starch, lactose and gelatin
Eighteenth century	Industrial Revolution	Mass production of drugs necessitated consistent excipients for manufacturability and stability
Mid-nineteenth century	Start of semi-synthetic excipients	Development of cellulose derivatives such as methylcellulose for enhanced functionality
1900s	Standardisation in excipient use	Introduction of lactose and starch as primary diluents and binders in tablet formulations
1930s	Introduction of surfactants	Polysorbates and sorbitan esters were developed for emulsion stabilisation and solubility enhancement
1950s	Growth of synthetic polymers	Polymers like PEG and PVP were introduced for controlled drug release and solubility improvements
1960s–1970s	Controlled-release technology	Introduction of matrix-forming and coating agents like HPMC for sustained-release formulations
1980s	The rise of multifunctional and co-processed excipients	Improved compressibility and flow properties for high-speed manufacturing
1990s	Biodegradable polymers	Introduction of PLGA for controlled release and targeted delivery systems
2000s	Smart excipients	Development of pH-sensitive and temperature-responsive materials for site-specific drug delivery
2010s	Advanced excipients for biologics	Stabilisers like trehalose and mannitol are used for protein-based therapeutics and vaccines
2020s	Focus on sustainability and personalised medicine	Development of eco-friendly, biodegradable excipients and materials tailored to patient-specific formulations

a significant role in the performance of dosage forms, including the stability, release and efficacy of APIs.

The last 50 years of the twentieth century witnessed a new emphasis on the concept of multifunctional and co-process excipients. These innovations coalesced several desirable functions of excipients into one material with enhanced compression strength, flow properties and solubility with APIs. Co-processed excipients helped reduce formulation complexity and scale up manufacturing efficiency to meet the increasing production capacity requirement of faster throughput lines. Also, the improvement of regulatory science makes the enhanced quality standards for excipients to guarantee that they meet safety requirements and are equivalent to other batches. In modern times, functional changes in excipients have been induced by the requirement of enabling advanced drug delivery systems such as transdermal patches, nanoparticles and liposomal systems. Innovations involving the utilisation of biodegradable polymers such as poly(lactic-co-glycolic acid) (PLGA) have created important excipients for the controlled release and deliverance of small molecules and biologics. Freezing constituents and stabilisers like trehalose and mannitol became indispensable for retaining the stability of a number of protein therapeutics and vaccines. In addition, the intelligent excipients which are sensitive to the pH value and/or temperature have also allowed the site-specific drug release, improving therapeutic effectiveness with reduced side-effects in the systemic circulation. Moreover, due to the advancement of targeted drug delivery systems due to the concept of personalised medicines, the concept of smart excipients that are best suited for the specific patient with desired pharmacokinetics is in the pipeline. Furthermore, topical imperatives towards sustainability have increased the usage of environmentally friendly and biodegradable products towards the achievement of pharmaceutical practices. Excipients of the modern world have shifted from simple non-active components that are added to the recipe to inactive multifaceted systems that are viewed as critical components of complicated systems for the targeted delivery of drugs, underlining the significance of excipients in advancing pharmaceutical technology.

1.3 Regulatory Frameworks and Technological Advances in Excipients

Specific legal requirements governing excipients have been developed with a view of safeguarding the health of the public through standardisation of safety, quality and overall performance of these components. These regulations are important as excipients greatly affect the stability, solubility and effectiveness of the products of the formulation.

Excipients' regulation is a critical factor in compliance with safety, quality and excipient consistency in pharmaceutical products. Excipients thus are very important in the formulation and functionality of a pharmaceutical product no matter the fact that they may not possess a therapeutic effect on their own. Limits for excipients are set out by several national and international regulatory bodies such as the

US Food and Drug Administration (FDA), European Medicines Agency (EMA) and International Council for Harmonisation (ICH) that have set guidelines that could protect the general public from harm caused by unwanted interactions of excipients while encouraging the development of innovative drug delivery methods [13].

1.3.1 US Food and Drug Administration (FDA)

In the United States, the FDA has formally recognised excipients as the other components in formulations of drugs. These excipients are essential in controlled drug delivery influencing such factors as dissolution rate, solubility, stability and release characteristics, but do not possess pharmacological activity. The FDA provides regulatory oversight under several key frameworks:

- **Generally Recognized as Safe (GRAS):** Many excipients are GRAS, which implies that they have been widely used for a long time or there is some evidence that they are safe to use. The other excipients known as Generally Recognized as Safe excipients are not subjected to lengthy pre-market approvals but the manufacturer must consider that the use of such excipients does not go beyond certain thresholds [14].
- **Inactive Ingredient Guide (IIG):** The IIG, which is sustained by the FDA, contains names of excipients that are deemed safe to be used in the formulation of drugs. It gives manufacturers all including safety information of excipients, their allowed concentration and usage according to dosage forms.
- **New Excipients:** The FDA demands full safety information for those excipients which have not been widely used or not used at all in the drug products. This encompasses acute and chronic toxicity studies, degradation rates to assess the stability of the candidate excipient and human clinical data concerning the impact of the excipient on the final dosage form. The use of excipients in biologic products such as vaccines and protein therapy is slightly more complicated due to the sensitivity of the active protein to the conditions of the environment, temperature, pH and light. Increased regulatory scrutiny of excipients is maintained to guarantee that these components will not in any way, affect the stability or performance of the biologic drug [15].

1.3.2 European Medicines Agency (EMA)

The European Medicines Agency (EMA) from within the European Union (EU) is responsible for the regulation of excipients. The regulation of excipients is led mainly by the European Pharmacopoeia (Ph. Eur.), which describes quality requirements for excipients and their application in medicines. The Ph. Eur. contains a list of monographs of excipients that show the requirements for their identity, purity, quality and methods of test.

- **European Pharmacopoeia (Ph. Eur.):** Characterisation of substances used as excipients in medicinal products in Europe is well described in a reference document called Ph. Eur. It gives special test methods for determining whether excipients have the desirable quality and safety in order. For instance, in parenteral (injectable) formulations, the choice of the excipients must be of very high purity since the contaminant would be dangerous and would affect the stability and sterility of the final product.
- **Authorisation of New Excipients:** EMA declares that any new excipient or an excipient that is used in a novel drug product must get approval from the EMA. In one case the manufacturers must provide considerable information about the safety profile of the excipient, such as toxicological data, stability studies and interaction with the API. Excipients in biologic drugs are especially scrutinised for their safety as they may have an impact on active pharmaceutical ingredients and/or the patient. The EMA also ensures the chosen excipients for the drug product will not interfere with the mechanisms of drug release and stability of the drug substance, its efficacy and bioavailability.

1.3.3 International Council for Harmonisation (ICH)

The International Council for Harmonisation (ICH) is an organisation whose aim is to harmonise guidelines for pharmaceutical products in some of the world's key markets, such as the United States, Europe and Japan. ICH guidelines have played a significant role in standardising excipient testing and quality across these regions.

- **ICH Q6A and Q3C:** ICH guidelines Q6A on 'Specifications: Test Procedures and Acceptable Criteria for New Drug Substances and New Drug Products', and Q3C on 'Impurities: Guideline for New Substances and Products' both contain information on the maximum level of impurities in excipients and the testing protocols for quality of excipients. These guidelines ensure that excipients meet rigorous standards for purity and safety, and they help harmonise regulatory requirements globally.
- **Stability and Toxicology:** Among ICH guidelines, specific references concern stability and toxicology of the excipients, together with recommendations on how to evaluate the stability of the excipients with different stress conditions and on possible toxicological interactions of the excipients with the active ingredients. These guidelines are aimed at guaranteeing that copy substances do not cause the presence of toxic products in the production of the formulation or cause the destabilisation of the drug. These steps make up the International Council for Harmonisation, or ICH, whose mission is to synchronise these procedures applied in distinct geographical areas to reduce the complexities of gaining marketing approval while sustaining high standards of safe and efficient drugs.

1.3.4 Other Regional Regulatory Bodies

Beyond the FDA, EMA and ICH, several other regulatory agencies contribute to the global regulatory framework for excipients [16]:

- **World Health Organisation (WHO):** In terms of exposure, the WHO is instrumental in setting standards for excipients, especially concerning developing countries and generic products. WHO guidelines ensure that excipients will not be health hazardous to the population and will fulfil quality benchmarks to encourage the manufacture and availability of drugs in low-resource health systems. It also supports the use of appropriate excipients that make important medicines stable and safe.
- **Health Canada:** In Canada, excipients come under the directives of the Health Products and Food Branch (HPFB), the agency which befits the excipients according to the standards of Pharmacopoeia of Canada. Like in the European system, only excipients that conform to the set criteria of identity, purity and quality can be used in drugs in Canada.
- **Pharmaceuticals and Medical Devices Agency (PMDA):** In Japan, only the PMDA oversees excipients used in pharmaceutical products. To facilitate this, the agency operates in conjunction with the ICH to ensure that the standards set for excipients of medicines in Japan are as standard as possible internationally. The PMDA has also offered its recommendations on the suitability and safety of the excipients in compound preparations, especially those for biological products and other special category products.

1.3.5 Emerging Challenges and Future Directions

As the pharmaceutical industry continues to evolve, new challenges in excipient regulation are emerging. With the rise of biologics, gene therapies and personalised medicine, there is an increasing demand for specialised excipients that can meet the unique needs of these advanced therapies. Regulatory bodies are adapting their frameworks to address these challenges, including more stringent requirements for biocompatibility, biodegradability and targeted delivery [17].

1.3.5.1 Nanotechnology

There are new trends with nanoparticles and nano-based excipients in the system, especially for the administration of BCS class II drugs. Presently, there are no clear guidelines for the uses of nanomaterials, and the FDA and EMA are evaluating the risks and merits of these sophisticated excipient technologies.

- **Sustainability:** Pressure is growing for pharma companies to swap environmentally unfriendly excipients for more sustainable materials. Those which are renewable or have a lower ecological impact during their manufacturing, are now being regarded in the corridors of regulation [18].

1.3.5.2 Technological Advances Impacting Excipients

Excipients have achieved a new status in pharmaceutical sciences due to improved technology and are no longer just an inert carrier but a functional component in drug delivery, stability and acceptance. The latter breakthrough is concerned with nanotechnology-based excipients which have facilitated the fabrication of nanoscale pharmaceutical delivery systems like liposome, solid lipid nanoparticle (SLN) and polymeric micelles. These excipients enhance the solubility, bioavailability and delivery of the class of drugs labelled as poorly soluble or highly sensitive. New advances in material science have also identified smart excipients which can respond to physiological stimuli like pH, temperature or enzyme activity for the release of drugs utilising a controlled or site-specific manner. For instance, stimuli-responsive polymers enable the controlled drug delivery system in targeted tissues to minimise symptoms caused by a drug dispersed throughout the body and improve curative effect. The recent introduction of a class of increasingly popular pharmaceuticals—biologics, monoclonal antibodies and RNA-based drugs—has stimulated the need for excipients that can stabilise such complex structures. Previous to this, cryoprotective agents, stabilisers and surfactants are now standard, ensuring manufacturing, storage and administration stability and activity of the biologics. In addition, it has been possible to use 3D printing technologies to develop and customise the excipients to accurately design the drug delivery system like the tablets in terms of dissolution profile and multiple active ingredient combinations. Computational modelling and recently machine learning approaches to support formulation development are also presented and describe how new excipients can be successfully designed based on their prediction of physicochemical characteristics and interactions with APIs. Convenience factors have jumped up as another core consideration, whereby excipients themselves coming from renewable resources, or those being designed to biodegrade into the environment, are becoming a standard strategy. In addition, excipients as active components are increasingly incorporated into fixed-dose combination agents as well as into complicated preparations like implants as well as transdermal patches, hence calling for high-performance efficient materials that conform to regulatory measures. These are not only enhancing the functional utility of excipients, but are also solving problems relating to formulation, specific needs for patient treatment and disease and poverty-stricken regions of the world, thus proving itself to be a vital element in the progress of pharmaceutical sciences [19].

Table 1.2 presents a comparative analysis of regulatory guidelines for excipients across different regions, highlighting the varying approaches taken by regulatory authorities such as the FDA in the United States, the EMA in the European Union and the PMDA in Japan. These guidelines ensure that excipients meet the necessary safety, efficacy and quality standards required for pharmaceutical applications.

Table 1.2 Comparative analysis of regulatory guidelines on excipients

Aspect	United States (US)	European Union (EU)	Japan
Regulatory authority	Food and Drug Administration (FDA)	European Medicines Agency (EMA)	Pharmaceuticals and Medical Devices Agency (PMDA)
Key frameworks	Generally recognized as safe (GRAS), inactive ingredient guide (IIG)	European Pharmacopoeia (Ph. Eur.)	Japanese Pharmacopoeia
Approval for novel excipients	Requires extensive safety and toxicological data, stability studies and pre-market approval	Comprehensive evaluation including toxicology, compatibility and quality specifications	Detailed review involving safety, stability and compliance with global standards
Focus areas	Bioavailability enhancement, excipient interaction with APIs and biologics	Quality specifications, controlled release and innovative delivery systems	Harmonisation with ICH guidelines, advanced formulations and international standards
Harmonisation	Collaborates globally but has distinct GRAS/IIG systems	Relies on European Pharmacopoeia while aligning with ICH standards	Strong alignment with ICH and global practices

1.4 Natural Sources of Pharmaceutical Excipients

The use of naturally derived products for pharmaceutical excipients defines drug formulation, providing biocompatibility, tunable properties and renewability. These excipient categories are obtained from plant, animal and marine sources that play relevant functions, including binder, diluent, emulsifier and stabiliser functions. Examples include plant-origin cellulose, starch and gums, animal-origin gelatin and chitosan, magnesium and calcium stearate mineral-origin talc and calcium carbonate. Such origin also fits regulatory and patient concerns about using natural and environmentally friendly products instead of synthetic ones. Subsequent developments in purification and modifying processes have only served to expand the versatility of these natural excipients and thus are key components to contemporary drug formulation [17].

1.4.1 Plant-Based Excipients

Plant-based excipients play a very significant role in pharmaceutical sciences because they are natural, biocompatible and farinaceous in the formulation of drugs. These carriers ranging from cellulose, starch and alginates derived from plant materials play a range of roles including fillers, binders, disintegrants and coating to improve functionality and stability of drug delivery systems. The most abundant natural polymer is cellulose and although it is rarely found in tablet formulations as microcrystalline cellulose, it fulfils several functions including a binder and a diluent with good compressibility and mechanical strength. Its derivatives, HPMC and carboxymethyl cellulose, are used in controlled-release systems where they are used as the matrix former, which controls the rate of release of the drug. HPMC is also commonly employed in mucoadhesive film coatings to shield the APIs from the environment and to cloak bitter tastes. Starch, yet another commercially popular plant-derived excipient, is obtained from common plants like corn, potatoes and tapioca. Native starch acts as a binder and a disintegrant assisting strong tabletting strength during production and quick disintegration after administration, which improves drug bioavailability. Starches that have been modified for their structure include pregelatinised or cross-linked starches, which have better flow properties and higher compressibility than other starches—an important feature for high-speed tableting operations [20].

Alginates are naturally occurring calcium, magnesium and sodium salts of polysaccharides isolated from brown seaweeds, which are particularly useful in the development of gastroretentive systems and controlled-release matrices because the swellability of their hydrogels increases in the stomach acid environment and consequently prolongs the time of the drugs. Another advantage of alginates is that they show mucoadhesive behaviour that in turn increases drug availability in certain segments of the gastrointestinal tract. In addition to the functional diversification, these plant-based excipients are renewable, which puts them in a strategic place at a time when the industry values environmentally friendly products. The science of material processing has enhanced the use of plant-derived excipients in chemical as well as physically modified forms with new potential in nanotechnology, liposome and biodegradable implants. Furthermore, since the products meet some of the strict regulatory measures in various jurisdictions, their use predominates in pharmaceutical markets around the globe. Recognising that patient-focused formulation and incorporation of green chemistry aspects into finished products are key to the industry's further development, plant-based excipients remain the leaders of excipient trends, ensuring functionality–safety nexus [21, 22].

1.4.2 Marine and Animal-Derived Excipients

Over the past few years, marine and animal-derived excipients have been identified as essential components in the preparation of pharmaceutical formulations because of their special structures and functions to meet certain problems in the field of drug delivery. Marine-derived excipients including chitosan, a non-swellable and cationic polysaccharide, derived from the exoskeleton of crustaceans including shrimp and crabs are widely used in pharmaceutical formulations due to their bioadhesive, biodegradable and non-toxic properties. The mucoadhesive properties of chitosan are particularly outstanding for increasing drug permeation across the mucosal barrier. Because of its versatility, it has been widely used in pharmaceutical formulations for drug solubility and permeability enhancement in the forms of nanoparticles, hydrogels and films. Some of the areas that have benefited from this chitosan versatility include a formulation of wound healing where its antimicrobial as well as biocompatible characteristic promotes tissue regeneration. Another frequently reported marine excipient is carrageenan extracted from red seaweeds, which provides thickening, gelling and stabilising action mostly for topical and oral applications. Due to its ability to form stable hydrogels under certain conditions, it has a very important application in controlled-release systems [23, 24].

The animal-derived excipients like gelatin, obtained from the collagen of bovine or porcine origin, are inevitable in the formulation of modern drugs and functional food. Thus, gelatin offering thermoreversible gelling behaviour is the first preference for manufacturing hard and soft capsules for drug protection, formability and rapid dissolution of the encapsulated drug. For this reason, it is also used as a stabiliser in parenteral preparations and as a binder in tablets. Moreover, there is also another large group of substances to produce solid dosage forms—excipients, in particular, lactose, which is produced from milk and is used as a filler and diluent for regulation of powder flow and compression during the preparation of tabletting. Hydrophilic and viscoelastic hyaluronic acid obtained from animal sources has become popular in ophthalmology formulations, soft tissue fillers and wound healing agents [25].

However, there are also some drawbacks associated with the use of marine and animal-derived excipients. They have raw material variability issues, allergenicity issues and transmissible diseases like bovine spongiform encephalopathy (BSE) in gelatin. As a result of these concerns, there is a clear and comprehensive protocol to ensure proper purification of the excipients and their quality control. Conservational and ethical concerns over the sources of animal-derived materials and the essence of cultural differences have also propelled the search for non-animal substitutes and synthetically derived near-equivalents. However, the uniqueness of the application of marine and animal-derived excipients is still unprecedented and hence continues to be incorporated in the modern pharmaceutical formulation and development process through new extraction and chemical modification to further improve their performance and adaptability in new and diverse drug delivery systems [26].

1.4.3 Role of Biotechnology in Natural Excipients

With biotechnology, the area of natural excipients has changed dramatically, and new highly purified and multifunctional components for applications in the pharmaceutical industries have become available. Purified and functionalised natural excipients can now be developed using biochemical and molecular technologies including fermentation, enzymatic treatments and genetic manipulation to fulfil new roles required in current drug delivery systems. Microbial fermentation is an important biotechnological process which finds usage in the production of polysaccharides like xanthan gum and hyaluronic acid. These materials possess very high thickening, stabilising and viscoelasticity characteristics and are applicable across numerous preparations such as topicals, injectables and controlled delivery systems. In the same way, enzymatic processing makes it possible to achieve controlled chemical modification of natural polymers like starch and cellulose in terms of solubility, stability and compatibility with the other ingredients to be effectively used in modern drug delivery systems including hydrogels, nanoparticles and matrix tablets [27].

Biotechnology has even widened the prospect of natural polymers or excipients by synthesising bioengineered polysaccharides and proteins containing highly unique functional profiles. For instance, biotechnology has allowed for the synthesis of proteins such as gelatin and albumin with definite molecular mass and low immunogenicity, which was a major problem for substances of animal origin. Further, genetically modified microorganisms are being used for the synthesis of plant-derived excipients like pectin and alginate, thus avoiding dependency on agricultural sources and the consequent batch-to-batch variation in the quality of raw materials.

Sustainability in the manufacture of excipients is also facilitated by biotechnology. Biotechnological processes depend on renewable feedstock to bypass the pollutive means of conventional extraction and have strong, stable supply chains. Also, progress in analytical tools and techniques has made it possible to accurately define the physicochemical properties of the biotechnologically produced excipients to meet tough regulatory requirements set in various parts of the world and increase the popularity of these materials in international markets for drugs and medicinal products [28].

The improvements in the production of natural excipients through biotechnology have not only enhanced the performance and quality of these materials but also paved the way for designing excipients that meet the needs of advanced drug delivery techniques. For example, the biotechnology of polysaccharides is employed in stimuli-responsive systems for drug delivery, where the release of the drug is controlled by changes in pH, temperature or enzyme action. With the rise of precision medicine and greener production solutions, biotechnology is being seen as the next frontier for enhancing and creating new natural excipients in both the pharmaceutical and blue pharma markets. Table 1.3 provides a classification of natural excipients based on their sources and applications.

Table 1.3 Classification of natural excipients based on sources and applications

Source	Example excipients	Applications
Plant-based	Cellulose, starch, pectin, gums (e.g. guar gum, xanthan gum), alginates	Binders, disintegrants, thickeners, emulsifiers, film-forming agents
Marine-based	Chitosan, carrageenan, agar, alginates	Bioadhesive systems, controlled-release formulations, stabilisers
Animal-based	Gelatin, lactose, hyaluronic acid	Capsule shells, stabilisers, fillers, dermal fillers, ophthalmic formulations
Mineral-based	Talc, calcium carbonate, kaolin	Diluents, adsorbents, anti-caking agents
Microbial-based	Xanthan gum, dextran, hyaluronic acid (via fermentation)	Thickeners, stabilisers, viscosity modifiers, injectable systems

1.5 Current Trends in Excipients: Sustainability and Green Chemistry

The recent development of trends in excipients is dominated by questions of sustainable development and green chemistry for the pharmaceutical industry. While the world turns its attention towards environmental conservation and use of assets that are environmentally friendly, the advancement of excipients is experiencing the following significant changes.

1.5.1 Increasing Need for Eco-friendly Stabilisers [29]

The cost increases, consciousness of environmental problems, depletion of resources and improved stewardship in the drug manufacturing process have been the driving force of the demand for sustainable excipients. Due to the emergent sustainability regulations all over the world, the pharmaceutical industries are shifting their importance to have effective excipients that will act as performance enhancers for formulation and will also conform to global sustainability and ethical standards. This change is due to recent tendencies that are being implemented in the sphere of sustainable production of pharmaceuticals with the perspective of lower carbon footprint, overall waste reduction and the usage of renewable resources. Bioderived and biodegradable excipients have attracted a lot of attention because they are evaluated to be more eco-friendly than the commonly used excipients. Substances like cellulose derivatives, starch and gelatin have found wide application in formulations, primarily, because they are biodegradable and self-destruct in the long run without leaving behind a harsh impact on the environment. These excipients are also renewable, ranging from plants and therefore the transition to sustainability is commendable. Second, natural polymers of marine origin such as alginate and chitosan are finding more use in the formulation; these materials are biocompatible, biodegradable and offer other functional benefits in controlled drug delivery systems [30].

A second important aspect of sustainable excipients is the employment of green chemistry in the development of processes that do not harm the environment. Manufacturers are approaching excipient production more cleanly and efficiently through the use of water-based extraction and biocatalysis to eliminate the use of toxic solvents and achieve a reduced waste output. More so, biodegradable polymers like polylactic licentiate acid (PLA) and poly(lactic-co-glycolic acid) (PLGA) are used in the fabrication of drug delivery systems and replace the use of petroleum products. Moreover, there has been a significant effort to provide sustainable supply chains for raw materials. Increased focus is being placed by the pharmaceutical producers on sourcing the excipient ingredients from a sustainable agriculture point of view and fair trade certified sources. It not only pegs the extraction of resources in the environment but also guarantees that raw material is procured through the right means. For instance, starch and pectin used for formulation are now sourced from agricultural wastes or other food residuals, hence a circular economy approach in drug production [27].

Examples dealing with some of the biggest pharmaceutical companies support these trends towards sustainability. For its part, a major global supplier of excipients, badische anilin- und soda-fabrik (BASF), has been seen to have advanced its effort in providing excipients from renewable sources and with enhanced biodegradability. Another significant participant in the excipient market, Evonik, is also trying to create new excipients with better sustainability characteristics, for instance, biodegradable polyesters. However, the current world has seen leading corporations like Johnson & Johnson and GlaxoSmithKline incorporate sustainable business models to support their excipients' supply chain and minimise environmental degradation in production. Increases in sustainable excipients are therefore inevitable due to expanding ethical, environmental and legal concerns and consumer trends of corporate sustainability and green pharmaceutical products. This has led to metamorphosis within the pharmaceutical industry regarding the future of excipients in a sustainable context with considerable stakes in the advancement of technologies for the production of sustainable and efficient excipients.

1.5.2 Role of Green Chemistry in Excipients

Green chemistry is revolutionising pharma excipients through the acquisition of a green approach to chemical manufacturing focusing on sustainability, and environmental and health concerns. The concepts of green chemistry are focused on the minimisation of the creation and use of hazardous substances during the chemical processes, and, therefore, their application to excipient manufacturing has contributed to the development of processes which are more environmentally friendly, efficient and safe. Green chemistry has made one of the most valuable contributions to the formation of excipients—the use of recyclable materials. Many conventional excipients are based on synthetic polymers derived from petroleum products and can be environmentally hazardous when not biodegraded. Nevertheless, green chemistry permits the utilisation of biopolymers, the sources of which comprise cellulose, pectin, starch, chitosan and alginate. These materials are biodegradable and non-toxic, and therefore they lessen pollution and potential adverse effects on human health. Furthermore, competency in renewable raw material sources has a significant role in making excipient production more environmentally friendly. Polymeric examples include polysaccharides that are mostly derived from renewable resources such as plant or marine sources and hence do not exhaust raw materials. The application of these naturally derived excipients also meets the principles of green chemistry by approaching the challenges without relying on petrochemicals and avoiding waste of resources [31].

The other way that green chemistry influences excipients is through the manufacturing processes of the raw materials. Conventional dry-bonding excipient preparation includes the need for hazardous solvents and high-energy procedures responsible for pollution and extra cost. The principles of green chemistry involve the use of green solvents particularly water, ethanol as well as supercritical carbon dioxide to replace 'volatile organic compounds (VOCs)' and toxic chemicals. Such solvents are less toxic, are less dangerous and many of them can again be used, meaning that wastes produced are few and thus are not dangerous to the environment. Also, solvent-free methods such as hot-melt extrusion and solid-state reactions have gained acceptability and are used in the process, greatly reducing the environmental impacts for production.

Concerning the preparation of biocatalysts, biocatalysis of excipients is yet another real-life application of green chemistry. Sustainable synthesis routes are employed in the production of excipients such as xanthan gum, dextran and hyaluronic acid via microbial fermentation. These processes in most cases might call for lesser harsh environments, like being subjected to low temperatures and pressures, and in most cases yield comparatively fewer by-products hence having lower energy intensity and waste. We are also able to use renewable starting materials such as sugars in a process that is known as fermentation [32, 33].

In addition, it is of particular importance that green chemistry is involved in the task of increasing the efficiency of various processes connected with the production of excipients. Current strategies like enzyme catalysis and biotransformation are being used to design specific excipients, at the same time cutting short the costs of environmental and economic impact of the conventional chemical manufacturing techniques. The virtue of the green approach in improving drug solubility or incorporating controlled-release features represents another exciting area of green chemistry that is gaining popularity. Another growth area for the application of green chemistry is enhancing the release characteristics of a drug or developing stimuli-sensitive excipients which release their load on cue. More integration of green chemistry concepts in the creation of new and improved excipients, however, remains problematic. The changeover from synthetic excipients in drugs to bio-based and more sustainable ones is a significant undertaking that always demands capital expenditure on research and enhanced manufacturing systems. Also, although the bio-based excipients are available with improved frequency, those are relatively expensive than the petroleum-based excipients, which may be a bottleneck for industries in terms of expenditure. However, with time, as biotechnology continues developing causing improvements in production, the cost gap between regular and green excipients is projected to narrow [34, 35].

Hence the impact of green chemistry has been observed in tailoring the excipient industry where materials of higher sustainability, biodegradability and non-toxicity are being developed. The application of green energy sources and raw materials, green solvents, biocatalysis and other green chemical synthesis methods can significantly minimise the volume of environmental footprint from the manufacturing of excipients. With the current focus within the pharmaceutical industry towards environmental conservation, the use of green chemistry in formulation of excipients has an outlook as the regulatory authorities put pressure on industries to adopt sustainable practices [36].

1.5.3 Consumer-Driven Trends in Excipients

There are further trends driven by the consumers and impacting the pharmaceutical industry more and more: natural, clean-label, sustainable excipients. Due to the increase in consumption and awareness by the public, there are subtle trends whereby the consumer base demands natural and clean-label excipients for the formulation of the medicine. This phenomenon is being spurred by consumers' aversion to synthetic material with unknown potential health risks, and preference for food ingredients which they consider safer, green and healthy in contemporary tendencies towards health and wellness. Natural non-chemical additives such as cellulose, pectin chitosan and alginate from plant and sea origin findings are highly demanded. People consider these ingredients to be relatively free of chemical reprocessing and better fit to cope with natural life perspectives. Therefore, the use of such bio-based materials is gradually becoming popular among pharmaceutical manufacturers to meet the escalating trends in greener products. This has especially led to the use of natural excipients that are also considered biocompatible ingredients and where possible biodegradable. As a result, the use of biodegradable polymers like poly(lactic-co-glycolic acid) (PLGA) and polylactic licentiate acid (PLA) has received much attention because they are synthesised from bio-based feedstocks including corn and sugarcane instead of petroleum-based excipients.

An important force behind the changes in excipients has been the focus on clean-label products that contain no artificial additives, preservatives or synthetic chemicals. Today's customers pay special attention to the lists of ingredients avoiding food additives such as artificial colours, flavour and stabilisers. There is also more use of natural types of binders, fillers and stabilisers that include starch, microcrystalline cellulose and hydroxypropyl methyl cellulose because most are recognised as safe and have a relatively small, potentially neutral, impact on the environment. Through such excipients, manufacturers can meet the consumers' demands for products free from unnecessary components that are thought to be safer in line with natural health trends. Another consumer-determined factor is the current trend towards sustainability. With the increased understanding of the impact of a product's environment, customers require goods that benefit the earth as well as themselves. Natural and renewable sources to produce excipients are emerging, and consumers are increasingly demanding pharmaceutical products that have little or no negative impact on the environment. Excipients derived from plant or marine origin are considered even more environmentally friendly options in comparison with most of the existing excipients based on petroleum products. This trend has forced manufacturers into the implementation of green chemistry where excipient production processes do not involve the use of toxic solvents, generation of waste or the use of non-renewable resources.

In addition, there is an increased need for functional and individualised medicine to change the excipient market. With increased focus on patient-specific solutions in advanced pharma delivery, excipients have gained importance in delivery systems as never before. Customers are continuing to demand products with better performance, higher solubility or more controlled release. Accordingly, there is a trend of

Table 1.4 Future trends in excipients: green chemistry and sustainability innovations

Trend	Description	Examples
Biodegradable excipients	Increasing the use of excipients that break down safely in the environment, reducing pharmaceutical waste	Starch, cellulose, chitosan, alginates
Green chemistry in production	Adoption of eco-friendly production methods to minimise environmental impact by reducing waste and toxic by-products	Microbial fermentation for xanthan gum; enzymatic synthesis of polysaccharides
Natural and clean label excipients	Demand for excipients derived from natural sources and free from synthetic additives, aligning with consumer preferences	Plant-derived cellulose, pectin, hydroxypropyl methyl cellulose (HPMC)
Circular economy in excipients	Recycling and repurposing waste materials in excipient production to reduce reliance on virgin raw materials	Agricultural by-products for starch; food industry waste for pectin
Sustainable sourcing	Ethical sourcing of raw materials, ensuring excipients are derived from renewable resources with minimal ecological impact	Marine-based chitosan, renewable plant polysaccharides

increasing the manufacturing of a new generation of specialised excipients, including controlled-release polymers, solubility enhancers and nano-excipients, that enable the formulation of specific drugs with high levels of accuracy. That is why these innovations are reflective of movements made by consumers who favour medications that offer better results, fewer negative effects and regimens that are easier concerning dosing.

Another person-related trend about excipient use is the increasing concern with ethical sourcing and product transparency. Ethical considerations about sourcing and production of the ingredients are on the minds of a growing number of consumers. Specifically, this means pressure on the suppliers of excipients to employ traded and sustainably sourced inputs. This involves the employment of natural excipients derived from sustainable resources whose origin can be traced from the point of origin through the manufacturing cycle without compromising the rights of workers or the environment. Such trends exert pressure on the supply chain; as the authors point out, the quality of products such as ethically sourced excipients is becoming ever more important to consumers, as shown in Table 1.4.

1.6 Future Directions in Excipients

The main trends in excipients' development are aimed at increasing sustainability, rationalisation of delivery systems and the emerging requirements of the pharmacy. This has been further fuelled by continually rising consumer expectations for safer, greener and higher-performing excipients by the pharma and biotech industries, where excipients advance through innovations in biodegradable materials, green chemistry and sophisticated drug delivery systems. The continuous transition to naturally derived, more ecologically friendly types of excipients and the application of advanced biotechnology for further sophisticated production lines are likely to serve as the primary drivers towards these goals. Also, the growth of the targeted smart excipients for the individual approach in the treatment and controlled-release formulations will continue. These trends are shifting the market for excipients towards increased sustainability, efficiency and patient-centricity so that excipients are expected to not only perform their conventional functions but also meet the agenda-setting objectives of the pharmaceutical industry.

1.6.1 Nanotechnology and Excipients

Pharmaceutical excipients are being revolutionised by nanotechnology, with better delivery systems, bioavailability and target therapies. It allows fabricating nano-sized excipients such as liposomes, solid lipid nanoparticles (SLNs), polymeric nanoparticles and nanostructured lipid carriers (NLCs) by controlling materials at 1–100 nm scale. Thus, these nanocarriers enhance the solubility and stability and reduce burst release of the poorly water-soluble drugs besides enhancing tissue- and cell-specific targeting as opposed to the systemisation of side effects [37].

Nanotechnology also changes the classic excipients for polyfunctionality. For example, nanocrystalline cellulose and chitosan nanoparticles increase the stability, the dispersion surface of the compound and drug release. The method of functionalisation includes a narrowly targeted group, like the folic acid conjugated carriers for the cancer cells. Other systems such as nanoemulsions together with smart, self-assembling excipients go further in improving the drug stability and release under certain circumstances, particularly in oral, topical and parental formulations.

Long-term toxicity is demonstrated to be low with biodegradable nanocarriers, for instance, PLGA-based polymers, along with safety. However, with these advancements, there are still issues like the complexity of manufacturing these structures, cost of manufacture, toxicity of materials used and lack of standard regulation policies which hinder their general use. As new studies develop, it has given hope for the transformation of excipients using nanotechnology in future enhanced systems of drug delivery [38, 39].

1.6.2 Biocompatible Polymers and Smart Excipients

Biocompatible polymers, smart excipients, are heralding the new generation of drug delivery systems simply because they improve solubility, stability, bioavailability and targeted drug delivery and at the same time reduce side effects. Such excipients satisfy the requirements of contemporary drugs, which require accuracy, controllability and security. PLGA, PEG, polycaprolactone (PCL) and PLA are the commonly used polymers because they are biodegradable and compatible with human tissue. These polymers create nano-carriers, microspheres and hydrogels where they allow the release of the drug to be effected in an efficient and controlled process. For example, PLGA microspheres give controlled release and PCL for elongated release for a drug. Additional targeting ligands or stimuli response modification of functionalised polymers contribute towards the high therapeutic efficiency, for example, folate functionalised PLGA nanoparticles for cancer treatment [40, 41].

There are many novel capability characteristics such as smart excipients that respond to biological or environmental signals: pH-sensitive polymers to release drugs in specific gastrointestinal conditions or thermoresponsive polymers like poloxamers for transition states for injectable controlled-release systems. Conventional controlled-release technology includes stimuli-sensitive hydrogels like hyaluronic acid polymers to deliver the drug at the moisture or pH-sensitive excipients that release the drugs in the presence of enzymes; disease-responsive excipients are designed to target diseases such as cancer by liberating drugs at the site of the disease. Overall, both biocompatible polymers and smart excipients provide an innovative means of delivering drugs that are either poorly soluble or have unwanted side effects and toxicity. In time, and as future technology evolves, they will serve an even greater purpose in pharmaceutical science by delivering safer, more accurate and individualised treatments for patients [42, 43].

1.6.3 Challenges and Opportunities

New approach formulations in excipients used in pharmaceutical products are threatened by various challenges yet hold growth potential. One of the main challenges is the rather high level of regulatory enshrinement. Progressively with the change of status of excipients as active components in the drug delivery process, they attract more attention from the regulatory agencies. This is made even more difficult because there is no global standardisation of guidelines used in creating these guidelines, meaning that the safety standard, quality and performance may vary from one country to another. Such contradiction can slow down the generation and approval of new excipients, which slows down advancement in this field. Sustainability is the other emerging issue that will hardly reduce the effect of the pharmaceutical industry on the environment. The patients also highlight issues such as energy dependency on petroleum-based excipients and the production of hazardous waste. While special attention is paid to biodegradable and plant-derived polymers, these materials' increased production and affordable functionality are still an issue.

Some of the challenges include safety issues and toxicity. Several synthetic and even some natural excipients can cause undesirable effects depending on the dosage, and more so in those who are frail, such as children or the aged. This means that the risks arising from the use of nanotechnology in excipients are especially felt due to the unknown impacts of nanoparticles on the body thus requiring extensive safety assessments and standard testing methods on the nanoparticles in question. Second, the same costs of raw material, manufacturing and longer regulatory cycles place financial hurdles, especially for companies with limited resources, on excipient innovation. However, looking at the challenges faced in the field it also provides a lot of opportunities. It is presently possible to design stimulus-sensitive smart excipients, which change their properties in response to the environment such as pH, temperature or enzyme activity to offer better control and targeting of the active pharmaceutical ingredient. This capability proves most useful for precision medicine, where formulas can be customised by genetic and environmental parameters.

1.7 Conclusion

The pharmaceutical excipient development process is a transformation from basically serving as carriers in the early days of pharmacological practices to being multifunctional components in contemporary drug formulations. The natural products of honey, beeswax and plant-derived gums were used in stabilities and reinforcements in some of the preparative mixtures of herbal medications. As the sophistication of trade and chemistry increased, firms began to utilise more predictable bulk solids such as lactose, starch and gelatin, ultimately promoting drug formulation standardisation.

Semi-synthetic and synthetic excipients came to the limelight in the twentieth century, such as celluloses and polyethylene glycols for better solubility, stability and controlled-release properties. With the advancement of controlled release and targeted drug delivery systems, excipients creep from being passive players to being controllers of the kinetics and dynamics of API. This has been evidenced through newer smart excipients based on stimuli-sensitive properties such as pH and temperature, co-processed excipients that ease formulation processes, and the growing trend towards sustainability through biodegradable excipients. New technologies such as biotechnology and nanotechnology have played a major role towards the formulation of sophisticated and advanced drug delivery systems including biologics and nanomedicines. Current innovative developments have overcome some of the issues related to solubility, bioavailability and API stability, and at the same time meet the needs of a patient-oriented and sustainable healthcare system.

Excipients in pharmaceuticals have evolved from merely inactive fillers into vital components required to provide complex delivery systems. The changes recorded on their application and use show how excipient is indeed very central to improving the efficacy, safety and manufacturability of drugs leading to improved health outcomes for patients. In the future, the excipient's future will be shaped via new technologies such as nanotechnology, green chemistry and biotechnology amongst others. Taking care of the environmental effects, the company will also use sustainable and biodegradable polymers; using, on the other hand, smart excipients and biocompatible polymers will help create polymer drugs for personalised medicine and targeted therapies. As regulatory scores compute on these advancements, excipients still play a central role in dictating the next path of the advancement of drug formulation and delivery in supporting the mission of the pharmaceutical industry in addressing the multifaceted challenges of global healthcare.

Conflicts of Interest No conflict of interest.

References

1. Kalasz H, Antal I. Drug Excipients. Curr Med Chem. 2006;13:2535–63. https://doi.org/10.2174/092986706778201648.
2. Pifferi G, Restani P. The safety of pharmaceutical excipients. Il Farmaco. 2003;58:541–50. https://doi.org/10.1016/S0014-827X(03)00079-X.
3. Nguyen TTH, Jung SJ, Kang HK, et al. Production of rubusoside from stevioside by using a thermostable lactase from Thermus thermophilus and solubility enhancement of liquiritin and teniposide. Enzym Microb Technol. 2014;64–65:38–43. https://doi.org/10.1016/J.ENZMICTEC.2014.07.001.
4. Kumar S, Bhargava D, Thakkar A, Arora S. Drug carrier systems for solubility enhancement of BCS class II drugs: a critical review. Critical Rev Trade; Therapeutic Drug Carrier Sys. 2013;30:217–56. https://doi.org/10.1615/CRITREVTHERDRUGCARRIERSYST.2013005964.
5. Pilcer G, Amighi K. Formulation strategy and use of excipients in pulmonary drug delivery. Int J Pharm. 2010;392:1–19. https://doi.org/10.1016/j.ijpharm.2010.03.017.
6. Mansuri S, Kesharwani P, Jain K, et al. Mucoadhesion: a promising approach in drug delivery system. React Funct Polym. 2016;100:151–72. https://doi.org/10.1016/j.reactfunctpolym.2016.01.011.
7. Baldrick P. Pharmaceutical excipient development: the need for preclinical guidance. Regul Toxicol Pharmacol. 2000;32:210–8. https://doi.org/10.1006/RTPH.2000.1421.
8. Acharya S, Sahoo SK. PLGA nanoparticles containing various anticancer agents and tumour delivery by EPR effect. Adv Drug Deliv Rev. 2011;63:170–83. https://doi.org/10.1016/j.addr.2010.10.008.
9. Seoane-Viaño I, Januskaite P, Alvarez-Lorenzo C, et al. Semi-solid extrusion 3D printing in drug delivery and biomedicine: personalised solutions for healthcare challenges. J Control Release. 2021;332:367–89. https://doi.org/10.1016/j.jconrel.2021.02.027.
10. Elder DP, Kuentz M, Holm R. Pharmaceutical excipients — quality, regulatory and biopharmaceutical considerations. Eur J Pharm Sci. 2016;87:88–99. https://doi.org/10.1016/J.EJPS.2015.12.018.
11. Thoorens G, Krier F, Leclercq B, et al. Microcrystalline cellulose, a direct compression binder in a quality by design environment - a review. Int J Pharm. 2014;473:64–72. https://doi.org/10.1016/j.ijpharm.2014.06.055.
12. Pifferi G, Santoro P, Pedrani M. Quality and functionality of excipients. Il Farmaco. 1999;54:1–14. https://doi.org/10.1016/S0014-827X(98)00101-3.
13. Wang T, Jacobson-Kram D, Pilaro AM, et al. ICH guidelines: inception, revision, and implications for drug development. Toxicol Sci. 2010;118:356–67. https://doi.org/10.1093/TOXSCI/KFQ286.
14. Osterberg RE, See NA. Toxicity of excipients – a Food and Drug Administration perspective. Int J Toxicol. 2003;22:377–80. https://doi.org/10.1177/109158180302200507.
15. Osterberg RE, DeMerlis CC, Hobson DW, McGovern TJ. Trends in excipient safety evaluation. Int J Toxicol. 2011;30:600–10. https://doi.org/10.1177/1091581811423582.
16. Steinberg M, Borzelleca JF, Enters EK, et al. A new approach to the safety assessment of pharmaceutical excipients - the safety committee of the international pharmaceutical excipients council. Regul Toxicol Pharmacol. 1996;24:149–54. https://doi.org/10.1006/RTPH.1996.0120.
17. Kar M, Chourasiya Y, Maheshwari R, Tekade RK. Current developments in excipient science: implication of quantitative selection of each excipient in product development. Basic Fundamentals Drug Delivery. 2019;29–83 https://doi.org/10.1016/B978-0-12-817909-3.00002-9.
18. Devalapally H, Chakilam A, Amiji MM. Role of nanotechnology in pharmaceutical product development. J Pharm Sci. 2007;96:2547–65. https://doi.org/10.1002/JPS.20875.
19. Narang AS, Boddu SH. Excipient applications in formulation design and drug delivery; 2015. p. 1–681. https://doi.org/10.1007/978-3-319-20206-8.

20. Pal RS, Pal Y, Wal A, Wal P. Current review on plant based pharmaceutical excipients. Open Med J. 2019;6:1–5. https://doi.org/10.2174/1874220301906010001.
21. Awasthi R, Sharma B, Kulkarni GT. Plant derived polysaccharides as pharmaceutical excipients: an overview. In: Herbal medicine in India: indigenous knowledge, practice, innovation and its value; 2019. p. 191–204. https://doi.org/10.1007/978-981-13-7248-3_13.
22. Mandal SC, Chakraborty R, Sen S. Evidence based validation of traditional medicines: a comprehensive approach; 2021. p. 1–1151. https://doi.org/10.1007/978-981-15-8127-4.
23. Dehghani M, Taherizadeh MR, Homaei A. Marine origin bioactive peptides: novel advances in the therapeutic potential. In: Marine Biomaterials: Therapeutic Potential; 2022. p. 351–92. https://doi.org/10.1007/978-981-16-5374-2_11.
24. Sukumaran S, Nadarajan S. Marine peptides and their bioactive potentials: new horizon for biotherapeutics. In: Drugs from nature: targets, assay systems and leads; 2024. p. 153–78. https://doi.org/10.1007/978-981-99-9183-9_7.
25. Tyagi U, Sarma AK. Bioactive compounds from algae: extraction, purification, and applications. In: Value added products from bioalgae based biorefineries: opportunities and challenges; 2024. p. 259–88. https://doi.org/10.1007/978-981-97-1662-3_12.
26. Alamgir ANM. Therapeutic use of medicinal plants and their extracts: volume 2. 2018:74. https://doi.org/10.1007/978-3-319-92387-1.
27. Chi EY. Excipients used in biotechnology products. In: Pharmaceutical excipients: properties, functionality, and applications in research and industry; 2016. p. 145–98. https://doi.org/10.1002/9781118992432.CH4.
28. Koo OMY. Pharmaceutical excipients: properties, functionality, and applications in research and industry. In: Pharmaceutical excipients: properties, functionality, and applications in research and industry; 2016. p. 1–338. https://doi.org/10.1002/9781118992432.
29. Sheldon R. Introduction to green chemistry, organic synthesis and pharmaceuticals. In: Green chemistry in the pharmaceutical industry; 2010. p. 1–20. https://doi.org/10.1002/9783527629688.CH1.
30. Dunn PJ, Wells AS, Williams MT. Green chemistry in the pharmaceutical industry; 2010. https://doi.org/10.1002/9783527629688.
31. Kunieda H, Yamagata M. Mixing of nonionic surfactants at water-oil interfaces in microemulsions. Langmuir. 1993;9:3345–51. https://doi.org/10.1021/LA00036A005.
32. Sethi MK, Chakraborty P, Shukla R. Biocatalysis—a greener alternative in synthetic chemistry. RSC Catalysis Series 2018-January. 2017:44–76. https://doi.org/10.1039/9781782629993-00044.
33. Baker Dockrey SA, Doyon TJ, Perkins JC, Narayan ARH. Whole-cell biocatalysis platform for gram-scale oxidative dearomatization of phenols. Chem Biol Drug Des. 2019;93:1207–13. https://doi.org/10.1111/CBDD.13443.
34. Farrán A, Cai C, Sandoval M, et al. Green solvents in carbohydrate chemistry: from raw materials to fine chemicals. Chem Rev. 2015;115:6811–53. https://doi.org/10.1021/CR500719H.
35. Horváth IT, Anastas PT. Introduction: green chemistry. Chem Rev. 2007;107:2167–8. https://doi.org/10.1021/CR0783784.
36. Zimmerman JB, Anastas PT, Erythropel HC. Leitner W (2020) designing for a green chemistry future. Science. 1979;367:397–400. https://doi.org/10.1126/SCIENCE.AAY3060/ASSET/8997170C-4499-4F22-A12F-B46943929ED6/ASSETS/GRAPHIC/367_397_F3.JPEG.
37. Zhu Y, Ye J, Zhang Q. Self-emulsifying drug delivery system improve Oral bioavailability: role of excipients and Physico-chemical characterization. Pharm Nanotechnol. 2020;8:290–301. https://doi.org/10.2174/2211738508666200811104240.
38. Li C, Wang J, Wang Y, et al. Recent progress in drug delivery. Acta Pharm Sin B. 2019;9:1145–62. https://doi.org/10.1016/J.APSB.2019.08.003.
39. Lakkireddy HR, Urmann M, Besenius M, et al. Oral delivery of diabetes peptides — comparing standard formulations incorporating functional excipients and nanotechnologies in the translational context. Adv Drug Deliv Rev. 2016;106:196–222. https://doi.org/10.1016/J.ADDR.2016.02.011.

40. Badreldin M, Salas-Ambrosio P, Ayala M, et al. Synthesis of polypeptides by ring-opening polymerization: a concise review. Curr Org Chem. 2024;28:1154–63. https://doi.org/10.2174/0113852728274519240228105518.
41. Aundhia C, Parmar G, Talele C, et al. Light-responsive polymers: developments in drug delivery systems. Curr Org Chem. 2024;28:1179–89. https://doi.org/10.2174/0113852728307241240430055059.
42. Seth AK, Parmar G, Aundhia C, et al. Smart polymer systems: a futuristic approach to enhance therapeutic efficacy. Curr Org Chem. 2024;28:1164–78. https://doi.org/10.2174/0113852728305580240429100851/CITE/REFWORKS.
43. Mauro N, Utzeri MA, Drago SE, et al. Carbon Nanodots as functional excipient to develop highly stable and smart PLGA nanoparticles useful in cancer theranostics. Pharmaceutics. 2020;12:1012 12:1012. https://doi.org/10.3390/PHARMACEUTICS12111012.

Natural Polymers in Drug Delivery: An Overview

2

Sneha Agrawal, Ruchi Shivhare, and Entela Haloci

2.1 Introduction to Natural Polymers

Natural polymers are chemical compounds, which are formed by smaller units naturally. The word polymer is originally derived from the Greek language, and it means a substance built from a large number of simple molecules of the same kind, polymers "having many parts," which consist of the words polys "many" + meros "part" [1]. Polymers are also called giant molecules of high molecular weight or macromolecules, formed by linking or combining chemically many small molecules called monomers. Polymers differ from monomers in chemical, physical, and mechanical properties, so studying them is important to understand their advantages and disadvantages better. The formation of polymers can occur naturally or synthetically, and the process is called polymerization [2]. Polymerization occurs via various reaction mechanisms related closely to functional groups present in reacting compounds. Many authors have studied and classified natural polymers during the development of polymer science. The historical timeline of the authors involved in this science shows the importance and complexity of these macromolecules.

S. Agrawal (✉)
Bharati Vidyapeeth's College of Pharmacy, Navi Mumbai, Maharashtra, India
e-mail: sneha.agrawal@bvcop.in

R. Shivhare
Department of Pharmaceutical Chemistry, Dadasaheb Balpande College of Pharmacy, Nagpur, Maharashtra, India
e-mail: ruchishivhare@dbcop.org

E. Haloci
Department of Pharmacy, Faculty of Medicine, Universiteti I Mjekësisë, Tiranë, Albania
e-mail: entela.haloci@umed.edu.al

J. S et al. (eds.), *Innovative Pharmaceutical Excipients: Natural Sources*,
https://doi.org/10.1007/978-981-96-7959-1_2

Natural polymers, which refer to International Union of Pure and Applied Chemisrty (IUPAC), include protein, starch, cellulose, and DNA and make up most of the structures of living tissue. The classification of natural polymers is based on different models and can be made based on their origin, function, or chemical nature [3].

2.2 Classification of Natural Polymers

2.2.1 Natural Polymer Is Classified Based on Plant Origin and Can Be Derived from Plant, Animal, and Microbial Origins

1. The polymers derived from *plant materials* are primarily polysaccharides (carbohydrates) and some proteins, which include:
   (i) Cellulose: The most abundant natural polymer found in the cell walls of plants. It is a polymer of glucose.
   (ii) Starch: A polysaccharide made of glucose units, found in plants as a source of energy storage.
   (iii) Gum: Various natural gums, such as xanthan gum and guar gum, are polysaccharides used in food and industrial applications.
   (iv) Pectin: Found in fruits, and used as a gelling agent.
2. The polymers *derived from animals* include proteins and some polysaccharides, which include:
   (i) Proteins: These are polymers of amino acids, including silk, a protein produced by silkworms; wool, a protein produced by sheep; collagen, found in connective tissues like skin, tendons, and cartilage; and keratin, found in hair, nails, and feathers.
   (ii) Chitin: A polymer of N-acetylglucosamine, found in the exoskeletons of insects and crustaceans like shrimp and crabs.
3. The polymers *derived by microorganisms*, often used in industrial and medical applications, include:

   Polyhydroxyalkanoates (PHAs): Biodegradable plastics produced by bacteria, and used in packaging and medical applications.

   Dextran: A polysaccharide produced by certain bacteria, and used in medical applications like plasma expanders.

2.2.2 Natural Polymers Based on Function Are Classified as Follows

1. Structural Polymers: These provide structural support in organisms. Examples include

   cellulose (plants), chitin (insects, crustaceans), and collagen (animals).
2. Storage Polymers: These serve as energy storage in plants and animals. Examples include

starch (plants) and glycogen (animals).

3. Functional Polymers: These have specific biological functions. Examples include:
   (i) Proteins: Such as enzymes, antibodies, and hormones (e.g., insulin)
   (ii) Nucleic acids: DNA and RNA for genetic storage and expression

2.2.3 Natural Polymers Based on Chemical Nature Are Classified as Follows

1. Polysaccharides (carbohydrates): These natural polymers are composed of sugar monomers (saccharides). Examples include:
   (i) Cellulose: A linear polymer made from glucose monomers, the main structural component of plant cell walls
   (ii) Starch: A storage polysaccharide in plants comprising glucose units (amylose and amylopectin)
   (iii) Glycogen: A storage form of glucose in animals, similar to starch in plants
   (iv) Chitin: An N-acetylglucosamine polymer found in arthropod exoskeletons and fungi
2. Proteins: These polymers comprise amino acid monomers linked by peptide bonds. Examples include:
   (i) Collagen: Found in animal connective tissues
   (ii) Elastin: A protein found in connective tissues that allows them to return to their original shape after stretching
   (iii) Keratin: Found in hair, nails, and the outer layer of skin
   (iv) Casein: The main protein in milk
3. Nucleic acids: These polymers are made up of nucleotide monomers. They include:
   (i) DNA (deoxyribonucleic acid): Carries genetic information in cells
   (ii) RNA (ribonucleic acid): Involved in protein synthesis and gene expression

2.2.4 Natural Polymers Based on Polymerization Reaction

One classification is based on polymer structure and divides polymers into condensation and addition polymers. The other classification is based on the polymerization mechanism and divides polymerization into step and chain polymerizations [4]. Addition polymer: polyethylene, polypropylene, polyvinyl chloride; condensation polymer: polyurethane, polyester [5].

2.2.5 Based on Degradability

Biodegradable: polylactic acid, polyglycolic acid, polycaprolactone, polyanhydrides; non-biodegradable: polydimethyl siloxane, polyether urethane, ethyl cellulose

2.2.6 Nature of Polymer-Water Interaction

Hydrophobic Polymer: ethyl cellulose, polydiethyl siloxane; hydrophilic polymer: cCellulosic: MC, HPMC (hydroxypropyl methylcellulose), HPC, HEC; cCellulosic: sodium alginate, xanthan gum, chitosan; hydrogel material: cross-linked polyvinyl alcohol, polyethylene oxide, polyacrylamide

2.3 Natural Polymer Nomenclature

Polymer nomenclature is not very complicated. It can be named in one of two ways, such as source-based and structure-based. Meanwhile, some traditional names are still in use. Source-based nomenclature can be used when the monomer can be identified, and structure-based nomenclature can be used when the polymer structure is proven. Whatever method is used, all polymer names have the prefix poly, followed by enclosing marks around the rest of the name. Many details should be kept in mind for the right nomenclature of the polymers such as mono or polyvalent group, linear or non-linear structure, regular or not regular double bonds, and organic or inorganic chemical nature [6].

2.4 Historical Background and Evolution in Drug Delivery

Particular attention in drug design and delivery is given to natural polymers because of their biocompatibility and biodegradability, which means that they can be inverted to non-toxic products for all ecosystems. Referring to the National Library of Medicine/PubMed database, the historical timeline of their application in pharmaceutical formulation starts with the use of silicon rubber for prolonged drug therapy in 1964, followed by a very rapid increase of publishing information with a total of 15,392 count items in the range of 60 years (1964–2024), excluding biographies, lectures, letters, dictionary, duplicate studies, portraits, and news [7]. Meanwhile, the total number of items in PubMed that include natural polymers in drug development is around 7729, from which 779 articles and reviews are in the range of 60 years (1964–2024). The most intensive reporting years are the last 20 years (2004–2024). In 1987, another important researcher, Harborne, investigated and described them as important agents in different pharmaceutical industry processes. Within these dosage forms, polymeric materials have fulfilled different roles such as binders, matrix formers, or drug-release modifiers; film coating formers, thickeners, or viscosity enhancers; stabilizers; disintegrants; solubilizers; emulsifiers; suspending agents; gelling agents; and bioadhesives [8]. The use of natural polymers in drug delivery has developed significantly over the years, with different types of polymers and applications. Their development is dynamic and has changed over the years as shown in Table 2.1.

Table 2.1 Historic view of natural polymers in drug development implementation

Timeline	Natural polymer class: pharmaceutical uses
Pre-twentieth century: early use	(a) Crude natural material (acacia, gum arabic, and starch): traditional medicine for binding and stabilizing formulations (b) Gelatin and collagen: capsule coatings
1950–1970: moderate use	(a) Alginate: agent for healing and drug carriers (b) Chitosan: mucoadhesive and antimicrobial properties (c) Gelatine: capsules as encapsulation agents (d) Cellulose derivatives: tablets binders, coatings, and controlled-release matrices
1980–1990: advanced drug delivery period	(a) Hydrogels (alginate, gelatine, agarose): controlled release (b) Chitosan derivatives: transdermal, oral, and nasal delivery systems (c) Polysaccharides: dextran, hyaluronic acid as specific targets (d) Cyclodextrins: use of natural polymer for encapsulating drugs into nanoparticles, to improve solubility and bioavailability
2000–2009: targeted drug delivery	(a) Polysaccharides: dextran, hyaluronic acid as specific targets (b) Stimuli-responsive polymer: delivery systems to pH, temperature, and enzymes (c) Mucoadhesive systems: chitosan and its derivative for oral, buccal, and nasal delivery (d) Nanogels: natural polymer-based nanoparticles for controlled and sustained release
2010–2019: biodegradable systems	(a) Biodegradable polymers: biopolymers such as alginate, collagen, and hyaluronic acid for sustainably and biocompatibility (b) Bioengineered polymers: PEGylated polymers used for specific receptor-mediated targeting (c) Drug-eluting implants: silk polymers (fibroin and fibrin) form implantable system (d) Tissue engineering: natural polymers in scaffolds combined with drug delivery systems
2020 onward: person	(a) Gene and protein delivery: natural polymers like chitosan and hyaluronic acid used for delivering mRNA, siRNA, and other systems (b) Microbiome-targeted delivery: polymers like pectins and inulin for drug delivery modulating microbiota (c) 3D drug delivery systems: natural polymers for 3D personalized dosage forms
Future of the use of natural polymer	(a) Artificial intelligence and the natural polymers formulation (b) Regenerative medicine: natural polymer used in delivering growth factors in stem cells or tissue repair (c) Polymers and vaccines: natural polymers can be applied in some type of vaccines

2.5 Types of Natural Polymers Used in Drug Delivery Based on Chemical Properties

2.5.1 Polysaccharides

(i) *Chitosan*: It is one of the most utilized polysaccharides derived from glucosamine and N-acetylglucosamine. It is sourced from chitin, which is present in the exoskeletons of crustaceans. This polymer is characterized by its non-toxic, biodegradable, and biocompatible properties, making it safe for the environment, pets, wildlife, and humans. The therapeutic potential of chitosan is attributed to its antioxidant, binding, and antimicrobial activities. Its diverse biological properties have garnered significant interest in scientific research and its applications across the pharmaceutical, cosmetic, aquaculture, and agricultural sectors. In the pharmaceutical domain, chitosan serves several purposes: (a) it acts as a biocompatible and biodegradable agent, ensuring safety and minimizing toxicity and drug accumulation within the body; (b) it functions as a mucoadhesive agent, enhancing drug absorption by adhering to mucosal surfaces and facilitating sustained release; (c) it serves as a controlled-release agent, which decreases the frequency of drug administration; (d) it acts as a nanoparticle (NP) agent, encapsulating drugs to protect them from degradation and improve bioavailability; and (e) it functions as a gene delivery agent, encapsulating DNA/RNA for gene therapy and aiding in cellular delivery. The applications of chitosan are evident in oral drug delivery, topical delivery, nasal drug delivery, and antitumoral therapy [9, 10].

(ii) *Alginate*: Extracted from seaweed, alginate is used to create hydrogels for sustained drug release. It can be cross-linked to form a gel that holds drugs in a controlled manner. Advantages of alginate in drug delivery are because they are biocompatible, biodegradable, and easy to be chemically modified, but they have weak points to be managed in drug delivery as they are sensitive to pH and ionic strength and can have degradation processes in different rates. Alginates are used in different drug delivery agents for controlled release, oral drugs, gene delivery, wound healing, and ophthalmic drugs, as well as in injectable systems, cell encapsulation, and vaccine delivery [11].

(iii) *Hyaluronic acid*: A naturally occurring polysaccharide found in connective tissues, skin, and eyes, which is used for targeted drug delivery, particularly for ocular and joint-related drug delivery applications. Hyaluronic acid is an agent with useful properties such as biocompatible, biodegradable, hydrophilic, mucoadhesive, and anti-inflammatory. It is applied in different processes of drug delivery as oral drugs, gene delivery, wound healing, ophthalmic drugs, etc. The advantages of hyaluronic acid are natural origin, versatility, target action, and compatibility with organisms, and the challenges are the stability, controlled-release systems, and their cost [12].

(iv) *Starch*: It is a natural polysaccharide that is widely used in the pharmaceutical industry due to its biocompatibility, abundance, and ease of modification. It is derived from plants such as corn, potatoes, and rice. Starch's unique properties make it suitable for various drug delivery applications, ranging from oral formulations to controlled-release systems. Starch properties are similar to other polymers of the family [13].

2.5.2 Proteins

(i) *Gelatine*: Derived from animal collagen, gelatine is often used for creating drug delivery systems such as microspheres, nanoparticles, and hydrogels. Gelatine is well tolerated in organisms, and derived from collagen. Gelatine is a protein-based polymer known for its gelling properties, making it useful for creating drug delivery systems. It is applied to formulate microspheres, hydrogels, and films for controlled drug release [14].

(ii) *Albumin*: Serum albumin, especially human serum albumin (HSA), is used in drug delivery for targeted therapy, including cancer treatment, due to its ability to bind and transport. Albumin is a globular protein that is abundant in blood plasma, known for its biocompatibility and ability to bind and carry various molecules. It is used in different applications as liposome formulations to enhance targeted delivery, especially for cancer therapies, and in nanoparticles to improve drug stability and bioavailability [15].

(iii) *Silk fibroin*: Derived from silk, it is used in drug delivery systems, particularly for long-term controlled release and tissue engineering applications. Silk fibroin is a protein extracted from silk, known for its mechanical strength, biocompatibility, and biodegradability. It is applied for drug-loaded nanoparticles and hydrogels, controlled release of drugs and proteins, and gene delivery systems [16].

(iv) *Collagen*: A structural protein that is often used for controlled drug release in tissue regeneration and healing, such as in wound dressings. Collagen is biocompatible, meaning it is not harmful to the body, which makes it suitable for medical applications like drug delivery and tissue repair. Collagen is biodegradable, meaning it breaks down naturally in the body without causing harmful accumulation. This allows for the controlled release of drugs over time. Collagen is water-loving, making it ideal for use in hydrogels in the role of hydrophilic agent and other delivery systems that require a moist environment, such as for controlled-release formulations. Collagen supports cell adhesion, proliferation, and differentiation, which is important in tissue regeneration and drug delivery targeting specific tissues [17]. This review discusses the use of hydrogels, including collagen-based hydrogels, for controlled drug release and tissue engineering applications [18].

2.5.3 Nucleic Acids

DNA and RNA: These polymers are sometimes used in gene delivery systems, particularly in therapies for genetic diseases. They can be encapsulated in nanoparticles for targeted delivery. These polymers are used in gene therapy like cystic fibrosis, Duchenne muscular dystrophy, and hemophilia by delivering functional genes. Nucleic acids can be used to express tumor-suppressing genes or silence oncogenes, offering a targeted approach to cancer treatment. Also, mRNA vaccines, like those developed for COVID-19, instruct cells to produce antigens and stimulate an immune response. Nucleic acid-based therapies, such as siRNA or CRISPR, are being developed to target viral genomes and stop replication. In theory and in the long term, gene therapy can offer long-lasting therapeutic benefits by modifying the genetic material [19].

2.5.4 Lipids in Drug Delivery

Lipids play a key role in drug delivery systems due to their biocompatibility, biodegradability, and ability to encapsulate both hydrophobic and hydrophilic drugs. Key lipid-based systems include (a) *liposomes*, which are spherical lipid bilayers enclosing an aqueous core and are used for drug delivery, especially for anticancer, antiviral, and vaccine therapies; (b) *solid lipid nanoparticles*, which are solid lipid nanoparticles that encapsulate lipophilic drugs and are used for controlled release and improved stability of drugs; (c) *nanostructured lipid carriers*, which are a mix of solid and liquid lipids for enhanced drug loading and used for delivery of poorly soluble drugs and targeted release; (d) *micelles*, which are amphiphilic lipid molecules forming spherical structures that improve solubility of hydrophobic drugs; (e) *lipid emulsions*, which are lipids dispersed in water, stabilized by surfactants, that are used for parenteral delivery of lipophilic drug. The advantages of lipids in drug delivery are similar to the other natural polymers such as improved solubility, increased controlled release, and low toxicity but on the other hand, they should be stabilized, and improve the regulatory of application [20, 21].

2.6 Mechanism of Drug Delivery Using Natural Polymers

Drug delivery systems aim to deliver therapeutic agents to specific sites in the body at a controlled rate and concentration to achieve optimal therapeutic outcomes while minimizing side effects. The use of natural polymers in drug delivery systems has garnered significant attention due to their biocompatibility, biodegradability, and ability to be chemically modified for specific applications. Drug dissolution, diffusion through the polymer matrix, and release into the environment are the stages of the drug delivery mechanism employing natural polymers, as shown in Fig. 2.1, which illustrates the crucial processes in diffusion-controlled-release systems.

Natural polymers derived from biological sources include polysaccharides (alginate, chitosan) and proteins (gelatin, collagen). These materials act as carriers in a variety of drug delivery systems, allowing for regulated drug release via processes such as diffusion, swelling, erosion, and stimuli-responsiveness. An outline of the

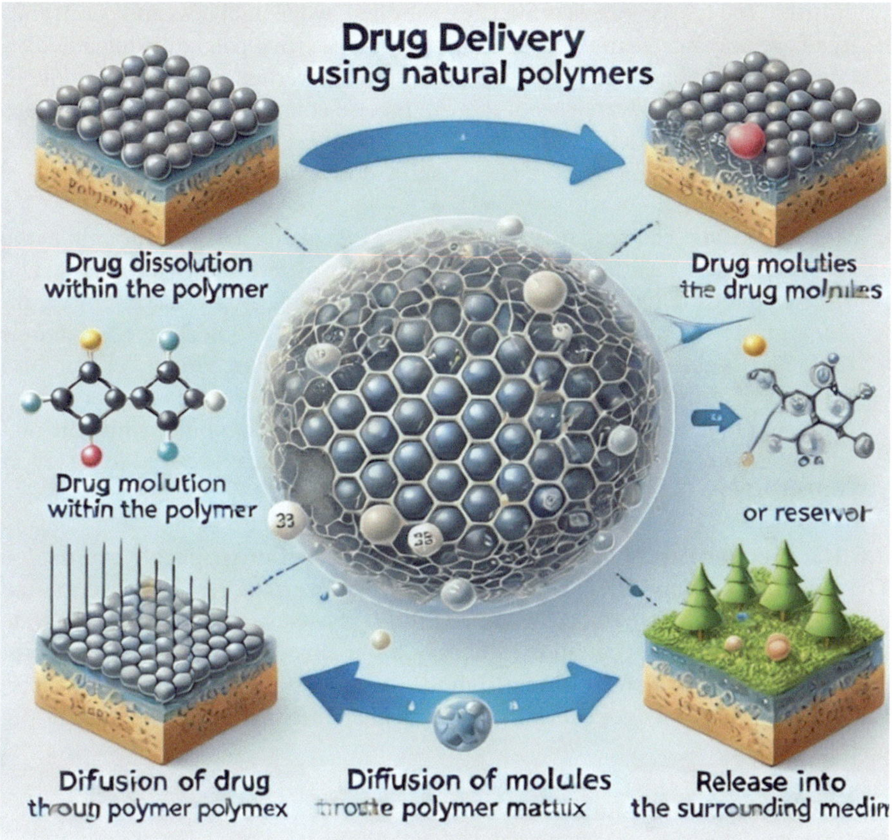

Fig. 2.1 Mechanism of drug delivery using natural polymers

mechanics underlying diffusion-controlled, swelling-controlled, erosion-controlled, and stimuli-responsive delivery systems may be found below:

2.6.1 Diffusion-Controlled Systems

The diffusion of drug molecules from a polymer matrix or reservoir to the surrounding environment is the basis of diffusion-controlled systems, which are commonly employed among drug delivery mechanisms. The release is controlled by Fick's laws of diffusion, which explain how molecules move from a higher concentration to a region of lower concentration.

2.6.1.1 Types of Diffusion-Controlled Systems

1. *Reservoir Systems*: In this system, the drug is enclosed within a core surrounded by a polymeric membrane. The polymer acts as a barrier through which the drug

diffuses. The release rate is primarily controlled by the thickness and permeability of the polymer membrane. For example, transdermal patches using ethylcellulose or chitosan as the polymeric barrier for delivering drugs like nicotine or fentanyl [22]. The advantages of this are precise control over drug-release rates and long-term drug delivery, while the disadvantages are that it requires precise membrane fabrication and contains risk of dose dumping if the membrane is damaged.

2. *Matrix Systems*: The drug in matrix systems is evenly distributed throughout the polymer matrix. The drug dissolves when the solvent enters the matrix and diffuses out through the matrix of polymers, causing drug release. For example, alginate hydrogels containing antibiotics for wound healing applications [23], and hydroxypropyl methylcellulose (HPMC)-based matrices for oral delivery of drugs like theophylline. The advantages are like simpler design compared to reservoir systems and lower risk of dose dumping, while the disadvantages are like limited control over the release rate compared to reservoir systems [24].

2.6.1.2 Mechanism of Drug Release in Diffusion-Controlled Systems

The migration of drug molecules through a polymer matrix or reservoir controls the drug-release mechanism in diffusion-controlled systems. As molecules migrate from areas of higher concentration to areas of lower concentration, the process complies with Fick's laws of diffusion. Both the system's design (reservoir or matrix) and the drug's and polymer's physicochemical characteristics affect the rate of drug release.

2.6.1.3 Steps in Drug Release

1. *Drug Dissolution*:

In both reservoir and matrix systems, the drug first dissolves in the polymer or surrounding 90 Dm33.edium if it is not already in a dissolved state. The solubility of the drug in the polymeric matrix significantly affects its release rate.

2. *Diffusion Through the Polymer*

The drug molecules diffuse through the polymer structure. In reservoir systems, this diffusion occurs across a semi-permeable polymeric membrane. In matrix systems, the drug diffuses through the interconnected pores or the swollen polymeric network.

3. *Drug Release to the Surrounding Environment*

Once the drug reaches the polymer surface, it is released into the surrounding environment, completing the diffusion process.

2.6.1.4 Factors Influencing Drug Release

1. *Polymer Properties*:

Molecular weight, cross-linking density, and porosity of the polymer affect drug diffusion. Example: Chitosan, with its adjustable degree of deacetylation, offers tunable diffusion characteristics [25].

2. *Drug Characteristics*:

Solubility, molecular size, and partition coefficient of the drug influence its mobility through the polymer. Hydrophilic drugs diffuse more readily in hydrophilic polymers like alginate [23].

3. *Environmental Factors*:

pH, ionic strength, and temperature in the surrounding medium can alter the diffusion rate. Some examples are alginate's swelling and diffusion properties are pH-sensitive, affecting drug release in gastrointestinal environments [26].

2.6.1.5 Natural Polymer Applications in Diffusion-Controlled Systems

Insulin and anticancer medications are delivered orally and transdermally using chitosan.

For instance, in cancer treatment, chitosan nanoparticles are used to release paclitaxel in a regulated manner [25]. A common ingredient in hydrogels for wound healing and probiotic encapsulation is alginate. One example is the use of alginate beads to treat diabetes by releasing metformin over an extended period. Gelatin is used in films and microspheres for medication administration in regenerative medicine and tissue engineering. For example, the release of growth factors during wound healing using gelatin-based microspheres [27].

In diffusion-controlled systems, the mechanism of drug release depends on how the properties of the drug, the polymer, and the environment interact. Diffusion-controlled systems provide exact control over drug-release kinetics by comprehend ing and refining these variables, guaranteeing steady therapeutic effects.

2.6.2 Swelling-Controlled Systems

Natural polymers' capacity to absorb water or biological fluids, swell, and then regulate the release of medications incorporated inside them is the basis for swelling-controlled drug delivery systems. Because of their ease of usage, biocompatibility, and capacity for regulated and prolonged drug release, these systems are extensively utilized.

2.6.2.1 Mechanism of Drug Delivery in Swelling-Controlled Systems

1. *Polymer Hydration and Swelling*:

The swelling-controlled polymer begins to absorb water when it comes into contact with bodily fluids or other aquatic environments. Because of this hydration, the polymer chains relax and lengthen, forming a gel-like structure [22].

2. *Drug Diffusion Through the Swollen Layer*:

The inflated polymer layer functions as a diffusion barrier, controlling the drug's movement from the polymer matrix to the surrounding medium. The swelling layer's thickness and properties dictate how quickly the drug is released.

3. *Continuous Swelling and Release*:

As swelling rises, more polymer regions get hydrated, allowing for the release of additional drug molecules. The amount of edema, the density of polymer cross-linking, and the medicine affect the release of drug molecules.

2.6.2.2 Factors Influencing Swelling-Controlled Drug Delivery

1. *Properties of polymers*:

The hydrophilicity, molecular weight, and cross-linking density of the polymer control the extent and rate of swelling. For instance, alginate, a naturally occurring polymer, has pH-dependent swelling properties that enable targeted medication release in particular gastrointestinal tract areas [23].

2. *Environmental Conditions*:

Factors including pH, ionic strength, and temperature of the surrounding media influence the swelling rate. For instance, chitosan is a naturally occurring polymer perfect for stomach-targeted medication administration since it swells more in acidic conditions because of its protonated amino groups [25].

3. *Drug Properties*:

The release rate of a drug is influenced by its solubility, molecular size, and diffusion coefficient.When hydrophilic polymers swell, drugs that are hydrophilic tend to diffuse more quickly.

2.6.2.3 Natural Polymer Examples in Swelling-Controlled Systems

Alginate is frequently used for the controlled release of proteins and growth factors because it creates hydrogels that swell in aqueous environments. Alginate hydrogels that deliver insulin to treat diabetes are one example. Chitosan is used to deliver drugs to specific sites in the gastrointestinal tract; it shows pH-sensitive swelling behavior. For instance, chitosan-based hydrogels for antibiotics with regulated release. A naturally occurring galactomannan polysaccharide, guar gum greatly swells when water is present. As an illustration, consider tablets made of guar gum for colon-targeted medication delivery [28].

2.6.2.4 Swelling-Controlled System Applications

Oral Drug Delivery: By allowing for the regulated release of medications in the gastrointestinal tract, polymers such as guar gum and alginate enhance bioavailability and lessen adverse effects.

Transdermal Drug Delivery: Long-lasting drug release can be achieved with patches made of swelling polymers.

Implantable Devices: Natural polymer-based hydrogels are utilized as implants for long-term medication administration, including in cancer treatment. Swelling-controlled drug delivery systems offer a versatile and effective approach for sustained and site-specific drug release. The ability to tailor polymer properties and respond to environmental conditions makes natural polymers ideal candidates for advanced drug delivery applications.

2.6.3 Erosion-Controlled Systems

Drug delivery methods with erosion control make use of natural polymers that erode or break down over time in biological environments. The rate of polymer erosion—which is frequently controlled by the polymer's characteristics and the surrounding environment—directly affects the release of the medicine. For long-term, site-specific delivery of drugs, these methods are especially beneficial.

2.6.3.1 Mechanism of Erosion-Controlled Drug Delivery

1. *Polymer Degradation*:

The polymer's interaction with biological fluids or water initiates the erosion process. Natural polymers break down by cleaving their molecular chains hydrolytically or enzymatically. For instance, when calcium ions or enzymes like alginase are present, alginate breaks down [23].

2. *Drug Release*:

Embedded drug molecules are released into the environment as the polymer breaks down. The kinetics of drug release are determined by the rate of erosion.

Drug release can occur in two ways:

I. Surface Erosion: A regulated release profile is produced as the polymer breaks down layer by layer.
II. Bulk Erosion: A quicker release results from the simultaneous degradation of the entire polymer structure.

3. *Complete Polymer Degradation*:

The polymer eventually erodes completely, leaving no trace of its presence. Implantable medication delivery systems benefit greatly from this feature.

2.6.3.2 Types of Erosion-Controlled Systems

1. *Surface-Eroding Systems*:

In these systems, the polymer degrades at the surface, maintaining its shape and volume during erosion. This leads to predictable drug release. Polyanhydrides and natural polymers like chitosan exhibit surface erosion in specific conditions [25].

2. *Bulk-Eroding Systems*:

The polymer absorbs water and erodes throughout its volume. This may result in less predictable drug-release kinetics. Natural polymers, such as gelatin, commonly degrade due to bulk erosion.

2.6.3.3 Factors Affecting Erosion-Controlled Drug Delivery

1. *Polymer Properties*: Factors including pH, ionic strength, and temperature of the surrounding media influence the swelling rate. For instance, chitosan is a naturally occurring polymer perfect for stomach-targeted medication administration since it swells more in acidic conditions because of its protonated amino groups [25].
2. *Environmental Conditions*: The pH, enzymatic activity, and ionic strength of the surrounding media all substantially affect erosion rates. For example, alginate erodes faster in low pH conditions due to carboxylic group protonation [26].
3. *Drug Properties*: The drug's solubility, distribution, and interaction with the polymer matrix all have an impact on how it is released during erosion.

2.6.3.4 Natural Polymer Examples in Erosion-Controlled Systems

Alginate: Ion exchange and enzymatic activities are the two ways in which erosion happens. It is extensively employed in hydrogels for long-term medication delivery. An illustration would be alginate-based microspheres that release antibiotics during wound healing [23]. *Chitosan*: Degrades enzymatically in the presence of lysozyme, making it perfect for targeted drug delivery. Chitosan nanoparticles are used for the controlled release of anti-inflammatory medicines [25]. *Gelatin*: Erodes enzymatically and commonly used in biodegradable films and microspheres. Gelatin microspheres are used for the release of chemotherapeutic agents in cancer therapy [24].

2.6.3.5 Applications of Erosion-Controlled Systems

- *Implantable Drug Delivery Devices*: Biodegradable implants that release drugs locally and continuously without the need for surgical removal are known as implantable drug delivery devices.

- *Oral Drug Delivery*: Drug release in the gastrointestinal system is regulated by erosion thanks to polymers like chitosan and alginate.
- *Tissue Engineering*: As eroding polymers break down, growth factors and medications are released, acting as scaffolds for cell proliferation.

Utilizing the inherent degradation characteristics of polymers like chitosan, alginate, and gelatin, erosion-controlled drug delivery systems offer a reliable mechanism for sustained and site-specific drug release, ensuring effective therapeutic outcomes while reducing side effects. Further investigation into the properties of polymers and erosion mechanisms will expand the use of these systems in contemporary medicine.

2.6.4 Stimuli-Responsive Delivery Systems (pH, Temperature, and Enzymes)

Site-specific and controlled drug release is made possible by stimuli-responsive drug delivery systems, which use natural polymers that can change their physical or chemical characteristics in response to certain environmental triggers like pH, temperature, or enzymatic activity. This reduces side effects and improves therapeutic efficacy.

2.6.4.1 Mechanisms of Stimuli-Responsive Delivery Systems

pH-Responsive Systems The mechanism by which natural polymers react to pH changes is as follows: ionizable polymers (such as carboxylic or amino groups) undergo protonation or deprotonation based on pH, which causes changes in polymer solubility, swelling, or degradation, releasing the .drug. Chitosan swells in acidic environments because of protonation of amino groups, which makes it appropriate for drug delivery to the stomach [25].

Temperature-Responsive Systems When the temperature varies, polymers go through phase transitions (sol-gel or gel-sol). The mechanism of drug release is made possible by temperature-responsive polymers, which have critical solution temperatures (higher or lower) over which they swell or collapse. For instance, gelation of gelatin hydrogels at body temperature results in regulated medication release for targeted treatments [24].

Enzyme-Responsive Systems Certain enzymes found in the target tissue or biological milieu cause the release of drugs. *Mechanism*: The medicine is released when the carrier is broken down by enzymatic cleavage of polymer chains. In one example, in conditions with high enzymatic activity, alginate breaks down when alginase enzymes are present, allowing for the release of drugs [23].

2.6.4.2 Features of Stimuli-Responsive Systems

1. *Specificity*: By limiting off-target effects, stimuli-responsive systems guarantee that drug release only takes place under particular circumstances.
2. *Versatility*: By combining pH and temperature responsiveness, natural polymers can be made to react to a variety of stimuli.
3. *Biodegradability*: The safety profile of natural polymers such as gelatin, alginate, and chitosan is improved by their breakdown into non-toxic metabolites.

2.6.4.3 Applications of Stimuli-Responsive Systems

- *Cancer Therapy*: By releasing medications into the acidic tumor microenvironment, pH-responsive systems improve localized delivery. For instance, doxorubicin is released by chitosan nanoparticles during the therapy of cancer [26].
- *Wound Healing*: In infected or injured tissues, enzyme-responsive hydrogels release growth factors or antibiotics. For example, alginate hydrogels can be used to treat wounds [24].
- *Oral Drug Delivery*: Drugs are released in the intestine after being protected in the stomach by pH-responsive polymers. Probiotics are delivered to the intestine using alginate beads [26].
- *Thermal-Triggered Systems*: Drug administration using thermoresponsive hydrogels in inflammatory regions with mild temperature increases. For example, gelatin-based hydrogels for the control of localized inflammation [23].

2.7 Advantages of Using Natural Polymers in Drug Delivery (Table 2.2)

Table 2.2 Advantages of natural polymers in drug delivery

Advantage	Explanation	Citation
Biocompatibility	Natural polymers are inherently biocompatible and cause minimal immune responses, making them suitable for biomedical applications	Lee and Mooney [11]
Biodegradability	They degrade into non-toxic byproducts in physiological environments, eliminating the need for surgical removal in implantable systems	Bhattarai et al. [25]
Sustained and controlled release	Polymers like alginate and chitosan provide sustained drug release by controlling diffusion, swelling, or erosion mechanisms	Kumar et al. [26]
Environmental responsiveness	Natural polymers can be tailored to respond to environmental stimuli such as pH, temperature, or enzymes, enabling site-specific drug delivery	Morris et al. [27]
Renewable and abundant	Derived from natural sources such as plants, algae, and animals, they are renewable and available in large quantities, making them cost-effective	Lee and Mooney [11]
Versatility	Natural polymers can be modified chemically or blended with other materials to enhance properties like mechanical strength and drug encapsulation	Bhattarai et al. [25]
Minimal toxicity	Their natural origin and similarity to biological macromolecules reduce the risk of cytotoxicity and adverse effects	Kumar et al. [26]
Targeted delivery	Functionalization of natural polymers allows for targeted drug delivery, ensuring high drug concentration at the target site and minimizing systemic side effects	Morris et al. [27]
Gel-forming ability	Many natural polymers, like alginate and gelatin, form hydrogels in aqueous environments, providing matrices for controlled and localized drug delivery	Lee and Mooney [11]
Compatibility with hydrophilic drugs	Hydrophilic natural polymers facilitate the encapsulation and controlled release of water-soluble drugs, enhancing therapeutic effectiveness	Bhattarai et al. [25]

2.8 Challenges and Limitations of Natural Polymers in Drug Delivery

2.8.1 Stability Issues: Temperature and pH Sensitivity

A popularly investigated polymer among stimuli-responsive polymers is temperature-sensitive smart polymer, besides other more sensitive polymers. It consists of two distinct forms: one with lower critical solution temperature (LCST) and one with upper critical solution temperature (UCST). In the context of drug delivery applications, LCST polymers are frequently preferred over UCST polymers due to the susceptibility of heat-sensitive drugs. Poloxamer 407 or poloxamer 188 is popular in situ as a gel.

Below the LCST, the polymer bonds hydrogen with the aqueous solution, a process governed by enthalpy. Conversely, above the LCST, the polymer exhibits hydrophobic characteristics, leading to its precipitation, driven by entropy. Light-responsive polymers are those whose phase transitions are stimulated by ultraviolet and visible lights. Notable examples of such polymers include azobenzene, stilbene, and triphenylmethane.

These compound's structures contain a chromophore group enabling them to absorb light energy that is subsequently converted into localized heat energy. This process results in an elevation of temperature, which in turn influences the swelling behavior of the hydrogel. Electro-sensitive polymers represent another category of materials that, as their name suggests, experience gel-sol transitions when subjected to an electric current. There are a significant number of ionizable groups within its chemical structure and response to pH changes. The pulsatile release of drug embedded in such type of polymer can be achieved when electric current is regulated for its intensity. The key component that facilitates the pH responsiveness of these polymers is the ionizable group, which is linked to the non-polar skeletal of the polymer chain. An electrostatic charge electrostatic repulsion between polymer chains may occasionally occur, leading to the extension or unfolding of these chains from a previously collapsed state. Conversely, a decrease in electrostatic charge leads to the opposite effect. Moreover, an increase in electrostatic charge is linked to a rise in hydrophilicity, which can enhance the polymer's solubility in aqueous solutions. Fluctuations in electrostatic charge can also promote self-assembly phenomena, influence de-swelling of formed vesicles. The fluctuation in pH initiates conformational or structural changes, which is associated with the polymer's pKa value [29].

The study of polymers as biomaterials is particularly compelling due to the physicochemical traits that allow them to respond to subtle changes in their environment. A particularly intriguing question is whether it is feasible to engineer these materials to absorb drugs at physiological pH and subsequently release them in more acidic environments, such as those found in lysosomes or within tumor microenvironments. Various factors, including the temperature of body cavities, the pH of bodily fluids, ionic concentrations across cellular membranes, the presence of tissue-specific enzymes, and the overexpression of particular receptors, contribute to our understanding of the body's unique compartments. This knowledge is essential for the design of polymers that respond to specific stimuli. Additionally, fluctuations in ambient pH can induce considerable changes in the physical and chemical characteristics of these polymers. The structural composition of polyelectrolytes includes either acidic or basic groups, which play a crucial role in their behavior in response to pH variations in their environment. A slight alteration in the ambient pH can prompt the ionization of the functional groups, resulting in a notable phase transition, either through dissolution or swelling. The protonation and unionization of acidic groups at low pH levels lead to a marked decrease in the volume of polyacids. As the pH rises, anionic polymers tend to swell. In cationic polymers, the ionization of basic groups from polybases increases with a decrease in pH [30].

These three-dimensional structures can adapt to pH changes in the gastrointestinal tract and cellular compartments through structural and functional modifications, including changes in surface activity, conformation, solubility, and configuration.

pH-sensitive hydrogels are made from polymers with ionizable acidic or basic groups that can accept or donate protons [31].

2.8.2 Batch-to-Batch Variability in Polymer Quality

2.8.2.1 Laboratory-Scale Production of Batch

Polymerization of various monomers creates polymeric nanoparticles (PNPs) that can self-assemble into structures measuring 10–100 nanometers. Natural polymers are biocompatible, biodegradable, and exhibit low toxicity, as they break down into oligomers and monomers that the body can meet and eliminate. However, some derivatives or cross-linking agents may resist complete degradation. PNPs can effectively trap, encapsulate, or attach drugs, including in forms like nanospheres, nanocapsules, and drug conjugates.

In general, there are two methods for producing NPs in nanotechnology [32]:

1. *The top-down methodology*

It is a widely used approach in nanotechnology that focuses on the reduction of bulk materials to achieve nanoscale dimensions. This process typically begins with larger pieces of material, which are systematically broken down through various techniques such as milling, etching, or lithography. An alternative "top-down" strategy can be employed when the initial precursor size exceeds the nanoscale. In this scenario, the process still involves the reduction of material, but the starting point is significantly larger than typical nanoscale dimensions.

2. *The bottom-up approach:*

It is a fundamental strategy in nanotechnology that focuses on constructing nanostructures from the ground up, starting with individual atoms or molecules. In the bottom-up approach, the assembly of these building blocks occurs sequentially, allowing for the creation of complex structures with precise control over their composition and properties. The process typically involves self-assembly, a phenomenon where atoms or molecules spontaneously organize into structured arrangements without external guidance. One of the primary applications of the bottom-up methodology is the synthesis of nanoparticles (NPs). These nanoparticles can be composed of a wide range of materials, including metals, semiconductors, and polymers, and their properties can be finely tuned by altering the size, shape, and composition of the constituent atoms or molecules.

Emulsification-Solvent Evaporation

Emulsification is a sophisticated technique employed in the field of nanotechnology, particularly for the creation of polymeric nanoparticles (PNPs). This process involves the effective dispersion of an insoluble solvent within an aqueous medium, achieved through the application of high shear stress. The rapid response time of the process allows for the quick production of nanoparticles, which is essential in various biomedical applications, including drug delivery systems. However, despite its benefits, the emulsification-solvent evaporation technique is not without its challenges. One significant hurdle is the need for standardization tailored to each specific drug. This requirement can complicate the development process, as different drugs may necessitate unique conditions for optimal nanoparticle formation.

Solvent Diffusion

This methodology involves the development of an oil-in-water (o/w) emulsion, which is achieved by utilizing a partially water-soluble solvent that contains both the polymer and the drug of interest. In addition, an aqueous solution that includes a surfactant is incorporated into the process. However, it is important to highlight that this method is specifically applicable to lipophilic drugs, which can pose limitations in terms of the range of drugs that can be effectively encapsulated.

However, a significant challenge associated with the emulsification-diffusion technique is its reliance on organic solvents, which, while facilitating the production of nanoparticles, also raises concerns regarding the environmental impact and safety of the solvents used. Overall, this methodology represents a promising approach for the development of advanced drug delivery systems.

Reverse Salting-out

The emulsification/solvent diffusion method differs from the salting-out technique, which separates a hydro-miscible solvent from an aqueous solution to form nanospheres. In this method, an oil-in-water (o/w) emulsion is created using a water-soluble polymer solvent like acetone or ethanol, combined with a gel, salting agent, and colloidal stabilizer in the aqueous phase. Saturating the aqueous phase reduces acetone solubility, allowing for o/w emulsion formation at room temperature with vigorous stirring. While perfect miscibility of organic solvents and water is not required, it simplifies the process. The resulting nanospheres range from 170 to 900 nm. For hydrophilic drugs, a double emulsion can encapsulate them within the polymer, starting with a water-in-oil (w/o) emulsion. For dual drug delivery of hydrophobic and hydrophilic compounds, biodegradable Janus particles can be used.

Nanoprecipitation

This method, commonly referred to as solvent transfer, is a sophisticated technique that leverages the properties of two miscible solvents to facilitate the encapsulation of various substances. The process begins with the dissolution of a polymer in an organic solvent that possesses the ability to mix with water. This initial step is crucial, as the choice of organic solvent can significantly impact the efficiency of the subsequent stages. Once the polymer is fully dissolved, the organic solvent is carefully introduced into an aqueous environment that contains a surfactant. The presence of the surfactant is essential, as it stabilizes the system and helps to prevent the aggregation of particles during the transition from organic to the aqueous phase. These nanocapsules can encapsulate a variety of substances, making them valuable in drug delivery applications. Notably, this technique excels in encapsulating lipophilic (fat-soluble) drugs, which can be effectively trapped within the nanocapsules due to their affinity for the organic solvent. However, it is important to note that the method has limitations when it comes to encapsulating hydrophilic (water-soluble) drugs. The inherent properties of hydrophilic substances make it challenging for them to be effectively incorporated into the nanocapsule structure, thereby restricting the versatility of this approach in certain applications.

2.8.2.2 Scale-up Production

The role of sampling is very important. NPs are inherently unstable and have a high aggregation tendency. The nanoscale subject has different physical properties from the corresponding bulk values. Sample measurements require a well-dispersed particle population corresponding to a statistically representative sample. The sample preparation is a key step of the measuring process and requires a great amount of work to minimize the agglomeration state, which impacts the measurement result of constituent nanoparticles negatively.

1. *Membrane extrusion*

The self-emulsification technique creates large, polydisperse emulsion droplets, while nanoporous membrane extrusion yields nanoparticles with better size uniformity and reproducibility. Variations in membrane extrusion can be achieved by adjusting pore size, pressure, flow rate, and fluid surface tension. There are four prevalent strategies for the extrusion of nanoporous membranes: (a) vesicle extrusion, (b) membrane emulsification, (c) precipitation extrusion, and (d) biological membrane extrusion. Precipitation extrusion is a unique method for producing nanoparticles (NPs). Nanoparticles (NPs) that are encapsulated by cell membranes represent a significant advancement in the field of nanomedicine, as they harness the natural biological properties of the cell membrane and its associated proteins. The cell membrane provides a protective barrier that helps shield the nanoparticles from the immune system, particularly from the reticuloendothelial system (RES), which is responsible for the clearance of foreign particles from the bloodstream.

2. *Low-flow microfluidics*

Recent advancements in microfluidics could greatly enhance nanoparticle (NP) synthesis and accelerate clinical trial progression. The flow rate of reactants in microchannels is crucial to the synthesis process. Pulsed micro-mixing is an effective active mixing strategy that includes a driving mechanism (PZT micropump) and a Y-type micromixer. In microfluidizer technology, high-pressure frontal collisions between two fluids create shear and cavitation forces. Microfluidics and microreactors improve control over the shape and size of organic nanomaterials, enhancing their stability, solubility, and targeted accessibility. The small volume and large surface area increase heat and mass transfer rates, reducing response times from hours to seconds, while the reactor's limited capacity enhances safety.

3. *Continuous flow systems*

Sustainable processing at a small scale has garnered significant attention from both industry and academia over the last ten years. This growing interest can be attributed to several key factors that highlight the advantages and transformative potential of small-scale sustainable practices in the field of chemical production. Firstly, small-scale sustainable processing offers the opportunity to improve the efficiency of synthesis processes. Traditional large-scale chemical production often involves significant resource consumption and waste generation. Flow reactors, which are integral to this microfluidic synthesis method, offer numerous advantages over traditional batch chemical processes.

4. *Spinning disk processing (SDP)*

Spinning disk processing (SDP) plays a crucial role in the investigation of fluid dynamics and the advancement of process intensification, particularly in the context of materials synthesis and nanoparticle production. This innovative technique addresses several challenges associated with scaling up production processes, which is often a significant hurdle in the transition from laboratory-scale experiments to industrial-scale manufacturing. By utilizing a rapidly rotating disk, SDP creates a unique environment that allows for enhanced control over various parameters, leading to improved outcomes in the synthesis of materials.

Particle Size

The dimensions of nanoparticles (NPs) significantly influence their behavior within in vivo systems, affecting their pharmacokinetics, biodistribution, and overall therapeutic efficacy. Nanoparticles with a size of less than 5 nm are rapidly cleared from the bloodstream by the kidneys via glomerular filtration, which impacts their circulation duration and limits their potential for prolonged therapeutic action. Consequently, NPs in this size range may not accumulate in target tissues effectively, which can hinder their intended therapeutic effects. However, the size-dependent clearance mechanisms also mean that larger NPs may face challenges related to biocompatibility and potential immunogenicity, which can influence their overall safety profile.

The methodology for assessing NP size can be categorized into two primary approaches:

1. *Dynamic light scattering (DLS)*: This technique measures the scattering of light by particles in suspension to determine their size distribution. DLS is particularly useful for characterizing the hydrodynamic diameter of nanoparticles in solutions, providing insights into their stability and aggregation behavior.
2. *Transmission electron microscopy (TEM)*: TEM provides high-resolution images of nanoparticles, allowing for direct visualization and measurement of their size and morphology. This method involves the transmission of electrons through a

thin sample, resulting in detailed images that can reveal structural characteristics at the nanoscale. Non-imaging techniques that rely on light scattering are employed for suspension measurements, yielding results that are both rapid and reproducible; however, they do not provide information regarding the shape of the particles.

Particle Shape and Morphology

The morphology of nanoparticles (NPs) plays a crucial role in determining their behavior and efficacy within the bloodstream. This includes how they interact with endothelial cells, their duration of circulation, their biodistribution throughout the body, and their cellular uptake by target tissues. A comprehensive study has highlighted a notable preference for spherical NPs, which have emerged as the most commonly utilized shape in various biomedical applications due to their favorable properties. However, it is essential to recognize that NPs can be engineered in a wide range of geometries, including but not limited to cubes, bars, disks, and stars.

Surface Properties

The electrical charge present on the surface of polymeric nanoparticles (PNP) plays a crucial role in determining their interaction with cellular membranes, particularly in the context of drug delivery and targeted therapies. The surface charge influences the nanoparticles' ability to penetrate cellular membranes, which is a critical factor in their effectiveness as delivery vehicles for therapeutic agents. Electrostatic interactions are particularly significant during the process of internalizing endocytotic nanoparticles. When nanoparticles are introduced into a biological environment, their surface charge can affect how they interact with the negatively charged phospholipid bilayers of cell membranes. Positive nanoparticles (NPs) exhibited greater harmful effects compared to their negative and neutral counterparts. Nanoparticles with a positive charge or elevated "like" charges demonstrated enhanced interaction forces with cellular structures. The presence of charged particles has implications for the integrity of the blood-brain barrier (BBB).

Drug Release

The profile of drug release is essential for ensuring consistency and quality performance across different batches, as well as for assessing the in vitro to in vivo correlation (IVIVC) of the medication. Each of these methodologies necessitates validation. Agitation plays a crucial role in in vitro release study, which are essential for evaluating the performance of pharmaceutical dosage forms. This agitation is necessary to ensure that the drug is evenly distributed throughout the release medium, thereby preventing the aggregation of dosage forms. Aggregation can lead to inaccurate results and misinterpretation of the drug's release profile, which is why maintaining a consistent and uniform environment is vital during these studies. The characteristics of the release media used in these studies are influenced by several factors. One of the primary factors is the solubility of the drug being tested.

Dissolution

Increased dissolution rates can significantly enhance drug absorption, a critical factor in the efficacy of pharmaceutical formulations. When drugs dissolve more readily, they can be absorbed more efficiently by the body, leading to improved therapeutic outcomes. This principle is particularly relevant in the context of nanoparticles, which are increasingly being utilized in drug delivery systems due to their unique properties and ability to enhance bioavailability. The dissolution of nanoparticles is not only important for drug absorption but also plays a crucial role in their transformation and potential toxicity within aquatic ecosystems.

Stability and Degradation

Controlling the particle size, morphology, stability, and size distribution of biosynthesized nanoparticles (NPs) presents significant challenges that are critical to the successful application of these materials in various fields, particularly in drug delivery systems. The intricate nature of nanoparticle synthesis means that even minor variations in the production process can lead to substantial differences in the final properties of the nanoparticles. Therefore, a thorough understanding of the factors influencing these characteristics is essential for optimizing NP formulations.

2.8.3 Challenges in Scale-up Production of Nanomedicine

Nanomedicine signifies a transformative advancement in the field of healthcare, representing a paradigm shift in how diseases are diagnosed, treated, and prevented. This innovative approach leverages the unique properties of nanomaterials, enabling a range of clinical applications that are currently accessible, including targeted drug delivery systems, advanced imaging techniques, and novel therapeutic agents. These applications not only enhance the efficacy of treatments but also minimize side effects, thereby improving patient outcomes and quality of life. The integrity of experiments in nanomedicine is paramount, as it is in all scientific disciplines. Meticulous documentation and thorough analysis of experimental data are essential to ensure the reliability and reproducibility of results. This rigor is particularly critical in research focused on nanotechnology, where the behavior and interactions of nanoparticles (NPs) can be influenced by a myriad of factors. Researchers must adhere to stringent protocols to validate their findings, as even minor deviations in experimental conditions can lead to significant variations in outcomes [33].

However, the production of nanoparticles must also consider economic and environmental factors. The synthesis of NPs can involve complex processes that may require expensive materials and energy-intensive methods. Therefore, it is imperative to evaluate the cost-effectiveness of these production techniques to ensure that they are viable for widespread clinical use. Additionally, the environmental impact of nanoparticle production cannot be overlooked.

2.8.3.1 Reproducibility

Reproducibility is essential for the practical applications of nanomaterials, as it ensures that the properties and behaviors of nanoparticles (NPs) can be consistently replicated across different batches and production methods. This consistency is crucial for various fields, including medicine, electronics, and environmental science, where the performance of nanomaterials can significantly impact the effectiveness of applications such as drug delivery systems, sensors, and catalysts.

However, the challenges associated with the reproducibility of nanoparticles arise from several interconnected factors related to their production, characterization, and delivery processes. Firstly, the synthesis of NPs often involves complex chemical reactions and physical processes that can be sensitive to a variety of parameters, including temperature, pressure, concentration, and reaction time. Even slight variations in these conditions can lead to significant differences in the size, shape, surface chemistry, and overall properties of the resulting nanoparticles. This variability can hinder the ability to produce NPs with the same characteristics consistently. Moreover, the characterization of nanoparticles presents its own set of challenges. Techniques used to analyze NPs, such as electron microscopy, dynamic light scattering, and spectroscopy, may have limitations in resolution or sensitivity, which can affect the accuracy of measurements. Inconsistent characterization methods can lead to discrepancies in reported properties, making it difficult to compare results across different studies or applications [32, 33].

2.8.4 Use of Natural Polymers in Gene Therapy and Vaccine Delivery

The field of gene therapy has made significant strides in treating diseases by delivering exogenous functional genes to target cells, thereby enabling gene silencing, expression, and editing. However, the susceptibility of nucleic acids to degradation by endonucleases and their elimination by the immune system poses considerable obstacles. As a result, there is an urgent requirement for safe and efficient gene delivery vectors to advance the clinical application of gene therapy. Although viral vectors are recognized for their high transfection efficiency, safety concerns limit their broader application. Similarly, while lipid nanoparticles show promise in clinical scenarios, they are hindered by issues of stability and storage. In this context, polymer cationic vectors (PCVs) have emerged as a focus of research due to their low cost, reduced immunogenicity, and adaptability. This article reviews the latest strategies aimed at enhancing the gene transfection efficiency of polymeric delivery systems and their applications in specific diseases. Despite these advancements, challenges in gene delivery with polymer vectors persist. This chapter concludes with insights into the future development of PCVs, which may aid in their design and application [34, 35]. The exploration of polymeric materials as vectors for gene delivery reveals a broad spectrum of goals and guiding principles. Key among these are the synthetic designs that adeptly adjust charge density, reduce toxicity, and refine structural characteristics to incorporate targeting ligands. The versatility of polymers is significantly enriched by the availability of pi-conjugated structures, which facilitate not only the fluorescent tracking of complexes but also the effective complexation and delivery of DNA/RNA. The emergence of innovative polymer designs, particularly when paired with cutting-edge delivery techniques, is expected to propel the advancement of enhanced therapeutic solutions. This evolution is likely to focus on the development of biocompatible polymers with degradable backbones that enable swift or triggered release of cargo, while maintaining an optimal balance between the degradability and stability of the polyplexes [36]. Employing physical techniques such as electroporation and sonoporation to aid polyplex transfection leads to improved targeting outcomes and a decrease in off-target toxicity. Recent gene delivery protocols that utilize polymers and plasmid DNA (pDNA) have been effectively translated into chemotherapy applications for multiple cancer types, showcasing low toxicity levels and robust biological activity of the gene products. Thus, the future holds significant potential for the application of polymer gene delivery systems in managing challenging conditions, including tumor growth, chronic inflammatory diseases (such as arthritis), and the repair of musculoskeletal tissues (cartilage and bone).

2.9 Regulatory and Safety Concerns

The regulatory framework may offer specific directives for demonstrating the bioequivalence between an innovator product and its generic counterpart, particularly in the context of complex medicinal formulations. However, pinpointing the critical attributes and developing in vitro/in vivo correlations (IVIVC) prove challenging due to intricate release properties and the absence of standardized compendial methods for in vitro release testing. For example, the rapid release observed in vitro may be obscured by the absorption phase occurring at the intramuscular injection site, or it may be mitigated by the formation of fibrous encapsulation resulting from the host's immune response or steric hindrance posed by the extracellular matrix. Consequently, in vitro assessments may possess limited efficacy in accurately predicting the in vivo performance of formulations that exhibit considerable variability in their release profiles.

In order to enhance our comprehension of the challenges associated with quality testing, we will examine the technological characteristics of polymer-based lipid drug delivery systems available in the market. This examination will be in accordance with the definitions outlined in the primary Pharmacopoeias, which consider various attributes for classification. Consequently, the requirements may vary depending on the specific monograph being referenced.

Poly(methyl vinyl ether-co-maleic anhydride) (P(VME/MA)), also known as Gantrez (G), is a biodegradable and biocompatible linear polymer with low toxicity and FDA approval. It is used in dental adhesives, tissue engineering scaffolds, and microneedle arrays for drug delivery. Its biocompatibility is further improved by carboxylic groups, and when combined with alcohol-containing macromolecules, it undergoes esterification, resulting in enhanced biomaterial properties.

The Department of Biotechnology (DBT) has submitted a draft titled "Guidelines for the Evaluation of Nanopharmaceuticals in India." To facilitate comprehensive discussions with stakeholders in the nanotechnology sector, the DBT has asked the Central Drugs Standard Control Organization (CDSCO) to publish the draft guidelines on its official website. Consequently, the draft guidelines for the evaluation of nanopharmaceuticals are provided for the feedback and recommendations of relevant stakeholders [34–37].

2.10 Future Perspective

The solvent transfer is a valuable technique in the field of nanotechnology, particularly for the development of nanocapsules. While it offers significant advantages for lipophilic drug encapsulation, ongoing research is needed to enhance its efficacy for hydrophilic compounds, thereby broadening its applicability in pharmaceutical formulations. The emulsification-solvent evaporation technique is a widely utilized method for the encapsulation of drugs within polymeric nanoparticles. Its non-toxic nature, rapid processing capabilities, and versatility make it an attractive option in drug delivery systems. However, the challenges of standardization, energy requirements, and the use of volatile solvents must be carefully addressed to optimize the process and ensure the stability and efficacy of the resulting pharmaceutical formulations. The electrical charge on the surface of PNP is a key determinant of their ability to penetrate cellular membranes and interact with immune cells.

2.11 Future Challenges

Understanding these electrostatic interactions is essential for optimizing the design of nanoparticles for targeted drug delivery and improving their performance in therapeutic applications. Agitation is a fundamental aspect of in vitro release studies, essential for preventing aggregation and ensuring accurate results. Controlling the particle size, morphology, stability, and size distribution of biosynthesized nanoparticles (NPs) presents significant challenges that are critical to the successful application of these materials in various fields, particularly in drug delivery systems. The intricate nature of nanoparticle synthesis means that even minor variations in the production process can lead to substantial differences in the final properties of the nanoparticles. Therefore, a thorough understanding of the factors influencing these characteristics is essential for optimizing NP formulations. The choice of release method, the solubility and stability of the drug, the sensitivity of the analytical techniques, and the specific testing procedures all play significant roles in shaping the characteristics of the release media and ultimately determining the success of the study. Understanding and controlling these factors is critical for the development of effective pharmaceutical formulations and for ensuring that they perform as intended in clinical settings. Furthermore, the specific procedures employed in the testing process can also affect the characteristics of the release media. This includes factors such as the temperature of the medium, the pH level, and the volume of the release medium used. Each of these parameters can influence the solubility and stability of the drug, as well as the overall kinetics of the release process. One of the primary factors affecting the properties of biosynthesized NPs is the choice of excipients. Excipients are substances formulated alongside the active ingredients in a drug product, and they can significantly influence the physical and chemical stability of nanoparticles. The selection of appropriate excipients can enhance the solubility, bioavailability, and overall efficacy of the nanoparticles, while also ensuring that they remain stable over time. Additionally, the methods employed in the production of NP-based systems, such as green synthesis techniques, chemical reduction, or physical methods, can also impact the size, shape, and distribution of the nanoparticles. Each method has its own set of advantages and limitations, and the choice of method must align with the desired characteristics of the final product. Ultimately, a comprehensive understanding of the factors influencing the particle size, morphology, stability, and size distribution of biosynthesized nanoparticles, along with rigorous stability assessments, is essential for advancing the field of nanomedicine and ensuring the successful translation of these technologies into clinical applications. A significant challenge to progress in this interdisciplinary field is the variability in the published literature regarding the types of characterizations conducted and the experimental data presented. This inconsistency can arise from several factors, including differences in research methodologies, variations in the parameters measured, and the diverse analytical techniques employed by different research groups. Additionally, the field's interdisciplinary nature means that researchers from different backgrounds—such as chemistry, physics, engineering, and biology—may use distinct terminologies and frameworks for their analyses. This can create barriers to communication and collaboration, further complicating efforts to establish a cohesive understanding of the subject matter.

Replication studies are vital for resolving inconsistencies among systems and accurately assessing their differences and similarities. For example, 60% of identified protein types displayed low affinity for nanoparticles (NPs). Of the remaining 40%, 60% of binding interactions were affected by surface charge, while 40% were associated with specific surface groups. Our analysis also revealed that only 5 of the 20 most common serum proteins were linked to the NPs studied.

References

1. Silva ACQ, Silvestre AJD, Vilela C, Freire CSR. Natural polymers-based materials: a contribution to a greener future. Molecules. 2021;27(1):94. https://doi.org/10.3390/molecules27010094.
2. Gowariker VR, Viswanathan NV, Shreedhar J. Polymer science. New Delhi: New Age International; 2005. p. 1–1.
3. Kumar A, Saini S. Natural polymers and their applications in drug delivery: a review. Asian J Pharm Sci. 2013;8(4):215–25.
4. Odian G. Principles of polymerization. 4th ed. Wiley; 2004. p. 1–835. https://unpa.edu.mx/~aramirez/Principles%20of%20polymerization.pdf

5. Kaur R, Kaur S. Role of polymers in drug delivery. J Drug Deliv Ther. 2014;4(3):32–6. https://doi.org/10.22270/jddt.v4i3.826.
6. IUPAC. Compendium of chemical terminology (the "gold book"). 2nd ed. Blackwell Scientific Publications; 2008. https://doi.org/10.1351/goldbook.
7. Folkman J, Long DM. The use of silicone rubber as a carrier for prolonged drug therapy. J Surg Res. 1964;4:139–42. https://doi.org/10.1016/s0022-4804(64)80040-8.
8. Guo J, Skinner GW, Harcum WW, Barnum PE. Pharmaceutical applications of naturally occurring water soluble polymers. PSTT. 1998;1:254–61.
9. Rinaudo M. Chitin and chitosan: properties and applications. Prog Polym Sci. 2006;31(7):603–32.
10. Liu H, Piao H. Chitosan and its derivatives in drug delivery systems. J Pharm Pharmacol. 2013;65(8):1111–22. https://doi.org/10.1111/jphp.12069.
11. Lee KY, Mooney DJ. Alginate: properties and biomedical applications. Prog Polym Sci. 2012;37(1):106–26. https://doi.org/10.1016/j.progpolymsci.2011.06.00.
12. Zhu X, et al. Hyaluronic acid-based nanocarriers for targeted drug delivery. Mol Pharm. 2016;13(5):1705–14.
13. Patel DK, Mishra A. Starch and starch-based drug delivery systems: a review. Int J Pharm Sci Res. 2014;5(3):836–43.
14. Jung SH, Lee SH, Kwon IC. Gelatin-based drug delivery systems. J Control Release. 2006;114(2):151–60.
15. Gupta A, Wang W. Albumin-based drug delivery systems: a review. J Control Release. 2011;148(1):91–105.
16. Jin H, et al. Silk fibroin nanoparticles as drug carriers for cancer treatment. Int J Nanomedicine. 2013;8:1955–64.
17. Li W, Mooney DJ. Designing hydrogels for controlled drug delivery. Nat Rev Mater. 2016;1(12):16071. https://doi.org/10.1038/natrevmats.2016.71.
18. Xie J, Wang CH. Collagen-based drug delivery systems. J Control Release. 2014;190:140–53. https://doi.org/10.1016/j.jconrel.2014.06.020.
19. Seneviratne SL, Sudhakar A. Nucleic acid-based drug delivery for gene therapy. Adv Drug Deliv Rev. 2020;156:128–45. https://doi.org/10.1016/j.addr.2020.05.005Th.
20. Zhang Y, Cheng X. Nucleic acid-based therapies in the treatment of genetic disorders: current status and future directions. Mol Ther Nucleic Acids. 2019;17:250–67. https://doi.org/10.1016/j.omtn.2019.04.009.
21. Barenholz Y. Liposome application: problems and prospects. Nanomedicine. 2012;8(6):70–6. https://doi.org/10.1016/j.nano.2012.03.005.
22. Torchilin VP. Lipid-core micelles for targeted drug delivery. Curr Drug Discov Technol. 2005;2(1):15–23.
23. Siepmann J, Peppas NA. Modeling of drug release from delivery systems based on hydroxypropyl methylcellulose (HPMC). Adv Drug Deliv Rev. 2001;48(2–3):139–57. https://doi.org/10.1016/S0169-409X(01)00112-0.
24. Lee KY, Mooney DJ. Alginate: properties and biomedical applications. Prog Polym Sci. 2012;37(1):106–26. https://doi.org/10.1016/j.progpolymsci.2011.06.003.
25. Bhattarai N, Gunn J, Zhang M. Chitosan-based hydrogels for controlled, localized drug delivery. Adv Drug Deliv Rev. 2010;62(1):83–99. https://doi.org/10.1016/j.addr.2009.07.019.
26. Kumar MN, Muzzarelli RA, Muzzarelli C, Sashiwa H, Domb AJ. Chitosan chemistry and pharmaceutical perspectives. Chem Rev. 2004;104(12):6017–84. https://doi.org/10.1021/cr030441b.
27. Morris ER, Nishinari K, Rinaudo M. Gelation of alginate and chitosan. Carbohydr Polym. 2011;84(1):95–102. https://doi.org/10.1016/j.carbpol.2010.12.008.
28. Singh A, Sharma PK, Garg VK, Garg G. Review on alginate-based hydrogel for drug delivery applications. Int J Pharm Pharm Sci. 2008;10(2):30–6.
29. Patel MM, Patel SL, Patel MT. Natural polymers in colon-specific drug delivery. J Sci Ind Res. 2007;66:869–73.

30. Muhammad FM, Chatterjee B, Jaffri JM. pH-responsive polymers in drug delivery. Res J Pharm Technol. 2018;11(11):5115–22. https://doi.org/10.5958/0974-360X.2018.00934.
31. Coşman BP, Bucătariu SM, Constantin M, Fundueanu G. Temperature/pH-sensitive double cross-linked hydrogels as platform for controlled delivery of metoclopramide. Gels. 2022;8(12):824. https://doi.org/10.3390/gels8120824.
32. Herdiana Y, Wathoni N, Shamsuddin S, Muchtaridi M. Scale-up polymeric-based nanoparticles drug delivery systems: development and challenges. OpenNano. 2022;7(2022):100048. https://doi.org/10.1016/j.onano.2022.100048.
33. Draft guidelines for evaluation of nanopharmaceuticals in India government of India directorate general of health services central drugs standard control organization (new drugs division) (2019). https://cdsco.gov.in/opencms/resources/UploadCDSCOWeb/2018/UploadPublic_NoticesFiles/newdrunoti7march.pdf
34. Guidelines for evaluation of nanopharmaceuticals in India. Government of India New Delhi (2019). https://dbtindia.gov.in/sites/default/files/uploadfiles/Guidelines_For_Evaluation_of_Nanopharmaceuticals_in_India_24.10.19.pdf
35. Yijia L, Ruizhen T, Jiayun X, Yingping Z, Tingting W, Junqiu L. Recent developments of polymeric delivery systems in gene therapeutics. Polym Chem. 2024;15:1908–31. https://doi.org/10.1039/D4PY00124A.
36. Benner B, Scarberry L, Suarez-Kelly LP, Duggan MC, Campbell AR, Smith E, Lapurga G, Jiang K, Butchar JP, Tridandapani S, Howard JH, Baiocchi RA, Mace TA, Carson WE 3rd. Generation of monocyte-derived tumor-associated macrophages using tumor-conditioned media provides a novel method to study tumor-associated macrophages in vitro. J Immunother Cancer. 2019;7(1):140. https://doi.org/10.1186/s40425-019-0622-0.
37. Yan Y, Xiong H, Zhang X, Cheng Q, Siegwart DJ. Systemic mRNA delivery to the lungs by functional polyester-based carriers. Biomacromolecules. 2017;18(12):4307–15. https://doi.org/10.1021/acs.biomac.7b01356.

Extraction and Characterization of Natural Polymers for Pharmaceutical Use

3

Pragya Sharma, Smita Jain, Shagufta Khan, and Meghraj Suryawanshi

3.1 Introduction

Natural polymers, derived from renewable sources, have emerged as pivotal materials in pharmaceutical applications due to their exceptional biocompatibility, biodegradability, and sustainability. The rising need for eco-friendly alternatives to synthetic materials had driven extensive research into their extraction, characterization, and optimization. This chapter delves into the extraction and analysis of key natural polymers, including cellulose, pectin, chitosan and alginate, highlighting their versatile applications in drug delivery systems, excipients, and biomedical devices.

Polymers formed by the polymerization of monomers exhibit as diverse array of properties tailored for specific functions. In nature, polymers such as cellulose, proteins, alginate, starch, chitin, and chitosan are ubiquitously distributed across plants, animals, and microorganisms. Among these, cellulose is the most abundant biopolymer, primarily sourced from plant cell walls, but also bacteria and algae. Its crystalline structure, resulting from intermolecular hydrogen bonding, imparts remarkable

P. Sharma (✉)
Amity Institute of Pharmacy, Amity University Rajasthan, Jaipur, Rajasthan, India

S. Jain
Department of Pharmacy, School of Chemical Sciences & Pharmacy, Central University Rajasthan, Kishangarh, India

S. Khan
Department of Pharmaceutical Chemistry, College of Pharmacy, Jazan University, Jazan, Kingdom of Saudi Arabia

M. Suryawanshi
Department of Pharmaceutics, Sandip Institute of Pharmaceutical Sciences (SIPS), Affiliated to Savitribai Phule Pune University (SPPU, Pune), Nashik, Maharashtra, India

J. S et al. (eds.), *Innovative Pharmaceutical Excipients: Natural Sources*,
https://doi.org/10.1007/978-981-96-7959-1_3

mechanical strength and thermal stability. However, limitations such as low solubility and lack of thermoplasticity necessitate chemical or physical modifications to enhance its utility in drug formulations. Methods like alkali treatment, bleaching, and hydrolysis have been employed to isolate and refine cellulose, yielding forms such as microcrystalline cellulose with applications in controlled drug release and composite materials.

Pectin, a negatively charged polysaccharide predominantly extracted from citrus peels and apple pomace, is another promising polymer due to its gelling properties and ability to form biocompatible films. Extraction techniques, including acid-mediated, enzymatic and advanced methods like microwave-assisted and subcritical water extraction, have been optimized to enhance pectin yield and purity. Similarly, chitosan, derived from the deacetylation of chitin, is widely sourced from crustacean shells and fungal cell walls. Its antimicrobial properties and solubility in acidic solutions make it a valuable component in wound dressings and novel drug delivery systems. Extraction methods range from traditional chemical approaches involving demineralization and deproteination to eco-friendly enzymatic and fermentation-based techniques.

Alginate, predominantly obtained from brown algae, is a versatile polysaccharide known for its mucoadhesive properties and gel-forming capabilities. The extraction process involves proton exchange, neutralization, and purification steps to yield sodium alginate, which is extensively used in pharmaceutical formulations for sustained drug release. These natural polymers, when subjected to advanced characterization techniques such as Fourier-transform infrared spectroscopy (FTIR), nuclear magnetic resonance (NMR) spectroscopy, and scanning electron microscopy (SEM), demonstrate structural and functional attributes essential for pharmaceutical applications.

The significance of these biopolymers lies not only in their intrinsic properties but also in their contribution to sustainable development goals. By replacing synthetic materials, natural polymers mitigate environmental impact while enabling the design of safer and more effective pharmaceutical products. This chapter underscores the importance of optimizing extraction methods to preserve polymer integrity and maximize yield, thereby paving the way for innovative applications in drug formulation and delivery. Through a comprehensive understanding of their extraction, modification, and characterization, natural polymers hold the potential to revolutionize the pharmaceutical industry, aligning it with modern sustainability objectives.

3.2 Extraction Techniques of Natural Polymers (Cellulose, Pectin, Chitosan, and Alginate)

Polymers are made up of several tiny molecules joined together to form a single big molecule. Polymers are made from small molecules called monomers. While building polymers are composed of monomer units that link together in a way that provides a tiny molecule, usually water, during each reaction, expanding polymers are composed of monomer units that are purposefully joined together. In nature, polymers are widely distributed. The bodies of plants, animals, and human beings include a variety of natural polymers, such as cellulose, proteins, alginate, starch, gelatine, chitin, and chitosan. To enhance their performance in a variety of applications, these polymers can be separated using different extraction methods, then improved and altered [1].

3.2.1 Cellulose

Cellulose is potentially the most abundant chemical on Earth, as it makes up the majority of plant cell walls and nearly half of the biomass in photosynthetic organisms. Traditional sources of cellulose include a variety of materials such as cotton, linen, wood, hemp, jute, kenaf, sugar beet cereal straws, and flax. Additionally, cellulose can be obtained from other organisms, including certain bacteria (e.g., Acetobacter), algae (such as Valonia and Microdictyon), and marine animals belonging to the Ascites family [2].

The chemical composition of cellulose shows three hydroxyl groups per anhydroglucose unit (AGU), except at the terminal ends. These hydroxyl groups work together to form intra- and inter-hydroxyl hydrogen bonds, creating a supramolecular polymer with crystalline (ordered) regions [3]. Amorphous patches, formed by randomly arranged cellulose molecules, link the ordered crystalline areas [4]. The cellulose structure significantly influences its reactivity. Surface hydroxyl groups on cellulose molecules can establish both intramolecular and intermolecular hydrogen bonds [5]. These hydrogen bonds between polymeric chains largely determine their physical and chemical characteristics, potentially enhancing their linear integrity. However, cellulose has inherent limitations, including poor crease resistance, limited solvent solubility, and lack of thermoplasticity. To enhance cellulose properties, controlled chemical or physical surface modification is required [6].

Generally, the component of natural cellulose produced through biosynthesis cannot be processed like a synthetic polymer [7]. A proposed solution involves modifying cellulose molecules by attaching foreign groups to their surface. This approach allows for alterations to the cellulose's surface chemistry without impacting crucial inherent properties of the drug, such as self-assembly, controlled dispersion within various matrix polymers, and enhancement of particle–particle and particle–matrix adhesion strengths. Many drug delivery systems utilize traditional cellulose and its analogues, including cellulose ethers, esters, and oxycellulose. Several naturally occurring cellulose ethers exist, such as sodium carboxymethyl cellulose (CMC), methylcellulose (MC), ethylcellulose (EC), hydroxypropyl cellulose (HPC), and hydroxypropyl methylcellulose (HPMC) [8].

3.2.2 Cellulose Extraction

Cellulose can be extracted mechanically or chemically into nano or microforms [9]. Cellulosic fiber extraction is typically carried out in three main steps: pre-hydrolysis treatment, which opens the matrix by using mineral acid or alkali. One of the most important steps in producing crystalline cellulose from various sources is pre-treatment, which involves removing or changing substances like lignin that obstruct the crystallization process. Biological, physical, and chemical processes can all be used for pre-treatment. Using an alkali, such as NaOH, to cook the fiber is known as pulping. The last stage, bleaching, which can be done with a variety of substances, including H_2O_2, extracts the pure bleached form of cellulose. Hydrolysis with sulfuric acid in a reflux medium is done to increase the purity of the cellulose that is produced [10].

Numerous studies on the extraction of cellulose from various sources have been carried out in 2023, for instance, leftover celery pulp was used to extract cellulose microfibers, as reported by Abzan and colleagues [11]. Two chemo-mechanical methods, including bleaching with $NaClO_2$ and NaOCl, were used in this study to extract pure cellulose from leftover celery pulp (Apium graveolens var. dulce). The diameters of the untreated cellulose fibers ranged from 100 to 150 μm, while the treated ones ranged from 10 to 15 μm, according to optical microscopy and morphology studies. The results also demonstrated the high product stabilities under mechanical and thermal conditions. Cellulose nanocrystals were extracted from hardwood pulp and reported by Chen et al. [12].

Following the oxidation of these cellulose fibers with ammonium persulfate, the reaction suspension was centrifuged to produce two types of cellulose nanocrystals. The cellulose nanocrystals were categorized as 1 and 2 based on their reliance on ultrasonication. Higher surface charge, crystallinity, thermal stability, shorter length, narrower length distribution, and marginally wider width were all displayed by cellulose nanocrystal 1. The highest yields were 42.65% and 34.11%. In order to create cellulose or materials based on cellulose, Freixo et al. utilized sugarcane bagasse, a significant by-product of the sugarcane industry, as feedstock [13]. Pretreatment techniques of lignocellulosic feedstocks were employed for the composition of the complex cellulose–lignin hemicellulose matrix in order to obtain more purified cellulose. Due to its high hemicellulose extraction selectivity, autohydrolysis was actually employed as a cost-effective fractionation technique. Without the use of chemical agents, this process used hot water (typically between 160 °C and 240 °C) and pressures between 10 and 35 bar. Bleaching, alkaline sulfite extraction and the impact of numerous conditions were assessed. A bleaching step after autohydrolysis at 170 °C produced the best results in terms of crystallinity (73.4%) and cellulose purity (77.8%).

Cellulose consists of glucopyranosyl units connected by 1 → 4 bonds. It forms monoclinic, rod-like crystals with a long b-axis composed of chains aligned parallel to the fiber direction. These chains are believed to be folded, enabling the formation of intrachain hydrogen bonds between O-4 and O-6, as well as O-3 and O-5. The parallel chains are stabilized by intermolecular hydrogen bonds along the a-axis, while hydrophobic interactions occur along the c-axis. Crystalline regions account for 60% of the original cellulose structure. These sections are separated by amorphous gel areas that can crystallize upon dehydration. These areas are thought to contain acid- or alkali-sensitive linkages. When hydrolyzed, these bonds lead to the production of microcrystalline cellulose. This partially depolymerized cellulose product has a molecular weight of 30–50 kD and lacks a fibrous structure, yet remains insoluble in water [14].

3.2.3 Cellulose Modification

The growing popularity of cellulose macro- and nanofibers stems from their remarkable properties, including high strength, stiffness, biodegradability, and renewability, as well as their use in composite fabrication. Research into the application of cellulose nanofibers for composite production is a relatively recent field. The outstanding mechanical, thermal, and biodegradation properties of cellulose macro- and nanofibers make them excellent candidates for strengthening composite materials. Due to the inherent water-loving nature of cellulose fibers, increasing their surface roughness is essential for developing composites with enhanced characteristics [15].

3.2.4 Source of Pectin

Pectin, a complex negatively charged polysaccharide, is commonly extracted from the central layer of plant cell walls, particularly from fruits and vegetables. The primary commercial sources for industrial pectin extraction are citrus peels (predominantly lemon, orange, and lime due to their high pectin content) and apples. However, numerous other potential sources exist, including various agricultural by-products such as apple pomace, banana skins, grapefruit rinds, sisal remnants, pumpkin, tomato, grape, mango, watermelon rind, sunflower heads, carrots, and pulp from coffee and potato processing [16, 17].

3.2.5 Methods for Pectin Extraction

Several common methods that are employed to extract pectin from diverse sources can be seen in (Fig. 3.1). The extraction of pectin is influenced by several factors in the production process. These include the concentration of acid (when utilized), the temperature or power applied, the ratio of solids to liquids, and the length of time for extraction. These variables impact various characteristics of the resulting pectin, such as its molecular weight, chemical makeup, level of purity, color, and the extent of esterification [18]. Agro-industrial residues must be pre-treated before pectin can be extracted. Despite being optional, this technique has an impact on the product's quality and quantity. With the purpose of increasing the effectiveness of extraction, the dregs are primarily cleansed via washing (washing with water, acid washing, or alcohol washing). After that, the drying stage is performed. Moreover, powdering the material further increases the surface area and improves extraction [16, 19].

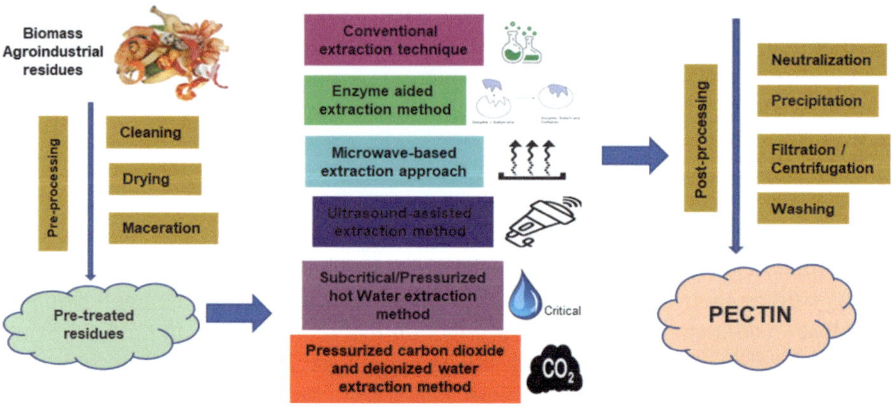

Fig. 3.1 An overview of the process for extracting pectin

Acid-Mediated or Conventional Extraction The extraction process employs acidic conditions and elevated temperatures (pH 1.5–3.0, 75–100 °C) for a duration of up to 3 h. The procedure involves heating the raw materials in acidic buffers to convert protopectin into pectin and dissolve the resulting pectin. Traditionally, potent and highly corrosive mineral acids such as HNO3, HCl, and H2SO4 have been widely utilized. However, these acids present challenges, including high extraction costs and environmental concerns. To address these issues, milder food-grade organic acids like acetic, citric, oxalic and tartaric acids have been introduced. Notably, citric acid, functioning as an effective chelating agent, demonstrated the highest pectin yield (14.34%) when extracting from potato pulp [18–20].

In order to release pectin, the cell wall matrix is broken down and degraded using enzymes in the enzyme-assisted extraction method. Protopectinases, cellulases, proteases, α-amylase, β-glucanase, hemicellulases, and xylanases are the most commonly used enzymes. Environmental friendliness, high yield, high selectivity, low temperature, low energy consumption, and the avoidance of harmful chemicals like acids are just a few of the many benefits of this process. However, this method is more expensive and tedious, and enzymes are sensitive to pH, temperature, and chemical compositions [18].

When the dielectric materials are exposed to microwave radiation, heat is produced using the microwave-based approach. Compared to the traditional method, it saves more time and has a shorter extraction period. But longer irradiation times or higher microwave power could result in pectin degradation and decreased production [18]. The ultrasound-assisted extraction method uses sound waves between 20 and 40 kHz to remove pectin. There were several reasons why the raw materials were exposed to ultrasonic radiation. Actually, ultrasonic vibrations create pressure changes that create microbubbles, or microjets, which break down cell membranes. Compared to conventional extraction techniques, this procedure has several benefits, including a shorter extraction time, a higher yield, and environmental friendliness [19].

Subcritical water has been used gradually in recent years to extract pectin. Pressurized hot water, also known as subcritical water, is water maintained in a liquid state at temperatures exceeding 100 °C (the typical boiling point) through the application of pressure. This form of water can be utilized to facilitate the extraction of pectin from agricultural and industrial by-products. Water in its liquid form at temperatures ranges from 100 °C to 374 °C (the critical point) is known as subcritical water. This extraction technique offers advantages over traditional acid-based methods, being more environmentally friendly, cost-effective, and less hazardous. However, higher temperatures from a particular range may result in pectin degradation (based on the pectin content in the source materials, as well as components that are sensitive to heat) [16]. Conversely, a novel extraction method utilizing pressurized carbon dioxide and deionized water can create acidic conditions in water and produce carbonate ions, which naturally chelate calcium ions. Although this technique is still in its early stages, it shows considerable promise due to its low environmental impact [16]. Furthermore, the objective of post-extraction processing is to eliminate or neutralize chemicals that have mixed with the pectin, as well as to remove or replace the buffer/solvent before subsequent use. Following filtration of the extract, the solubilized pectin should be precipitated by combining it with alcohol (60% ethanol) [20].

3.2.6 Source and Extraction Techniques of Chitin and Chitosan

Chitin is existent in an immense range of sources. Most amply it is found in the crustacean's shells and marine exoskeleton (like crabs, shrimp, crustaceans, arthropods, krill, cephalopods, prawns, shrimps, lobsters, and mollusks), multilayered extracellular matrix of arthropod, and cell walls of fungi, yeasts, and insects. Marine-derived raw materials are utilized for the large-scale production of chitin and chitosan [21, 22]. Conversely, in current eons, unconventional sources like biomass of microbes and insects, mushroom/fungal, and cell walls have gathered heightened attention. The origin of chitin, the parameters of the reaction, and the extent of deacetylation are critical determinants influencing the molecular weight of chitosan as well as its functional characteristics. Chitin and chitosan are procured utilizing both chemical and biological methodologies, and at times, a synergistic application of these techniques may be employed [23, 24]. Chitin and chitosan are extracted from crustacean shell debris, mollusks, insects, and fungus using the traditional (chemical extraction) approach, which involves high temperatures (>100 °C) and alkaline and acidic solvents (NaOH >40% wt) over an extended length of time (Fig. 3.2). The most popular method and one that is frequently used commercial purposes is the chemical process. Three reaction steps are involved in the preparation of chitosan using this method: (1) demineralization, in which the minerals, primarily $CaCO_3$, in the shell are eliminated by treating it with strong acids (H_2SO_4, HNO3, HCl, and HCOOH) for 2–3 h while stirring (2) deproteination

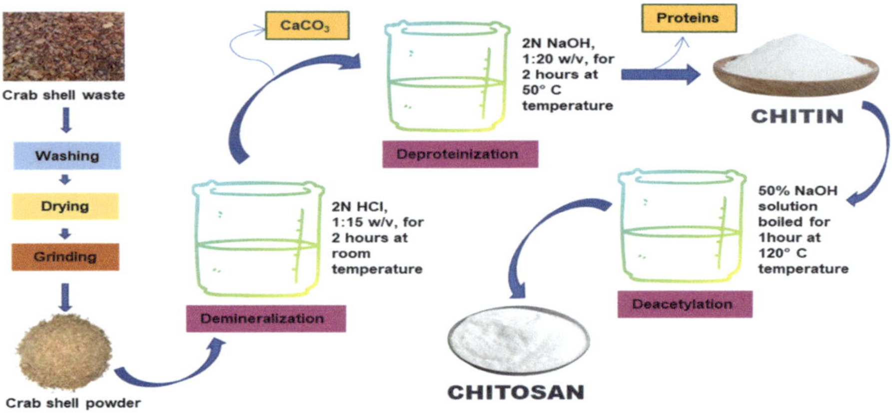

Fig. 3.2 An overview of the process for extracting chitin and chitosan

involves the application of chemical agents such as sodium hydroxide (NaOH) and potassium hydroxide (KOH) to disrupt the chemical bonds between proteins and chitin. This process facilitates the removal of proteins and other organic materials, excluding chitin, at temperatures ranging from 65 °C to 100 °C and (3) deacetylation—amide compounds are subjected to hydrolytic processes in both acidic and alkaline environments. Hisham and colleagues employed a two-stage process involving acid and alkaline treatments to extract chitin and chitosan from shrimp shells. The deproteinization of shrimp shells was accomplished using NaOH, while demineralization was carried out using hydrochloric acid at concentrations of 1%, 2%, 3%, 4%, and 5%. FTIR analysis revealed that the extracted chitin was in the α-chitin isomorph form. The surface morphology of chitin became more pronounced as the acid concentration increased. This chemical method effectively produced chitosan with a 65% degree of deacetylation and a rod-like microstructure. The resulting chitosan was soluble in 2% (w/v) acetic acid [25].

In addition to the chemical methodology, environmentally sustainable techniques encompassing microbial fermentation, microwave-assisted techniques, and enzymatic processes are available for the synthesis of chitosan. In fact, bio-based techniques have been developed to address the issue of chemical methods, which alter the physicochemical properties of chitosan. Acid is used in a demineralization process to extract the $CaCO_3$ in a shell in enzyme-assisted extraction. Milder temperatures (25–59 °C) are used to carry out the enzyme-based deproteinization and deacetylation reactions [23]. Microorganisms and fish intestines are common sources for enzyme extraction. Utilizing enzymes that prevents the uneven deacetylation and molecular weight reduction covers examples like chitinases, carbohydrases, and proteases. Although this method offers the advantage of extraction under gentle conditions, it faces numerous challenges for large-scale industrial production. The primary concern is the high cost of enzymes used in deproteinization and deacetylation processes, which stands in stark contrast to the more economical chemical reagents. In a study conducted by Hongkulsup and colleagues [26], the researchers demonstrated the use of enzymatic methods to extract chitin from the exoskeletons of Litopenaeus vannamei shrimp. The impact of various process parameters, including the ratio of enzyme to substrate and duration of incubation, was investigated to assess process efficiency and yield outcomes. Shells that underwent deproteinization were subjected to treatment with 0.5 M solutions of lactic and acetic acids. For both acids, the demineralization process showed improvement (duration: 20 min), reaching peak values of 73.85% and 65.66% for lactic and acetic acids, respectively. Lactic acid achieved a higher degree of demineralization due to its stronger acidic nature (pKa = 3.86) compared to acetic acid (pKa = 4.76). The findings revealed that demineralization rates of 98.64 and 97.57% were attained for lactic and acetic acids, respectively, under the following conditions: temperature of 25 °C, duration of 20 minutes, shells-to-lactic acid ratio of 1:1.1 w/w, and shells-to-acetic acid ratio of 1:1.2 w/w.

To address the issue of expensive enzymes, researchers have proposed an alternative approach using microbial fermentation. This method is divided into two categories: lactic acid fermentation, where microbial strains produce lactic acid for demineralization, and non-lactic acid fermentation, which utilizes other organic acids. Although chemical methods offer advantages such as faster processing times and less complex manufacturing procedures, they have drawbacks. These include producing chitosan with lower molecular weight and higher deacetylation degrees, requiring harmful or caustic chemicals like HCl, and generating unwanted by-products. In contrast, biological methods yield chitosan with higher molecular weight and superior mechanical properties. Nian Tan and colleagues documented the extraction of chitin from seafood by utilizing sugars obtained from fruit waste streams [27]. The co-fermentation of shrimp waste was conducted using two bacterial strains: Lactobacillus plantarum subsp. plantarum ATCC 14917 and Bacillus subtilis subsp. subtilis ATCC 6051. Chitin was successfully extracted from the prawn shells following their proteinization and demineralization. Utilizing waste substrate glucose supplementation for fermentation maximized waste utilization while lowering costs. As carbon sources, sugarcane molasses, light corn syrup, red and white grape pomace, apple, pineapple, and banana peels, as well as potato, mango, and sweet potato peels were analyzed and contrasted. At 0.89 g, the chitin that was extracted from fermentation using potato peel had the highest dry weight. A 6.85 C/N ratio, 72.90% deacetylation degree, and 95.54% crystallinity degree were obtained when chitin was recovered using red grape pomace. The microwave-aided extraction technique offers numerous benefits, such as low operational costs, brief heating durations, energy efficiency, straightforward process control, and selective heating capabilities. This method has been utilized for extracting polysaccharides like chitosan, while also minimizing the need for chemicals. Various factors, including solvent concentration, solid-to-liquid ratio, and reaction time, influence the molecular weight (MW) and degree of deacetylation, making it essential to optimize these conditions. Chitosan produced through microwave heating exhibited a higher molecular weight compared to that obtained via conventional heating methods, while maintaining similar crystallinity, structures, and morphologies [23].

3.2.7 Sources and Extraction of Alginate

Two major sources of alginates are seaweed (covers the cell wall of brown marine algae like *Phaeophyceae* family) and bacteria (resembling *Azotobacter* and *Pseudomonas*). Their commercial extraction is done through algae sources, although microbial fermentation is possible for manufacture. Numerous sources of brown marine algae are *Ascophyllum nodosum*, *Laminaria hyperborea*, and *Macrocystis pyrifera* [28–30].

3.2.8 Extraction of Sodium Alginate and Other Forms of Alginates from Seaweeds

General Extraction Procedure The process of extracting alginates (Fig. 3.3) from brown algae involves several steps. First, the raw algae are mechanically harvested and then dried, with the exception of *M. pyrifera*, which can be used in wet conditions. Next, the dried material is treated with mineral acids to remove counter ions through a proton exchange.

Following this, alginate is extracted by solubilizing the insoluble fractions of alginic acid via a neutralization process, using an alkali agent such as NaOH or Na2CO3. To separate the precipitates, centrifugation or flotation is utilized. The precipitates are then treated with alcohols or mineral acids to directly extract sodium (Na)-alginate. This step helps eliminate impurities, such as the homopolysaccharides fucoidan and laminarin.

An alternative method for extracting alginate involves using barium (Ba) or calcium (Ca) ions to produce a stable alginate gel. Finally, pure Na-alginate is obtained by treating the material with alkaline solutions.

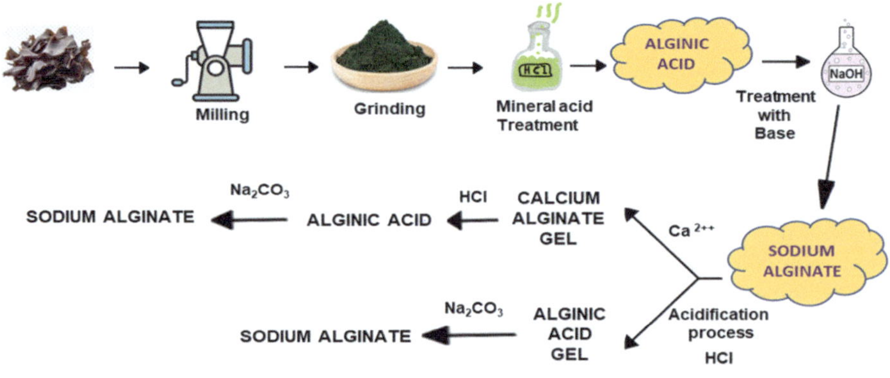

Fig. 3.3 The diagrammatic representation elucidating the procedural flow for the isolation of alginates from algal sources

3.3 Characterization of the Natural Polymer

Polysaccharides make up the majority of the mucilages and gums found in wild plants. Elucidating the structure, physicochemical characteristics, physicomechanical properties, and drug delivery capabilities of mucilages and gums in great detail is necessary to move the application of these substances from the bench to the pilot scale. Additionally, it is of the utmost importance to differentiate between each natural polymer and the others. It is possible to demonstrate the one-of-a-kind qualities of a natural polymer and enhance its utilization in medicinal applications by all-encompassing characterization [31].

3.3.1 Determination of the Structural Composition of Natural Polymers

In most cases, the identification of the polymer is accomplished by identifying the structure of the polymer. The chemical structure of polymers and the chain conformations of polymers are the factors that determine their qualities. There is a large variety of methods that may be used to characterize and categorize a polymer systematically. These methods are accessible to completely explicate the structure and chain conformation of a polymer. The polymeric structures can be classified and identified through the utilization of techniques such as Fourier transform infrared spectroscopy, solid-state nuclear magnetic resonance (NMR) (one and two dimensions), solid-state NMR, Raman spectroscopy, gas chromatography (GC), GC-mass spectroscopy (GC-MS), and high-performance liquid chromatography. Static and dynamic light dispersion, viscosity analysis, circular dichroism analysis, atomic force microscopy (AFM), single-molecule force spectroscopy based on AFM, fluorescence correlation spectroscopy, and nuclear magnetic resonance (NMR) spectroscopy are some of the techniques that may be used to shed light on the conformations of polysaccharide chains in solutions [31].

3.3.2 Determination of the Sugar Ingredients

For the purpose of identifying the sugar components, the gum or mucilage is hydrolyzed with diluted acids. Chromatographic methods, such as size exclusion chromatography with multi-angle laser light scattering, and gas chromatography (GC), are able to separate the numerous monosaccharides that are present in gum and mucilage. Elucidation of the structure may also be accomplished by the use of NMR spectroscopy and MS. In the field of polymer chemistry, the process of monosaccharide-monosaccharide bonding is best understood via the use of methylation, periodate, and lead tetra-acetate oxidation [32].

3.3.3 Polymorphism Analysis and Comparison

Changing the structure of the polymer may also result in a change in the degree of crystallinity of the polymer, which in turn has an additional impact on the characteristics of the polymer, including its solubility, stability, and drug release. Gel electrophoresis, differential scanning calorimetry (DSC), wide-angle X-ray diffraction (XRD), and powder X-ray diffraction (XRD) are some of the analytical methods that have been used in the process of determining polymorphism. In spite of this, the X-ray diffraction method is the most common method that is used in order to ascertain the degree of crystallinity that a polymer offers. There is a correlation between the degree of crystallinity in the resulting spectrum and the characteristic peaks. Indicative of the polymer's complete amorphous existence is the absence of distinctive peaks in the spectrum [33]. Amorphous or semi-crystalline polymers make up the vast bulk of naturally occurring polymers. While the polymer, which has a high crystallinity in its natural state, is more soluble, the opposite is also true [34].

3.3.4 The Determination of Polydispersity Index and Molecular Weight

Due to the heterogeneity and polydispersity characteristic of polysaccharides, it may be difficult to determine the molecular weights of polysaccharides. It is possible to see polysaccharide fractions of mucilage and gums by the use of preparative chromatography techniques such as scale exclusion chromatography (SEC), gas chromatography (GC), and viscometer rheometric method, which are often used for the assessment of molecular weight and polydispersity [35]. Due to the fact that it is a direct representation of molecular weight, the assessment of viscosity provides a concise understanding of the molecular weight. Using chromatographic methods, such as size exclusion chromatography (SEC), in conjunction with multi-angle laser light scattering results in a more rapid and dependable process [36]. Summers et al. have recently advocated for the use of an advanced polymer chromatography technique, which offers a higher resolution of polymer distribution while simultaneously reducing the amount of time required to assess molecular weight distribution. On the other hand, it seems that the technique is likewise recommended for samples of polymer with a low molecular weight.

3.3.5 Natural Polymers' Surface Features and Their Properties

The surface morphology of a polymer or its derivative can have an effect on the rate of drug release from the polymer as well as the processes involved [37]. The screening of pharmaceuticals is accomplished by the alteration of the surface of a polymer. It is also possible for the shape of a polymer, whether it is a natural or modified polymer, to influence the degree and duration of its circulation in the bloodstream, the interaction between polymers and drugs, nano, or microparticles and cells, or the encapsulation of living cells [38]. Therefore, it is essential to have a comprehension of the morphology of natural polymers, which includes aspects such as the form, size, and surface chemistry. This may be accomplished via the utilization of scanning tunnel microscopy, transmission electron microscopy (TEM), scanning electron microscopy (SEM), and atomic force microscopy (AFM). TEM gives a two-dimensional picture of a material, while SEM provides a three-dimensional perspective. Atomic force microscopy (AFM) goes even farther by revealing the molecular surface and mechanical characteristics of the substance. With the help of the inner ring display, transmission electron microscopy (TEM) demonstrated that the particle in question is a nanocapsule, and scanning electron microscopy (SEM) confirmed this with the hollowness display [39]. Polymer surface characterization may be improved by the use of technologies such as atomic force microscopy (AFM) and Raman spectroscopy. The high-resolution data of both chemical and morphological structures are supplied in addition to the topographical, electromagnetic, thermal, and near-field optical characteristics that have been described on the molecular scale [40].

3.4 Thermal Behavior Study of the System

The differential scanning calorimetry (DSC) is often used to identify the physical and chemical changes that occur in the polymer during the heat processes. The polymer is subjected to a variety of temperatures, both of which contribute to the identification of the glass transition, crystallization, and decomposition characteristics of the polymer. The behavior of polymers is distinct both above and below the temperatures at which they undergo the glass transition [41]. A further factor that influences the mechanical characteristics of the polymer is the temperature at which it transitions. Due to the fact that the thermal activity of polymers is unique, it is expected that each polymer would exhibit a different thermogram profile. Dilatometry, thermally stimulated current spectroscopy, and dynamic mechanical thermal analysis are some of the other methods that are used in the process of investigation into the thermal behavior of polymeric materials [42].

3.5 An Investigation of the Rheological Behavior of Natural Polymer

The viscoelastic nature of the polymer is yet another attribute that is determined by the molecular weight of the material. Several factors are influenced by the viscoelastic qualities of the material, including the strength of the substance, its stiffness, the release of the medicine, its suspend ability, and distribution [43]. However, the rheometer can measure not only the viscosity and flux of the polymer, but also other parameters, such as the effect of shear on the flow property of the polymer, the effect of different frequencies or temperatures on polymers, and it assists in gaining an understanding of the polymer's ability to behave as a viscous fluid, paste, gel, or a three-dimensional network [44].

3.6 Conclusion

Natural polymers, such as cellulose, pectin, chitosan and alginate, hold immense promise as sustainable alternatives to synthetic materials in pharmaceutical applications. Their exceptional biocompatibility, biodegradability, and unique physicochemical properties position them as ideal candidates for various uses, including drug delivery systems, excipients, and biomedical devices. The optimization of extraction techniques, including solvent extraction, enzymatic treatments and advanced methods like microwave-assisted processing, ensures the preservation of polymer structure and functionality while maximizing yield. Characterization through advanced techniques like FTIR, NMR spectroscopy, and SEM validate their structural integrity and suitability for medicinal applications.

Beyond their inherent properties, these polymers contribute significantly to the development of eco-friendly pharmaceutical solutions, addressing the growing demand for sustainable materials. Their applications in enhancing drug delivery, improving therapeutic efficacy, and reducing environmental impact underscore their transformative potential. This chapter demonstrates the critical role of extraction and characterization in unlocking the full potential of natural polymers. By advancing our understanding of their properties and refining their applications, these biopolymers can pave the way for safer, more effective, and environmentally conscious pharmaceutical products. The continued exploration of natural polymers aligns with modern sustainability goals, fostering innovation in healthcare while minimizing ecological footprints.

Acknowledgments The authors are grateful to Prof. Amit Jain, Vice-Chancellor, Amity University Rajasthan, Jaipur, Rajasthan, India, for providing all the necessary resources in the completion of this chapter.

References

1. Dávid A, Lewandrowski KU, Josten C, Ekkernkamp A, Clasbrummel B, Muhr G. Surgical correction of talipes equinovarus following foot and leg compartment syndrome. Foot Ankle Int. 1996;17(6):334–9. https://doi.org/10.1177/107110079601700607.
2. Pérez S, Mazeau K. Conformations, structures, and morphologies of celluloses. In: Polysaccharides: Structural diversity and functional versatility. 2nd ed; 2005. https://doi.org/10.1201/9781420030822.
3. Allen AW, Cook JG, Armbruster MJ. Habitat suitability index models: pronghorn. Washington, DC: Western Energy and Land Use Team, Division of Biological Services, Research and Development, Fish and Wildlife Service, US Department of the Interior; 1984.

4. Kulasinski K, Keten S, Churakov SV, Derome D, Carmeliet J. A comparative molecular dynamics study of crystalline, paracrystalline and amorphous states of cellulose. Cellulose. 2014;21(3):1103–16. https://doi.org/10.1007/s10570-014-0213-7.
5. Koch U, Popelier PLA. Characterization of C-H-O hydrogen bonds on the basis of the charge density. J Phys Chem. 1995;99(24):9747–54. https://doi.org/10.1021/j100024a016.
6. Hebeish A, Guthrie JT. The chemistry and technology of cellulosic copolymers. Heidelberg: Springer; 1981. ISBN 3-540-10164-0.
7. Roy D, Semsarilar M, Guthrie JT, Perrier S. Cellulose modification by polymer grafting: a review. Chem Soc Rev. 2009;38(7):2046. https://doi.org/10.1039/b808639g.
8. Farrán A, Cai C, Sandoval M, Xu Y, Liu J, Hernáiz MJ, Linhardt RJ. Green solvents in carbohydrate chemistry: from raw materials to fine chemicals. Chem Rev. 2015;115(14):6811–53. https://doi.org/10.1021/cr500719h.
9. Trache D, Tarchoun AF, Derradji M, Hamidon TS, Masruchin N, Brosse N, Hussin MH. Nanocellulose: from fundamentals to advanced applications. Front Chem. 2020;8:392. https://doi.org/10.3389/fchem.2020.00392.
10. Chopra L, Manikanika. Extraction of cellulosic fibers from the natural resources: a short review. Mater Today Proc. 2022;48:1265–70. https://doi.org/10.1016/j.matpr.2021.08.267.
11. Abzan N, Abbasian A, Jonoobi M, Ghasemi I. Cellulose microfiber extraction from leftover celery pulp: chemomechanical treatments, structural, morphological, and thermal characterization. Int J Biol Macromol. 2023;253(Pt 3):126834. https://doi.org/10.1016/j.ijbiomac.2023.126834.
12. Chen Z, Xie Z, Jiang H. Extraction of the cellulose nanocrystals via ammonium persulfate oxidation of beaten cellulose fibers. Carbohydr Polym. 2023;318:121129. https://doi.org/10.1016/j.carbpol.2023.121129.
13. Freixo R, Casanova F, Ribeiro AB, Pereira CF, Costa EM, Pintado ME, Ramos ÓL. Extraction methods and characterization of cellulose fractions from a sugarcane by-product for potential industry applications. Ind Crop Prod. 2023;197:Article 116615. https://doi.org/10.1016/j.indcrop.2023.116615.
14. Belitz HD, Grosch W, Schieberle P. Food chemistry. 4th ed. Berlin: Springer-Verlag; 2009. 1070p.
15. Kalia S, Dufresne A, Cherian BM, Kaith BS, Avérous L, Njuguna J, Nassiopoulos E. Cellulose-based bio- and nanocomposites: a review. Int J Polym Sci. 2011;Article ID 837875:35. https://doi.org/10.1155/2011/837875.
16. Roy S, Priyadarshi R, Łopusiewicz Ł, Biswas D, Chandel V, Rhim JW. Recent progress in pectin extraction, characterization, and pectin-based films for active food packaging applications: a review. Int J Biol Macromol. 2023;239:124248. https://doi.org/10.1016/j.ijbiomac.2023.124248.
17. Huang J, Zhihong H, Lingping H, Li G, Yao Q, Yaqin H. Pectin-based active packaging: a critical review on preparation, physical properties and novel application in food preservation. Trends Food Sci Technol. 2021;118(Part A):167–78. https://doi.org/10.1016/j.tifs.2021.09.026. ISSN 0924-2244.
18. Roman-Benn A, Contador CA, Li M-W, Lam H-M, Ah-Hen K, Ulloa PE, Ravanal MC. Pectin: an overview of sources, extraction and applications in food products, biomedical, pharmaceutical and environmental issues. Food Chem Adv. 2023;2:100192. https://doi.org/10.1016/j.focha.2023.100192. ISSN 2772-753X.
19. Singhal S, Hulle NR. Citrus pectins: structural properties, extraction methods, modifications and applications in food systems—a review, vol. 2. Appl Food Res; 2022. p. 100215.
20. Sun RR, Niu Y, Li M, Liu Y, Wang K, Gao Z, Wang Z, Yue T, Yuan Y. Emerging trends in pectin functional processing and its fortification for synbiotics: a review. Trends Food Sci Technol. 2023;134:80–97. https://doi.org/10.1016/j.tifs.2023.03.004.

21. Huq T, Khan A, Brown D, Dhayagude N, He Z, Ni Y. Sources, production and commercial applications of fungal chitosan: A review. J Bioresour Bioprod. 2022;7(2):85–98. https://doi.org/10.1016/j.jobab.2022.01.002.
22. Azmana M, Mahmood S, Hilles AR, Rahman A, Arifin MAB, Ahmed S. A review on chitosan and chitosan-based bionanocomposites: promising material for combatting global issues and its applications. Int J Biol Macromol. 2021;185:832–48. https://doi.org/10.1016/j.ijbiomac.2021.07.023.
23. Mohan K, Ganesan AR, Ezhilarasi PN, Kondamareddy KK, Rajan DK, Sathishkumar P, Rajarajeswaran J, Conterno L. Green and eco-friendly approaches for the extraction of chitin and chitosan: a review. Carbohydr Polym. 2022;287:119349. https://doi.org/10.1016/j.carbpol.2022.119349.
24. Hisham F, Akmal MM, Ahmad F, Ahmad K, Samat N. Biopolymer chitosan: potential sources, extraction methods, and emerging applications. Ain Shams Eng J. 2024;15(2):102424.
25. Hisham F, Maziati Akmal MH, Ahmad FB, Ahmad K. Facile extraction of chitin and chitosan from shrimp shell. Mater Today Proc. 2021;42(Part 5):2369–73. https://doi.org/10.1016/j.matpr.2020.12.329. ISSN 2214-7853.
26. Hongkulsup C, Khutoryanskiy VV, Niranjan K. Enzyme assisted extraction of chitin from shrimp shells (Litopenaeus vannamei). J Chem Technol Biotechnol. 2015;91(5):1250–6. https://doi.org/10.1002/JCTB.4714.
27. Tan YN, Lee PP, Chen WN. Microbial extraction of chitin from seafood waste using sugars derived from fruit waste-stream. AMB Express. 2020;10(1):17. https://doi.org/10.1186/s13568-020-0954-7.
28. Hasnain, M. S., Jameel, E., Mohanta, B., Dhara, A. K., Alkahtani, S., & Nayak, A. K. (2020). Alginates: sources, structure, and properties. doi:https://doi.org/10.1016/b978-0-12-817640-5.00001-7.
29. da Silva, T. L., Vidart, J. M. M., da Silva, M. G. C., Gimenes, M. L., & Vieira, M. G. A. (2017). Alginate and sericin: environmental and pharmaceutical applications. InTech. doi: https://doi.org/10.5772/65257.
30. Draget KI, Nordga CTA. Handbook of hydrocolloids. Sawston: Woodhead Publishing; 2021. p. 805–29.
31. Yang L, Zhang L. Chemical structural and chain conformational characterization of some bioactive polysaccharides isolated from natural sources. Carbohydr Polym. 2009;76:349–61.
32. Prajapati VD, Jani GK, Moradiya NG, Randeria NP. Pharmaceutical applications of various natural gums, mucilages and their modified forms. Carbohydr Polym. 2013;92:1685–99.
33. Vinod VT, Sashidhar RB, Suresh KI, Rao R, Suresh B. Morphological, physico-chemical and structural characterization of gum kondagogu (*Cochlospermum gossypium*): A tree gum from India. Food Hydrocoll. 2008;22:899–915.
34. Shekunov BY, York P. Crystallization processes in pharmaceutical technology and drug delivery design. J Cryst Growth. 2000;211:122–36.
35. Harding SE, Varum K, Stokke BT, Smidsrod O. Molecular weight determination of polysaccharides. Adv Carbohydr Anal. 1991;1:63–144.
36. Zhang M, Zhang L, Cheung PC, Dong J. Fractionation and characterization of a polysaccharide from the sclerotia of Pleurotus tuber-regium by preparative size-exclusion chromatography. J Biochem Biophys Methods. 2003;56:281–9.
37. Summers M, O'Leary M. High speed, high-resolution analysis of low molecular weight polymers using the advanced polymer chromatography (APC) system. Polym Anal Appl. 2013;5:9–12.
38. Champion JA, Katare YK, Mitragotri S. Particle shape: A new design parameter for micro and nanoscale drug delivery carriers. J Control Release. 2007;121:3–9.
39. Caldorera-Moore M, Guimard N, Shi L, Roy K. Designer nanoparticles: incorporating size, shape and triggered release into nanoscale drug carriers. Expert Opin Drug Deliv. 2010;7:479–95.

40. Eronen P, Osterberg M, Jaaskelainen A. Effect of alkaline treatment on cellulose supramolecular structure studied with combined confocal Raman spectroscopy and atomic force microscopy. Cellulose. 2009;16:167–78.
41. Di Lorenzo ML. The crystallization and melting processes of poly (L-lactic acid). Macromol Symp. 2006;234:176–83.
42. Omelczuk MO, McGinity JW. The influence of polymer glass transition temperature and molecular weight on drug release from tablets containing poly (DL-lactic acid). Pharm Res. 1992;9:26–32.
43. Dong Y, Ruan Y, Wang H, Zhao Y, Bi D. Studies on glass transition temperature of chitosan with four techniques. J Appl Polym Sci. 2004;93:1553–8.
44. Jones DS, Woolfson AD, Brown AF. Textural, viscoelastic and mucoadhesive properties of pharmaceutical gels composed of cellulose polymers. Int J Pharm. 1997;151:223–33.

Plant-Derived Excipients: Applications in Drug Formulation

4

Jatin Kumar, Neha Kanojia, Jatindernath,
Manoj Dalabehera, and Prarit Chandel

4.1 Introduction

Excipients are components used in the manufacturing of medicinal products. They can be employed as a disintegrant, binder, or to modify pH. Excipients might be natural, synthetic, or semi-synthetic. In the pharmaceutical business, the use of excipients generated from plants is rapidly gaining traction resulting in new opportunities to address present drug delivery problems [1]. Natural materials are increasingly being used as excipients in formulation systems because they have relative advantages over synthetic polymers. Natural polymers including mucilage, gum and pectin are examples of natural excipients. Mango peels, banana peels, mango gum, dehydrated banana powder, guar gum, gum karaya, starch, ispaghula husk, chitosan, *Lepidium sativum* mucilage and fenugreek seed mucilage are a few sources of plant-derived natural polymers for pharmaceutical formulations [2]. Currently, more than 40% of medications approved by the US Food and Drug Administration (FDA) contain chemicals originating from plants. Products and materials made from plants are already becoming more and more well-liked. Numerous studies have already highlighted the potential functions and applications of bio polysaccharides generated from plants in the treatment of numerous illnesses during the last few decades. Since bio polysaccharides made from plant

J. Kumar (✉) · N. Kanojia · Jatindernath · P. Chandel
Chitkara University School of Pharmacy, Chitkara University,
Baddi, Himachal Pradesh, India
e-mail: prarit1055.bpharm21@chitkarauniversity.edu.in

M. Dalabehera
Department of Pharmaceutics, Uttaranchal Institute of Pharmaceutical Sciences,
Uttaranchal University, Deharadun, Uttarakhand, India

Chitkara College of Pharmacy, Chitkara University, Rajpura, Punjab, India
e-mail: manojdalabehera@myyahoo.com

J. S et al. (eds.), *Innovative Pharmaceutical Excipients: Natural Sources*,
https://doi.org/10.1007/978-981-96-7959-1_4

sources make up a large portion of the human diet, they are accepted as safe and non-toxic substances to eat [3]. Plant-derived excipients have become popular high molecular weight biopolymers with a number of great advantages over various conventional synthetic polymers. These advantages include broad accessibility due to the availability of plant-based natural sources, sustainable production, the availability of simple and affordable extraction techniques, aqueous solubility, the ability to swell in aqueous media, non-toxicity and biodegradability [4]. The ease of cultivation and harvesting, which offers a steady supply of plant materials, is the primary driver of the increased interest in using a range of excipients obtained from plants. These excipients are often photosynthetic by-products that are present in many plant parts, including leaves, pods, stems, grain, seeds, corms, rhizomes, roots and bark exudates. Plant-based excipients are important because they have the potential to offer sustainable substitutes for their traditional synthetic counterparts, supporting the pharmaceutical industry's dedication to environmentally friendly operations [5]. Investigating the scientific underpinnings of plant-based excipients is crucial in order to formulate formulations that satisfy both medicinal needs and environmental sustainability objectives, given the growing demand for natural and biodegradable ingredients. The purpose of this chapter is to give a thorough analysis of plant-based excipients used in medication formulation and their range of uses [6]. Plant-based excipient classification, the necessity for herbal excipients, particular examples of plant-derived excipients, commonly used plant-based materials, application obstacles, successful implementations and conclusion are all covered in this book chapter [7].

4.2 Classification of Plant-Derived Excipients

Polymers, gums and mucilage, waxes and oils and other plant-derived substances are all included in the category of excipients derived from plants. These natural excipients are essential to contemporary pharmaceutical formulations due to their non-toxic nature, biodegradability and affordability. Their many uses in improving medication delivery systems are highlighted by recent studies [8].

4.2.1 Polymers

Plant-based natural polymers are useful excipients because they can alter drug release characteristics and are biocompatible. For example, alginate, which is derived from brown algae, is used as a matrix in wound dressings and sustained-release tablets. Its efficacy in encapsulating curcumin to increase its bioavailability in therapeutic applications was demonstrated by recent research in 2024. Similarly, a crucial polymer for regulated drug delivery that improves drug permeability and stability is chitosan, which is frequently modified from plant-based sources. Alternatives to gelatin derived from plants, including soy protein-based polymers, are becoming more popular due to their biodegradable and environmentally beneficial qualities, especially in soft capsule formulations [9].

4.2.2 Gums and Mucilage

Gums and mucilages derived from plants are essential for enhancing drug solubility and therapeutic results because of their hydrophilic and sticky properties. The sap of the Astragalus species is used to make tragacanth gum, which is frequently used as a suspending agent in oral formulations. Its capacity to stabilise paediatric lipophilic medication solutions was demonstrated in a recent study conducted in 2024. Another noteworthy example is carrageenan, which is taken from red algae and used in gel formulations for topical medication delivery. Its versatility in pharmaceutical applications was demonstrated by the effective application of its antibacterial qualities in the development of therapies for skin infections [10].

4.2.3 Waxes and Oils

Plant-derived oils and waxes are vital excipients for emulsifying and stabilising medicinal formulations, while being rarely talked about. Because of its hydrophobic qualities, carnauba wax is frequently used as coatings for controlled-release tablets. When mixed with plant oils, it can regulate drug release rates, according to a 2024 study. The use of coconut oil, a lipid derived from plants, in nano-emulsions specifically, in the delivery of lipophilic medications like itraconazole has been thoroughly investigated. It is a priceless excipient due to its inherent antibacterial action and capacity to increase bioavailability [11].

4.2.4 Other Plant-Derived Compounds

This group comprises plant-based excipients with specific uses, like thickening, gelling and stabilising medication formulations. Corn and cassava are two common sources of starch, which is utilised in granulation and as a super disintegrant in tablet formulations. Targeted medication delivery systems using modified starch nanoparticles were the subject of recent studies in 2024. Another significant excipient used in colon-targeted drug administration is pectin, a polysaccharide that is taken from citrus fruits. Research has shown how well it shields probiotics from the hostile gastric environment, preserving their vitality in the intestinal tract [12].

4.3 Need of Herbal Polymers

The synthetic polymers used in the current pharmacological system are poisonous, non-biodegradable and incompatible with human anatomy. Furthermore, the synthetic synthesis of polymers necessitates the chemical synthesis of particular monomeric components, which drives up the cost of production. Synthetic polymer-based drug delivery methods can exhibit problems with physiological and biological compatibility [13]. Natural polymer-based drug delivery systems (DDS) are thought to be better than synthetic polymer-based ones because (i) Biodegradability: All living things, including bacteria, plants and animals, develop natural polymers. Because of their natural occurrence, they are easily broken down, metabolised and digested by enzymes, leaving no residue behind. Therefore they are readily broken down and processed by the body. (ii) Biocompatibility: Because biopolymers are derived from monomers that are frequently found in living things, they are both structurally and chemically compatible with the human body and does not show any toxicity. (iii) Economic: The cost of producing, purifying and processing them is decreased because the raw material is conveniently accessible and derived from natural resources. (iv) Eco-friendly: Under environmental conditions, by-products and other leftovers from the natural polymer processing industries decompose more readily. Furthermore, little to no additional processing is needed to enable its environmental integration. (v) Easy availability: They can be obtained through the spread of their sources and are found everywhere. Large-scale production of plant-based polymers is achieved by growing plant crops under natural and regulated circumstances [14].

4.4 Examples of Plant-Derived Excipients

Excipients produced from plants are essential in pharmaceutical formulations because of their multifunctional qualities, biocompatibility and biodegradability. Due to their many advantages, such as lower toxicity and cost-effectiveness, these natural polymers are becoming more and more preferred over synthetic ones. Specific instances and uses of these excipients originating from plants are described in the sections that follow. Various plant-based natural excipients that are often utilised in medication formulations are shown in the Fig. 4.1. It illustrates their distinct functions in

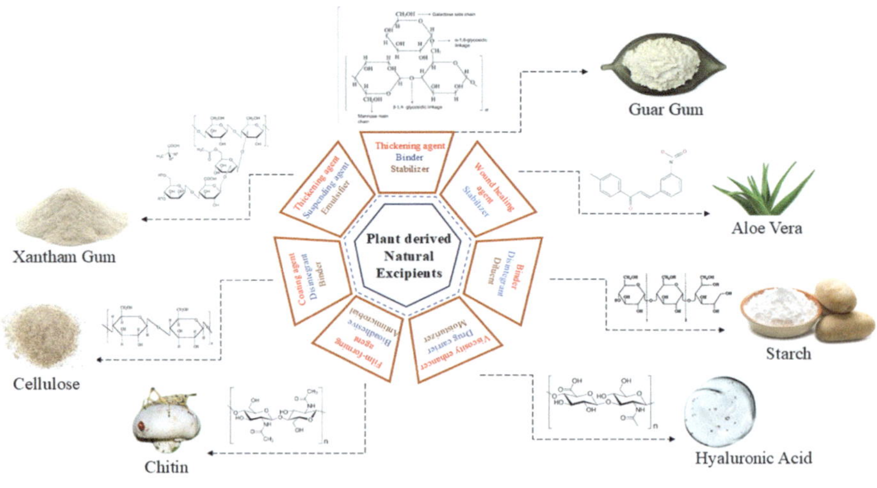

Fig. 4.1 Examples of various plant-derived natural excipients in drug formulation

pharmaceutical applications and provides examples along with their chemical structures. This comprehensive visual connects excipient sources, structures and functionalities in an integrated, easy-to-understand the key aspect of this book chapter.

4.4.1 Plant Gums

Guar gum is a hydrophilic, non-ionic polymer that is both biocompatible and biodegradable. It forms hydrogels, films and nanoparticles that enable targeted drug delivery, making it a common component of drug delivery systems. Its eco-friendliness and ease of availability make it a perfect fit for pharmaceutical applications, improving patient safety and adherence. Gum Xanthan Xanthomonas campestris ferments to create xanthan gum, a polysaccharide with thickening and stabilising qualities. It is used in a variety of formulations to increase stability and viscosity, which make it appropriate for emulsions and suspensions [15]. Because it is non-toxic and adaptable to certain formulation requirements, xanthan gum can improve the bioavailability of active compounds.

4.4.2 Mucilage

Polysaccharides like glucomannan, which are found in aloe vera, give it a gel-like consistency. It improves the bioavailability of active compounds by stabilising and thickening a variety of formulations [5, 8]. It is appropriate for topical and oral treatments due to its inherent healing and anti-inflammatory qualities. The soluble fibre found in psyllium husk is mostly made up of mucilage. It helps with drug release and absorption by acting as a bulking agent and is used in formulations for gastrointestinal health. Its capacity to gel and absorb water aids in regulating drug release and enhancing patient compliance [16].

4.4.3 Polysaccharides

The structural element of plant cell walls is cellulose, the most prevalent organic polymer on the planet. Cellulose is found in materials such as cotton, wood, hemp and flax, and it also comes from marine life and bacteria like Acetobacter. In terms of chemistry, cellulose is made up of glucose units that have three hydroxyl groups per anhydro-glucose unit (AGU) [17]. These groups create hydrogen connections both within and between molecules. Its characteristics, like stiffness and limited solubility, are influenced by the crystalline and amorphous zones formed by these interactions. Despite its lack of thermoelectricity, cellulose performs better in drug delivery when its surface is modified. Its derivatives, including hydroxypropyl methylcellulose (HPMC), ethyl cellulose (EC) and methylcellulose (MC), are essential in pharmaceutical formulations because they enhance solubility, stability and controlled drug release [18].

4.4.3.1 Starch

Another important excipient obtained from plants is starch, which is mostly utilised for its ability to store energy. Amylose (20–25%) and amylopectin (75–80%) make up its mixture, which varies depending on the source (e.g. maize, wheat and potatoes). The structures of these polysaccharides are different; amylopectin is branching, whereas amylose is linear. In drug delivery systems, starch is crucial due to its widespread availability and biocompatibility, particularly for controlled-release formulations and immediate-release tablets. Its derivatives, such as cross-linked and pregelatinised starches, improve its disintegration and swelling qualities, thereby broadening its pharmacological uses [19].

4.4.3.2 Glycogen

Glycogen is another kind of carbohydrate that is used to store energy. Though more compacted and branching, it resembles other storage polysaccharides like amylopectin. Glycogen, a form of glucose, is typically found in the cytoplasm of animal cells. Despite being necessary for physiological metabolism, glycogen is not used in industry and is solely mentioned here for completeness [20].

4.4.3.3 Chitin

Animals provide chitin, a linear polysaccharide with acetyl and amino groups inside its unit that is highly hydrophobic. It is soluble in specialised solvents such dimethylacetamide with 5% lithium chloride and chloro-alcohols in aqueous mineral acid solutions, but insoluble in water and typical organic solvents. It is another common structural polysaccharide that can be found in the spores and mycelia of fungi [21].

4.4.3.4 Hyaluronic Acid

One kind of glycoprotein, referred to as mucopolysaccharides or mucins, is hyaluronic acid. It is more commonly present in developing embryos and is one of the few polysaccharides identified in vertebrate tissue compared to other polysaccharides like heparin. These are proteins that have covalent bonds with polysaccharides. Among the others are proteoglycans. Often referred to as hyaluronan, hyaluronic acid is a straight-chain polymer with a molecular mass of roughly 7 × 106 g/mol that is often firmly bonded to proteins via hydrogen bonds like those seen in water. Isolation and extraction are therefore challenging processes. It can be used in surgical procedures in conjunction with alginate to promote wound healing, among other biological uses [22, 23].

4.4.4 Waxes

Because of their distinct chemical composition and range of applications, waxes derived from both plants and insects are frequently employed as excipients in the food, cosmetic and pharmaceutical sectors. Long-chain fatty acids, esters, hydrocarbons and alcohols make up the majority of these naturally occurring, biodegradable compounds, which help explain their emulsifying, water-resistant and protecting qualities [24].

4.4.5 Carnauba Wax

One of the hardest and highest-melting natural waxes, with a melting point of 82–86 °C, is made from the leaves of the Copernicia cerifera palm in Brazil. Hydrocarbons, 10–30% free fatty alcohols, 3–6% free fatty acids and 62–85% fatty esters make up its chemical makeup [25]. Carnauba wax is frequently used in oil structuring to increase the stability of formulations based on oil because of its inertness and stability. It gives tablets a glossy, protective covering in the pharmaceutical business by acting as a polishing agent. It is utilised in food-grade polishes and edible coatings and has been approved as a food additive (E903) and given GRAS classification in the United States. In addition to their usefulness in a variety of formulations, these plant and insect-derived waxes are prized for being environmentally benign and biodegradable, which satisfies the growing need for sustainable excipients in the cosmetics and pharmaceutical sectors. Their adaptability keeps spurring innovation in the advancement of cutting-edge medication delivery systems and eco-friendly packaging options [26, 27].

4.5 Various Applications of Plant-Derived Excipients in Drug Formulation

Natural excipients derived from plants have become more important in pharmaceutical formulations, improving the efficacy and safety of medicinal medicines.

4.5.1 Binders, Disintegrants and Coating Agents

Binders, disintegrants and coating agents are examples of excipients that are crucial for altering drug release patterns and enhancing bioavailability. Natural polymers, such as cellulose derivatives and starch, are frequently employed as disintegrants to facilitate medication release after administration and as binders to preserve tablet integrity [28]. Furthermore, plant-based substances like zein and gum arabic are used as coating agents because they provide controlled release mechanisms and shield delicate medications from deterioration [29]. Natural gelling agents, such as chitosan and alginate, are added to gels and creams for topical use to enhance drug absorption and stability. Furthermore, while reducing adverse effects, plant protein nanoparticles made from substances like soy and zein enable tailored drug delivery. Additionally, natural polymers aid in the creation of sustained-release systems, which guarantee the slow release of medications over long periods of time, maximising therapeutic results [30]. Notwithstanding the obvious advantages, issues like the unpredictability of natural sources and legal restrictions can prevent these excipients from being widely used in pharmaceutical applications.

4.5.2 Thickening and Gelling Agents

Plant-derived polymers are essential to medicinal formulations, especially as thickening and gelling agents in topical treatments. Cellulose, a valuable polymer derived from plant cell walls, is highly regarded for its non-toxicity and its ability to perform various functions in drug delivery to regulate drug release and contributions to the stability and smoothness of products. Starch, which comes from a variety of plants, improves topical formulations' rheological qualities and affects how drugs are released [31]. The thickening and stabilising properties of guar gum, which is derived from guar beans, enhance the consistency of products. Because of its effect on stability and viscosity, carrageenan, which is derived from red seaweed, is frequently included in gel compositions. Likewise, microbial xanthan gum is praised for improving consistency by thickening and stabilising [32]. Another polymer generated from seaweed, agar-agar, enhances product spread ability, texture and structural integrity. The creation of well-tolerated topical drugs is greatly aided by these plant-based polymers, which offer efficient and sustainable substitutes for synthetic compounds.

4.5.3 Emollients and Lipid Components

The exceptional emollient qualities of plant-based oils and waxes make them highly prized since they greatly improve the moisturising and protecting benefits of topical treatments. These organic ingredients soothe inflammation, encourage hydration and rebuild the lipid barrier of the skin [33]. Rich in fatty acids and antioxidants, olive oil is well acclaimed for its capacity to enhance skin moisture. A common ingredient in skincare, coconut oil is well-known for its many uses and has antibacterial, anti-inflammatory and moisturising properties. Beeswax ensures formulation stability by acting as a thickening in addition to its emollient properties. In a similar vein, shea butter, which is praised for its deep moisturising and anti-inflammatory qualities, improves skin hydration and promotes the function of the skin's barrier [34]. These plant-derived emollients work together to form a foundation for topical solutions that are both mild and effective, and that are designed to meet a variety of skin needs.

4.5.4 Active Agents and Extracts

Rich in bioactive components with therapeutic qualities, plant-based extracts provide mild yet efficient remedies for a range of skin issues. Because of its well-known calming and anti-inflammatory properties, chamomile extract is especially useful for treating sensitive or irritated skin [35]. With its bioactive ingredients, aloe vera extract, which is well-known for its relaxing and wound-healing qualities, is a flexible ingredient in topical formulations that supports the general health of the skin. Rich in antioxidants, green tea extract offers antiaging and anti-inflammatory properties as well as defence against oxidative damage [36]. Calendula extract promotes skin regeneration and repair because of its potent anti-inflammatory and wound-healing qualities. By adding these natural extracts to topical formulations, skincare solutions that are both efficient and appropriate for a variety of skin conditions can be created [37].

4.5.5 Healing Agents and Wound Care Components

Due to their antibacterial, anti-inflammatory and regenerative properties, plant-based chemicals have gained widespread recognition for their role in wound care. This makes them gentle and effective choices for treating skin injuries. The strong antibacterial and anti-inflammatory qualities of myrrh resin set it apart and are essential for promoting tissue regeneration. Its ability to promote wound healing and aid in the repair of injured skin is further highlighted by its distinct chemical makeup [38].

4.5.6 Fragrance and Aromatic Components

Fragrances made from plants, which are praised for their sensual appeal, also have amazing medicinal properties. Known for its peaceful scent, lavender oil has antimicrobial and calming properties that help heal wounds and encourage relaxation. Similar to this, frankincense resin has anti-inflammatory and antioxidant qualities that make it particularly useful for topical applications and the treatment of inflammatory skin diseases. It also has a unique scent. These organic substances have a variety of uses, combining strong health benefits with a pleasing scent [39, 40].

4.6 Role of Excipients in Noval Dosage Forms

Natural plant-derived excipients are becoming increasingly popular as sustainable, biocompatible substitutes for synthetic ones in the development of sophisticated pharmaceutical formulations [7, 41]. In addition to improving medication stability and delivery, these plant-based compounds offer affordable, non-toxic alternatives that work with a variety of drug delivery systems (DDS). Because of their many uses, they are essential in contemporary pharmaceutical formulations such as hydrogels, nanoparticles and controlled-release systems [42].

4.6.1 Controlled Release Agents

In drug delivery systems, plant-derived natural polymers offer great promise as controlled release agents because of their low toxicity, biocompatibility and biodegradability. According to a recent study, these polymers, which include chitosan, alginate and other plant gums, can be engineered to build systems that improve drug stability and release profiles, hence enhancing therapeutic efficacy and patient adherence [43, 44]. In controlled-release formulations, polymers such as hydroxypropyl methylcellulose (HPMC) and ethyl cellulose are employed to adjust the active pharmaceutical ingredient (API's) release rate, guaranteeing a long-lasting therapeutic impact. However, a number of obstacles stand in the way of their development and commercialisation.

4.6.2 Polysaccharides as Versatile Delivery Agents

Because of their hydrophilic, gel-forming and biodegradable qualities, polysaccharides such as pectin, alginate and guar gum are essential parts of DDS. Their use in controlled-release, targeted and sustained formulations is made possible by their structural variety. Pectin, which is derived from plant cell walls, is frequently used in medication delivery systems that target the colon [45]. By decreasing solubility in gastric fluids and permitting colon-specific medication release, modifications such as the production of calcium salt or combining with ethyl cellulose improve its effectiveness. Its versatility in different DDS is demonstrated by its application in matrix tablets, transdermal patches and microspheres for ocular drug delivery. Brown seaweed is the source of alginate, which is used to create hydrogels, microspheres and nanoparticles for long-term medication delivery [46, 47]. It is appropriate for encapsulating bioactive substances due to its ability to cross-link with divalent ions such as calcium, guaranteeing extended and precise drug release. Guar gum is excellent for colon-specific administration and is well-known for its biodegradability and sensitivity to colonic bacteria. It provides delayed and site-specific drug release and is commonly used in matrix tablets and compression-coated formulations [22, 48].

4.6.3 Gums and Their Multifunctional Roles

Because of their ability to bind, suspend and gel, natural gums including xanthan gum, gum acacia and karaya gum are frequently employed in DDS. Xanthan gum, a microbial polysaccharide, is an essential component of hydrophilic matrices. It uses gel barrier creation and swelling to regulate medication release [49, 50]. Its ability to work with other polymers, such as hydroxypropyl methylcellulose, improves its performance in formulations with modified release. Acacia gum is a useful component of pellet formulations and monolithic osmotic tablet systems (MOTS) due to its emulsifying and suspending qualities, which provide consistent drug dispersion and controlled release. Similar to this, karaya gum is a viable excipient for buccal medication administration because of its exceptional mucoadhesive qualities when employed in sustained-release formulations [50, 51].

4.6.4 Hydrogels for Controlled Drug Delivery

Hydrogels offer a three-dimensional network that may swell in water and regulate the release of encapsulated medications; plant-based hydrogels have completely changed controlled drug delivery. Examples include xanthan gum hydrogels and gels based on alginate, which are utilised in products such as ocular inserts and transdermal patches [52, 53].

4.6.5 Plant-Derived Volatile Oils

Plant-derived volatile oils are used in transdermal drug delivery systems (TDDS) to improve penetration. Plant-derived volatile oils are used in transdermal drug delivery systems (TDDS) to enhance skin penetration. For example, carvone and limonene are employed to increase the permeability of active drugs and to ensure controlled, sustained release, while menthol facilitates consistent plasma concentrations and improves the transdermal absorption of medications such as nimodipine [54, 55].

4.6.6 Flavouring and Colouring Agent

By masking bad tastes and giving dosage forms a pleasing look, flavouring and colouring compounds derived from plants increase patient compliance. Natural dyes like henna and turmeric, as well as essential oils like menthol and peppermint, are frequently employed in medicinal formulations [56].

4.7 Role in Enhancing Solubility, Solubility and Bioavailability

Excipients are important for preserving the API's physical and chemical stability. While chelating compounds like ethylenediaminetetraacetic acid (EDTA) guard against instability brought on by metal ions, antioxidants like ascorbic acid and tocopherols stop oxidative deterioration. Excipients such as cyclodextrins, surfactants (like polysorbates) and solubilising agents (like polyethylene glycol (PEG)) can help poorly soluble medications become more soluble and, as a result, more bioavailable [28, 57]. The low water solubility of many active pharmaceutical ingredients (APIs) may restrict their bioavailability. Through a variety of methods, excipients can greatly improve these APIs' solubility and, in turn, their bioavailability [58]. Surfactants that lower surface tension and speed up the dissolution of poorly soluble medications include sodium lauryl sulphate and polysorbates (Tween 80). By developing inclusion complexes with APIs, solubilisers such as cyclodextrins and polyethylene glycol (PEG) increase the solubility of APIs in water. By altering the solvent environment, cosolvents including ethanol, propylene glycol and glycerine make hydrophobic medications more soluble [59, 60]. Phospholipids and block copolymers are examples of amphiphilic excipients that create micelles in aqueous solutions, encasing hydrophobic medications within their cores to improve solubility and bioavailability. The drug's dissolution rate and bioavailability can be increased by dispersing it in a carrier matrix using excipients like HPMC and polyvinyl pyrrolidone (PVP) to form solid dispersions [61, 62].

4.8 Therapeutics Applications

Excipients derived from plants are becoming more and more popular for their therapeutic uses in chronic illnesses. These natural compounds, which include alkaloids, terpenoids, extracellular vesicles, phytomedicines and bioactive peptides, present encouraging therapeutic and preventative options. Both contemporary scientific validation and traditional knowledge support their effectiveness, underscoring their potential to improve healthcare outcomes [63, 64].

4.8.1 Phytomedicines

Plant-derived substances known as phytomedicines are extensively utilised in conventional medical systems and have been shown to have therapeutic potential by contemporary research. They have a variety of biological functions, including as immunomodulatory, anti-inflammatory and antioxidant actions [65]. They frequently work in concert to improve the effectiveness of treatments for long-term conditions like diabetes and cardiovascular illnesses. Phytomedicines are accessible, affordable and easily extracted using straightforward techniques like infusions and decoctions, which make them useful for applications in global healthcare. An evidence-based strategy for utilising traditional knowledge is provided by their incorporation into contemporary medicine [66, 67, 68].

4.8.2 Bioactive Peptides

Short amino acid sequences having strong therapeutic potential, plant-derived bioactive peptides (PDBPs) have antihypertensive, antibacterial, anticancer and antioxidant qualities. Because of their ethical and ecological advantages, they are becoming more and more favoured over peptides obtained from animals due to their sustainable sourcing from grains, legumes and seeds. PDBPs are appropriate for disorders including diabetes and hypertension because they have demonstrated promise in the treatment of chronic diseases, specifically in lowering oxidative stress, regulating blood pressure and enhancing glucose metabolism [9].

4.8.3 Extracellular Vesicles

Nanoparticles called plant-derived extracellular vesicles (PDEVs) are released by plants and contain bioactive substances such as proteins, lipids and RNAs. These vesicles are extremely adaptable in therapeutic applications due to their anti-inflammatory, anticancer, and antiaging characteristics. They are also prospective drug carriers for therapeutic usage due to their inherent biocompatibility and capacity to transport bioactive compounds to target cells. PDEVs are becoming a viable and cutting-edge platform for improving precision medicine in the treatment of chronic disorder [10, 69].

4.8.4 Terpenoids

Terpenoids, often referred to as isoprenoids, are a vast class of substances generated from plants that have a variety of biological properties, such as antiviral, anti-inflammatory and anticancer properties. While some terpenoids, such as paclitaxel, have been developed into clinically authorised medications, others are being studied for their potential to treat chronic inflammation and neurological disorders [70]. Terpenoids' natural abundance and therapeutic diversity make them attractive candidates for the development of new drugs to treat chronic disorder, even in the face of obstacles like low bioavailability [71, 72].

4.8.5 Alkaloids

Alkaloids are nitrogen-containing plant substances that have a variety of pharmacological properties, such as analgesic, anti-inflammatory and neuroprotective actions. With substances like vincristine and berberine already being utilised in cancer treatment and diabetes treatment, respectively, they show great promise for treating inflammatory and neurodegenerative illnesses [73]. Alkaloids are a vital resource for medication development because of their capacity to target illnesses of the central nervous system and penetrate the blood-brain barrier, while issues like toxicity and standardisation still need to be resolved [74].

4.9 Challenges and Limitations

Despite its benefits such as non-toxicity, affordability and biocompatibility, plant-derived excipients encounter several obstacles that prevent them from being widely used in pharmaceuticals [75, 76]. Microbial contamination, which can happen during processing in uncontrolled situations and jeopardise their safety, is a big worry. Furthermore, the impact of environmental conditions on plant sources causes significant variation in the quality of these polymers; variations in region, species, climate and harvesting time result in compositions that are inconsistent. Their performance is also impacted by physical constraints such as viscosity drop, uncontrolled hydration and temperature instability, as well as uneven hydration rates. The physiology and development rate of plants dictate the production process's inherent slowness, which cannot be intentionally accelerated. Furthermore, there is a chance of contamination due to plants' natural absorption of heavy metals. Regulatory obstacles also make the clearance process more difficult because their quality varies due to laxer criteria than those of synthetic polymers. Another barrier is market rivalry, where synthetic polymers predominate because of their well-established production methods and consistent performance indicators [77]. These problems are made worse by variations in plant materials, a lack of standardised analytical procedures and a lack of regulatory frameworks. This emphasises the necessity of thorough pharmacopoeial monographs, sophisticated analytical techniques and strict quality control. The potential of plant-derived excipients in pharmaceutical applications can be increased by addressing these constraints with creative strategies like grafting and enhanced regulatory standards [1, 78].

4.10 Conclusion

Excipients generated from plants have great potential to improve drug formulations. Their natural source addresses important drawbacks of synthetic polymers by guaranteeing biocompatibility, biodegradability and decreased toxicity. The creation of novel drug delivery methods, such as hydrogels, nanoparticles and controlled-release formulations, is made possible by these adaptable excipients, which improve drug solubility, stability and bioavailability. Applications include thickening agents, medicinal enhancers, binders and disintegrants. Their wider acceptance is being facilitated by continuous research and standardisation initiatives, notwithstanding obstacles like variability, microbial contamination and regulatory barriers. Plant-derived excipients can greatly contribute to sustainable and effective healthcare solutions by utilising their economical and environmentally beneficial qualities.

References

1. Pockle RD, Masareddy RS, Patil AS, et al. A comprehensive review on pharmaceutical excipients. Ther Deliv. 2023;14(7):443–58.
2. Pal D, Sahu P, Nayak AK, et al. Plant polysaccharides as emulsifiers in pharmaceutical emulsions. In: Plant polysaccharides as pharmaceutical excipients, vol. 1; 2023. p. 125–47.
3. Izah SC, Ogidi OI, Ogwu MC, et al. Historical perspectives and overview of the value of herbal medicine. In: Herbal medicine phytochemistry: applications and trends, vol. 10. Cham: Springer; 2024. p. 3–35.
4. Getahun MJ, Kassie BB, Alemu TS. Recent advances in biopolymer synthesis, properties, & commercial applications: a review. Process Biochem. 2024;
5. Guru PR, Kar RK, Nayak AK, et al. A comprehensive review on pharmaceutical uses of plant-derived biopolysaccharides. Int J Biol Macromol. 2023;1(233):123454.

6. Gericke M, Amaral AJ, Budtova T, et al. The European Polysaccharide Network of Excellence (EPNOE) research roadmap 2040: advanced strategies for exploiting the vast potential of polysaccharides as renewable bioresources. Carbohydr Polym. 2024;15(326):121633.
7. Hulke Y, Bhange M. Naturally occurring excipients and their pharmaceutical and engineering applications. In: AIP conference proceedings, vol. 3188, no. 1. AIP Publishing; 2024.
8. Mam NM, Bias S, Manasi M. Review on herbal excipients. RJPP. 2023;15(2):118–24.
9. Sun L, Liu J, He Z, et al. Plant-derived as alternatives to animal-derived bioactive peptides: a review of the preparation, bioactivities, structure–activity relationships, and applications in chronic diseases. Nutrients. 2024;16(19):3277.
10. Alzahrani FA, Khan MI, Kameli N, et al. Plant-derived extracellular vesicles and their exciting potential as the future of next-generation drug delivery. Biomol Ther. 2023;13(5):839.
11. Parekh SS, Parmar GR, Kanojiya D et al Unlocking the therapeutic potential of terpenoids: a roadmap for future medicine. Pharm Res 2024;16(4).
12. Sirin S, Dolanbay SN, Aslim B, et al. Role of plant derived alkaloids as antioxidant agents for neurodegenerative diseases. Health Sci Rev. 2023;1(6):100071.
13. Al-Roujayee AS, Hilaj E, Deepak A, et al. Alginate-based systems: advancements in drug delivery and wound healing. Int J Polym Mater. 2024;15:1–29.
14. Saadh MJ, Hsu CY, Mustafa MA, et al. Advances in chitosan-based blends as potential drug delivery systems: a review. Int J Biol Macromol. 2024;4:132916.
15. Behrooznia Z, Nourmohammadi J. Polysaccharide-based materials as an eco-friendly alternative in biomedical, environmental, and food packaging. Giant. 2024;6:100301.
16. Virmani T, Kumar G, Virmani R, et al. Xanthan gum-based drug delivery systems for respiratory diseases. In: Natural polymeric materials based drug delivery systems in lung diseases. Singapore: Springer; 2023. p. 279–95.
17. Chegu Krishnamurthi M, Tiwari S, Veera Bramhachari P, et al. Exploitation of marine-derived multifunctional biomaterials in biomedical engineering and drug delivery. In: Marine bioactive molecules for biomedical and pharmacotherapeutic applications. Singapore: Springer; 2024. p. 231–50.
18. Agnihotri JA. Milestones in the technology of modified starch in pharmaceutical formulation. J Nat Prod. 2024;
19. Babarabie M, Sardoei AS, Jamali B, et al. Carnauba wax-based edible coatings retain quality enhancement of orange (Citrus sinensis cv Moro) fruits during storage. Sci Rep. 2024;14(1):4133.
20. Baig MS, Karade SK, Ahmad A, et al. Lipid-based nanoparticles: innovations in ocular drug delivery. Front Mol Biosci. 2024;17(11):1421959.
21. Roy S, Priyadarshi R, Łopusiewicz Ł, et al. Recent progress in pectin extraction, characterization, and pectin-based films for active food packaging applications: a review. Int J Biol Macromol. 2023;1(239):124248.
22. Benalaya I, Alves G, Lopes J, et al. A review of natural polysaccharides: sources, characteristics, properties, food, and pharmaceutical applications. Int J Mol Sci. 2024;25(2):1322.
23. de Paula MC, Carvalho SG, Silvestre AL, et al. The role of hyaluronic acid in the design and functionalization of nanoparticles for the treatment of colorectal cancer. Carbohydr Polym. 2023;3:121257.
24. Dayan A. Excipients and solvents: the plant sources of excipients and solvents. In: Modern medicines from plants. CRC Press; 2023. p. 336–44.
25. Fitzgibbons EM. Characterizing and manipulating Sorghum Kernel Wax as a possible alternative for Carnauba Wax (Master's thesis, University of Minnesota).
26. Dimakopoulou-Papazoglou D, Giannakaki F, Katsanidis E. Structural and physical characteristics of mixed-component oleogels: natural wax and monoglyceride interactions in different edible oils. Gels. 2023;9(8):627.
27. Alexander I, Sodri A, Mizuno K. The effect of different natural waxes to hydrophobic properties of starch-based biodegradable foams. J Environ Sustain. 2023;6(1):86–101.
28. Goel R, Bhardwaj S, Bana S. Pharmaceutical excipients. In: Dosage forms, formulation developments and regulations. Academic Press; 2024. p. 311–48.

29. Malik MK, Bhatt P, Kumar T, et al. Significance of chemically derivatized starch as drug carrier in developing novel drug delivery devices. J Nat Prod. 2023;13(6):40–53.
30. Pires PC, Damiri F, Zare EN, et al. A review on natural biopolymers in external drug delivery systems for wound healing and atopic dermatitis. Int J Biol Macromol. 2024;19:130296.
31. Singh NK, Baranwal J, Pati S, et al. Application of plant products in the synthesis and functionalisation of biopolymers. Int J Biol Macromol. 2023;15(237):124174.
32. Guru PR, Kar RK, Nayak AK, et al. A comprehensive review on pharmaceutical uses of plant-derived biopolysaccharides. Int J Biol Macromol. 2023;233:123454.
33. Abdalla S, Aroua MK, Gew LT. A comprehensive review of plant-based cosmetic oils (virgin coconut oil, olive oil, argan oil, and jojoba oil): chemical and biological properties and their cosmeceutical applications. ACS Omega. 2024;9(44):44019–32.
34. Selwyn A, Govindaraj S. Study of plant-based cosmeceuticals and skin care. S Afr J Bot. 2023;1(158):429–42.
35. Agrawal R, Jurel P, Deshmukh R, et al. Emerging trends in the treatment of skin disorders by herbal drugs: traditional and nanotechnological approach. Pharmaceutics. 2024;16(7):869.
36. Dubey P. Tea catechins as potent antioxidant and anti-inflammatory agents: possibilities of drug development to promote healthy aging. In: Plant bioactives as natural panacea against age-induced diseases. Elsevier; 2023. p. 253–69.
37. Ferreira SM, Gomes SM, Santos L. A novel approach in skin care: by-product extracts as natural UV filters and an alternative to synthetic ones. Molecules. 2023;28(5):2037.
38. Qu S, Yu S, Ma X, et al. Medicine food homology plants promote periodontal health: antimicrobial, anti-inflammatory, and inhibition of bone resorption. Front Nutr. 2023;15(10):1193289.
39. Mensing J. Scent therapy: scents for More Joie De vivre: working with primal perfumes in scent-assisted therapy a look into the future at what else might be possible for increasing well-being. In: Beautiful SCENT: the magical effect of perfume on well-being. Berlin/Heidelberg: Springer; 2023. p. 223–53.
40. Natarajan H, Siddhan P. Herbal infusions: weaving nature's pharmacy into textiles. In: Textile materials for good health and wellbeing: design and applications. Singapore: Springer Nature; 2024. p. 309–35.
41. George A, Shrivastav PS. Plant polysaccharides as excipients in oral drug delivery. In: Plant polysaccharides as pharmaceutical excipients. Elsevier; 2023. p. 215–47.
42. Gali L, Pirozzi A, Donsì F. Biopolymer-and lipid-based carriers for the delivery of plant-based ingredients. Pharmaceutics. 2023;15(3):927.
43. Song X, Ji M, Shu X, et al. Drug delivery system loaded with plant-derived natural products for dental caries prevention and treatment. J Mater Chem B. 2025;
44. Mondal A, Nayak AK, Chakraborty P, et al. Natural polymeric nanobiocomposites for anti-cancer drug delivery therapeutics: a recent update. Pharmaceutics. 2023;15(8):2064.
45. Tudu M, Samanta A. Natural polysaccharides: chemical properties and application in pharmaceutical formulations. Eur Polym J. 2023;7(184):111801.
46. Kareemi AF, Likhitkar S. Applications and advancements of polysaccharide-based nanostructures for enhanced drug delivery. Colloids Surf B: Biointerfaces. 2024;30:113883.
47. Paliya BS, Sharma VK, Sharma M, et al. Protein-polysaccharide nanoconjugates: Potential tools for delivery of plant-derived nutraceuticals. Food Chem. 2023;1(428):136709.
48. Kumar J, Singh G. Mapping the copolymer research knowledge domain in lung cancer. Indian Drugs. 2024;61(8)
49. Veerapandian M, Ramasundaram S, Jerome PN, et al. Drug delivery application of functional nanomaterials synthesized using natural sources. J Funct Biomater. 2023;14(8):426.
50. Deep A, Rani N, Kumar A, et al. Prospective of natural gum nanoparticulate against cardiovascular disorders. Curr Chem Biol. 2019;13(3):197–211.
51. Shirwaikar A, Shirwaikar A, Sl P, et al. Herbal excipients in novel drug delivery systems. Indian J Pharm Sci. 2008;70(4):415.

52. Velghe C. Oral controlled drug delivery systems, optimization of release patterns and elucidation of release mechanisms (Doctoral dissertation, Université du Droit et de la Santé-Lille II).
53. Hou Y, Ma S, Hao J, et al. Construction and ion transport-related applications of the hydrogel-based membrane with 3D nanochannels. Polymers. 2022;14(19):4037.
54. Chelu M. Hydrogels with essential oils: recent advances in designs and applications. Gels. 2024;10(10):636.
55. Isopencu GO, Covaliu-Mierlă CI, Deleanu IM. From plants to wound dressing and transdermal delivery of bioactive compounds. Plan Theory. 2023;12(14):2661.
56. Pore AV, Bais SK, Shinde DN. Review on herbal monograph preparation.
57. Deshpande RD, Shah DS, Gurram S, et al. Formulation, characterization, pharmacokinetics and antioxidant activity of phloretin oral granules. Int J Pharm. 2023;15(645):123386.
58. Nyamba I, Sombie CB, Yabre M, et al. Pharmaceutical approaches for enhancing solubility and oral bioavailability of poorly soluble drugs. Eur J Pharm Biopharm. 2024;21:114513.
59. Ding Y, Zhao T, Fang J, et al. Recent developments in the use of nanocrystals to improve bioavailability of APIs. Wiley Interdiscip Rev Nanomed Nanobiotechnol. 2024;16(2):e1958.
60. Sachdeva S, Singh H, Singh J. Enhancing dissolution and bioavailability: a review on co-processed superdisintegrants in pharmaceutical formulations. JDDT. 2024;14(8)
61. Attia MS, Elshahat A, Hamdy A, et al. Soluplus® as a solubilizing excipient for poorly water-soluble drugs: recent advances in formulation strategies and pharmaceutical product features. J Drug Deliv Technol. 2023;1(84):104519.
62. Rusdin A, Mohd Gazzali A, Ain Thomas N, et al. Advancing drug delivery paradigms: polyvinyl pyrrolidone (PVP)-based amorphous solid dispersion for enhanced physicochemical properties and therapeutic efficacy. Polymers. 2024;16(2):286.
63. Feng Y, Pan M, Li R, et al. Recent developments and new directions in the use of natural products for the treatment of inflammatory bowel disease. Phytomedicine. 2024;13:155812.
64. Liu H, Liang J, Ye X, et al. The potential role of extracellular vesicles in bioactive compound-based therapy: a review of recent developments. Crit Rev Food Sci Nutr. 2023;63(31):10959–73.
65. Ghosh D, Datta N, Banerjee D, et al. Current trends on phytochemicals toward herbal medicine development. In: Herbal medicine phytochemistry: applications and trends. Cham: Springer; 2023. p. 1–26.
66. Qassadi FI, Zhu Z, Monaghan TM. Plant-derived products with therapeutic potential against gastrointestinal bacteria. Pathogens. 2023;12(2):333.
67. Chaachouay N, Zidane L. Plant-derived natural products: a source for drug discovery and development. Drugs Drug Candidates. 2024;3(1):184–207.
68. Verma N, Kumar J, Thapa K, et al. Role of copolymers in lung cancer. In: Block co-polymeric nanocarriers: design, concept, and therapeutic applications. Singapore: Springer; 2023. p. 211–26.
69. Xu Z, Xu Y, Zhang K, et al. Plant derived extracellular vesicles (PDEVs) in nanomedicine for human disease and therapeutic modalities. J Nanobiotechnol. 2023;21(1):114.
70. Terpenoids, often referred to as isoprenoids, are a vast class of substances generated from plants that have a variety of biological properties, such as antiviral, anti-inflammatory, and anticancer properties.
71. Maswal M, Pandita M, Bashir S. Terpenes, terpenoids and steroids: properties, biosynthesis and functions. In: Steroids and their medicinal potential. Bentham Science; 2023. p. 1–38.
72. Mohammadi-Cheraghabadi M, Hazrati S. Terpenoids, steroids, and phenolic compounds of medicinal plants. In: Phytochemicals in medicinal plants: biodiversity, bioactivity and drug discovery; 2023. p. 105.
73. Xavier V, Spréa R, Finimundy TC, Heleno SA, Amaral JS, Barros L, Ferreira IC. Terpenes. In: Natural secondary metabolites: from nature, through science, to industry. Cham: Springer; 2023. p. 107–56.

74. Kaur S, Kaur G, Rana R, et al. Therapeutic impact of alkaloids in neurological diseases: a promising key molecule. In: E3S Web of conferences, vol. 556. EDP Sciences; 2024. p. 01004.
75. Christi VE, Chowdary JS, Kumar SS, et al. Therapeutic values of pyridine molecules from natural sources: a comprehensive review. J Pharm Res Int. 2024;36(5):48–64.
76. Lamichhane S, Sahariah BJ, Das B, et al. Herbal drug standardization: a systemic review. JDDT. 2023;13(4):149–53.
77. Wang H, Chen Y, Wang L, et al. Advancing herbal medicine: enhancing product quality and safety through robust quality control practices. Front Pharmacol. 2023;25(14):1265178.
78. Putro JN, Soetaredjo FE, Lunardi VB, et al. Polysaccharides gums in drug delivery systems: a review. Int J Biol Macromol. 2023;21:127020.

5 Marine Polymers: Novel Sources for Drug Delivery Systems

Mirunalini Gobinath, L. Priyanka Dwarampudi, Shanmugam Ramaswamy, and Naina Mohamed Pakkir Maideen

5.1 Introduction

A drug which is commonly known as the active pharmaceutical ingredient (API) is a substance intended for use in diagnosis, cure, mitigation, treatment, or prevention of disease as per the US Food and Drug Administration (FDA, recognized in official pharmacopoeia). At some point, due to the body's defense mechanisms, certain medications become either resistant or inactive during a period. Thus, to overcome the existing issues with an API, drug delivery systems (DDS) were introduced [1]. These include several techniques to deliver the API effectively to the individual by increasing the concentration of the drug. The common goal of DDS would be to deliver in adaptive, biocompatible, and sustained or controlled manner. Using DDS, it is compatible to fabricate a dosage form which is safer, efficacious, and less toxic. DDS also

M. Gobinath
Department of Pharmacognosy and Phytopharmacy, JSS College of Pharmacy, JSS Academy of Higher Education & Research, Ooty, Tamil Nadu, India
e-mail: mirunalini.phd@jssuni.edu.in

L. P. Dwarampudi (✉)
JSS College of Pharmacy, JSS Academy of Higher Education & Research, Ooty, Tamil Nadu, India
e-mail: lalithapriyanka25@jssuni.edu.in

S. Ramaswamy
TIFAC CORE HD, JSS College of Pharmacy, JSS Academy of Higher Education & Research, Ooty, Tamil Nadu, India
e-mail: shanmugamr@jssuni.edu.in

N. M. P. Maideen
College of Medicine, Mohammed Bin Rashid University of Medicine and Health Sciences, Dubai, UAE

Dubai Health, Dubai, UAE

J. S et al. (eds.), *Innovative Pharmaceutical Excipients: Natural Sources*, https://doi.org/10.1007/978-981-96-7959-1_5

improves stability parameters and bioavailability of the main drug in specific cases such as when the API is sensitive to environmental exposure like light, moisture, and temperature. DDS is said to improve patient compliance by masking some organoleptic properties of API, such as appearance, taste, and odor. Thus, incorporating DDS is essential, but in some cases the excipients or additives that are used in these delivery systems do not have the characteristics like biocompatibility, stability, and nontoxic [2]. As a result, opting for natural products that serve as biopolymers is an ideal choice in avoiding such conditions. Polymers also aid in masking therapeutic drugs that are either hydrophilic or lipophilic in nature. They are often incorporated in the formulation of nanoparticles, microneedles, implants, hydrogels, and various nanotechnology-based formulations that help in time-sense release of drugs [3].

Biopolymers are covalently bonded polymeric molecules that are purely derived from natural origin. They have more complex structures than synthetic polymers and thus produce viable results. In addition, polymers derived from natural sources are environmentally friendly, as they are biodegradable when compared with synthetic polymers. Based on the source it is made of, biopolymers are classified into many types. Some instances are acacia gum, gelatin, starch, alginate, chitosan, collagen, etc. [4]. Among these those that are derived purely from marine origin are known as marine polymers. They are beneficial in drug delivery systems in recent advancements of research. These biopolymers are directly derived from the ocean, which is an incredible source on the earth. Almost 70–80% of the earth is made up of oceans, and various biodiversity exist, which serve as useful natural resource in multiple ways [5]. Marine species which have undergone several environmental insults have evolved to produce viable metabolites that have multifaceted benefits over existing polymers. Apart from being biocompatible, they are nontoxic, easy stimuli-responsive, and biodegradable polymers with various added characteristics. Surface modification is also easy in the case of marine polymers and is incorporated as coating materials in nanoparticles. Thus, this book chapter deals with marine polymers which have been served as a novel source in pharmaceutical drug delivery systems [6]. Figure 5.1 represents various uses of marine polymers in drug delivery and advanced therapies.

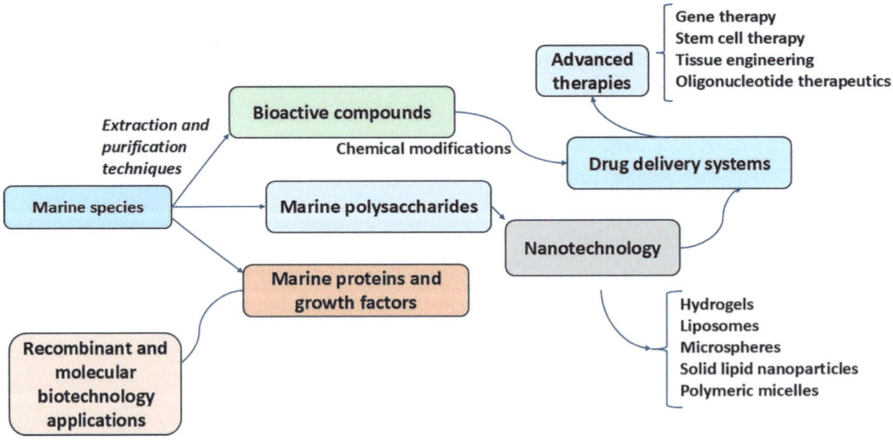

Fig. 5.1 Various marine polymers and their use in drug delivery [7]

5.2 Classification of Marine Polymers

Marine polymers can be derived from various biological sources such as marine algae, crustaceans, invertebrates, microorganisms, fish, and sponges. Based upon their origin and chemical structure, marine polymers are classified as follows:

(a) **Marine polysaccharides from algae**

Polymers that are derived from marine have tremendous properties; due to their structural complexity, it is difficult to distinguish the exact activity of polysaccharides. Some instances are monosaccharides with heterogeneous glycosidic bonds, high-branched heteropolymers with different functional groups in their carbon skeleton, and their composition of structure and function as observed in Gyrodinium impudicum, which is a dinoflagellate that yields homogalactan and β-glucan. These variations in characteristics differ based on harvesting time, place, and species strains [8].

5.2.1 Alginates

From the marine brown algae group Phaeophyceae, alginates are naturally formed. It is composed of uronic acid monomers, i.e., alpha-L-guluronic and beta-D-mannuronic acid, which are the essential structures in natural polysaccharides. Alginates are linear anionic marine-based polymers. Brown algae are usually composed of various organic and inorganic compounds like minerals, carbohydrates, and others [9]. Thus, to obtain alginates, it must undergo simple isolation techniques to obtain pure compound. Thus, the extraction procedure of brown algae biomass includes successive steps such as a) pre-extraction process by washing the algae sample, b) filtering and acid treatment using hydrochloric acid, which converts alginate to alginic acid by removing ions, and c) further alkali neutralization using NaOH or $NaHCO_3$ for converting alginic acid into sodium alginate [10]. Additional steps like further purification and filtration are done to remove extra debris. When ethanol or isopropanol is used as a solvent, sodium alginate is precipitated. Then, further precipitation steps are done to validate the higher pure compound, and final steps include drying and milling [11].

Some important key properties of alginates are gel formation, water absorption, and biocompatibility, due to which they are implemented in industrial applications, including food, pharmaceutical and biomedical industries. In addition, alginates are good gelling agents; at pH more than 6 they form gel spheres by ionotropic gelation using cations like calcium, zinc, or barium. These spheres of gels could be incorporated into encapsulation of various unstable drugs in pharmaceutical industry and nanotechnology as they are low-toxic and immunogenic [12].

Microencapsulation is a technique employed to mask the unwanted characteristics of drug especially that are of protein in nature because they are quickly degraded or denatured in the pH and environment of the stomach. Furthermore, alginates have their biological roles in providing structural support, flexibility, and mechanical strength [13].

5.2.2 Ulvan

Ulvan is a marine polysaccharide derived from green seaweed Ulva species, which has a high nutritional value and contains ulvanobiuronic acid three-sulfate type A and B, glucuronic acid, iduronic acid, sulfated rhamnose, xylose, and a few quantities of galactose. This composition of the monosaccharide and the sulfation degree varies with several factors like species variation, cultivation methods used, harvested season, growth environment, and extraction techniques performed to obtain biopolymers of *Ulva* sp. [14].

These polysaccharides have various biological and pharmacological activities such as anti-inflammatory, antioxidant, anticoagulant, anticancer, immune modulation, and lipid-lowering properties. Thus, they are implemented in various biomedical and health industries. Unique features of ulvan include gel formation, high water solubility, pH stability, temperature sensitivity, due to the presence of sulfate content, rare sugars, and several functional groups and structural branching patterns [15].

5.2.3 Fucan

These are sulfated polysaccharides which are composed of L-fucose sulfated units, which often exist in two forms: fucoidans and fucans. The chemical composition of these polymers includes α-L-fucopyranose, with sulfation positions in C-chain 2, 3, and 4. These contain additional components like acetyl groups, neutral sugars like galactose, mannose, xylose, protein residues, and uronic acids like glucuronic and galacturonic acid. Fucoidans are highly negative in charge due to the presence of sulphate groups in them [16].

Structural variations are seen based on which source fucoidans have been extracted, and which consist of simple and complex fucans. Some instances of marine sources where fucoidan polysaccharides are extracted are Brown seaweeds like *Undaria pinnatifida*, Laminaria species, Fucus species, *Ascophyllum nodosum*, and marine invertebrates like sea cucumbers, echinoderms, and sea urchins [17]. Extraction techniques such as acid extraction, enzymatic, hot water extraction, and newer technologies like microwave and ultrasound-assisted extraction, followed by purification steps such as ethanol precipitation and various chromatographic methods, are performed to validate and isolate polysaccharide [18].

5.2.4 Agar

Agar, which is obtained from a complex of saccharide molecules from certain red algae species such as Gracilaria, Gelidium, and Pterocladia, has phycocolloid nature and is composed of uronate, pyruvate, and sulphate substances. Agar or agarose has tremendous properties like reversible gelling process without degradation of their characteristics in heat and mechanical treatment. The solubility of agar differs with temperature. They are insoluble in cold water; thus, solubility range exists in boiling water. They form thermally reversible gels without the requirement of any oxidizing agents or acidic media, when dissolved in hot water and cooled gradually. This formation of gel is due to the repeated hydrogen bond between agar molecules [19]. When agar and glutaraldehyde are cross-linked, superabsorbent hydrogels are formed. Another important application of agar gel is the protein impurities. Since agar is ionic in nature, it permits the formation of complexes with protein structure and thus they are used in food industries for removal of proteins using centrifugation or filtration [20].

Other biomedical and health applications are as a DDS, in which pharmaceutical graded agar has viscous consistency. In radiology, it is used as suspending agent for solutions like barium sulphate. In microbiology, agar plays a crucial role as a bacterial culture media. Agarose or agar beads could be used as sustainable release dosage forms for hydrophilic drugs [21].

5.2.5 Carrageenan

It is a sulfated polysaccharide, which is obtained by extracting from red seaweed species *Rhodophyceae*. They are hydrocolloids in nature and are composed of calcium sulfate, magnesium, potassium, sodium, and esters of galactose. It has applications for the food industry, cosmetics like moisturizers, and DDS in pharmaceutical industries [22]. Due to their better gel texture characteristics, carrageenan is preferred mostly in food production based on rigidity, softness, firm, elasticity, and non-gelling agent. It is also used in nonfood industries such as pharmaceuticals, printing, textile industries, and cosmeceuticals. Apart from these industries, combination of chitosan and carrageenan nanoparticles has tremendous benefits in regenerative medicine and tissue engineering [23].

(b) **Marine polymers from fish**

Natural polymers that are derived from fish and other related species are categorized into many, depending on the nature of the biopolymer. Some studies have found that biopolymers from fish can also be obtained from fish waste, which leads to biorefinery processing. Some instances of marine polymers from fish are chitin, chitosan, collagen, gelatin, etc. [24].

5.2.6 Marine Collagen

Collagen is the most essential structural protein of the body, which is present in the extracellular matrix in a fibrous architecture. It is crucial in maintaining proper structural integrity of tissues, skeleton, and elasticity of the skin. Thus, it is abundant in mammals and composed of around 30–40% of bones, cartilage, and skin [25]. Based upon the amino acid sequences, morphological characteristics, and physiological properties, almost 29 types of collagens are identified. Among them, the triple-helix structured collagen type 1 called tropocollagen is the most common. It is structurally made up of lengthy amino acid sequences with three domains of H-terminal, C-terminal, and N-terminal, which assemble to form network of fibrils and is fused together to form collagen fibers [26].

Collagen has been used in various industries like in pharmaceuticals, food and cosmetics, medical devices, and thus the demand for it rose exponentially. Despite its use, many ethical considerations such as animal cruelty, religious prohibitions, limited resources, potential immune adverse reactions from land animals are observed. On the other hand, marine collagen possesses physiochemical properties that are quite similar to mammalian collagen, with added advantages like less risk in infectious disease transmission, reduced denaturation time and temperature, simple extraction techniques, and mild inflammatory response [27]. Methods like ultrasound-assisted, physical-aided, enzymatic, and acidic techniques are used for extracting and separating collagen. Some processes like deep eutectic solvent extraction, extrusion-hydro-extraction, and supercritical fluid extraction are environment friendly and have less pollution causing methods [28]. During the extraction process, these covalent and hydrogen bonds which stabilize the triple helix matrix are broken. This leads to an essential byproduct known as gelatin. Some instances of marine sources from which marine collagen are obtained are crustaceans, fish, marine sponges, sea cucumbers, mollusks, and jellyfish [29].

5.2.7 Gelatin

Partial degradation of collagen gives rise to gelatin with molecular weight ranging from 15 to 250 kDa. This process is through the hydrolysis of collagen by partial thermal degradation. Though gelatin is denatured, it has similar amino acid composition as collagen. Type A and B are the two types of gelatins that are obtained by pre-treatment techniques such as acid or alkaline conditions and are classified based on their isoelectric point ranges [30]. In gelatin production, enzyme pretreatment methods using specific peptide bonds are also performed, which result in polypeptide with various conformational changes due to extraction and pretreatment techniques. High molecular weight fractions such as β-chains and γ-chains are covalently linked to α-chain dimers and trimers. Aquatic sources like fish, jellyfish, and marine invertebrates are used as an alternative to traditional gelatin, which are derived from porcine skin and bovine hides. These marine gelatins also have properties like antioxidants, immune regulation, antimicrobial, antihypertensive, mineral chelating, and wound healing activity. Drawbacks are due to limited sources; marine gelatins hard to produce at a commercial scale, and they also have pungent odor, which requires proper optimization in extraction procedures [31].

5.2.8 Chitin and Chitosan

These are cationic mucopolysaccharide natural polymers obtained abundantly in walls of shellfish and other species with shells like crabs. This neutral solid compound contains n-acetyl-glucosamine monomers structure. Chitins and chitosans are differentiated to two types, alpha and beta, which are found in shrimps, crabs, and squids, respectively. Among these two types, beta is more sensitive. They have properties like antimicrobial, anticancer, antioxidant, and wound-healing characteristics [32]. Features like biocompatible, nontoxic, biodegradable, gelling properties, and permeability through membranes make them a better choice for biomedical and health applications. They can be easily degraded by human lysozymes and hence used as beads, films, membranes, microparticles, nanofibers, nanoparticles, and scaffolds. Some of the methods to prepare these polymers are complex coacervation, cross-linking, co-precipitation, emulsification, ionotropic gelation, polyelectrolyte complexes, and solvent evaporation methods. Thus, chitin and chitosan have a wide range of applications in nanomedicine [32].

5.2.9 Fish Oils

These are significant in various biomedical applications for decades due to their multifaceted properties. Main ingredients that are present in natural fish oils are omega (ω)-3-poly-unsaturated-fatty-acids (PUFA), docosahexaenoic acid (DHA), and eicosapentaenoic acid (EPA). These are essential fatty acids and cannot be produced de novo in human beings and thus are acquired through diet or supplementation of fish oils. The desirable physio-chemical properties such as amphiphilic nature makes fish oils as compatible with both hydrophobic and lipophobic drugs, thus providing better biodegradability and biocompatibility. Fish oil components also have intrinsic immune modulatory and anti-inflammatory properties that have synergistic effect with certain category of drugs [33]. PUFAs and other fish oils are used in drug delivery systems like emulsions, liposomes, nano-liposomes, solid-lipid nanoparticles, nano-structured lipid carriers, and self-emulsifying drug delivery systems. Challenges like unpleasant or pungent odor and taste, and shortest shelf-life are faced with fish oils; to overcome these issues using pulse electrospray and microencapsulation methods, Alfatama et al. have formulated alginate microspheres that are incorporated with fish oils. In this study, they have utilized wave-based drop on demand jetting electrospray to prepare controlled and proportionate size of microspheres. Characterization techniques are done to validate the microspheres and thus enhancing efficacy, stability, and palatability of fish oils using these alginate microspheres [34].

(c) **Microorganisms-Derived Polymers**

Microbial polymers are referred to as sustainable alternatives for conventional polymers due to their indispensable characteristics such as biodegradability.

Conventionally, these polymers from microorganisms are produced via fermentation of carbohydrate-rich foods, resulting in complex compounds. Biopolymers are classified based upon their backbone structure. Some instances of biopolymers from microbes are polysaccharides, peptides polynucleotides like de-oxy ribonucleic acids. Microorganism-based polymers could probably be extracted from bacteria, fungi, and algae, which give rise to xanthan gum, chitin, and alginate type of polymers [35].

5.2.10 Bacteria-Derived Polymers

These biodegradable polymers that are hydrophilic in nature are commonly produced by bacteria such as *Corynebacterium, Bacillus, Streptococcus, Mycobacterium, and Acetobacter*, which produce polyphosphates as a complex compound. Bacterial-derived polymers possess significant characteristics like anti-inflammatory, anticancer, and antimicrobial properties, due to which they are used in several industries and biomedical applications. In addition, other important benefits are their biodegradability, nontoxic, non-immunogenic, and anti-inflammatory activities. Thus, they could be used as a replacement for conventional nondegradable plastics [36].

Challenges that could be faced in the preparation of bacterial-based polymers are higher, depending on the versatility of microorganisms, mechanisms, materials, enzymes, and reagents that are used during the process. Exopolysaccharides are usually secreted by the bacterial cell wall or could be synthesized within cell wall surface proteins or via enzymes. These are located in the extracellular membrane in an amorphous slimy form, which is more abundant in prokaryotes especially in microalgae in comparison to yeasts and fungi [37]. For example, xanthan gum, which is a polyhydroxyalkanoate (PHAs), that are produced by bacteria like *Pseudomonas putida* and *Pseudomonas mendocina*, and capsular polysaccharides (CPS), which forms a protective layer around the bacteria [38].

5.2.11 Fungi-Derived Polymers

These polymers are made up of chitin that are extracted from mushrooms and later turned into fibers. These have self-healing properties which is why they are used in wound healing. Biocomposites from fungal mycelium are nontoxic and inert, which could be used in insulating and packaging [39].

(d) **Marine Polymers from Invertebrates**

These include crustaceans, shrimps, marine sponges, crabs, and other species of marine origin. These polymers contain bioadhesive substances, which have a significant role in biomedicals and as biomaterials for drug delivery. Organisms like sandcastle worms are crucial examples of bioadhesive properties, whereas mollusks like oysters and clams, starfish, and sea urchins which belong to echinoderms and crustaceans of certain types are extensively used for protein-based polymers. Marine invertebrates have been extensively studied for their biological activities and bioactive compounds such as from cnidarians, sponges, echinoderms, mollusks, and ascidians [40].

These invertebrates are water dwelling organisms that have a vast diversity, which not only differ in their structure but also in physiology, leading to a remarkable utilization by collagenous tissue. In addition, since marine invertebrates produce bioadhesives, they have more wet-locking and moisturizing properties and can be used in various industries and cosmeceuticals. In tissue repair and engineering, these polymers act as wound sealants and improve adhesiveness [41].

5.3 Drug Delivery Systems Using Marine Polymers

DDS are formulations or devices that assist in loading or introducing the therapeutic agent into the body. In order to improve the flow rate, safety, and efficiency of drug delivery, principles of nanotechnology could be incorporated into the traditional DDS. They are often used in the formulation of nanoparticles, microneedles, implants, hydrogels, and various nanotechnology-based formulations that help in time-sense release of drugs [2]. Table 5.1 gives some examples of marine polymer-based novel drug delivery systems.

Polymers play a significant role in drug delivery systems due to their several beneficial physicochemical properties including adaptability, biocompatibility, and controlled release of the API. These are essential in the field of nanotechnology for targeted drug delivery and various applications. Polymers that are used in nanotechnology comprise in two natures: biodegradable or nonbiodegradable and ideally be nontoxic and nonreactive, making them suitable for various therapeutics agents. Among these, marine polymers have created more attention in the domain of drug delivery because of their convincing properties [42].

5.3.1 Hydrogels

Hydrogel preparations have a few ideal characteristic requirements, namely biocompatibility, biodegradation and ability for mass production, and marine polymers possess all three of these characters making them ideal to produce hydrogels and incorporating various drugs in this drug delivery system. The topmost candidate that has been used in drug delivery for a long time now is alginate, commonly available in its salt form sodium alginate, the incorporation of alginate comes with the challenge of its low mechanical strength. Therefore, the strategies such as cross-linking by using calcium chloride, covalent cross-linking, and synthetic polymer grafting are usually followed, upon which the calcium chloride method is cheaper, easier, and less complex making it the most favored choice of strength enhancement. Alginate hydrogels have shown uses in the delivery of API in case of encapsulation and controlled release in acidic environment of gastrointestinal tract (GIT). Having

Table 5.1 Novel drug delivery systems with marine polymers

SL	Polymer composition	Details	References
1.	Polylactic acid (PLA) with chitosan	Delivered ovalbumin (OVA) for immunization with improved response of antibodies than that of the standard ones	[50]
2.	PLA, Polyvinyl pyrrolidone (PVP), and chitosan	Used in the delivery of inactivated influenza virus with a 16-week lasting effect with higher efficacy than Intramuscular (IM) dosing	[51]
3.	Chitosan alone	Successfully delivered hepatitis B antigens with a 1.5 times higher antibody count	[52]
4.	Silicon with chitosan	Successfully delivered polio antigens in vivo	[53]
5.	Chitosan or hyaluronic acid or gelatine	Used to deliver insulin for the treatment of diabetes	[54]
6.	PVA and PVP with a combination of chitosan	Delivery of 1 methyl D and L tryptophan with a combination of anti-programmed death-1/programmed death-ligand 1 (PD-1/L1) antibody for melanoma immunotherapy resulted in higher efficacy and targeted delivery	[55]
7.	Chitosan with graphene	Smart drug delivery of cephalexin to provide antibacterial effect	[56]
8.	Chitosan with beta-sodium glycerophosphate	Delivery of levonorgestrel for contraception resulting in higher penetration and hence enhancement of bioavailability	[57]
9.	Chitosan with Transcutol®	Use of sibutramine for obesity treatment	[58]
10.	Chitosan and graphene integrated with quantum dots	Use of lidocaine/BSA for macromolecules delivery due to improved electrical conductivity	[59]
11.	Alginate with trehalose Bacillus Calmette-Guérin (BCG)	Successfully delivered BCG antigens for BCG immunization	[60]
12.	HA with transferosomes	Delivery of doxorubicin for cancer therapy by accumulating the drug in lymph nodes leading to increased bioavailability	[61]
13.	Hyaluronic acid alone	Delivery of EV71 virus-like particles for EV71 virus immunization	[62]
14.	Hyaluronic acid with polycaprolactone	Use of canine influenza virus antigens for the development of canine influenza vaccine	[63]
15.	Oligo-HA with ethanol	Use of artemether for the treatment of collagen induced arthritis	[64]
16.	Hyaluronic acid integration with sucrose as well as dextran along with PVA and PVP	Use of diclofenac sodium for macromolecules delivery and for controlling Inflammation control	[65]

(continued)

Table 5.1 (continued)

SL	Polymer composition	Details	References
17.	Methacrylated hyaluronic acid with irgacure	Delivery of antiangiogenic monoclonal antibody and diclofenac for treating eye diseases	[66]
18.	Hyaluronic acid with imigran and methylene blue	Delivery of sumatriptan for migraine treatment	[67]
19.	Hyaluronic acid with fluorescein amine	Use of humanized monoclonal Immunoglobulin G1 (IgG1) antibody for protein delivery	[68]
20.	Hyaluronic acid with trehalose and PVP	Use of human parathyroid hormone for peptide delivery	[69]
21.	Hyaluronic acid with poly(lactic-co-glycolic acid) (PLGA) and silicon	Use of magnesium ascorbyl phosphate for vitamin delivery	[70]
22.	Hyaluronic acid with 4-arm-polyethylene glycol (PEG)-amine	Delivery of quininib for retinal angiogenesis and vascular permeability	[71]
23.	Hyaluronic acid alone	Delivery of uricase or horseradish peroxidase for the treatment of hyperuricemia	[72]
24.	Hyaluronic acid alone	Use of siRNA for nucleic acid delivery or abnormal scar treatment	[73]
25.	Alginate with mesoporous silica	Used as a sustained drug delivery system for indomethacin, a sparingly water-soluble drug	[74]
26.	Alginate nanohydroxyapatite	Delivered bacteriophages to inhibit orthopedic implant infection-related issues	[75]
27.	Calcium carbonate compound with alginate	Showed successful entrapment of bacteriophages	[76]
28.	Alginate salt form	Delivered VEGF in an injectable hydrogel delivery system	[77]
.29.	Alginate	Controlled delivery of curcumin and resveratrol in prostate cancer	[78]

a superior moisture retaining capacity, alginates are also a good choice for preparation of topical drug delivery systems in tissue engineering and wound healing via scaffolds [43, 44].

Chitosan is the second most used polymer in hydrogel type drug delivery, because of its excellent bioadhesion and natural antimicrobial properties. Chitosan is cross-linked by physical and chemical bonding, especially by using ionic interactions and hydrogen bonding methods. In chemicals, glutaraldehyde cross-linking is a popular choice. Chitosan hydrogels in drug delivery are generally used in wound dressings and pH responsive drug releasing in tissue engineering systems. On the other hand, potassium is used as the cross-linking ion for carrageenan; carrageenan also can show thermal gelation and polymerization with other polymerization enhanced its use as a thermos sensitive gel drug delivery system as well as sustained and targeted release drug deliveries. Apart from drug delivery, carrageenan is also used in the food industry as a thickening agent. Sourced from brown seaweeds, fucoidan is a hydrogel synthesizing polymer that has enhanced physical cross-linking, but it has capabilities to form chemical as well as enzymatic cross-linkage. The biggest use of fucoidan in drug delivery in hydrogels is in oncology; it is used in targeted and controlled API release in cancer tissue [43].

5.3.2 Microneedle Drug Delivery

In the recent decade, microneedle (MN) arrays have increasingly become one of the sought after transdermal drug delivery systems for their high penetration yet noninvasive manner, precision delivery of active pharmaceutical ingredients (APIs). They utilize small micron size pore creation in the epidermis via usage of micro-sized needle structures. These are prepared in cylindrical, conical, spike shaped, candle shaped, star and octagonal shapes, pyramid or even rocket shaped, having hollow, solid, coated, dissolving or hydrogel forming character [45]. These are used in therapeutics as well as cosmetics and diagnostic tools. One of the most adaptable classes of materials is polymers, which are also commonly utilized in the manufacturing of MNs systems [46].

Polymers have acceptable strength, cost, and availability in addition to their biocompatible and biodegradable qualities. Drug loading in large quantities is simple with polymeric MNs, and breakdown profiles vary according to their composition, swelling characteristics, and reactions to physical and biological stimuli [47].

Among all the available polymeric materials, marine originated ones are highly sought after due to their ease of bulk production and excellent drug delivery ability when used in MN. There have been numerous studies that point towards the use of marine polymers in MN for a wide range of applications. Table 5.1 summarizes the major research in different polymers used in drug delivery [48].

5.3.3 Polymeric Microspheres

Marine-derived microspheres are obtained using an array of different techniques. Both hydrogelation and cross-linking of marine polymers are good approaches; the different techniques are microfluidic emulsification, bath emulsification, EHD (electro-hydrodynamic) techniques spraying method, photolithography, etc. [43, 49].

5.4 Advantages and Challenges of Marine Polymers in Drug Delivery

Salient advantages of marine polymers over other natural biopolymers are their tremendous benefits including biocompatibility and biodegradability, which make them ideal for pharmaceutical and biomedical applications [79]. They are nontoxic and nonreactive in nature. Marine polysaccharides have unique structural features that could produce therapeutic activities like anticancer, anti-inflammatory, and antioxidants, which could be incorporated as synergistic agents for certain pharmacological roles. Marine polymers like carrageenan and alginates have high water absorption properties, gel strength capacity, and tunable viscosity, which make them ideal to form hydrogels that are suitable for cosmetic, food, and pharmaceutical applications [80]. Chitosan-based biopolymers have strong metal ion chelating properties, which make them suitable for wastewater treatment by removal of heavy metal ions. Marine polymers are also eco-friendly, which could be better alternatives for existing plastics to reduce environmental and biodiversity issues like resource depletion and pollution. These marine polysaccharides have a wide range of significance in drug delivery systems and targeted therapies; for instance, chitosan/alginate/fucoidan is being already used in nanotechnology and marine polymer science [81].

Challenges that are faced with marine polymers in drug delivery systems include stability issues; in case of marine polymers structural instability is seen due to variations in physiological conditions like pH, temperature, and enzymatic reactions. Thus, in turn it reduces the drug-loading capacity as in chitosan, which is susceptible for degradation via hydrolytic acidic environments causing drug release before reaching the targeted site. Thus, it is difficult to make it for oral drug delivery [82]. Another example for stability issues are alginate hydrogels in the presence of divalent cations, undergoing ionotropic gelation, thus leading to premature drug release due to ionic changes. Possible solutions to overcome these problems are hybrid polymer composites like alginate gelatin beads and chitosan polyvinyl alcohol (PVA) that provide resilience under normal physiological conditions. To enhance the mechanical stability of marine polymers, it can be incorporated into chemical cross-linking with noncytotoxic glutaraldehyde or genipin [83].

Poor solubility in some biological conditions is observed, in case of chitosan which is poorly soluble in basic and neutral pH due to 6.5 and above pKa value. Fucoidan is another example which is a sulfated polysaccharide with poor solubility in aqueous solutions due to intermolecular H-bonding. These conditions limit their potential for systemic drug delivery. To overcome these problems, chemical derivatization to improve solubility in various pH conditions might help [84]. Since these are novel natural products, limited knowledge on long-term effects is known which might not give us a clear idea about biodegradation and biocompatibility challenges. Concerns like inflammatory responses caused by degradation of polymer byproducts, accumulation into vital organs and variability in metabolism could alter the toxic profiles. To overcome these issues, extensive in vitro and in vivo assessment and computational toxicology models should be done to estimate degradation of biopolymers, systemic toxicity, and long-term effects before clinical studies. This could also reduce possible drug interactions, which are mainly caused due to altered bioavailability and pharmacokinetics [85].

The presence of charges also determines the therapeutic action of some drugs, for example, the cationic nature of chitosan, which interferes with negative charged drugs that cause reduced bioavailability and unpredicted absorption of certain small molecule drugs. On the other hand, fucoidan with strong anionic nature interacts with peptides and proteins, and as a result, it affects the stability and therapeutic action of the drug. Solutions like using predictive computational tools, rational polymer design and surface modification could prevent these unnecessary electrostatic interactions. Other issues like limited production possibilities due to variation in marine biomass sources, high processing costs, biodiversity and environmental concerns, complexity in modifications, and hurdles in ethical and regulatory authorities are some challenges that are faced while using marine polymers [86, 87].

5.5 Recent Advancements and Conclusions

Nevertheless, in the past decades, research interest and perspectives in the field of natural polymer discovery have been increasing to identify novel biocompatible materials and eco-friendly production processes to improve the conventional drug delivery systems. In this view, when compared to artificial polymers, natural products-based biopolymers show high grades of biocompatibility and easy modification processes. In future perspectives, mixing natural polymers with synthetic compounds makes it more complex hybrid nanostructures.

Marine polymers have always been an integral part of conventional and novel drug delivery systems due to their wide availability, biocompatibility, degradation profile, responsive to stimuli like pH, light, and temperature. These polymers are suitable raw materials for the creation of increasingly complicated loading devices with a possible regulated release as well. Devices that can shield various bioactive substances, including proteins and nucleic acids, can be made in a variety of ways and in a variety of geometries, including hydrogels, particles, and capsules. Stimuli-responsive polymers can be used to control the release of bioactive substances through a variety of ways, facilitating either a rapid or prolonged release. The usage of these materials may improve encapsulation efficiency and encourage the protection of a bioactive agent because they are frequently biocompatible and biodegradable.

Systems that can regulate the release of therapeutic chemicals to treat genetic illnesses are already available today. Some issues, like effective targeted delivery, precise control over the release rate to fit within a therapeutic window, and adaptability to more patient-friendly administration routes, still need to be resolved despite the pharmaceutical industry's extensive knowledge and widespread use of marine polysaccharides. To improve the extraction and purification of marine biopolymers, as well as the synthesis of their chemical modification and processing into the several potential matrices shapes, more research will be necessary.

References

1. Park H, Otte A, Park K. Evolution of drug delivery systems: from 1950 to 2020 and beyond. J Control Release. 2022;342:53–65.
2. Ezike TC, Okpala US, Onoja UL, Nwike CP, Ezeako EC, Okpara OJ, et al. Advances in drug delivery systems, challenges and future directions. Heliyon [Internet]. 2023 [cited 2025 Mar 12];9(6). Available from: https://www.cell.com/heliyon/abstract/S2405-8440(23)04696-0

3. Liechty WB, Kryscio DR, Slaughter BV, Peppas NA. Polymers for drug delivery systems. Annu Rev Chem Biomol Eng. 2010;1:149–73.
4. Atanase LI. Micellar drug delivery systems based on natural biopolymers. Polymers. 2021;13(3):477.
5. Galasso C, Ruocco N, Mutalipassi M, Barra L, Costa V, Giommi C, et al. Marine polysaccharides, proteins, lipids, and silica for drug delivery systems: a review. Int J Biol Macromol. 2023;253:127145.
6. Banday AH, ul Azha N, Farooq R, Sheikh SA, Ganie MA, Parray MN, et al. Exploring the potential of marine natural products in drug development: a comprehensive review. Phytochem Lett. 2024;59:124–35.
7. Cardoso MJ, Costa RR, Mano JF. Marine origin polysaccharides in drug delivery systems. Mar Drugs. 2016;14(2):34.
8. Lee YE, Kim H, Seo C, Park T, Lee KB, Yoo SY, et al. Marine polysaccharides: therapeutic efficacy and biomedical applications. Arch Pharm Res. 2017;40(9):1006–20.
9. Lai J, Azad AK, Sulaiman WMAW, Kumarasamy V, Subramaniyan V, Alshehade SA. Alginate-based encapsulation fabrication technique for drug delivery: an updated review of particle type, formulation technique, pharmaceutical ingredient, and targeted delivery system. Pharmaceutics. 2024;16(3):370.
10. Shaikh MAJ, Gupta G, Afzal O, Gupta MM, Goyal A, Altamimi ASA, et al. Sodium alginate-based drug delivery for diabetes management: a review. Int J Biol Macromol. 2023;236:123986.
11. Dalal SR, Hussein MH, El-Naggar NEA, Mostafa SI, Shaaban-Dessuuki SA. Characterization of alginate extracted from Sargassum latifolium and its use in Chlorella vulgaris growth promotion and riboflavin drug delivery. Sci Rep. 2021;11(1):16741.
12. Abka-khajouei R, Tounsi L, Shahabi N, Patel AK, Abdelkafi S, Michaud P. Structures, properties and applications of alginates. Mar Drugs. 2022;20(6):364.
13. Wani SUD, Ali M, Mehdi S, Masoodi MH, Zargar MI, Shakeel F. A review on chitosan and alginate-based microcapsules: mechanism and applications in drug delivery systems. Int J Biol Macromol. 2023;248:125875.
14. Anisha GS, Augustianath T, Padmakumari S, Singhania RR, Pandey A, Patel AK. Ulvan from green macroalgae: bioactive properties advancing tissue engineering, drug delivery systems, food industry, agriculture and water treatment. Bioresour Technol Rep. 2023;22:101457.
15. Tziveleka LA, Pippa N, Ioannou E, Demetzos C, Roussis V. Development of ulvan-containing liposomes as antibacterial drug delivery platforms. J Funct Biomater. 2022;13(4):186.
16. Barbosa JDS, Sabry DA, Silva CHF, Gomes DL, Santana-Filho AP, Sassaki GL, et al. Immunostimulatory effect of sulfated Galactans from the green seaweed Caulerpa cupressoides var. flabellata. Mar Drugs. 2020;18(5):234.
17. Zahariev N, Katsarov P, Lukova P, Pilicheva B. Novel Fucoidan pharmaceutical formulations and their potential application in oncology – a review. Polymers. 2023;15(15):3242.
18. George A, Shrivastav PS. Fucoidan, a brown seaweed polysaccharide in nanodrug delivery. Drug Deliv Transl Res. 2023;13(10):2427–46.
19. Zhang L, Ye S, Chen F, Xiao Q, Weng H, Xiao A. Super absorbent glutaric anhydride-modified agar: structure, properties, and application in biomaterial delivery. Int J Biol Macromol. 2023;231:123524.
20. Abel S, Jule LT, Gudata L, Nagaraj N, Shanmugam R, Dwarampudi LP, et al. Preparation and characterization analysis of biofuel derived through seed extracts of Ricinus communis (castor oil plant). Sci Rep. 2022;12(1):11021.
21. Navarrete-León C, Doherty A, Strimaite M, Bear JC, Olivo A, Endrizzi M, et al. Nanoparticle contrast agents for dark-field X-ray imaging. Nano Lett. 2025;25(3):1036–42.
22. Marunganathan V, Kumar MSK, Kari ZA, Giri J, Shaik MR, Shaik B, et al. Marine-derived κ-carrageenan-coated zinc oxide nanoparticles for targeted drug delivery and apoptosis induction in oral cancer. Mol Biol Rep. 2024;51(1):89.
23. Geyik G, Işıklan N. Design and fabrication of hybrid triple-responsive κ-carrageenan-based nanospheres for controlled drug delivery. Int J Biol Macromol. 2021;192:701–15.

24. Kumar M, Jha A, Mishra B. Marine biopolymers for transdermal drug delivery. In: Jana S, Jana S, editors. Marine biomaterials: drug delivery and therapeutic applications [Internet]. Singapore: Springer Nature; 2022 [cited 2025 Mar 12]. p. 157–207. Available from: https://doi.org/10.1007/978-981-16-4787-1_6.
25. Nuerjiang M, Bai X, Sun L, Wang Q, Xia X, Li F. Size effect of fish gelatin nanoparticles on the mechanical properties of collagen film based on its hierarchical structure. Food Hydrocoll. 2023;144:108931.
26. Farooq S, Ahmad MI, Zheng S, Ali U, Li Y, Shixiu C, et al. A review on marine collagen: sources, extraction methods, colloids properties, and food applications. Collagen Leather. 2024;6(1):11.
27. Shaik MI, Kadir ANA, Sarbon NM. Physicochemical and thermal properties of pepsin- and acid-soluble collagen isolated from the body wall of sea cucumbers (Stichopus hermanni). J Food Sci. 2024;89(1):320–9.
28. Xiao XC, Lin D, Cao KY, Sun LC, Chen YL, Weng L, et al. Properties of Pacific white shrimp (*Litopenaeus vannamei*) collagen and its degradation by endogenous proteinases during cold storage. Food Chem. 2023;419:136071.
29. Smith IP, Domingos M, Richardson SM, Bella J. Characterization of the biophysical properties and cell adhesion interactions of marine invertebrate collagen from Rhizostoma pulmo. Mar Drugs. 2023;21(2):59.
30. Ahmad MI, Li Y, Pan J, Liu F, Dai H, Fu Y, et al. Collagen and gelatin: structure, properties, and applications in food industry. Int J Biol Macromol. 2024;254:128037.
31. Ranasinghe RASN, Wijesekara WLI, Perera PRD, Senanayake SA, Pathmalal MM, Marapana RAUJ. Functional and bioactive properties of gelatin extracted from aquatic bioresources – a review. Food Rev Int. 2022;38(4):812–55.
32. Baharlouei P, Rahman A. Chitin and chitosan: prospective biomedical applications in drug delivery, cancer treatment, and wound healing. Mar Drugs. 2022;20(7):460.
33. Tao L, Wang P, Zhang T, Ding M, Liu L, Tao N, et al. Preparation of multicore millimeter-sized spherical alginate capsules to specifically and sustainedly release fish oil. Food Sci Human Wellness. 2023;12(2):397–406.
34. Alfatama M, Doolaanea AA, Choukaife H. Fish oil alginate microspheres produced via wave-based drop-on-demand jetting electrospray: improving stability and palatability. Food Biosci. 2024;61:104852.
35. Sachin K, Karn SK. Microbial fabricated nanosystems: applications in drug delivery and targeting. Front Chem [Internet]. 2021 [cited 2025 Mar 12];9. Available from: https://www.frontiersin.org/journals/chemistry/articles/10.3389/fchem.2021.617353/full
36. Wu F, Liu J. Decorated bacteria and the application in drug delivery. Adv Drug Deliv Rev. 2022;188:114443.
37. Li Z, Wang Y, Liu J, Rawding P, Bu J, Hong S, et al. Chemically and biologically engineered bacteria-based delivery systems for emerging diagnosis and advanced therapy. Adv Mater. 2021;33(38):2102580.
38. Mezzina MP, Manoli MT, Prieto MA, Nikel PI. Engineering native and synthetic pathways in pseudomonas putida for the production of tailored Polyhydroxyalkanoates. Biotechnol J. 2021;16(3):2000165.
39. Cord-Landwehr S, Moerschbacher BM. Deciphering the ChitoCode: fungal chitins and chitosans as functional biopolymers. Fungal Biol Biotechnol. 2021;8(1):19.
40. Hassan J, Faiyazuddin M, Uddin MJ. Marine drug delivery systems. In: Handbook of research in marine pharmaceutics. Apple Academic Press; 2025.
41. Romano G, Almeida M, Varela Coelho A, Cutignano A, Gonçalves LG, Hansen E, et al. Biomaterials and bioactive natural products from marine invertebrates: from basic research to innovative applications. Mar Drugs. 2022;20(4):219.
42. Ege H, Ege ZR, Gunduz O. Marine-derived materials for the development of advanced drug delivery systems. In: Maia FR, Oliveira JM, Reis RL, editors. Handbook of the extracellular matrix: biologically-derived materials [Internet]. Cham: Springer; 2024 [cited 2025 Mar 12]. p. 1157–71. Available from: https://doi.org/10.1007/978-3-031-56363-8_55.

43. Lin X, Wang J, Wu X, Luo Y, Wang Y, Zhao Y. Marine-derived hydrogels for biomedical applications. Adv Funct Mater. 2023;33(6):2211323.
44. Maity C, Das N. Alginate-based smart materials and their application: recent advances and perspectives. Top Curr Chem Cham. 2021;380(1):3.
45. Sharma S, Hatware K, Bhadane P, Sindhikar S, Mishra DK. Recent advances in microneedle composites for biomedical applications: advanced drug delivery technologies. Mater Sci Eng C. 2019;103:109717.
46. Donnelly RF, Larrañeta E. Microarray patches: potentially useful delivery systems for long-acting nanosuspensions. Drug Discov Today. 2018;23(5):1026–33.
47. Gill HS, Prausnitz MR. Coated microneedles for transdermal delivery. J Control Release. 2007;117(2):227–37.
48. Tarbox TN, Watts AB, Cui Z, Williams RO. An update on coating/manufacturing techniques of microneedles. Drug Deliv Transl Res. 2018;8(6):1828–43.
49. Wang H, Zhao Z, Liu Y, Shao C, Bian F, Zhao Y. Biomimetic enzyme cascade reaction system in microfluidic electrospray microcapsules. Sci Adv. 2018;4(6):eaat2816.
50. Chen MC, Huang SF, Lai KY, Ling MH. Fully embeddable chitosan microneedles as a sustained release depot for intradermal vaccination. Biomaterials. 2013;34(12):3077–86.
51. Chen F, Yan Q, Yu Y, Wu MX. BCG vaccine powder-laden and dissolvable microneedle arrays for lesion-free vaccination. J Control Release. 2017;255:36–44.
52. Zhu DD, Zhang XP, Yu HL, Liu RX, Shen CB, Zhang WF, et al. Kinetic stability studies of HBV vaccine in a microneedle patch. Int J Pharm. 2019;567:118489.
53. van der Maaden K, Sekerdag E, Schipper P, Kersten G, Jiskoot W, Bouwstra J. Layer-by-layer assembly of inactivated poliovirus and N-Trimethyl chitosan on pH-sensitive microneedles for dermal vaccination. Langmuir. 2015;31(31):8654–60.
54. Chen BZ, Ashfaq M, Zhang XP, Zhang JN, Guo XD. In vitro and in vivo assessment of polymer microneedles for controlled transdermal drug delivery. J Drug Target. 2018;26(8):720–9.
55. Yang P, Lu C, Qin W, Chen M, Quan G, Liu H, et al. Construction of a core-shell microneedle system to achieve targeted co-delivery of checkpoint inhibitors for melanoma immunotherapy. Acta Biomater. 2020;104:147–57.
56. Gaware SA, Rokade KA, Bala P, Kale SN. Microneedles of chitosan-porous carbon nanocomposites: stimuli (pH and electric field)-initiated drug delivery and toxicological studies. J Biomed Mater Res A. 2019;107(8):1582–96.
57. Yao G, Quan G, Lin S, Peng T, Wang Q, Ran H, et al. Novel dissolving microneedles for enhanced transdermal delivery of levonorgestrel: In vitro and in vivo characterization. Int J Pharm. 2017;534(1):378–86.
58. Castañeda PS, Chávez JJE, Hipólito EAM, Delgado CLD, Reyes VHA. Microneedles and Transcutol® as transdermal penetration enhancers of sibutramine formulated in a transdermal patch. Rev Cuba Farm. 2013;47(3):289–99.
59. Justin R, Román S, Chen D, Tao K, Geng X, Grant RT, et al. Biodegradable and conductive chitosan–graphene quantum dot nanocomposite microneedles for delivery of both small and large molecular weight therapeutics. RSC Adv. 2015;5(64):51934–46.
60. Arshad MS, Fatima S, Nazari K, Ali R, Farhan M, Muhammad SA, et al. Engineering and characterisation of BCG-loaded polymeric microneedles. J Drug Target. 2020;28(5):525–32.
61. Yang H, Wu X, Zhou Z, Chen X, Kong M. Enhanced transdermal lymphatic delivery of doxorubicin via hyaluronic acid based transfersomes/microneedle complex for tumor metastasis therapy. Int J Biol Macromol. 2019;125:9–16.
62. Zhu Z, Ye X, Ku Z, Liu Q, Shen C, Luo H, et al. Transcutaneous immunization *via* rapidly dissolvable microneedles protects against hand-foot-and-mouth disease caused by enterovirus 71. J Control Release. 2016;243:291–302.
63. Choi IJ, Kang A, Ahn MH, Jun H, Baek SK, Park JH, et al. Insertion-responsive microneedles for rapid intradermal delivery of canine influenza vaccine. J Control Release. 2018;286:460–6.
64. Qiu Y, Li C, Zhang S, Yang G, He M, Gao Y. Systemic delivery of artemether by dissolving microneedles. Int J Pharm. 2016;508(1):1–9.

65. Hou A, Quan G, Yang B, Lu C, Chen M, Yang D, et al. Rational design of rapidly separating dissolving microneedles for precise drug delivery by balancing the mechanical performance and disintegration rate. Adv Healthc Mater. 2019;8(21):1900898.
66. Than A, Liu C, Chang H, Duong PK, Cheung CMG, Xu C, et al. Self-implantable double-layered micro-drug-reservoirs for efficient and controlled ocular drug delivery. Nat Commun. 2018;9(1):4433.
67. Ito Y, Kashiwara S, Fukushima K, Takada K. Two-layered dissolving microneedles for percutaneous delivery of sumatriptan in rats. Drug Dev Ind Pharm. 2011;37(12):1387–93.
68. Mönkäre J, Pontier M, van Kampen EEM, Du G, Leone M, Romeijn S, et al. Development of PLGA nanoparticle loaded dissolving microneedles and comparison with hollow microneedles in intradermal vaccine delivery. Eur J Pharm Biopharm. 2018;129:111–21.
69. Kim HK, Lee SH, Lee BY, Kim SJ, Sung CY, Jang NK, et al. A comparative study of dissolving hyaluronic acid microneedles with trehalose and poly(vinyl pyrrolidone) for efficient peptide drug delivery. Biomater Sci. 2018;6(10):2566–70.
70. Kim Y, Bhattaccharjee SA, Beck-Broichsitter M, Banga AK. Fabrication and characterization of hyaluronic acid microneedlesto enhance delivery of magnesium ascorbyl phosphate into skin. Biomed Microdevices. 2019;21(4):104.
71. Galvin O, Srivastava A, Carroll O, Kulkarni R, Dykes S, Vickers S, et al. A sustained release formulation of novel quininib-hyaluronan microneedles inhibits angiogenesis and retinal vascular permeability in vivo. J Control Release. 2016;233:198–207.
72. Hao Y, Li H, Cao Y, Chen Y, Lei M, Zhang T, et al. Uricase and horseradish peroxidase hybrid CaHPO4 nanoflower integrated with transcutaneous patches for treatment of hyperuricemia. J Biomed Nanotechnol. 2019;15(5):951–65.
73. Wang M, Han Y, Yu X, Liang L, Chang H, Yeo DC, et al. Upconversion nanoparticle powered microneedle patches for transdermal delivery of siRNA. Adv Healthc Mater. 2020;9(2):1900635.
74. Hu L, Sun C, Song A, Chang D, Zheng X, Gao Y, et al. Alginate encapsulated mesoporous silica nanospheres as a sustained drug delivery system for the poorly water-soluble drug indomethacin. Asian J Pharm Sci. 2014;9(4):183–90.
75. Barros JAR, de Melo LDR, da Silva RAR, Ferraz MP, de Azeredo JCVR, Pinheiro VMDC, et al. Encapsulated bacteriophages in alginate-nanohydroxyapatite hydrogel as a novel delivery system to prevent orthopedic implant-associated infections. Nanomed Nanotechnol Biol Med. 2020;24:102145.
76. Colom J, Cano-Sarabia M, Otero J, Aríñez-Soriano J, Cortés P, Maspoch D, et al. Microencapsulation with alginate/CaCO3: a strategy for improved phage therapy. Sci Rep. 2017;7(1):41441.
77. DeVolder R, Antoniadou E, Kong H. Enzymatically cross-linked injectable alginate-g-pyrrole hydrogels for neovascularization. J Control Release. 2013;172(1):30–7.
78. Saralkar P, Dash AK. Alginate nanoparticles containing curcumin and resveratrol: preparation, characterization, and in vitro evaluation against DU145 prostate cancer cell line. AAPS PharmSciTech. 2017;18(7):2814–23.
79. Mukherjee C, Varghese D, Krishna JS, Boominathan T, Rakeshkumar R, Dineshkumar S, et al. Recent advances in biodegradable polymers – properties, applications and future prospects. Eur Polym J. 2023;192:112068.
80. Moniz T, Costa Lima SA, Reis S. Marine polymeric microneedles for transdermal drug delivery. Carbohydr Polym. 2021;266:118098.
81. Yang C, Ding X, Yang C, Shang L, Zhao Y. Marine polymers-alginate/chitosan composited microcapsules for wound healing. Chem Eng J. 2023;456:140886.
82. Safarzadeh Kozani P, Safarzadeh Kozani P, Hamidi M, Valentine Okoro O, Eskandani M, Jaymand M. Polysaccharide-based hydrogels: properties, advantages, challenges, and optimization methods for applications in regenerative medicine. Int J Polym Mater Polym Biomater. 2022;71(17):1319–33.

83. Uma Thanu Krishnan Neela N, Szewczyk PK, Karbowniczek JE, Polak M, Knapczyk-Korczak J, Stachewicz U. Improving stability and mechanical strength of electrospun chitosan-polycaprolactone scaffolds using Genipin cross-linking for biomedical applications. Macromol Rapid Commun:2400869.
84. Caetano PA, do Nascimento TC, Fernandes AS, Nass PP, Vieira KR, Maróstica Junior MR, et al. Microalgae-based polysaccharides: Insights on production, applications, analysis, and future challenges. Biocatal Agric Biotechnol. 2022;45:102491.
85. Patel AK, Vadrale AP, Singhania RR, Michaud P, Pandey A, Chen SJ, et al. Algal polysaccharides: current status and future prospects. Phytochem Rev. 2023;22(4):1167–96.
86. Santinon C, Ochi D, Beppu MM, Vieira MGA. Chemical modifications in the structure of seaweed polysaccharides as a viable antimicrobial application: a current overview and future perspectives. Algal Res. 2022;66:102796.
87. Liu T, Ren Q, Wang S, Gao J, Shen C, Zhang S, et al. Chemical modification of polysaccharides: a review of synthetic approaches, biological activity and the structure–activity relationship. Molecules. 2023;28(16):6073.

Polysaccharides from Natural Sources: Functional Properties and Applications

6

Neha Minocha

6.1 Introduction

Complex carbohydrates or polysaccharides are large macromolecules comprising repetitive chains of monosaccharide units that are joined to together by glycosidic connections. These naturally occurring macromolecules are ubiquitous in nature with wide structural variety. Polysaccharides have critical roles in biological systems, including energy storage and the support of structural integrity in living organisms. They have been of interest in medicines, nutraceuticals, and biotechnology due to their distinctive physicochemical characteristics and biocompatibility [1]. Polysaccharides (polymers of ten or more monosaccharides which are often linked by O-glycosidic linkages) adaptability results from the difference in monosaccharide components, the type of glycosidic linkages, and the degree of branching in the structure of polysaccharide. As a result, these elements improve their characteristic performance of solubility, viscosity, and mechanical strength, making them applicable in different applications [2].

6.1.1 Natural Sources of Polysaccharides

Polysaccharides are found in nature in many sources, such as plants, animals, algae, fungi, and even microbes. Polysaccharides from every source of course exhibit unique structures and activities as they have ecological and biological importance.

(i) *Plant-derived polysaccharides*: Polysaccharides are a major source from plants. Cellulose is the main organic polymer on Earth and it is a fundamental

N. Minocha (✉)
School of Medical and Allied Sciences, K.R. Mangalam University, Gurugram, Haryana, India

J. S et al. (eds.), *Innovative Pharmaceutical Excipients: Natural Sources*,
https://doi.org/10.1007/978-981-96-7959-1_6

constituent of the plant cell walls and therefore offers the plant cell walls structural stability. Polysaccharide starch is produced by plants as an energy reserve, and is found in cereals, tubers, and legumes. Natural thickening and gelling agents in the food business are usually pectins and hemicelluloses, both from fruits and seeds [3, 4]. Other gums, such as gum arabic and guar gum and their derivatives, are collected for emulsifying and stabilizing activity [5].

(ii) *Animal-derived polysaccharides*: Polysaccharides are synthesized by other animals in lesser amounts than plants. An energy reserve in liver and muscle tissues, the animal equivalent of starch, is glycogen, a branching polysaccharide. Connective tissues rely on the preservation of skin moisture and joint lubrication, which is completely dependent on hyaluronic acid. Dermatan sulfate and chondroitin sulfate are sulfated polysaccharides which give cartilage and skin structural support [6].

(iii) *Algal polysaccharides*: Marine algae are a unique rich source of some polysaccharides bearing various physicochemical features. Carrageenan and agar are gelling agents extensively used in the food and pharmaceutical sectors, synthesized by red algae. Alginate, a polysaccharide that is well known for its good biocompatibility, is utilized in wound dressings, drug delivery, and tissue engineering, namely brown algae [7].

(iv) *Fungal and microbial polysaccharides*: Polysaccharides from fungi and microbes are attractive because they are bioactive. Fungal cell walls and crustacean exoskeletons are the sources for the precursor of chitosan, a biopolymer with antibacterial and wound healing properties. Immunostimulatory properties of fungal exopolysaccharides, i.e., β-glucans from *Saccharomyces cerevisiae* as well as *Aspergillus*, Xanthan gum and dextran are especially used as stabilizers and emulsifiers in food and cosmetic industry [8].

(v) *Polysaccharides from alternative sources*: Some polysaccharides are obtained from unconventional origins—lichenin from exotic plants like tamarind and fenugreek, gum-like compounds from the seeds of exotic plants, lichen-derived lichenin, and possible applications in medicines and nutraceuticals. Natural biochemistry is based upon polysaccharides, with an incredibly broad biological and economic range of uses. Their categorization and natural origins tell us their structural and functional diversity, which has profound impacts on human health and technology. The polysaccharide landscape is augmented with natural sources and shows the potential for sustainable and renewable production [9, 10].

6.2 Structural Diversity of Polysaccharides

Polysaccharides are some of the most structurally diverse macromolecules of nature and hence are utilized as naturally occurring biopolymers with extensive biological roles and in numerous industrial applications. The diversity of these glucans arises from differences in monosaccharide composition, glycosidic linkage, chain branching, and molecular weight. The different arrangement of polysaccharides makes them structural, energy reserve and bioactive compounds, depending on what they are [11, 12]. Polysaccharides are structurally diverse and relevant to versatile and important biological and commercial applications. Spectrum of polysaccharide qualities includes bioactive sulfated glycosaminoglycans, and crystalline rigidity of cellulose, with each fulfilling a specific biological requirement and environmental tolerance. As their structural classification becomes clearer, we understand better the structure of natural macromolecules and develop new applications in medicine, food science, and materials engineering [13–15].

6.2.1 Structural Characteristics That Enhance Diversity

The structural diversity of polysaccharides is attributable to various factors:

(i) Glycosidic linkages: Substantial differences in the properties of polysaccharides (solubility and crystallinity) are due to the type (α or β) and location (i.e. $1 \rightarrow 4$, $1 \rightarrow 6$) of glycosidic bonds [16, 17].
(ii) Branching: Glycogen and amylopectin, like other branched polysaccharides, are more compact and soluble than linear polysaccharides such as cellulose. Solubility, reactivity, and functional characteristics of the polysaccharide are influenced by the extent and configuration of branching.
(iii) Degree of polymerization: The length of the polysaccharide chain dictates its molecular weight and mechanical characteristics.
(iv) Chemical alterations: Natural modifications, including sulfation (as in heparin) and acetylation (as in chitin), add to functional diversity while increasing bioactivity [18, 19].

6.2.2 Categories of Polysaccharides

Homopolysaccharides and heteropolysaccharides are two types of polysaccharides, and additional polysaccharide classifications depend on functions as structural polysaccharides or storage polysaccharides.

A. **Homopolysaccharides**

The polymeric framework of homopolysaccharides consists of repeating one type of monosaccharide. Their composition is not so simple, and the dissimilarities in glycosidic linkages and chain configuration give rise to large structural variety [20, 21].

- *Cellulose*: The most common polysaccharide on Earth is cellulose and it is the major component of plant cell walls as a structural element. A linear homopolysaccharide of β-($1 \rightarrow 4$)-linked D glucose units which form elongated chains is capable of hydrogen bonding with adjacent chains. As a result of this, highly crystalline microfibrils are created, giving plant tissues their abilities of tensile strength and stiffness [22]. As a water insoluble structural material, cellulose is a durable material that is resistant to enzymatic breakdown.

- *Starch*: It is the primary energy storage carbohydrate in plants, composed of two glucose-derived polymers: amylose and amylopectin [23, 24].
- *Amylose*: It is linear, and mostly consists of α-(1 → 4)-linked D-glucose units, which form helical structures. In contrast, amylopectin contains extensive branching consisting of α-(1 → 4) glucose chains and α (1 → 6) branch sites. Starch structural variability is responsible for increased water solubility and enzymatic accessibility and is therefore a key dietary carbohydrate [21].
- *Chitin*: The exoskeletons of arthropods, and cell walls of fungi, are structural homopolysaccharides of chitin. The structure consists of elongated strands similar to cellulose, formed by β-(1 → 4)-linked N-acetylglucosamine units. The stiffening and insolubility of chitin that occurs via chemical modification of the backbone augment its use in nature for protective and supportive functions [25].
- *Glycogen*: In animals and fungus, glycogen is the principal storage polymer. The structure is extensively branched and has α-(1 → 4) linked glucose chains with α-(1 → 6) branching sites occurring every 8 to 12 glucose units. This elaborate branching allows the onset of glycogen irreverently mobilized for energy production under metabolic increases, such as in muscle and liver tissue [17, 26].

B. **Heteropolysaccharides**

Heteropolysaccharides consist of two or more distinct monosaccharide types, which may be organized in linear or branching structures. Their structural intricacy and functional adaptability are essential to their biological functions [27, 28].

- *Hyaluronic acid*: Hyaluronic acid is a linear heteropolysaccharide present in the extracellular matrix of connective tissues. It consists of recurring disaccharide units of D-glucuronic acid and N-acetylglucosamine connected by β-(1 → 3) and β-(1 → 4) linkages. Hyaluronic acid demonstrates exceptional hydrophilicity and viscoelasticity, rendering it vital for tissue hydration, joint lubrication, and wound healing [29–31].
- *Heparin and heparan sulfate*: These are sulfated heteropolysaccharides consisting of alternating glucosamine and uronic acid units. Heparin, mostly located in mast cells, serves as a powerful anticoagulant by binding to antithrombin and obstructing blood coagulation. Heparan sulfate, structurally analogous to heparin but exhibiting less sulfation, is integral to cell signaling and adhesion [32–34].
- *Pectins*: Pectins are heteropolysaccharides of plant origin that are abundant in galacturonic acid residues. They are primarily linear but may incorporate side chains. Pectins possess significant hydrophilicity and gel-forming properties, rendering them vital in food processing and as dietary fibers that enhance gut health [35].
- *Hemicelluloses*: Hemicelluloses are branching heteropolysaccharides present in plant cell walls, commonly linked with cellulose and lignin. They com-

prise an array of sugars including xylose, mannose, arabinose, and galactose. In contrast to cellulose, hemicelluloses exhibit amorphous characteristics and are more readily hydrolyzed, enhancing the flexibility of plant cell walls [36, 37].

- *Exopolysaccharides (EPS):* A multitude of microbes synthesize exopolysaccharides, including xanthan gum and gellan gum, which are excreted into their surroundings. These polysaccharides are essential for biofilm development, microbial adhesion, and defense against environmental stressors. Their structural diversity results from the amalgamation of several sugar residues and side groups, including acetyl or pyruvyl groups [38].

6.3 Extraction and Purification Methods of Natural Polysaccharides

Extraction and purification operations represent the fundamental procedures for the isolation of natural polysaccharides from their matrices and their purification, as well as for the determination of the purity and bioactivity of these macromolecules. They are primarily extracted from plants, fungi, algae, and microorganisms and obtain polysaccharides of varying degrees of extraction that determine degrees of polymeric yields and structures and usefulness [39, 40]. It is a challenging process of extracting and purifying polysaccharides from natural sources, preserving their integrity while avoiding proteins, phenolics, and pigments impurities. Ease of use and scalability are the reasons why numerous traditional technologies have been used, while eco-friendly and sustainable practices are receiving greater focus in the mitigation of environmental damage. Although progress has been made, the purifying process remains challenging for the most part, as polysaccharides and other biomolecules rarely behave as simple molecules [41–43].

6.3.1 Traditional Extraction Methods

The most common methods for extraction of polysaccharide were the conventional procedures. They extract polysaccharides from the source by physical, chemical or enzymatic processes.

1. *Hot water extraction* (HWE): It is one of the most common ways to go. Biomass is heated with water (80–100 °C), so polysaccharides dissolve out into the aqueous phase. Medicinal plant, fungus (e.g., β-glucans), and seaweed (e.g., agar, alginate) derived optimal polysaccharides [44, 45].
2. *Acid-assisted extraction*: It uses dilute acid (e.g., hydrochloric acid) to disrupt plant or algal cell walls. It shows particular strong extraction efficiency for pectins and hemicelluloses [46, 47].

3. *Alkaline-assisted extraction*: This method makes use of bases (e.g., sodium hydroxide) to release lignin or cell wall components to release cellulose or hemicelluloses. It is common in plant-based polysaccharides such as cellulose [48].
4. *Enzymatic extraction*: It is decomposition of the cellular matrix through the activity of enzymes such as cellulases, pectinases, or hemicelluloses; it is selective liberation of polysaccharides. Fragile polysaccharides are preserved in their structural integrity, and bioactive polymers are used [49, 50]. Different extraction techniques for polysaccharides with principle involved, applications, and advantage are tabulated in Table 6.1.

Table 6.1 Extraction techniques for natural polysaccharides

Technique	Principle	Applications	Advantages	Limitations
Hot water extraction	Heat solubilizes polysaccharides into the aqueous phase	Medicinal plants, fungi, seaweeds (e.g., agar)	Simple, scalable, cost-effective	Long extraction time, high energy use
Acid-assisted extraction	Acids disrupt cell walls, releasing polysaccharides	Pectins, hemicelluloses	High efficiency for certain polysaccharides	Can degrade sensitive structures
Alkaline-assisted extraction	Bases break down lignin and release polysaccharides	Cellulose, hemicellulose	Effective for lignin-containing biomass	Harsh conditions may damage polysaccharides
Enzymatic extraction	Enzymes selectively degrade cell components to release polysaccharides	Bioactive polysaccharides (e.g., pectin)	Gentle on sensitive polysaccharides	High cost of enzymes
Ultrasound-assisted extraction	Cavitation breaks cell walls, enhancing polysaccharide release	Marine algae, plant polysaccharides	Fast, energy-efficient, preserves bioactivity	Limited penetration for dense materials
Microwave-assisted extraction	Rapid heating disrupts cell walls, facilitating extraction	Plant, fungal, and algal polysaccharides	Short processing time, low energy	Equipment cost can be high
Supercritical fluid extraction	Supercritical CO_2 solubilizes polysaccharides in combination with ethanol	Bioactive polysaccharides, especially medicinal	Environmentally friendly, no residues	Requires specialized equipment
Ionic liquid extraction	Ionic liquids dissolve biomass to release polysaccharides	Cellulose, hemicellulose	Minimal environmental impact	Expensive and less studied

6.3.2 Green and Sustainable Extraction Approaches

This requires conventional methods, which include use of toxic solvents, high consumption of energetic resources, and chemical degradation. By increasing efficiency, decreasing waste, and preserving the structural integrity of polysaccharides, thus, green and sustainable extraction techniques have emerged as eco-friendly alternatives [51, 52].

1. *Ultrasound-assisted extraction* (UAE): The release of polysaccharides occurs after cavitation bubbles are generated and rupture the cell walls by ultrasound waves. Lowest extraction times, lowest temperatures, and preserved bioactivity are the benefits [53, 54].
2. *Microwave-assisted extraction* (MAE): Using microwaves to fast heat the sample disrupts cellular architecture and allows for a favorable extraction of polysaccharides. It reduces energy use for plant and algal polysaccharides [51, 55].
3. *Supercritical fluid extraction* (SFE): It employs supercritical CO_2 as a solvent, alone or in combination with ethanol, to extract bioactive polysaccharides. It is an extremely sustainable, but highly specialized equipment needed approach, free of toxic leftovers [51].
4. *Ionic liquid extraction*: Polysaccharides in lignocellulosic biomass are depolymerized in ionic liquids with minimal environmental impact. It is effective for the extraction of cellulose and hemicellulose [52]. Green and sustainable approaches of polysaccharide extraction are represented on Table 6.2, while those of environmental impact, cost, and percentages yield of each technique are tabulated in Table 6.3.

6.3.3 Challenges in Polysaccharide Purification

The procedure is intrinsically difficult due to the complexity, heterogeneity, and high molecular weight of polysaccharides, which frequently occur as combinations of structurally distinct macromolecules with varying degrees of branching, charge density, and molecular conformations. Polysaccharides, complex, heterogeneous, and of high molecular weight are commonly found as mixtures of structurally different macromolecules with varying degrees of branching, charge density, and molecular conformations, making the procedure intrinsically difficult [7, 62]. The co-extraction of contaminants including proteins, phenolic compounds, pigments, and salts is a major problem, leading to disruption of further downstream applications and bioactivity evaluations. Efficient simultaneous purification and concentration of polysaccharides with selective removal of contaminants remain difficult, because conventional purification techniques (such as ethanol precipitation, ultrafiltration, dialysis, and various chromatographic methods including ion exchange [63], size exclusion, and affinity chromatography) do not provide the necessary selectivity to remove the contaminant(s), while maintaining the polysaccharide yield or functionality.

Table 6.2 Comparison of green and sustainable extraction techniques for polysaccharides

Extraction method	Principle	Advantages	Limitations	Example	Ref.
Enzyme-assisted extraction (EAE)	Uses specific enzymes to hydrolyze cell walls and release polysaccharides	– Mild conditions prevent degradation – High specificity and purity – Reduces solvent and energy use	– High enzyme cost – Requires process optimization	Pectin extraction from citrus peels using pectinase	[56]
Microwave-assisted extraction (MAE)	Microwave energy disrupts cell walls by rapid heating, enhancing solvent penetration	– Fast and efficient – Reduces solvent use – Preserves polysaccharide integrity	– Requires power and solvent optimization – High initial equipment cost	Alginate extraction from brown seaweed with 30% higher yield	[57]
Ultrasound-assisted extraction (UAE)	Ultrasound waves generate cavitation bubbles, disrupting cell walls	– Faster extraction with less energy – Low temperatures preserve structure – Reduces solvent use	– Ultrasound waves may degrade polysaccharides – Requires frequency and duration optimization	Fucoidan extraction from seaweed with high purity	[58]
Supercritical fluid extraction (SFE)	Uses supercritical CO_2 as a solvent to dissolve and extract polysaccharides	– No toxic solvents – No residual solvent in product – Highly selective extraction	– High pressure and equipment cost – Limited for hydrophilic polysaccharides	Arabinoxylans from wheat bran with improved bioactivity	[59]
Subcritical water extraction (SWE)	Uses hot pressurized water (100–374 °C) as a green solvent	– No toxic chemicals – Efficient for various polysaccharides – Enhances bioavailability	– High temperature may degrade polysaccharides – Requires pressurized equipment	Chitosan extraction from shrimp shells with reduced acid use	[60]
Deep eutectic solvent extraction (DES)	Uses eco-friendly solvents made from natural compounds to dissolve polysaccharides	– Nontoxic and biodegradable – Highly tunable for different polysaccharides – Low-cost and energy-efficient	– Solvent recovery needs optimization	Pectin from fruit peels with higher yield and reduced waste	[61]

Table 6.3 Comparison of different factors associated with green extraction methods for polysaccharides

Method	Solvent use	Energy efficiency	Yield and purity	Cost	Environmental impact
EAE	Low	High	High	Moderate	Low
MAE	Low	High	High	High	Low
UAE	Low	High	High	Moderate	Low
SFE	None (CO_2)	High	High	High	Very low
SWE	None	Moderate	Moderate to high	High	Low
DES	None	Moderate	High	Low	Very low

Purification of oligogalacturonides is complicated by the presence of bound proteins and polyphenols requiring additional procedures including enzymatic hydrolysis or activated carbon treatment; this can produce structural modifications or diminished bioactivity. Polydispersity and solubility variability pose additional problems as polysaccharides exhibit heterogeneity in solubility profiles depending on the molecular weight and charge distribution. Filtration and centrifugation of high viscosity polysaccharides, for example hyaluronic acid and alginate, are difficult, leading to process efficiencies losses and increased operation costs [64]. In addition, the use of aggressive chemical pretreatments, such as acid hydrolysis or strong alkaline solutions, for deproteinization and decolorization can also lead to depolymerization, which will alter the functional properties of polysaccharides and oblige them for use in timid biological systems. Supply demand reliability is a serious concern since variation in raw material sources from batch to batch, as well as seasonal and extraction condition fluctuations, prevents the standardization of purification processes at large scale industrial manufacturing [65]. Green purification techniques, such as membrane filtration, supercritical fluid fractionation and deep eutectic solvents, are viable because they decrease reliance on potentially hazardous reagents and minimize structural degradation. However, still, scalability, cost efficiency, and regulatory approval still remain big challenges. Solutions to these problems call for an integrated approach that combines the handling of sophisticated purification methods, process improvements as well as real time analytical instruments such as nuclear magnetic resonance (NMR) spectroscopy, mass spectrometry, and chromatographic fingerprinting to ensure consistent, pure, and functional stable polysaccharides for various applications [66–68].

6.4 Functional Properties of Polysaccharides

Polysaccharides are biopolymers composed of long chains of monosaccharide units linked by glycosidic bonds. Their functional properties determine their applications in food, pharmaceuticals, biomaterials, and biomedicine. These properties can be broadly classified into physicochemical, biological, and rheological properties, which influence their role as stabilizers, emulsifiers, bioactive agents, and structural materials [69–71].

6.4.1 Physicochemical Properties of Polysaccharides

The physicochemical properties of polysaccharides are diverse, leading to differences in their solubility, stability, interaction with other biomolecules and functionality in various applications. Because their solubility, viscosity, and gelation behavior are critically dependent on the molecular weight and structural conformation (linear, branched), they remain useful—especially for protecting proteins and assisting transport of cellular chemicals. Water solubility is enhanced by the presence of hydrophilic groups, whereas hydrophobic interactions promote aggregation and film formation [72]. They also possess high water holding capacity that makes them vital to moisture retention applications (hydrogels and food stabilizers). Furthermore, due to their ion exchange properties, these resins may interact with metal ions affecting their bioavailability and functional activity in drug delivery. Polysaccharides undergo chemical modifications which improve their bioavailability and tunable properties (improved solubility, biodegradability, and mechanical strength) [73]. Some properties and their effects include:

- Molecular weight and structure: Polysaccharides have different molecular weight and different structure (linear or branched). Hyaluronic acid has a high molecular weight, which gives it high viscosity and water retention capacity.
- Solubility: Cold water is not very soluble to starch but when heated, starch swells and gelatinizes to give a thick viscous solution in hot water.
- Water-holding and swelling capacity: Due to the large amounts of water absorbed by polysaccharides, these are ideal for forming hydrogels as well as controlled drug delivery. Hydrogels derived from alginate and xanthan gum, which hold in moisture, are used for wound healing as well as food stabilization.
- Thermal stability and gelation: Some polysaccharides are thermally gelled while others are thermally stable. Carrageenan and agar form very strong gels which are used in food and biomedical applications.
- Emulsification and film-forming ability: A few polysaccharides also behave as emulsifiers and film formers and improve the stability of emulsions and edible films. The excellent emulsifying properties of gum arabic rendered its use in beverage emulsions very common [74–76].

6.4.2 Biological Properties of Polysaccharides

Given their important role in biological systems via their antioxidant activity (scavenging of free radicals and amelioration of oxidative stress to reduce cellular damage and age-related illnesses), polysaccharides are deemed especially vital. In addition, numerous polysaccharides such as fucoidan and β glucans activated macrophages, improved cytokine synthesis, and altered inflammatory responses, making them particularly amenable to use in cancer immunotherapy and vaccine adjuvants. Polysaccharides with antitumor and anticancer properties act in the immune response to kill malignancies either directly via the suppression of tumor cell proliferation or by indirectly through the enhancement of an antitumor cell immune response [77]. Cancer suppressive polysaccharides of *Ganoderma lucidum* and *Lentinus edodes* have been reported. In addition, they have an antiviral and antibacterial characteristic that makes them strong contenders for the treatment of infectious diseases. Carrageenan and heparin, and other sulfated polysaccharides, can block viral entrance and replication, inhibiting herpes simplex virus and influenza [78].

Prebiotics are polysaccharide in nature and selectively enhance the proliferation of beneficial gut microbiota, such as bifidobacteria and lactobacilli, thus improving health of the gut, improving digestion, and strengthening immunity. Mucosa stimulating and indigestible polysaccharides (e.g., inulin and arabinoxylans) are able to adjust microbiota equilibrium and protect against intestinal inflammation. In addition to that, they have been extensively researched for their hypoglycemic and antidiabetic properties due to the fact that polysaccharides can regulate blood glucose by blocking α-glucosidase enzymes, improve insulin sensitivity, and delay nutrient absorption, which is a very important thing in terms of diabetes potential treatment [79, 80]. In addition, polysaccharides have important functions in the wound healing and the tissue regeneration because of its biocompatibility, the moisture retension properties, and the ability to stimulate cell growth. Hyaluronic acid and chitosan (polysaccharide) derived biomaterials have been widely applied in regenerative medicine, hydrogel wound dressings, and drug delivery systems. They are applied in the treatment of cardiovascular ailments and prevention of blood clots on account of their anti-inflammatory and anticoagulant characteristics of especially in heparin and dextran derivatives. Polysaccharides exhibit a long list of biological features and promise as multifunctional biopolymers in medicines, functional foods, and biomaterials, and thus they are a key area of current biomedical study [81–83].

6.4.3 Rheological Properties of Polysaccharides

As a high molecular weight biopolymer, polysaccharides have unique and different rheological properties, which therefore make polysaccharides indispensable in various industrial, pharmacological, and biological applications. The study of material flow and deformation under varied stress, temperature, and concentration situations is called rheology. Polysaccharides must be analyzed with consideration of their rheological properties which have an impact on texture, stability, and functionality of goods. Polysaccharides are linear polysaccharides characterized by long chains of repeating monosaccharide units joined by glycosidic linkages. The molecular architecture, branching degree, and functional groups of these biopolymers control their level of solubility, viscosity, elasticity, and gelation potential [75].

Polysaccharides have rheological properties influenced by molecular weight, concentration as well as the environment. Polysaccharides with a larger molecular weight tend to be more viscous at lower concentrations than do lower molecular weight polysaccharides, such as xanthan gum, guar gum, and alginates. Because of their larger molecule size, flow resistance is increased, and the result is this [82]. On the other hand, maltodextrins which are low molecular weight polysaccharides have lower viscosities, and are more fluid. Polysaccharides are very dependent on molecular structure, especially with respect to branching and the presence of such functional groups as hydroxyl and carboxyl groups, in their interaction with water and other substances, ultimately affecting their rheological behavior.

In aqueous solutions, polysaccharides often display pseudoplastic behavior. This implies that their viscosity decreases with increasing shear rate, a property especially important for many practical applications. At low shear rates, long polysaccharide chains are originally entangled and randomly oriented, but at higher-rate shear, their alignment with the force field leads to reduced internal friction and decreasing viscosity, an attractive property for products that require smooth flow at processing or application yet maintain structure and viscosity at low shear [84]. Shear thinning action assists in smooth pouring of foods where they would be in food preparations such as sauces and dressings that also allow for preservation of viscosity during storage. Shear thinning formulations could ensure effortless injection or administration of a drug delivery system in the first instance, but turn to become highly viscous once present in the body and eventually release the drug within their predetermined delivery window. In addition, a strong thixotropicity was exhibited by many polysaccharides, reflecting a decrease in their viscosity under shear stress, followed by a gradual return to original fullness of skeleton when shear stress is relieved. This is a critical feature in formulations where long-term stability is required or where ease of flow or spread is needed under certain conditions. Ease of application with uniformity retention after application is the main rationale for using a thixotropic compound in coatings and ink and in personal care products.

Rheologically, gel formation is one of the most significant processes among polysaccharides. Under different temperature, ionic strength, and pH conditions, polysaccharides such as agar, carrageenan, pectin, and gellan gum produce gels. Cooling a molten agar to a temperature below 37 °C causes it to form a gel whose strength of gelation is influenced dramatically by concentration, as well as by ionic environment. Gels formed by carrageenan are dependent on the presence of divalent cations, such as calcium ions. These gel forming capabilities are exploited for applications in the food, pharmaceuticals, and biotechnology industries. In drug delivery systems, polysaccharide-based gels can encapsulate and deliver pharmaceuticals in such a way as to create prolonged therapeutic benefits. Polysaccharide gels must possess certain gel strength, flexibility, and heat reversibility, and yet be able to remain unchanged for years or even decades, and upon encountering moisture, return to their original form. The rheological properties of polysaccharides are significantly affected by temperature. There are certain polysaccharides which, e.g., gelatin and agar, show thermoreversible gelation; i.e., they change between liquid and gel state on cooling and vice versa on heating. For instance, in tissue engineering, this property is particularly advantageous for use where cells or medications are contained within temperature-sensitive gels and release can be controlled by temperature fluctuations inside the body. Also, the shear modulus (an elastic modulus) of polysaccharide-based materials can also be tailored. The rigidity or flexibility of the polysaccharides to be used can be modified according to the desired application by varying concentration or by the addition of cross-linking agents. They are especially relevant to the formulation of hydrogels for biomedical application where mechanical strength, porosity, and swelling characteristics are critical factors to allow for suitable administration of drugs, healing of wounds, or scaffolding of tissue. Ionic strength and pH also affect the capacity of polysaccharides to behave in different ways with water to create gels. Maintenance of the molecular interaction within polysaccharide chains, required for specific viscosity and gelation behaviors, is dependent on sufficient amount of salt or variations of pH [85, 86]. The percentage of acid and calcium ions causing pectin to create gels, the gelation process can be controlled by changing these variables. The rheology of polysaccharides may be manipulated to engineer materials with desired flow properties, gel strengths, and mechanical properties, making them adaptable and valuable commodities in several industries.

6.5 Polysaccharides in Food Applications

It is known that polysaccharides are important in the food industry and they play an important part in improving the texture, stability, and overall product quality. These are the naturally occurring macromolecules having different sources such as plant, animal or microbial source, and the sugar units linked by glycosidic linkage having different dietary composition. The complexity of polysaccharide chemical structures (including the capacity to form gels, thicken liquids, stabilize emulsions, and affect food texture) ensures that polysaccharides are utilized as key ingredients in a broad range of food categories, including beverages, baked goods, dairy, and processed foods [87].

In the food industry, polysaccharides are almost exclusively employed as thickeners or gelling agents. Common polysaccharides with gel forming properties include agar, pectin, carrageenan, and gelatin and are used to make jellies, jams, gummy candies, and puddings. The ability of polymers to gelate depends strongly on their chemical structures, the presence of some ions, e.g., calcium for pectin and carrageenan, and environmental conditions such as pH and temperature [88]. Agar sets solid when it is cooled from a melted state; gelatin requires the water to be cooled first. Incorporating these gels not only strengthens the appeal of these food products from the consumer's perspective, but they also confer the desired firmness and consistency necessary for their stability as foods. Polysaccharides are often used as stabilizers and emulsifiers in the food processing, especially applications involving both water and oil, such as in salad dressings, mayonnaise, and ice cream. Emulsions contain polysaccharides which help assure consistent distribution of oil droplets in water and prevent phase separation. Salad dressings and sauces can be made more stable and more viscous by using two major polysaccharides: guar gum and xanthan gum, both from microbes. The polysaccharides are viscous, smooth, and serve to suspend particles such as spices or herbs in the liquid preventing the particles from settling [89, 90].

Polysaccharides other than their functions in thickening and stabilizing can give products a different texture and provide specific mouthfeel sensations. Polysaccharides containing glucose units are extensively used in the food sector due to their textural attributes on the food products. Gelatinization is a process whereby starch granules absorb water at elevated temperatures and swell and so it can be used as thickening agents in soups, gravies, sauces, and puddings. The starch type used—corn, potato, or rice—dictates how the finished product will feel and behave as a viscosity. Physical or chemical modification of starches creates modified starches, which are stable, freeze and thaw resilient, and more acid resistant, all of which are important in processed meals. Food products with polysaccharides as dietary fibers have both functional and nutritional benefits [91]. By increasingly being integrated in food formulations to serve improving digestive health, maintaining gut microbial composition, and lowering blood sugar levels, they are dietary fibers, nondigestible polysaccharides. For example, fibers found in inulin from chicory root and resistant starch in legumes and whole grain products added in breakfast cereals, snacks, and dairy products to increase their health benefits are included. These fibers offer advantages such as modulating bowel movements, preventing chronic diseases like cardiovascular disease, and aiding in hunger control and weight management. Polysaccharides exhibit gelling, thickening as well as emulsifying characteristics essential for the formulation of low or reduced calorie foods [92]. Polysaccharides can be used by food makers to make items to recreate the creamy mouth feel and texture of higher fat alternatives without the added calories. Frequently used to provide the appealing creamy texture associated with fat laden foods in low fat dairy products and in salads, xanthan gum and carrageenan are commonly used in these products [93]. The use of these polysaccharides helps meet future customer demand for healthier food alternatives without degrading the food sensory properties. Polysaccharides also demonstrate innovative use for food packaging, as well as further application. Polysaccharide films and coatings are being investigated as possible substitutes for conventional plastic packaging. Polysaccharide-based films like chitosan, starch, or alginate films are biodegradable films that protect food from contamination, extend shelf life, and reduce waste. In addition, polysaccharide coatings can serve as barriers to moisture, oxygen, and light in order to maintain the freshness and quality of food goods [94]. At the same time as an increased demand for clean label ingredients, there has been a transition toward the use of natural polysaccharides in food compositions in recent years. Polysaccharides can be used as natural and safe alternative for synthetic thickeners, stabilizers, and emulsifiers, replacing consumers' search for products with less artificial additives. Regulatory authorities have designated numerous polysaccharides, including guar gum, agar, and pectin as generally recognized as safe (GRAS) for their use in food at standard levels of consumption. Polysaccharides are important food industry components used for texture modification, stability, emulsification, and fiber enrichment. Diversity, their natural abundance and functional attributes make them applicable in various food formulations to incorporate them in developing healthier, more sustainable, superior food products [95, 96].

6.6 Polysaccharides in Pharmaceutical and Biomedical Applications

Due to distinctive physicochemical properties, biocompatibility, biodegradability, and adaptability, polysaccharides have been the source of interest in the pharmaceutical and biomedical fields. They are naturally occurring biopolymers, meaning that there is such a large array of uses they can either be as drug delivery systems, tissue engineering, etc. or they can be used in wound healing or diagnostic instruments. Because of their nontoxic ability to interact with biological systems for developing innovative medicines and creating improved medical devices and biomaterials, they provide exemplary prospects for that purpose [97, 98]. Polysaccharides are used in this context for their ability to form gels, encapsulate pharmaceuticals, enhance cell adhesion, and release medicinals in a regulated fashion. The pharmaceutical sector is largely employed practically in the formulation of drug delivery systems using polysaccharides. In contrast, polysaccharides, such as alginate, chitosan, pectin and hyaluronic acid, have been well characterized to encapsulate and transport pharmaceuticals in a regulated manner. Polysaccharides for use in drug delivery have advantages, including intrinsic biocompatibility that makes induced adverse immune reactions less likely, and the potential for facile modification to satisfy specific needs. Polysaccharide-based hydrogels can be engineered to allow for continuous (or regulated) release of medication over extended periods of time, thereby improving therapeutic efficacy while reducing delivery frequency. Nanoparticles and microparticles based on polysaccharides were engineered for the encapsulation of hydrophobic medicines improving solubility, stability, and bioavailability of insoluble substances [99]. Pharmaceutical and biomedical use of polysaccharides is illustrated in Fig. 6.1.

Among the most extensively studied polysaccharides in the field of drug administration, chitosan is produced from chitin and has cationic properties thus allowing interaction with negatively charged biological membranes. This characteristic further expands chitosan's ability to promote drug absorption across mucosal barriers, including the gastrointestinal tract. Additionally, chitosan is being used in preparation of nano and microencapsulation systems to protect pharmaceuticals from degradation and to deliver at preselected sites [100, 101]. Versatility in the development of drug delivery systems with customizable release profiles arises from the ability to alter chitosan's characteristics through changes in its molecular weight, degree of deacetylation, and cross-linkage. Thus, regulated drug delivery utilizes polymer chain network, or polysaccharide-based hydrogels, that can retain large amounts of water. Such hydrogels, in response to external stimuli such as pH, temperature, or ionic strength, can be tailor made for targeted release of medication to specific locations in the body. One particularly attractive advantage of pectin-based hydrogels

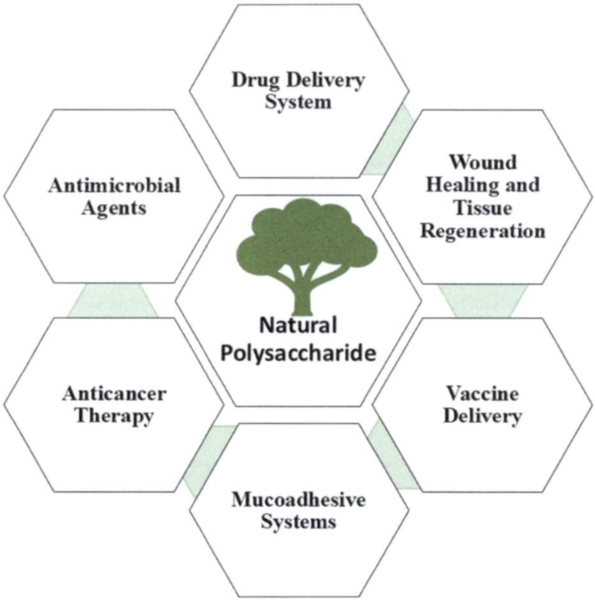

Fig. 6.1 Pharmaceutical and biomedical use of polysaccharides

for protein and peptide delivery is their predicated gelation under acidic conditions typical of the stomach, or with the presence of specific ions. Additionally, temporal degradation of polysaccharide hydrogels can be engineered to provide for the drug-controlled release and reducing the risk of overdosing [102].

Apart from administration in drugs, polysaccharides play a vital role in tissue engineering and regenerative medicine. Due to their polysaccharide-based nature, biomaterials can serve as scaffolds for tissue development and regeneration in wound healing and in bone repair. Well studied for its abilities to improve cell migration and proliferation, hyaluronic acid (polysaccharide naturally found in connective tissues) is a desirable candidate for wound healing applications. In the form of a sponge or hydrogel, it can be included to create conditions that are conducive for the tissue regeneration, or to accelerate healing [103]. Scaffolds for tissue engineering of cartilage and bone have been fabricated using alginate, a product obtained from brown seaweed. Offering an appropriate milieu for cell adhesion, proliferation, and differentiation, its capacity to generate hydrogels that may replicate the mechanical properties of a tissue makes it a potential biocompatible material [104]. Further, polysaccharides may be employed in the making of bioactive coatings for medical devices for enhancing biocompatibility as well as for aiding in tissue integration. Coatings based on polysaccharides can be applied on implants, such as stents or prostheses, to facilitate their embrittlement with lateral tissues and prevent infection or inflammation. Inhibition of bacterial adherence can be achieved with polysaccharide coatings such as chitosan or hyaluronic acid, which reduce the likelihood of device associated illnesses [105, 106].

Polysaccharides have been used in wound healing. The most efficacious dressings that polysaccharides provide include those of alginate or chitosan that sustain a moist environment, absorb exudates, and provide barrier against germs. The frequent use of these dressings in wound management, burns, and surgical incisions is well established. It has long been known that alginate is able to absorb huge volumes of fluid and form a gelling [107], thus facilitating the creation of a tissue regenerative layer. In addition, chitosan has already been known to offer antibacterial effects which can prevent infection at wound sites, and therefore help to speed up the healing process. It is also a further intriguing domain of research of polysaccharides, their application in diagnostics. Biosensors have been created where polysaccharides are used as carriers to increase the detection of disease indicators and infections by taking advantage of their ability to bind some analytes [108]. Polysaccharide-based materials can be used in diagnostic assays to trap antibodies, proteins, or other biomarkers associated with diseases like cancer, infection, or metabolic disorders. These materials can also be used to prepare nanoparticles for imaging, by coupling targeting ligands to drive the nanoparticles toward specific cells or tissues, for noninvasive imaging of disease locations. Much interest has been gathered in polysaccharides as carriers for vaccines, in particular within nanoparticle-based vaccine delivery systems. Dextran and chitosan can serve as encapsulating polysaccharides for encapsulating antigens in order to stabilize them and dictate the site for their immune cell targeting. Vaccines encapsulated into polysaccharide nanoparticles in vaccines may increase an immunological response by enhancing presentation of the antigen to the immune system and controlling release of the antigen at the immunization site. Implementation pharmacological and biological fields have made polysaccharides so adaptable and biocompatible. Polysaccharides offer many advantages that improve the efficacy, safety, and patient outcomes of therapeutic interventions, such as the administration of medication, tissue engineering, wound healing, diagnostics, and vaccine delivery [109].

6.7 Polysaccharides in Cosmetic and Personal Care Products

Due to the multifunctionality, biocompatibility, and sustainability of naturally occurring polysaccharides, these biopolymers have been increasingly attractive to the cosmetic and personal care industry. Sourced from plants, algae, fungi, and microorganisms, these macromolecules serve as thickeners, emulsifiers, moisturizes, stabilizers, and bioactive components. Additionally, their ability to retain moisture, form hydrogels, and interact with other biomolecules makes them vital for the development of skincare, haircare, and personal care products [110–112]. Table 6.4 lists the principal polysaccharides used in cosmetics. These biopolymers offer a sustainable and robust alternative to synthetic chemistry for hydration, anti-aging, rheology control, and also as antibacterial agent. Emerging trends of polysaccharide use in cosmetics are illustrated in Fig. 6.2.

Table 6.4 Polysaccharides in cosmetic industries

Polysaccharide	Source	Function in cosmetics	Ref.
Hyaluronic acid	Animal (*rooster combs*), microbial fermentation	Intense hydration, antiaging, wound healing	[123]
Xanthan gum	Bacterial fermentation (*Xanthomonas campestris*)	Thickener, stabilizer, texture enhancer	[124, 125]
Chitosan	Shellfish exoskeleton, fungal sources	Antimicrobial, film-forming, skin protection	[126]
Alginate	Brown algae	Moisturizer, emulsion stabilizer, film-forming	[127]
Carrageenan	Red algae	Gelling agent, texture enhancer, skin conditioner	[128]
Beta-glucans	Oats, fungi, yeast	Skin repair, anti-inflammatory, wound healing	[129]
Aloe vera gel (Acemannan)	Aloe vera plant	Soothing, anti-inflammatory, hydration	[130]
Fucoidan	Brown seaweed	Antiaging, antioxidant, collagen stimulation	[131]
Guar gum	Guar beans	Thickener, viscosity modifier	[132]

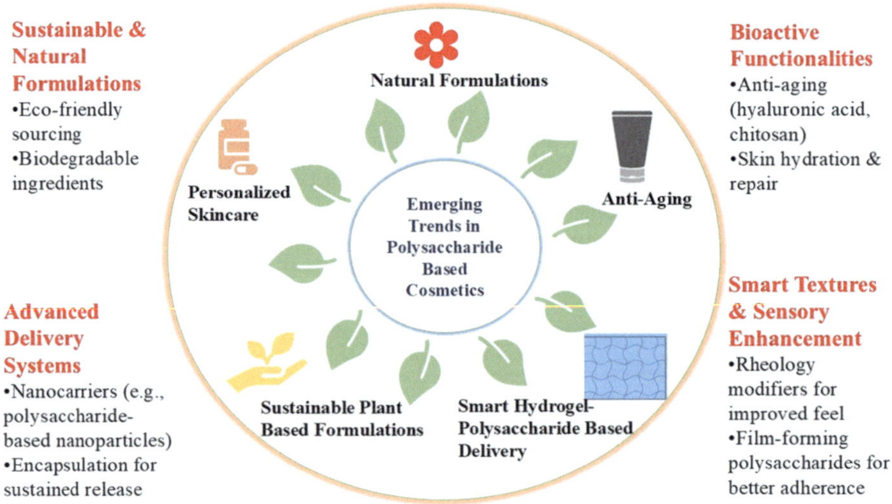

Fig. 6.2 Emerging polysaccharide trends in cosmetics

6.7.1 Key Functions of Polysaccharides in Cosmetics

- Moisturization and hydration: Polysaccharides are renowned for their outstanding water holding power. Preserving skin hydration also involves cutting out transepidermal water loss (TEWL) by forming a protective barrier. An important skincare polysaccharide widely used is hyaluronic acid, a glycosaminoglycan that is famed for its remarkable ability to retain hydration (it can hold on to 1000 x its weight in water). Additionally, alginates and carrageenans, prepared from seaweed, will hydrate the skin and develop film forming characteristics, leaving skin with improvement in its smoothness and elasticity [113, 114].

- Film formation and dermal protection: There are polysaccharides that make thin, permeable coatings over the skin to protect from environmental contaminants, irritants, and UV-related damage [115]. Of these, chitosan, a biopolymer from chitin, shows best film forming and antibacterial characteristics, rendering it excellent for antipollution skincare products [116]. Oats and fungi beta-glucans create a protective skin-based barrier offering anti-inflammatory and wound healing benefits.
- Rejuvenation and dermal restoration: Polysaccharides help with antiaging formulas as they help to give the skin suppleness, stimulate the production of collagen, and reduce oxidative stress. In addition to hydrating your skin, hyaluronic acid accelerates tissue regeneration and shrinks the appearance of wrinkles [117]. Brown algae induced polysaccharide sulfated, which has been shown to also facilitate fibroblast division and mitigate matrix metalloproteinase (MMP) activity to prevent against early ageing of the skin [118].
- Rheological modification and texture improvement: Many cosmetic compositions must be tightly controlled in viscosity and texture to provide a good user experience. Natural thickeners and stabilizers include polysaccharides like xanthan gum, guar gum, and cellulose derivatives which improve spreadability and consistency of creams, gels, and serums. Pseudoplastic properties of their application provide a smooth application which fits the occurrence of mechanical stress during utilization.
- Calming and anti-inflammatory attributes: Botanical and microbiological polysaccharides are calming and anti-inflammatory, making them useful ingredients in formulations for sensitive skin. Abundance of acemannan in aloe vera gel provides quick relief of irritation and erythema. Beta-glucans, as they are known for their immunomodulatory effects, have an astringent, anti-inflammatory property used for helpful in post-procedure skin care [119, 120].
- Antimicrobial and preservative properties: Intrinsic antibacterial properties of certain polysaccharides reduce the requirement for synthetic preservatives. Because of its antibacterial and antifungal properties, chitosan is an effective natural substitute for preservative systems in cosmetic formulations. It also enhances product efficacy by increasing the bioavailability of medicinal compounds [121, 122].

6.8 Polysaccharides in Industrial and Environmental Applications

As multifunctional biopolymers polysaccharides are of great importance in many different industrial and environmental domains on account of biodegradability, biocompatibility, and functional characteristics [133, 134]. These macromolecules are from natural sources and are vital for making such sustainable materials, for wastewater treatment, for bioenergy generation as well as environmental remediation. As a result of their capacity to generate hydrogels, absorb pollutants, and act as a bio-based substitute for synthetic polymers, they have become essential materials in the development of sustainable industrial practices [135].

6.8.1 Industrial Utilization of Polysaccharides

(i) *Bioplastics and packaging materials*: There has been a trend of escalating demand for biodegradable and environmentally friendly substitutes for petroleum-based plastics, turning polysaccharides as possible alternatives for bioplastic manufacturing. Biodegradable films, coatings and packaging materials are produced from starch, cellulose, chitosan, and alginate. Because thermoplastic starch (TPS) forms superior films, its use in the manufacture of biodegradable plastic films is extensive [136]. However, it requires plasticizers, such as glycerol, to contribute to flexibility and mechanical strength. The interest in cellulose derivatives, like cellulose acetate and nanocellulose, comes from their outstanding mechanical characteristics, transparency, and biodegradability for use in sustainable packaging. Chitosan is sourced from chitin and has shown antibacterial activity making it a critical element in food packaging prolonging shelf life and avoiding microbiological infection [137].

(ii) *Textile industry and fiber alteration*: Textile processing, fiber modification, and fabric finishing are all dependent on polysaccharides. Cotton fiber is an extensively used fiber derived from natural polysaccharides. In addition, cellulose derivatives, e.g., carboxymethyl cellulose (CMC) and hydroxypropyl cellulose (HPC), are used as thickeners and fillers in the printing and dyeing textiles. As visco-regulating agents and being non-toxic, alginate derived from brown seaweed is used as a thickener in textile printing pastes [138, 139].

(iii) *Paper and pulp industry*: Polysaccharides help improve paper manufacturing and processing by improving paper strength, water resistance, and printability. In papermaking, it performs as binder and coating in aid of making a better finish with strength. Fiber adhesion is enhanced by the use of natural hemicellulose as a natural adhesive and CNFs as a synergistic reinforcement to add mechanical properties along with reduction in the reliance of synthetic additives [140, 141].

(iv) *Adhesives and binders*: Bio-based adhesion and binders are provided by natural polysaccharides and used in various industrial applications. Its large film-forming and bonding characteristics make starch-based adhesives of choice in the manufacture of corrugated boards, book board, and wood adhesives. Chitosan-based binders are used for biomedical applications as well as biodegradable coatings because of their nontoxic and antibacterial properties [136].

6.8.2 Environmental Utilizations of Polysaccharides

Efficient bioadsorbents of polysaccharides are found to be used for the extraction of heavy metals, organic contaminants, and colors from industrial effluent. Polysaccharides are extensively used for removal of lead, cadmium, and arsenic from contaminated water because of strong affinity for metal ions. Polysaccharides produce gels and porous substances that can enclose contaminants and perform well for handling wastewater [142]. Chemically modified starch increases adsorption capacity toward the removal of synthetic colors from textile effluents. Microbial bioremediation of polluted soil and water requires polysaccharides. EPS produced by microorganisms increase the bioavailability of residues of contaminants, such as hydrocarbons or pesticides, or heavy metals, and are involved in their microbial degradation. By encapsulation of oil-decomposing microorganisms into alginate hydrogels, the microorganisms are augmented in their efficacy to degrade petroleum contaminants in marine environments [143].

Polysaccharides function as essential feedstocks for the manufacture of biofuels and biogas, facilitating the shift toward sustainable energy sources. Lignocellulosic biomass, consisting of cellulose and hemicellulose, undergoes enzymatic hydrolysis to yield fermentable sugars for bioethanol synthesis. Corn and cassava starch undergo fermentation to produce ethanol, aiding in the development of sustainable fuel alternatives [144]. Chitosan is investigated as an ingredient in anaerobic digestion systems to enhance microbial activity and increase methane production. Polysaccharides improve soil quality and promote plant growth by functioning as biofertilizers, soil conditioners, and moisture retainers. Chitosan functions as a biostimulant by promoting plant growth, enhancing disease resistance, and fostering beneficial microbial activity in the soil [145, 146]. Alginate is utilized for water retention in arid areas to enhance soil moisture retention and diminish irrigation requirements. Cellulose-derived hydrogels are designed to function as slow-release fertilizers, enhancing nutrient accessibility for crops [147].

6.9 Challenges and Future Perspectives

Polysaccharides are extensively used in pharmaceuticals, food, cosmetics, and biomaterials, because of their varied applications. Nonetheless, various obstacles impede their widespread application and clinical implementation. Extraction and purification of natural polysaccharides are particularly difficult. However, conventional extraction methods such as solvent extraction, enzyme assisted extraction, and ultrasonic extraction usually require great amounts of energy input, long extraction duration, and usage of aggressive chemicals. In addition, purification is still limited, due to the structural complexity of polysaccharides. They can also be influenced by their concentration by impurities such as proteins, phenolic chemicals, and residual solvents. A great effort has to be given for developing sustainable, cost-effective [148], high yield technologies for their extensive implementation. Their structural heterogeneity is a notable concern. Natural polysaccharides differ from synthetic polymers in that they have differences in molecular weight, branching configurations, and functional groups depending upon origin and environmental factors. This lack of predictability poses a problem in sustaining batch to batch uniformity which is a prerequisite for use in pharmaceutical and biological applications. For accurate structural elucidation and quality control, the analysis requires advanced characterization techniques, such as nuclear magnetic resonance (NMR) spectroscopy, mass spectrometry, and chromatography. Hence, another challenge is the limited functional modification options constrained by structural variability. Although bioactivity is intrinsic to all polysaccharides, polysaccharides often exhibit less functional quality than usually desired in an industrial or medical application. Typically chemical, enzymatic, and physical modifications are used to improve solubility, stability, and bioavailability. However, modifying the site-specifically is infeasible since they tend to disrupt fundamental structure; thus, advancement of sophisticated modification techniques, such as green chemical approaches and nanotechnology assisted modifications, is necessary to attain site specific alteration without compromising fundamental structure. Many polysaccharides show low aqueous or organic solvent solubility rendering their use in direct applications as medicines and biomaterials untenable. Reduced solubility is realized due to elevated molecular weight, hydrogen bonding, and crystallinity. Additionally, their bioavailability is often hindered due to limited uptake and enzymatic breakdown within biological systems. Strategies for formulation, including nanoencapsulation, hydrophilic alterations, and conjugation with bioactive chemicals, are being investigated to address these limitations [149].

A further significant difficulty pertains to their stability and shelf-life. Environmental degradation of natural polysaccharides such as hydrolysis, oxidation, and microbial contamination of polysaccharides is of importance. In pharmaceutical and food applications, their storage, shelf life, and functional efficacy are impacted by this instability. Shelf life extending measures such as lyophilization, regulated storage temperatures, and stabilization agent addition are also important [150]. Consequently, regarding their potential applications, natural polysaccharides are regulated by the challenges of their complicated and wide composition. Regulatory approval depends upon standardized techniques for extraction, characterization and safety evaluation. In addition, for commercialization they must adhere to international standards, such as Food and Drug Administration (FDA) and European Medicines Agency (EMA) [151].

These difficulties need to be tackled by future research using environmentally friendly and sustainable extraction technology. Recent development of supercritical fluid extraction, microwave assisted extraction, and deep eutectic solvent-based procedures are being used as ecofriendly methods for traditional methods. This allows optimizing solvent use, reducing the energy expenditure and maximizing the production rate with a concurrent preservation of bioactivity of polysaccharides [152]. Improvements in analytical methods such as high resolution mass spectrometry, cryo-electron microscopy, and machine learning-based structure prediction will greatly increase our understanding of polysaccharide architectures. This will enable the very tailored application of their characteristics to specific applications in drug delivery, tissue engineering, and biomaterials. Polysaccharide-based materials are also dependent on creating intelligent and functionalized biomaterials. Because they can change their properties when interacting with a pH, temperature, or other external information, these materials are very good candidates for applications in controlled medication release and in medicine. Moreover, these hybrid systems are intriguing in that incorporation of polysaccharides with nanomaterials, such as carbon nanotubes, graphene and metallic nanoparticles, increases the functions of these hybrid systems. Customized medicine makes it possible to customize polysaccharides for personalized medicinal uses. The development of 3D bioprinting and biofabrication methodologies will permit the generation of tailored polysaccharide-based scaffolds for tissue regeneration and wound healing as well as organ engineering. Additionally, they are next expected to be applied in immunomodulatory therapy and to targeted drug delivery. Artificial intelligence (AI) and computational modeling are becoming an important part of polysaccharide research [153, 154]. Predictive models can, in art, utilize AI to optimize extraction methods, to elucidate structure-activity connections, and to enhance formulation tactics. This will speed up the search for new polysaccharides with enhanced bioactivity and industrial relevance in this combination. Much needed effort should be directed at achieving worldwide regulatory uniformity to facilitate commercialization. Academic efforts will be paired with business and regulatory agency inputs to establish standardized protocol for polysaccharide synthesis and utilization [155]. Moreover, enhancing output through bioreactor-based fermentation and sustainable sourcing methods will provide a reliable supply for industrial applications.

6.10 Conclusion

A unique category of biomolecules that is derived from a natural source includes the polysaccharides whose functional characteristics vary and have been widely used. With their intrinsic biocompatibility, biodegradability, and adjustable physicochemical properties, they are extremely relevant to many industries, including pharmaceuticals, food, cosmetics and biomedical engineering. This structural diversity complexity presents issues with extraction, purification, repeatability, and regulatory compliance, and an ongoing need for advances in analytical technologies and processing developments. The multiple functionalities of natural polysaccharides are one of the most attractive properties of these materials. Water solubility, viscosity, gelation, and film forming abilities allow their wide application in the stabilization of food, administration of medication, and the production of biomaterials. Additionally, these properties biological make the probiotic strains multispecialty uses toward improve health. Polysaccharides have been shown to have bioactivities establishing them as attractive candidates for functional foods, medicines as well as for therapeutic formulations for disease prevention and management.

Despite these benefits, however, many impediments still remain. Unlike synthetic polymers, the variability in diversity of natural polysaccharides makes functional performance difficult to characterize and/or standardize in industrial and clinical applications. In addition, their bioavailability and stability must be improved through the use of sophisticated formulation techniques, such as nanoencapsulation, chemical modification, and hybrid material design. Future strategies for addressing the present limits and boosting the efficiency of polysaccharide-based products include sustainable methods of extraction, green chemical approaches, and AI-driven prediction models.

In the future, the combined use of polysaccharides with advanced technologies will carry a great potential in the use in the biological and industrial fields by means of 3D bioprinting, nanotechnology, artificial intelligence, etc. Next-generation drug delivery systems, wound healing scaffolds, and tissue engineering platforms are being advanced by intelligent polysaccharide-derived biomaterials that can respond to physiological stimuli. Additionally, polysaccharides are poised as key components facilitating the development of green alternatives to synthetic materials of increasingly high consumer demand for natural, ecofriendly and health improving products. Polysaccharides will persist in influencing the future of medicine, food science, and biomaterials through interdisciplinary collaboration and novel methodologies, fostering a more sustainable and health-oriented society.

References

1. Yu Y, Shen M, Song Q, Xie J. Biological activities and pharmaceutical applications of polysaccharide from natural resources: A review. Carbohydr Polym. 2018;183:91–101.
2. Mohammed ASA, Naveed M, Jost N. Polysaccharides; classification, chemical properties, and future perspective applications in fields of pharmacology and biological medicine (A review of current applications and upcoming potentialities). J Polym Environ. 2021;29:2359–71.
3. Yang X, Li A, Li X, Sun L, Guo Y. An overview of classifications, properties of food polysaccharides and their links to applications in improving food textures. Trends Food Sci Technol. 2020;102:1–15.
4. Torres FG, Troncoso OP, Pisani A, Gatto F, Bardi G. Natural polysaccharide nanomaterials: an overview of their immunological properties. IJMS. 2019;20:5092.
5. Le XT, Rioux L-E, Turgeon SL. Formation and functional properties of protein–polysaccharide electrostatic hydrogels in comparison to protein or polysaccharide hydrogels. Adv Colloid Interf Sci. 2017;239:127–35.
6. Li M, Jia X, Yao Q, Zhu F, Huang Y, Zeng X-A. Recent advance for animal-derived polysaccharides in nanomaterials. Food Chem. 2024;459:140208.
7. Patel AK, Vadrale AP, Singhania RR, Michaud P, Pandey A, Chen S-J, et al. Algal polysaccharides: current status and future prospects. Phytochem Rev. 2023;22:1167–96.
8. Freitas F, Torres CAV, Araújo D, Farinha I, Pereira JR, Concórdio-Reis P, et al. Advanced microbial polysaccharides. In: Rehm B, Moradali MF, editors. Biopolymers for biomedical and biotechnological applications [internet]. 1st ed. Wiley; 2021. [cited 2025 Feb 3]. pp. 19–62. Available from: https://onlinelibrary.wiley.com/doi/10.1002/9783527818310.ch2
9. Díaz-Montes E. Polysaccharides: sources, characteristics, properties, and their application in biodegradable films. Polysaccharides. 2022;3:480–501.
10. Moreira JB, Vaz BDS, Cardias BB, Cruz CG, Almeida ACAD, Costa JAV, et al. Microalgae polysaccharides: an alternative source for food production and sustainable agriculture. Polysaccharides. 2022;3:441–57.

11. Kabir SF, Rahman A, Yeasmin F, Sultana S, Masud RA, Kanak NA, et al. Occurrence, distribution, and structure of natural polysaccharides. Radiation-processed polysaccharides [Internet]. Elsevier. 2022 [cited 2025 Feb 3]. pp. 1–27. Available from: https://linkinghub.elsevier.com/retrieve/pii/B9780323856720000052
12. Górska S, Maksymiuk A, Turło J. Selenium-containing polysaccharides—structural diversity, biosynthesis, chemical modifications and biological activity. Appl Sci. 2021;11:3717.
13. Dumitriu S, editor. Polysaccharides: structural diversity and functional versatility. Second Edition [Internet]. 0 ed. CRC Press; 2004. [cited 2025 Feb 3]. Available from: https://www.taylorfrancis.com/books/9781420030822
14. Elango B, Shirley CP, Okram GS, Ramesh T, Seralathan K-K, Mathanmohun M. Structural diversity, functional versatility and applications in industrial, environmental and biomedical sciences of polysaccharides and its derivatives—A review. Int J Biol Macromol. 2023;250:126193.
15. Vaaje-Kolstad G, Forsberg Z, Loose JS, Bissaro B, Eijsink VG. Structural diversity of lytic polysaccharide monooxygenases. Curr Opin Struct Biol. 2017;44:67–76.
16. Peesapati S, Sajeevan KA, Patel SK, Roy D. Relation between glycosidic linkage, structure and dynamics of α – and β -glucans in water. Biopolymers. 2021;112:e23423.
17. Chaudhary S, Jain VP, Jaiswar G. The composition of polysaccharides: monosaccharides and binding, group decorating, polysaccharides chains. Innovation in Nano-Polysaccharides for Eco-sustainability [Internet]. Elsevier; 2022 [cited 2025 Feb 3]. p. 83–118. Available from: https://linkinghub.elsevier.com/retrieve/pii/B9780128234396000052
18. Chen R, Xu J, Wu W, Wen Y, Lu S, El-Seedi HR, et al. Structure–immunomodulatory activity relationships of dietary polysaccharides. Curr Res Food Sci. 2022;5:1330–41.
19. Zvyagintseva TN, Usoltseva RV, Shevchenko NM, Surits VV, Imbs TI, Malyarenko OS, et al. Structural diversity of fucoidans and their radioprotective effect. Carbohydr Polym. 2021;273:118551.
20. Monsan P, Bozonnet S, Albenne C, Joucla G, Willemot R-M, Remaud-Siméon M. Homopolysaccharides from lactic acid bacteria. Int Dairy J. 2001;11:675–85.
21. Nabot M, Guérin M, Sivakumar D, Remize F, Garcia C. Variability of bacterial Homopolysaccharide production and properties during food processing. Biology. 2022;11:171.
22. Korakli M, Vogel RF. Structure/function relationship of homopolysaccharide producing glycansucrases and therapeutic potential of their synthesised glycans. Appl Microbiol Biotechnol. 2006;71:790–803.
23. Joseph TM, Sathian A, Joshy KS, Kar Mahapatra D, Haponiuk JT, Thomas S. Modifications of starch and its characterizations. Handbook of Natural Polymers, Volume 2 [Internet]. Elsevier; 2024 [cited 2025 Feb 3]. p. 23–48. Available from: https://linkinghub.elsevier.com/retrieve/pii/B9780323998567000240
24. Holtekjølen AK, Uhlen AK, Bråthen E, Sahlstrøm S, Knutsen SH. Contents of starch and non starch polysaccharides in barley varieties of different origin. Food Chem. 2006;94:348–58.
25. Li J, Goddard-Borger ED, Raji O, Saxena H, Solhi L, Mathieu Y, et al. Chitin-active lytic polysaccharide monooxygenases are rare in *Cellulomonas* species. Kelly RM, editor. Appl Environ Microbiol. 2022;88:e00968–22.
26. Kalidas C, Sangaranarayanan MV. Carbohydrates, their reactions, thermochemistry and energetics. Biophys Chem [Internet]. Cham: Springer Nature Switzerland; 2023 [cited 2025 Feb 3]. pp. 59–85. Available from: https://link.springer.com/10.1007/978-3-031-37682-5_3
27. De Vuyst L. Heteropolysaccharides from lactic acid bacteria. FEMS Microbiol Rev. 1999;23:153–77.
28. Li Y, Guo X, Zhong R, Ye C, Chen J. Structure characterization and biological activities evaluation of two hetero-polysaccharides from Lepista nuda: cell antioxidant, anticancer and immune-modulatory activities. Int J Biol Macromol. 2023;244:125204.
29. Hussain A, Zia KM, Tabasum S, Noreen A, Ali M, Iqbal R, et al. Blends and composites of exopolysaccharides; properties and applications: a review. Int J Biol Macromol. 2017;94:10–27.

30. Buckley C, Murphy EJ, Montgomery TR, Major I. Hyaluronic acid: A review of the drug delivery capabilities of this naturally occurring polysaccharide. Polymers. 2022;14:3442.
31. Sudha PN, Rose MH. Beneficial effects of hyaluronic acid. Advances in food and nutrition research [Internet]. Elsevier; 2014 [cited 2025 Feb 3]. p. 137–76. Available from: https://linkinghub.elsevier.com/retrieve/pii/B9780128002698000099
32. Rodén L, Ananth S, Campbell P, Curenton T, Ekborg G, Manzella S, et al. Heparin — an Introduction. In: Lane DA, Björk I, Lindahl U, editors. Heparin and Related Polysaccharides [Internet]. Boston, MA: Springer US; 1992 [cited 2025 Feb 3]. pp. 1–20. Available from: http://link.springer.com/10.1007/978-1-4899-2444-5_1
33. Saravanan R. Isolation of low-molecular-weight heparin/Heparan sulfate from marine sources. Advances in Food and Nutrition Research [Internet] Elsevier; 2014 [cited 2025 Feb 3]. pp. 45–60. Available from: https://linkinghub.elsevier.com/retrieve/pii/B9780128002698000038
34. Demeter F, Peleskei Z, Kútvölgyi K, Rusznyák Á, Fenyvesi F, Kajtár R, et al. Synthesis and biological profiling of seven heparin and Heparan Sulphate analogue Trisaccharides. Biomol Ther. 2024;14:1052.
35. Montilla A, Muñoz-Almagro N, Villamiel M. A new approach of functional pectin and pectic oligosaccharides: role as antioxidant and antiinflammatory compounds. Current Advances for Development of Functional Foods Modulating Inflammation and Oxidative Stress [Internet]. Elsevier; 2022 [cited 2025 Feb 3]. p. 105–20. Available from: https://linkinghub.elsevier.com/retrieve/pii/B9780128234822000261
36. Spiridon I, Popa VI. Hemicelluloses: major sources, properties and applications. Monomers, polymers and composites from renewable resources [Internet]. Elsevier; 2008 [cited 2025 Feb 3]. p. 289–304. Available from: https://linkinghub.elsevier.com/retrieve/pii/B9780080453163000132
37. Gao Y, Guo M, Wang D, Zhao D, Wang M. Advances in extraction, purification, structural characteristics and biological activities of hemicelluloses: A review. Int J Biol Macromol. 2023;225:467–83.
38. Potnis AA, Raghavan PS, Rajaram H. Overview on cyanobacterial exopolysaccharides and biofilms: role in bioremediation. Rev Environ Sci Biotechnol. 2021;20:781–94.
39. Huang Y, Chen H, Zhang K, Lu Y, Wu Q, Chen J, et al. Extraction, purification, structural characterization, and gut microbiota relationship of polysaccharides: a review. Int J Biol Macromol. 2022;213:967–86.
40. Wang W, Tan J, Nima L, Sang Y, Cai X, Xue H. Polysaccharides from fungi: a review on their extraction, purification, structural features, and biological activities. Food Chemistry: X. 2022;15:100414.
41. Luan F, Ji Y, Peng L, Liu Q, Cao H, Yang Y, et al. Extraction, purification, structural characteristics and biological properties of the polysaccharides from Codonopsis pilosula: A review. Carbohydr Polym. 2021;261:117863.
42. Yang W, Huang G. Extraction methods and activities of natural glucans. Trends Food Sci Technol. 2021;112:50–7.
43. Nai J, Zhang C, Shao H, Li B, Li H, Gao L, et al. Extraction, structure, pharmacological activities and drug carrier applications of Angelica sinensis polysaccharide. Int J Biol Macromol. 2021;183:2337–53.
44. Mozammil Hasnain SM, Hasnain MS, Nayak AK. Natural polysaccharides. Natural Polysaccharides in Drug Delivery and Biomedical Applications [Internet]. Elsevier; 2019 [cited 2025 Feb 4]. pp. 1–14. Available from: https://linkinghub.elsevier.com/retrieve/pii/B9780128170557000017
45. Tang Y, Xiao Y, Tang Z, Jin W, Wang Y, Chen H, et al. Extraction of polysaccharides from *Amaranthus hybridus* L. by hot water and analysis of their antioxidant activity. PeerJ. 2019;7:e7149.
46. Wan C, Jiang H, Tang M-T, Zhou S, Zhou T. Purification, physico-chemical properties and antioxidant activity of polysaccharides from Sargassum fusiforme by hydrogen peroxide/ascorbic acid-assisted extraction. Int J Biol Macromol. 2022;223:490–9.

47. Shi Y, Zhu A, Shen L. Optimization of acid assisted extraction process of foxtail millet polysaccharides and its antioxidant activity. Int Agrophys. 2020;34:141–9.
48. Jia Y, Gao X, Xue Z, Wang Y, Lu Y, Zhang M, et al. Characterization, antioxidant activities, and inhibition on α-glucosidase activity of corn silk polysaccharides obtained by different extraction methods. Int J Biol Macromol. 2020;163:1640–8.
49. Lin L, Zhao L, Deng J, Xiong S, Tang J, Li Y, et al. Enzymatic extraction, purification, and characterization of polysaccharides from *Penthorum chinense Pursh* : natural antioxidant and anti-inflammatory. Biomed Res Int. 2018;2018:1–13.
50. Song Y-R, Sung S-K, Jang M, Lim T-G, Cho C-W, Han C-J, et al. Enzyme-assisted extraction, chemical characteristics, and immunostimulatory activity of polysaccharides from Korean ginseng (Panax ginseng Meyer). Int J Biol Macromol. 2018;116:1089–97.
51. Mena-García A, Ruiz-Matute AI, Soria AC, Sanz ML. Green techniques for extraction of bioactive carbohydrates. TrAC Trends Anal Chem. 2019;119:115612.
52. Morais ES, Lopes AMDC, Freire MG, Freire CSR, Coutinho JAP, Silvestre AJD. Use of ionic liquids and deep eutectic solvents in polysaccharides dissolution and extraction processes towards sustainable biomass valorization. Molecules. 2020;25:3652.
53. Sun C, Wang G, Sun J, Yin J, Huang J, Li Z, et al. A new method of extracting Polygonatum sibiricum polysaccharide with antioxidant function: ultrasound-assisted extraction-deep eutectic solvents method. Food Secur. 2023;12:3438.
54. Wang N, Li Q. Study on extraction and antioxidant activity of polysaccharides from radix Bupleuri by natural deep eutectic solvents combined with ultrasound-assisted enzymolysis. Sustain Chem Pharm. 2022;30:100877.
55. Gomez L, Tiwari B, Garcia-Vaquero M. Emerging extraction techniques: Microwave-assisted extraction. Sustainable Seaweed Technologies [Internet]. Elsevier; 2020 [cited 2025 Feb 4]. p. 207–24. Available from: https://linkinghub.elsevier.com/retrieve/pii/B9780128179437000081
56. Gligor O, Mocan A, Moldovan C, Locatelli M, Crișan G, Ferreira ICFR. Enzyme-assisted extractions of polyphenols—A comprehensive review. Trends Food Sci Technol. 2019;88:302–15.
57. Magnusson M, Yuen AKL, Zhang R, Wright JT, Taylor RB, Maschmeyer T, et al. A comparative assessment of microwave assisted (MAE) and conventional solid-liquid (SLE) techniques for the extraction of phloroglucinol from brown seaweed. Algal Res. 2017;23:28–36.
58. Thao My PL, Sung VV, Dat TD, Nam HM, Phong MT, Hieu NH. Ultrasound-assisted extraction of Fucoidan from Vietnamese Brown seaweed *Sargassum mcclurei* and testing bioactivities of the extract. ChemistrySelect. 2020;5:4371–80.
59. Santana ÁL, Meireles MAA. Valorization of cereal byproducts with supercritical technology: the case of corn. PRO. 2023;11:289.
60. Mohan K, Ganesan AR, Ezhilarasi PN, Kondamareddy KK, Rajan DK, Sathishkumar P, et al. Green and eco-friendly approaches for the extraction of chitin and chitosan: A review. Carbohydr Polym. 2022;287:119349.
61. Turan O, Isci A, Yılmaz MS, Tolun A, Sakiyan O. Microwave-assisted extraction of pectin from orange peel using deep eutectic solvents. Sustain Chem Pharm. 2024;37:101352.
62. Garcia-Vaquero M, Rajauria G, O'Doherty JV, Sweeney T. Polysaccharides from macroalgae: recent advances, innovative technologies and challenges in extraction and purification. Food Res Int. 2017;99:1011–20.
63. Tang W, Liu D, Yin J-Y, Nie S-P. Consecutive and progressive purification of food-derived natural polysaccharide: based on material, extraction process and crude polysaccharide. Trends Food Sci Technol. 2020;99:76–87.
64. Zhang X, Li L, Fung H, Chen N, Shan P, Zhou Y, et al. Critical review of the criterion of polysaccharide purity. Carbohydr Polym. 2025;352:123187.
65. Xiong Q, Song Z, Hu W, Liang J, Jing Y, He L, et al. Methods of extraction, separation, purification, structural characterization for polysaccharides from aquatic animals and their major pharmacological activities. Crit Rev Food Sci Nutr. 2020;60:48–63.

66. Persin Z, Stana-Kleinschek K, Foster TJ, Van Dam JEG, Boeriu CG, Navard P. Challenges and opportunities in polysaccharides research and technology: the EPNOE views for the next decade in the areas of materials, food and health care. Carbohydr Polym. 2011;84:22–32.
67. Liao B, Zheng J, Xia C, Chen X, Xu Q, Duan B. The potential, challenges, and prospects of the genus spirulina polysaccharides as future multipurpose biomacromolecules. Int J Biol Macromol. 2023;253:127482.
68. Zhang P, Tan J, Wang W, Zhang J, Gong H, Xue H. Extraction, separation, purification, chemical characterizations, and biological activities of polysaccharides from Chinese herbal medicine: A review. Starch Stärke. 2022;74:2200114.
69. Shen Y, Hong S, Singh G, Koppel K, Li Y. Improving functional properties of pea protein through "green" modifications using enzymes and polysaccharides. Food Chem. 2022;385:132687.
70. Cui SW, Wang Q. Cell wall polysaccharides in cereals: chemical structures and functional properties. Struct Chem. 2009;20:291–7.
71. Meng X, Luosang D, Meng S, Wang R, Fan W, Liang D, et al. The structural and functional properties of polysaccharide foulants in membrane fouling. Chemosphere. 2021;268:129364.
72. Cui R, Zhu F. Ultrasound modified polysaccharides: A review of structure, physicochemical properties, biological activities and food applications. Trends Food Sci Technol. 2021;107:491–508.
73. Wang X, Wang Z, Shen M, Yi C, Yu Q, Chen X, et al. Acetylated polysaccharides: synthesis, physicochemical properties, bioactivities, and food applications. Crit Rev Food Sci Nutr. 2024;64:4849–64.
74. Alba K, Kontogiorgos V. Techniques for the chemical and physicochemical characterization of polysaccharides. Handbook of Hydrocolloids [Internet]. Elsevier; 2021 [cited 2025 Feb 4]. p. 27–74. Available from: https://linkinghub.elsevier.com/retrieve/pii/B9780128201046000267
75. Cui SW. Food carbohydrates: chemistry, physical properties, and applications. Hoboken: Taylor and Francis; 2010.
76. Chen Y, Zhang N, Chen X. Structurally Modified Polysaccharides: Physicochemical Properties, Biological Activities, Structure–Activity Relationship, and Applications. J Agric Food Chem 2024;72:3259–3276.
77. Yoshida T. Synthesis of polysaccharides having specific biological activities. Prog Polym Sci. 2001;26:379–441.
78. Ullah S, Khalil AA, Shaukat F, Song Y. Sources, extraction and biomedical properties of polysaccharides. Food Secur. 2019;8:304.
79. Liu J, Willför S, Xu C. A review of bioactive plant polysaccharides: biological activities, functionalization, and biomedical applications. Bioact Carbohydr Diet Fibre. 2015;5:31–61.
80. Pereira L. Biological and therapeutic properties of the seaweed polysaccharides. IBR [Internet]. 2018 [cited 2025 Feb 4];2. Available from: https://journals.ke-i.org/index.php/ibr/article/view/1762
81. Roger O, Colliec-Jouault S, Ratiskol J, Sinquin C, Guezennec J, Fischer AM, et al. Polysaccharide labelling: impact on structural and biological properties. Carbohydr Polym. 2002;50:273–8.
82. Polysaccharides: properties and applications. Beverly, Massachusetts: Scrivener Publishing. 2021.
83. Aravamudhan A, Ramos DM, Nada AA, Kumbar SG. Natural polymers. Natural and synthetic biomedical polymers [Internet]. Elsevier; 2014 [cited 2025 Feb 4]. pp. 67–89. Available from: https://linkinghub.elsevier.com/retrieve/pii/B9780123969835000041
84. Arab K, Ghanbarzadeh B, Karimi S, Ebrahimi B, Hosseini M. Gelling and rheological properties of a polysaccharide extracted from Ocimum album L. seed. Int J Biol Macromol. 2023;246:125603.
85. Dimopoulou M, Alba K, Sims IM, Kontogiorgos V. Structure and rheology of pectic polysaccharides from baobab fruit and leaves. Carbohydr Polym. 2021;273:118540.

86. Mikušová V, Ferková J, Žigrayová D, Krchňák D, Mikuš P. Comparative study of polysaccharide-based hydrogels: rheological and texture properties and ibuprofen release. Gels. 2022;8:168.
87. Muthukumar J, Chidambaram R, Sukumaran S. Sulfated polysaccharides and its commercial applications in food industries—A review. J Food Sci Technol. 2021;58:2453–66.
88. Xie L, Shen M, Wang Z, Xie J. Structure, function and food applications of carboxymethylated polysaccharides: A comprehensive review. Trends Food Sci Technol. 2021;118:539–57.
89. Bilal M, Gul I, Basharat A, Qamar SA. Polysaccharides-based bio-nanostructures and their potential food applications. Int J Biol Macromol. 2021;176:540–57.
90. Manzoor A, Dar AH, Pandey VK, Shams R, Khan S, Panesar PS, et al. Recent insights into polysaccharide-based hydrogels and their potential applications in food sector: a review. Int J Biol Macromol. 2022;213:987–1006.
91. Roy S, Malik B, Chawla R, Bora S, Ghosh T, Santhosh R, et al. Biocompatible film based on protein/polysaccharides combination for food packaging applications: a comprehensive review. Int J Biol Macromol. 2024;278:134658.
92. Costa JAV, Lucas BF, Alvarenga AGP, Moreira JB, De Morais MG. Microalgae polysaccharides: an overview of production, characterization, and potential applications. Polysaccharides. 2021;2:759–72.
93. Wang Z, Wang L, Yu X, Wang X, Zheng Y, Hu X, et al. Effect of polysaccharide addition on food physical properties: A review. Food Chem. 2024;431:137099.
94. Zhu F. Polysaccharide based films and coatings for food packaging: effect of added polyphenols. Food Chem. 2021;359:129871.
95. Nasrollahzadeh M. Biopolymer-based metal nanoparticle chemistry for sustainable applications: volume 2: applications. Amsterdam, Netherlands: Elsevier; 2021.
96. Patel AR, Baldi A, Verma DK, Sandhu KS, Garcia S, editors. Biotechnical processing in the food industry: new methods, techniques, and applications. Boca Raton, FL, USA Abingdon, Oxon, UK: CRC Press; 2021.
97. Khodadadi Yazdi M, Seidi F, Hejna A, Zarrintaj P, Rabiee N, Kucinska-Lipka J, et al. Tailor-made polysaccharides for biomedical applications. ACS Appl Bio Mater. 2024;7:4193–230.
98. Schuerch C. Biomedical Applications of Polysaccharides. In: Gebelein CG, Carraher CE, editors. Bioactive Polymeric Systems [Internet]. Boston, MA: Springer US; 1985 [cited 2025 Feb 4]. p. 365–86. Available from: http://link.springer.com/10.1007/978-1-4757-0405-1_14
99. Nayak AK, Hasnain MS, editors. Natural polysaccharides in drug delivery and biomedical applications. London: Academic, an Imprint of Elsevier; 2019.
100. Priya S, Batra U, R.N. S, Sharma S, Chaurasiya A, Singhvi G. Polysaccharide-based nanofibers for pharmaceutical and biomedical applications: A review. Int J Biol Macromol. 2022;218:209–24.
101. Arokiarajan MS, Thirunavukkarasu R, Joseph J, Ekaterina O, Aruni W. Advance research in biomedical applications on marine sulfated polysaccharide. Int J Biol Macromol. 2022;194:870–81.
102. Jabeen N, Atif M. Polysaccharides based biopolymers for biomedical applications: A review. Polymers Adv Techs. 2024;35:e6203.
103. Tchobanian A, Van Oosterwyck H, Fardim P. Polysaccharides for tissue engineering: current landscape and future prospects. Carbohydr Polym. 2019;205:601–25.
104. Nayak AK, Ahmed SA, Tabish M, Hasnain MS. Natural polysaccharides in tissue engineering applications. Natural Polysaccharides in Drug Delivery and Biomedical Applications [Internet] Elsevier; 2019 [cited 2025 Feb 4]. p. 531–548. Available from: https://linkinghub.elsevier.com/retrieve/pii/B9780128170557000236
105. Rajalekshmy GP, Lekshmi Devi L, Joseph J, Rekha MR. An overview on the potential biomedical applications of polysaccharides. Functional Polysaccharides for Biomedical Applications [Internet] Elsevier; 2019 [cited 2025 Feb 4]. pp. 33–94. Available from: https://linkinghub.elsevier.com/retrieve/pii/B9780081025550000029
106. Maiti S, Jana S. Functional polysaccharides for biomedical applications. Duxford: Woodhead publishing; 2019.

107. Shariatinia Z. Pharmaceutical applications of natural polysaccharides. Natural Polysaccharides in Drug Delivery and Biomedical Applications [Internet]. Elsevier; 2019 [cited 2025 Feb 4]. pp. 15–57. Available from: https://linkinghub.elsevier.com/retrieve/pii/B9780128170557000029
108. Iacob A-T, Drăgan M, Ionescu O-M, Profire L, Ficai A, Andronescu E, et al. An overview of biopolymeric electrospun nanofibers based on polysaccharides for wound healing management. Pharmaceutics. 2020;12:983.
109. Nosrati H, Khodaei M, Alizadeh Z, Banitalebi-Dehkordi M. Cationic, anionic and neutral polysaccharides for skin tissue engineering and wound healing applications. Int J Biol Macromol. 2021;192:298–322.
110. Ahsan H. The significance of complex polysaccharides in personal care formulations. J Carbohydr Chem. 2019;38:213–33.
111. Savary G, Grisel M, Picard C. Cosmetics and Personal Care Products. In: Olatunji O, editor. Natural Polymers [Internet]. Cham: Springer International Publishing; 2016 [cited 2025 Feb 5]. pp. 219–61. Available from: http://link.springer.com/10.1007/978-3-319-26414-1_8
112. Semenzato A, Costantini A, Baratto G. Green polymers in personal care products: rheological properties of tamarind seed polysaccharide. Cosmetics. 2014;2:1–10.
113. Yao Y, Xu B. Skin health promoting effects of natural polysaccharides and their potential application in the cosmetic industry. Polysaccharides. 2022;3:818–30.
114. Albuquerque PBS, De Oliveira WF, Dos Santos Silva PM, Dos Santos Correia MT, Kennedy JF, Coelho LCBB. Skincare application of medicinal plant polysaccharides — A review. Carbohydr Polym. 2022;277:118824.
115. De Oliveira AC, Morocho-Jácome AL, De Castro Lima CR, Marques GA, De Oliveira Bispo M, De Barros AB, et al. Cosmetics applications. Microalgae [Internet] Elsevier; 2021 [cited 2025 Feb 5]. p. 313–338. Available from: https://linkinghub.elsevier.com/retrieve/pii/B9780128212189000104
116. Wu Q, Cheng N, Fang D, Wang H, Rahman F-U, Hao H, et al. Recent advances on application of polysaccharides in cosmetics. J Dermatl Sci Cosmetic Technol. 2024;1:100004.
117. Tafuro G, Costantini A, Baratto G, Francescato S, Busata L, Semenzato A. Characterization of Polysaccharidic associations for cosmetic use: rheology and texture analysis. Cosmetics. 2021;8:62.
118. Goddard ED. Principles of polymer science and technology in cosmetics and personal care. Milton: Taylor & Francis Group; 1999.
119. Arora S, Singh D, Rajput A, Bhatia A, Kumar A, Kaur H, et al. Plant-based polysaccharides and their health functions. FFHD. 2021;11:179.
120. Resende DISP, Ferreira M, Magalhães C, Sousa Lobo JM, Sousa E, Almeida IF. Trends in the use of marine ingredients in anti-aging cosmetics. Algal Res. 2021;55:102273.
121. Balkrishna A, Agarwal V, Kumar G, Gupta AK. Applications of bacterial polysaccharides with special reference to the cosmetic industry. In: Singh J, Sharma D, Kumar G, Sharma NR, (eds). Microbial Bioprospecting for Sustainable Development [Internet]. Singapore: Springer Singapore; 2018 [cited 2025 Feb 5]. pp. 189–202. Available from: http://link.springer.com/10.1007/978-981-13-0053-0_9
122. Plucinski A, Lyu Z, Schmidt BVKJ. Polysaccharide nanoparticles: from fabrication to applications. J Mater Chem B. 2021;9:7030–62.
123. Juncan AM, Moisă DG, Santini A, Morgovan C, Rus L-L, Vonica-Țincu AL, et al. Advantages of hyaluronic acid and its combination with other bioactive ingredients in cosmeceuticals. Molecules. 2021;26:4429.
124. Kanlayavattanakul M, Lourith N. Natural Polysaccharides for Skin Care. In: Oliveira J, Radhouani H, Reis RL (eds). Polysaccharides of Microbial Origin [Internet]. Cham: Springer International Publishing; 2021 [cited 2025 Feb 5]. pp. 1–23. Available from: https://link.springer.com/10.1007/978-3-030-35734-4_46-1
125. Baptista S, Freitas F. Bacterial polysaccharides: cosmetic applications. In: Oliveira J, Radhouani H, Reis RL, editors. Polysaccharides of microbial origin [internet]. Cham:

Springer International Publishing; 2021 [cited 2025 Feb 5]. pp. 1–42. Available from: https://link.springer.com/10.1007/978-3-030-35734-4_45-1.
126. Guzmán E, Ortega F, Rubio RG. Chitosan: A promising multifunctional cosmetic ingredient for skin and hair care. Cosmetics. 2022;9:99.
127. Gawade RP, Chinke SL, Alegaonkar PS. Polymers in cosmetics. Polymer science and innovative applications [Internet]. Elsevier; 2020 [cited 2025 Feb 5]. pp. 545–65. Available from: https://linkinghub.elsevier.com/retrieve/pii/B9780128168080000172
128. Shafie MH, Kamal ML, Zulkiflee FF, Hasan S, Uyup NH, Abdullah S, et al. Application of carrageenan extract from red seaweed (Rhodophyta) in cosmetic products: a review. J Indian Chem Soc. 2022;99:100613.
129. Sousa P, Tavares-Valente D, Amorim M, Azevedo-Silva J, Pintado M, Fernandes J. Β-Glucan extracts as high-value multifunctional ingredients for skin health: a review. Carbohydr Polym. 2023;322:121329.
130. Sadgrove NJ, Simmonds MSJ. Pharmacodynamics of *Aloe vera* and acemannan in therapeutic applications for skin, digestion, and immunomodulation. Phytother Res. 2021;35:6572–84.
131. Yuan M, Wang J, Geng L, Wu N, Yang Y, Zhang Q. A review: structure, bioactivity and potential application of algal polysaccharides in skin aging care and therapy. Int J Biol Macromol. 2024;272:132846.
132. Madni A, Khalid A, Wahid F, Ayub H, Khan R, Kousar R. Preparation and applications of guar gum composites in biomedical, pharmaceutical, food, and cosmetics industries. CNANO. 2021;17:365–79.
133. Raza ZA, Munim SA, Ayub A. Recent developments in polysaccharide-based electrospun nanofibers for environmental applications. Carbohydr Res. 2021;510:108443.
134. Souza MAD, Vilas-Boas IT, Leite-da-Silva JM, Abrahão PDN, Teixeira-Costa BE, Veiga-Junior VF. Polysaccharides in agro-industrial biomass residues polysaccharides. 2022;3:95–120.
135. Orejuela-Escobar L, Gualle A, Ochoa-Herrera V, Philippidis GP. Prospects of microalgae for biomaterial production and environmental applications at biorefineries. Sustain For. 2021;13:3063.
136. Gamage A, Thiviya P, Liyanapathiranage A, Wasana MLD, Jayakodi Y, Bandara A, et al. Polysaccharide-based bioplastics: eco-friendly and sustainable solutions for packaging. J Compos Sci. 2024;8:413.
137. Aleksanyan KV. Polysaccharides for biodegradable packaging materials: past, present, and future (brief review). Polymers. 2023;15:451.
138. Wang Z, Xu C, Qi L, Chen C. Chemical modification of polysaccharides for sustainable bioplastics. Trends Chem. 2024;6:314–31.
139. Reichert CL, Bugnicourt E, Coltelli M-B, Cinelli P, Lazzeri A, Canesi I, et al. Bio-based packaging: materials, modifications. Indus Appl Sustai Polymers. 2020;12:1558.
140. Zhao X, Wang Y, Chen X, Yu X, Li W, Zhang S, et al. Sustainable bioplastics derived from renewable natural resources for food packaging. Matter. 2023;6:97–127.
141. Martins BA, De Albuquerque PBS, De Souza MP. Bio-based films and coatings: sustainable polysaccharide packaging alternatives for the food industry. J Polym Environ. 2022;30:4023–39.
142. Bochek AM. Prospects for use of polysaccharides of different origin and environmental problems in processing them. Fibre Chem. 2008;40:192–7.
143. Maciel JV, Durigon AMM, Souza MM, Quadrado RFN, Fajardo AR, Dias D. Polysaccharides derived from natural sources applied to the development of chemically modified electrodes for environmental applications: a review. Trends Environ Analytical Chem. 2019;22:e00062.
144. Arias DM, Ortíz-Sánchez E, Okoye PU, Rodríguez-Rangel H, Balbuena Ortega A, Longoria A, et al. A review on cyanobacteria cultivation for carbohydrate-based biofuels: cultivation aspects, polysaccharides accumulation strategies, and biofuels production scenarios. Sci Total Environ. 2021;794:148636.

145. Picos-Corrales LA, Morales-Burgos AM, Ruelas-Leyva JP, Crini G, García-Armenta E, Jimenez-Lam SA, et al. Chitosan as an outstanding polysaccharide improving health-commodities of humans and environmental protection. Polymers. 2023;15:526.
146. Bakshi PS, Selvakumar D, Kadirvelu K, Kumar NS. Chitosan as an environment friendly biomaterial—a review on recent modifications and applications. Int J Biol Macromol. 2020;150:1072–83.
147. Selvaraj S, Chauhan A, Dutta V, Verma R, Rao SK, Radhakrishnan A, et al. A state-of-the-art review on plant-derived cellulose-based green hydrogels and their multifunctional role in advanced biomedical applications. Int J Biol Macromol. 2024;265:130991.
148. Biranje SS, Sun J, Shi Y, Yu S, Jiao H, Zhang M, et al. Polysaccharide-based hemostats: recent developments, challenges, and future perspectives. Cellulose. 2021;28:8899–937.
149. Bushra R, Ahmad M, Seidi F, Qurtulen SJ, Jin Y, et al. Polysaccharide-based nanoassemblies: from synthesis methodologies and industrial applications to future prospects. Adv Colloid Interf Sci. 2023;318:102953.
150. Ali SS, Alsharbaty MHM, Al-Tohamy R, Naji GA, Elsamahy T, Mahmoud YA-G, et al. A review of the fungal polysaccharides as natural biopolymers: current applications and future perspective. Int J Biol Macromol. 2024;273:132986.
151. Manikandan V, Min SC. Roles of polysaccharides-based nanomaterials in food preservation and extension of shelf-life of food products: a review. Int J Biol Macromol. 2023;252:126381.
152. Guo Q, Zhang M, Mujumdar AS. Progress of plant-derived non-starch polysaccharides and their challenges and applications in future foods. Comp Rev Food Sci Food Safe. 2024;23:e13361.
153. Mahendiran B, Muthusamy S, Sampath S, Jaisankar SN, Popat KC, Selvakumar R, et al. Recent trends in natural polysaccharide based bioinks for multiscale 3D printing in tissue regeneration: a review. Int J Biol Macromol. 2021;183:564–88.
154. Teixeira MC, Lameirinhas NS, Carvalho JPF, Silvestre AJD, Vilela C, Freire CSR. A guide to polysaccharide-based hydrogel bioinks for 3D bioprinting applications. IJMS. 2022;23:6564.
155. Kaith A, Jain N, Kaul S, Nagaich U. Polysaccharide-infused bio-fabrication: advancements in 3D bioprinting for tissue engineering and bone regeneration. Materials Today Commun. 2024;40:109429.

Exploring Proteins and Peptides as Natural Excipients

7

Nikunjkumar Parekh, Meghana N. Patel, Manish N. Nandpal, and Kamal Dua

Abbreviations

GI Tract	Gastrointestinal tract
PEG	Polyethylene glycol
$t_{1/2}$	Half life
HSA	Human serum albumin
kDa	Kilo Dalton
GA	Gum Arabic
FDA	Food and Drug Administration
DNA	Deoxyribonucleic acid
RNA	Ribonucleic acid
PBS	Phosphate buffered saline
CPPs	Cell-penetrating peptides
DDSs	Drug delivery systems
PDC	Peptide-drug conjugates
THPs	Tumor-homing peptides
CTPs	Cell-targeting peptides
PTX	Paclitaxel

N. Parekh (✉) · M. N. Patel · M. N. Nandpal
Department of Pharmaceutical Chemistry and Analysis, Ramanbhai Patel College of Pharmacy, Charotar University of Science and Technology (CHARUSAT), Anand, Gujarat, India
e-mail: nikunjparekh.ph@charusat.ac.in

K. Dua
Graduate School of Health, Discipline of Pharmacy, University of Technology Sydney, Sydney, NSW, Australia
e-mail: Kamal.Dua@uts.edu.au

J. S et al. (eds.), *Innovative Pharmaceutical Excipients: Natural Sources*,
https://doi.org/10.1007/978-981-96-7959-1_7

DOX	Doxorubicin
TME	Tumor microenvironment
SEC	Size-exclusion chromatography
DLS	Dynamic light scattering
mAbs	Monoclonal antibodies
HPLC	High-performance liquid chromatography
EMA	European Medicines Agency
API	Active pharmaceutical ingredient
NLCs	Nanostructured lipid carriers
Nm	Nanometer

7.1 Introduction

Amino acids are the fundamental units for building peptide chains, and the bonding forces between two or more peptide chains form protein structures. Proteins are essential components of the human body, functioning as enzymes, receptors, and transporters in various cellular mechanisms. It is particularly interesting to explore how these proteins and peptides are utilized in drug delivery systems (DDS) to achieve diverse goals, such as targeted delivery, improved bioavailability, reduced toxicity of parent compounds, and enhanced biological activity of drugs.

Excipients, typically supportive components in pharmaceutical formulations do not possess pharmacological effects like active drugs. However, many peptides have inherent pharmacological properties, such as antimicrobial activity, angiotensin-converting enzyme (ACE)-inhibitory action, and antioxidant characteristics, which can add value to pharmaceutical formulations as excipients. This chapter provides an overview of proteins and peptides, their chemistry, collection, and purification techniques, analysis, and applications in pharmaceutical formulations as excipients [1].

On the other hand, these structures are challenging to handle and store over long periods, and thus, this chapter will also discuss reported modifications to improve their stability, drug-carrying capacity, and pharmacological effects.

Currently, protein and peptide drugs are primarily administered through the intravenous route or via the subcutaneous route due to their instability and poor GI (Gastrointestinal) tract absorption when taken orally. To overcome these challenges, carrier systems, enzyme inhibitors, absorption enhancers, and stability enhancers are required. Transdermal delivery faces challenges such as absorption limitations due to proteolysis by protease, opsonization through microphages, and agglutination. Chemical modifications to the structure of proteins and peptides can address these drug delivery issues [2, 3].

In modern drug delivery systems, transferrin, serum albumin, recombinant proteins as well as virus capsids are utilized as carriers for active pharmaceutical ingredients (APIs) and biomolecules. Polymeric protein matrixes offer significant advantages for drug delivery systems. To develop polymer protein conjugated formulation, polyethylene glycol (PEG) is the polymer of choice. Other polymers, such as polylactic-co-glycolic acid, N-(2-hydroxypropyl) methacrylamide copolymer, polyvinyl pyrrolidone, and polyglutamic acid, have also been explored for conjugation with protein moieties. Most drug delivery systems are designed to target specific diseases, including cancer, diabetes, immunomodulation, neurodegenerative diseases, and antimicrobial activity. For the improvement in pharmaceutical formulation, there is a need of chemical modifications like altering the chemical structure, functionality or stability of natural proteins and peptides to serve as excipients. These scientific strategies are essential in optimizing drug delivery systems, enhancing therapeutic efficacy and ensuring patient compliance. Plenty of approaches have been developed, each contributing uniquely to the advancement of pharmaceutical science.

Chemical modifications represent a fundamental strategy, with pegylation and acylation being prominent techniques. Pegylation involves the covalent attachment of polyethylene glycol (PEG) to proteins or peptides. This modification improves solubility, enhances stability, and reduces immunogenicity, thereby prolonging the therapeutic effect. A notable example is pegylated interferon, widely recognized for its extended half-life in medical treatments. In contrast, acylation incorporates fatty acid chains into proteins or peptides, increasing their lipophilicity. This enhancement improves absorption and enables the formation of micelles, which are critical for sustained activity. Insulin detemir, modified with a fatty acid, exemplifies this approach by offering prolonged glucose-lowering effects [4].

Physical modifications further expand the utility of proteins and peptides in drug formulations. Coacervation, a process that utilizes proteins such as gelatin or casein, encapsulates active pharmaceutical ingredients (APIs) to provide controlled-release properties. This technique protects APIs from degradation and enhances delivery efficiency. Nanoparticle formation, another physical modification, leverages proteins like albumin to create nanoparticles. These serve as carriers for poorly soluble drugs, significantly improving their bioavailability. For instance, albumin-based nanoparticles are utilized in ABRAXANE®, a formulation of paclitaxel, to optimize drug delivery [5].

Enzymatic modifications provide an additional layer of functionality. Proteins can be cross-linked using enzymes such as transglutaminase, which enhance their mechanical strength and thermal stability. This approach is particularly useful in the development of pharmaceutical formulations requiring robust structural integrity. Stabilized hydrogels, commonly employed in wound healing, are a practical example of enzymatic cross-linking [6].

Functionalization of proteins and peptides involves the conjugation of ligands or surface modifications to enhance their properties. Conjugation with ligands such as sugars, targeting molecules, or fluorescent tags improves targeted delivery, cellular uptake, or imaging capabilities. Glycosylated excipients, for example, enhance cellular uptake in therapeutic applications. Surface modifications, on the other hand, involve the addition of hydrophilic or hydrophobic groups to optimize interactions with drugs and delivery systems [7–9].

Structural engineering employs advanced techniques like recombinant technology and cyclization to refine protein and peptide properties. Recombinant technology uses genetic engineering to produce proteins with enhanced stability, altered functionality, or specific binding properties. A prime example is the development of tailored human serum albumin (HSA) variants with improved drug-binding capabilities. Cyclization, another innovative approach, enhances peptide stability by protecting them from enzymatic degradation, thus increasing their bioavailability [10].

Derivatives of proteins and peptides, such as hydrolysates and microspheres, also find extensive applications in pharmaceutical formulations. Partial hydrolysis of proteins like whey or collagen yields smaller peptides with emulsifying or stabilizing properties, improving formulation consistency. Proteins like gelatin or silk fibroin can be modified into microspheres, which are employed in sustained drug-release systems [11, 12].

Cross-linking and gel formation techniques round out the arsenal of modification strategies. Proteins such as fibrin or collagen can be chemically cross-linked to form scaffolds. These structures are integral to controlled-release drug delivery systems and serve as excipients in various formulations, providing both structural and functional benefits.

Through these diverse strategies, the pharmaceutical industry continues to harness the potential of natural proteins and peptides. By optimizing their properties, scientists are developing innovative solutions that advance drug delivery, therapeutic efficacy, and patient outcomes [13–15].

Protein and peptide derivatives serve crucial roles as excipients in modern pharmaceutical formulations, addressing key challenges in drug delivery. Stabilizers are employed to prevent aggregation or degradation of biopharmaceuticals, ensuring the efficacy and shelf-life of sensitive molecules. For instance, modified gelatin is utilized in monoclonal antibody (mAb) formulations to maintain stability. Solubilizers play an essential role in enhancing the solubility of poorly soluble drugs, enabling effective therapeutic delivery. Additionally, these derivatives function as delivery systems, acting as carriers for sustained or targeted drug release, thereby improving bioavailability and therapeutic outcomes. Protectants shield sensitive active pharmaceutical ingredients (APIs) during manufacturing and storage, preserving their integrity and potency [16].

The future of protein and peptide derivatives as excipients lies in innovative advancements that push the boundaries of drug delivery technologies. Smart drug delivery systems, which respond to stimuli such as pH or temperature, are poised to revolutionize precision medicine. Personalized medicine approaches, leveraging protein-based excipients tailored to individual patient needs, promise enhanced therapeutic efficacy. Protein-polymer hybrid systems are emerging as versatile platforms, combining the benefits of both materials for superior performance. Applications in gene and vaccine delivery highlight their potential in addressing complex medical challenges. Enzyme-based cross-linking techniques offer improved mechanical strength and functionality, while 3D-printed drug delivery devices provide customized and precise solutions. Furthermore, advancements in biodegradability and safety ensure minimal environmental impact and patient-friendly formulations. Multifunctional platforms integrating these features signify a new era in drug delivery, promising to address diverse clinical needs effectively [17–19].

Despite their significant potential in pharmaceutical formulations, protein and peptide derivatives as excipients face numerous challenges that hinder their broader application. Stability issues remain a primary concern, as these biomolecules are prone to aggregation, denaturation, or degradation under varying conditions, impacting their efficacy. The cost of production is another critical limitation, driven by the complexity of their synthesis and the need for stringent quality controls.

Immunogenicity presents a significant risk, as certain protein-based excipients can elicit immune responses, compromising safety and therapeutic outcomes. Complex manufacturing processes add to the difficulty, requiring specialized facilities and expertise, which can limit scalability and increase production timelines. Source dependency poses additional challenges, with ethical considerations, allergenic risks, and the availability of raw materials affecting consistent supply [20].

Batch-to-batch variability further complicates the use of these derivatives, as minor differences in production can lead to variations in performance or safety. Regulatory challenges also loom large, as stringent guidelines must be met to ensure the safety, efficacy, and quality of these excipients, adding to development timelines and costs. Additionally, the short shelf-life of many protein and peptide derivatives limits their practicality, necessitating advanced stabilization techniques.

Compatibility with active pharmaceutical ingredients (APIs) is another hurdle, as interactions between the excipient and the API can lead to reduced drug efficacy or stability. Addressing these challenges requires ongoing research and development, innovative manufacturing technologies, and robust regulatory frameworks to unlock the full potential of protein and peptide derivatives in drug delivery [21–23].

Below is a table listing some proteins used as drug delivery systems, along with their advantages and disadvantages [24–26]. Summary of the information on proteins used as a drug delivery system with their advantages and disadvantages is shown in Table 7.1.

Table 7.1 Proteins used as a drug delivery system with its advantages and disadvantages (Hong et al. [25])

Sr. No.	Protein	Advantages	Disadvantages
1	Human serum albumin	Non-immunogenicity High stability	Expensive cost
2	Gliadin	Non-immunogenicity Nontoxic	Rapid degradation speed Large particle size
3	Gelatin	Easy to make a bridge Safety	Compromised mechanical strength
4	Silk protein fibroin	Biocompatibility, high mechanical strength with good flexibility, suitable for various physical conditions	Sericin may cause immunogenic reactions
5	Legumin	Broad surface area Particle size is small Low susceptible to immunogenicity	Yield loss
6	Lipoprotein	Long $t_{1/2}$ (half-life) Inbuilt cell targeting property	Difficult to separate out native low-density lipoproteins
7	30Kc19	Stabilize the enzyme Penetrate the cell Increased cell growth and viability	Low nanoparticle size and yield when using only 30Kc19
8	Ferritin	Stable with different pH and temperature conditions	Expensive

Let us turn by turn understand the chemistry and the properties of proteins used as excipients for DDS.

7.2 Proteins and Peptides Used as Excipients for Drug Delivery System

7.2.1 Proteins Used as Excipients for Drug Delivery System

7.2.1.1 Human Serum Albumin

It is a globe-shape protein found in the blood circulation, composed of ~585 amino acids, with a molecular weight of 66 g/mol. It has three main units (I, II, III), each with subunit A and subunit B, and has two numbers of primary drug-attachment sites, Sudlow's site I and Sudlow's site II. HSA is well known for its stability across a wide pH range, high temperatures, and organic solvents, making it ideal for drug delivery applications. Its nanoparticles, formed using methods like desolvation, offer advantages such as biodegradability, non-toxicity, and excellent ligand-binding properties, enabling targeted drug delivery. Studies have demonstrated HSA nanoparticles' effectiveness in enhancing drug stability, solubility, and targeting, with diverse applications, including treating cancer and cardiovascular diseases [20, 25, 27–30].

7.2.1.2 Gliadin

It is a key wheat gluten protein, is derived through an extraction process using 70% ethanol, and is made up of single-chain polypeptides with molecular weights of 25–100 kDa. Its low solubility in water stems from hydrophobic interactions and disulfide bonds, making it ideal for controlled-release systems for lipophilic and amphiphilic drugs. Gliadin nanoparticles exhibit excellent biocompatibility, biodegradability, and mucosal adhesion, enabling applications in oral and localized drug delivery. Studies have demonstrated their effectiveness in eradicating *Helicobacter pylori* and treating cancer by enhancing drug stability, solubility, and targeting. On the other hand, gliadin nanoparticles are sensitive to environmental factors like pH, heat, and salt. To improve stability, protein-polysaccharide interactions, such as adding *Gum arabic* (GA), have been explored, enhancing their resilience to environmental stress while maintaining functionality [31–33].

7.2.1.3 Gelatin

Gelatin is a water-soluble polymeric form derived from hydrolyzed collagen and widely utilized in nanoparticle formulation due to its biocompatibility, US Food and Drug Administration (FDA) approval, and versatility. Gelatin is classified into type A (cationic, IEP 7–9) and type B (anionic, IEP 4.8–5) based on hydrolysis methods. It has a three-helix structure, enabling chemical modifications for drug delivery. Gelatin nanoparticles can encapsulate biologics like DNA, RNA, and proteins, but require cross-linking (e.g., with GA) to enhance mechanical strength and control decomposition. These nanoparticles show potential for targeted delivery, including bone regeneration and anticancer therapies. Surface modifications, such as Polyethylenimine cationization, improve cellular uptake, stability, and protein loading, making gelatin nanoparticles a promising drug and biopolymer carrier [34, 35].

7.2.1.4 Legumin

It is a principal storage protein of soybean seeds (*Pisum sativum L.*), which belongs to the 11S globulin family, with a molecular weight of 300–400 kDa and six subunits are full of sulfur-containing amino acids. Legumin nanoparticles, synthesized mainly via the coacervation method, exhibit bio-adhesiveness and high-interaction characteristics with biological surfaces. This process involves phase separation and chemical cross-linking (e.g., with glutaraldehyde) to form stable nanoparticles, typically ranging from 250 to 300 nm in size at neutral pH. These nanoparticles are stable in phosphate buffered saline (PBS) and display low immunogenicity. Despite their promising stability and functionality, challenges like low yield and further optimization remain for their biomedical application potential [36–38].

7.2.1.5 30Kc19-Protein (Silkworm Hemolymph)

The 30K protein class, derived from *Bombyx mori hemolymph*, includes proteins such as 30Kc6, 30Kc12, 30Kc19, 30Kc21, and 30Kc23, each with a molecular weight of approximately 30 kDa. All these proteins enhance cell growth, viability, and enzyme stability. Among them, 30Kc19 is notable for its cell-penetrating properties, attributed to the CPP Pep-c19 in its N-terminal α-helix domain.

Studies have utilized 30Kc19 to develop nanoparticles for intracellular delivery of enzymes like β-galactosidase (β-gal) and α-galactosidase (α-gal). These nanoparticles, prepared through the desolvation method and cross-linked with glutaraldehyde (GA), demonstrated enhanced drug activity and stability. Combining 30Kc19 with human serum albumin (HSA) at 50 wt% resulted in well-formed nanoparticles with a size range of 230–310 nm and high drug loading efficiency (80–95%). The particles exhibited controlled-release profiles, with β-gal releasing 30–50% within 24 hours and continuing up to 60% [39–41].

7.2.1.6 Lipoproteins

They are natural nanoparticles essential for the transport of fats inside the body. Their unique structure, comprising a triglyceride and cholesterol ester core encased by a layer of phospholipid embedded with polar-nonpolar apolipoproteins, makes them versatile carriers for drug delivery. Lipoproteins are categorized based on size, density, and function:

HDL (high-density lipoproteins): 7–13 nm
LDL (low-density lipoproteins): 22–27 nm
IDL (intermediate-density lipoproteins): 27–30 nm
VLDL (very low-density lipoproteins): 35–80 nm
Chylomicrons: 80–1200 nm

Lipoproteins are typically isolated using density-based ultracentrifugation techniques. They are highly desirable as drug transportation systems due to their biodegradable nature, non-immunogenicity, bioresorbable, and natural targeting capabilities. Compared to synthetic nanocarriers, lipoprotein nanoparticles exhibit prolonged circulation half-lives (48–72 h) and can encapsulate therapeutic agents such as active pharmaceutical ingredients and nucleic acids. Their surface can also be modified with ligands to enhance targeting specificity.

Lipoprotein nanoparticles have shown promise in treating conditions such as neurodegenerative diseases like Alzheimer's disease and other disorders, leveraging their ability to maintain stability, avoid immune detection, and deliver therapeutic payloads efficiently [42–44].

7.2.1.7 Ferritin

Ferritin is basically a globular protein with a molecular weight of 474 kDa, and stores iron in its 6-nm core surrounded by a polypeptide shell. Found in microorganisms, plants, and animals, it has two subunits, H and L, which complement each other in iron absorption. The H subunit oxidizes iron via a dinuclear ferroxidase site, while the L subunit aids mineralization. Ferritin nanoparticles, with modifiable outer and inner interfaces, are thermostable and pH-restorable. Liang et al. used H-ferritin nanocages loaded with doxorubicin (HFn-Dox) to target tumor cells via transferrin receptor 1, achieving higher drug concentrations, reduced toxicity, and significant tumor growth inhibition compared to conventional treatments [45–47].

7.2.2 Peptides Used as Excipients for Drug Delivery System

Peptides exhibit high receptor binding, lower immunogenicity, an elementary amino acid building blocks, and tunable molecular size, making them ideal for conjugation with drugs via anchoring to form peptide-drug conjugates (PDCs), which act as pro-drugs. Figure 7.1 describes the classification of peptides used in drug delivery system. PDCs are considered favorable drug delivery systems (DDSs) due to their capacity to enhance drug biocompatibility and physiological stability. Peptide-linked drug delivery systems are commonly employed to deliver therapeutic agents, such as anticancer drugs and nucleic acid-based therapies, as they not only reduce in vivo decomposition rate of drugs but also ensure targeted drug delivery, maintain drug concentration at the site of action, and extend the drug's half-life in the body [29, 30]. Table 7.2 list outs the examples of different peptides used as excipients.

Targeted drug delivery systems (DDSs) utilize peptides, such as cell-targeting peptides (CTPs) and tumor-homing peptides (THPs), cell-penetrating peptides, and stimuli-responsive peptides to improve drug efficacy and minimize side effects [27, 28, 48].

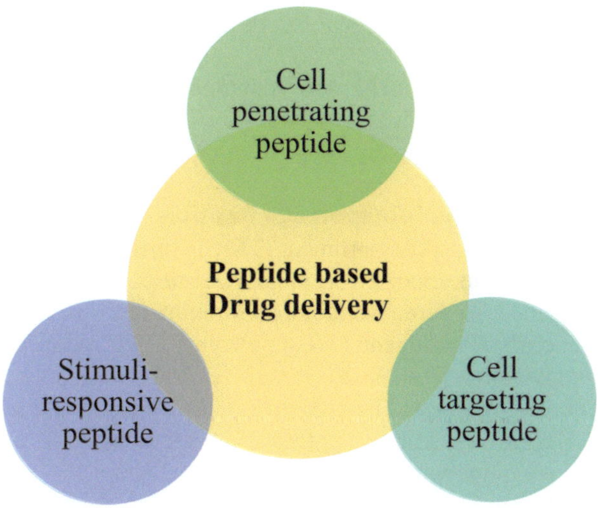

Fig. 7.1 Classification of peptides used in drug delivery system

Table 7.2 Examples of different peptides used as excipients (Guo et al. [30])

Cell-targeted peptides	Cell-penetrating peptides	Stimuli-responsive peptide
Angiopep-2 (ANG)	Xentry TAT-9	PEG-b-PLL/DMMA
THP F3	TAT-A1	RGD-ELP
WMW10	MGPE	mPEG-b-PDTG
TH-RGD (TR)	ANG-TAT	CPPT
Phycocyanin	Polyarginine	PLA-b-PEG-b-pHis

7.2.2.1 Cell-Targeted Peptides

CTPs, composed of 3 to 14 amino acids, specifically bind to receptors on targeted cells or tissues, facilitating precise drug delivery, particularly in cancer therapy. THPs can target tumor vasculature and cells, enhancing the penetration of cytotoxic drugs or nanocarriers into tumor tissues.

Examples include 177Lu-dotatate, a peptide-drug conjugate (PDC) used for gastroenteropancreatic neuroendocrine tumors (GEP-NETs), and F3-PEG-PLA nanoparticles used to deliver paclitaxel (PTX) to glioblastoma cells, demonstrating superior penetration depth compared to non-targeted nanoparticles. Additionally, F3-modified cisplatin nanoparticles have shown efficacy in inhibiting tumor angiogenesis in ovarian cancer models, highlighting the potential of peptide-based DDSs in precision cancer treatment.

7.2.2.2 Cell-Penetrating Peptides

CPPs are short peptides (8–30 amino acids) capable of crossing cell membranes without disrupting membrane integrity, making them effective carriers for delivering therapeutic agents like proteins, nucleic acids, and small molecule drugs. CPPs, such as TAT peptides, can transport drugs like doxorubicin (DOX) and paclitaxel (PTX) to targeted cells, inhibiting proliferation in human oral epidermoid carcinoma cells (KB). Figure 7.2 describes the cellular mechanism of peptide used as DDSs.

Additionally, TAT-siRNA conjugates have been utilized for gene therapy in human chronic myeloid leukemia (K562) cells, demonstrating their potential in drug and gene delivery applications.

7.2.2.3 Stimuli-Responsive Peptide

The tumor microenvironment (TME) consists of both internal and external factors that promote tumor growth and metastasis, influenced by surrounding tissues and the cancer's internal conditions. Stimulus-responsive drug delivery systems (DDSs) can react to endogenous (e.g., enzymes, pH, redox potential) or exogenous (e.g., temperature, light, heat) stimuli, causing changes such as charge rotation and decrease in a size. These changes enhance drug uptake, and internalization by cancer cells, and improve targeted delivery, enabling controlled and efficient drug release to malignant tissues.

Protein and peptide drug delivery offers numerous advantages across various administration routes. The buccal route ensures less pain and irritation during long-term treatments, with easy application and removal, enhancing patient comfort. The nasal route is convenient, provides rapid drug action, and bypasses first-pass metabolism. The rectal route allows for large drug doses, targets the lymphatic system, avoids gastrointestinal irritation, and is suitable for patients prone to nausea or vomiting. The pulmonary route offers a direct pathway to circulation, requires smaller doses, minimizes discomfort, and is effective even for lung disease patients. Lastly, the transdermal route enables controlled and sustained drug release, self-administration, reduced side effects, and bypasses hepatic first-pass metabolism, improving overall patient compliance [48–50].

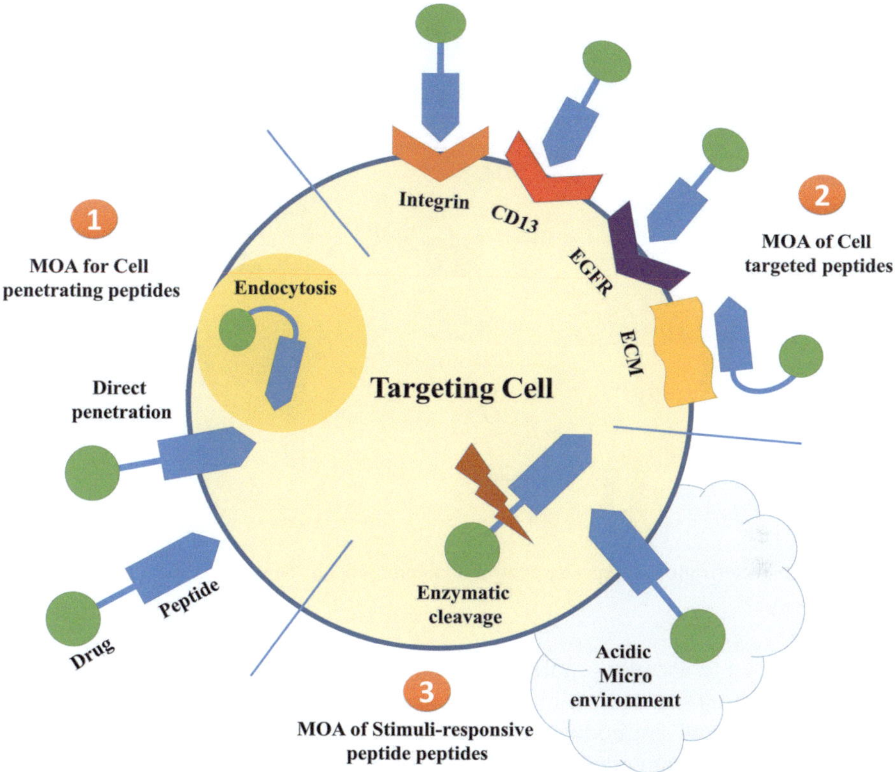

Fig. 7.2 Cellular mechanism of peptide used as drug delivery systems (DDSs). (Reproduced from Guo et al. [30])

7.3 Characterization and Quality Control of Natural Proteins/Peptides

Natural proteins and peptides need to be characterized for use as pharmaceutical excipients since their stability, molecular structure, and purity significantly impact their effectiveness as therapeutic agents. This characterization procedure includes physical, chemical, and biological assessments to validate their suitability for drug delivery applications [51, 52]. Figure 7.3 illustrates a brief overview of the methods utilized for the characterization and quality control of proteins and peptides.

7.3.1 Physical Characterization: Size, Shape, and Stability

Physical characterization focuses on analyzing the size, shape, stability, and interactions of proteins and peptides in various environments. Techniques like mass spectrometry (MS), gel electrophoresis, and size-exclusion chromatography

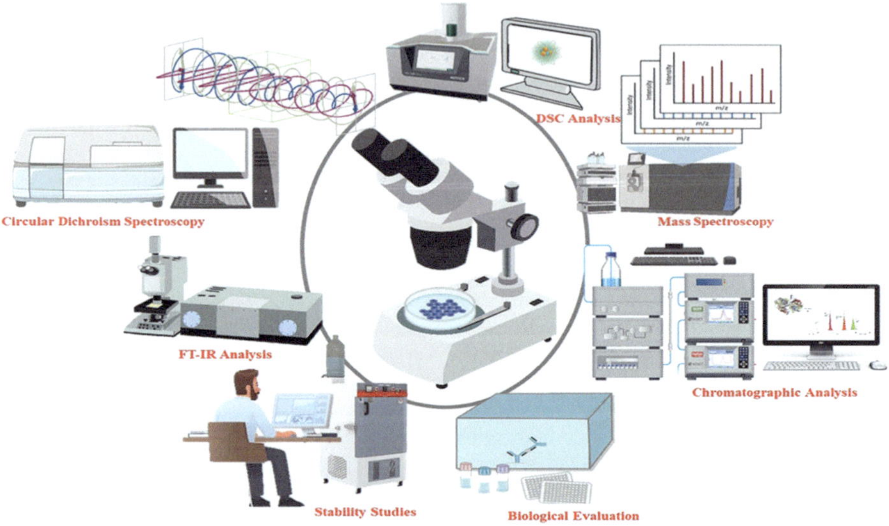

Fig. 7.3 Characterization and quality control of natural proteins/peptides

(SEC) are used to ascertain molecular weight and detect any modifications, truncations, or aggregates that could affect stability. For instance, in the characterization of monoclonal antibodies (mAbs), SEC is frequently used to evaluate the aggregation of protein molecules, which can hinder their therapeutic efficacy [53–58].

Dynamic light scattering (DLS) is utilized to assess nanoparticle compositions, which is essential for enhancing drug delivery capabilities. Atomic force microscopy (AFM) and cryogenic electron microscopy (cryo-EM) yield extensive information about protein shape and structural integrity. Cryo-EM has played an important role in structural investigations of viral proteins, such as the severe acute respiratory syndrome coronavirus 2 (SARS-CoV-2) spike protein, which aids vaccine development. Stability testing with differential scanning calorimetry (DSC) and thermogravimetric analysis (TGA) aids in determining thermal stability and denaturation under environmental stresses such as temperature and pH fluctuations [59–61]. Understanding protein-protein and protein-excipient interactions is critical, and advanced techniques such as isothermal titration calorimetry (ITC) and surface plasmon resonance (SPR) reveal binding thermodynamics and kinetics, which are required to predict formulation stability and release profiles. ITC investigations on albumin-based medication formulations have revealed important insights into drug molecules' binding characteristics, which influence their release mechanisms. Integrating these methodologies allows researchers to thoroughly examine the behavior of the proteins and peptides in pharmaceutical systems, ensuring their successful usage as excipients [62–64].

7.3.2 Chemical Characterization: Amino Acid Composition, Conformation, and Modifications

In chemical characterization, the primary, secondary, and tertiary structures of proteins and peptides are examined, as well as post-translational modifications (PTMs), which are essential for the stability and activity of these molecules. Finding the fundamental structure requires amino acid sequencing, which uses methods like tandem mass spectrometry (MS/MS) and Edman degradation to pinpoint particular residues and changes like phosphorylation, glycosylation, and acetylation. For instance, the fundamental structure of recombinant human insulin is confirmed using amino acid sequencing. Recombinant protein uniformity is guaranteed by accurate sequencing, which also facilitates batch-to-batch quality control [65, 66]. The goal of secondary structure analysis is to find random coils, β-helices, and β-sheets that affect protein stability.

To ensure structural integrity during formulation, Fourier transform infrared (FT-IR) spectroscopy utilizes amide bond signals to identify structural conformations, while circular dichroism (CD) spectroscopy provides real-time information on protein folding [67–69]. Sophisticated techniques like nuclear magnetic resonance (NMR) spectroscopy and X-ray crystallography are employed to assess tertiary structure and conformational stability. These techniques examine protein folding and stability at the atomic scale, crucial for understanding potential conformational alterations during formulation and storage. The structure of biologics, such as monoclonal antibodies, has been analyzed through NMR spectroscopy, for example, enhancing their design for greater efficacy [70–72].

Post-translational modifications (PTMs), such as acetylation, phosphorylation, and glycosylation, significantly influence protein stability and activity. For the identification and quantification of PTMs, mass spectrometry (MS), often combined with liquid chromatography (LC-MS), is considered the gold standard. To ensure that protein-based excipients are safe, effective, and immunogenic, these alterations need to be monitored carefully. For instance, the pharmacokinetics and immunogenicity of monoclonal antibodies are affected by glycosylation, which plays a crucial role in therapeutic applications. When used together, these techniques enable comprehensive chemical characterization, ensuring that proteins and peptides retain their intended properties and functions in therapeutic uses [73, 74].

7.3.3 Biological Characterization: Activity, Immunogenicity, and Biocompatibility

Evaluating the biological activity, immunogenicity, and biocompatibility of proteins and peptides necessitates biological characterization. The capability of proteins to perform their designated functions, like medication stabilization, controlled release, or targeting, is assessed through bioactivity assays, which include enzyme-linked immunosorbent assays (ELISA) and receptor binding assays. To evaluate the efficacy of antibody-drug conjugates (ADCs) in targeted cancer treatment, receptor-binding assays are crucial. Evaluating immunogenicity assesses the potential for immune system activation, which can lead to protein neutralization or hypersensitivity reactions. Alongside advanced technologies such as mass spectrometry for recognizing neoantigens or immunogenic regions, both in vitro (e.g., ELISA for detecting antibodies) and in vivo (e.g., using animal models) methods are utilized [75, 76].

Mass spectrometry is employed in the study of immunogenicity for therapeutic proteins like erythropoietin (EPO) to identify potential immunogenic epitopes. Biocompatibility testing ensures that proteins are safe, especially in systems that are implanted or injected. Testing for hemocompatibility, conducting cytotoxicity assays, and performing histological evaluations in animal models evaluate the impacts on the well-being of cells, tissues, and organs. In regenerative medicine, the potential of collagen-based scaffolds for tissue repair applications is validated by biocompatibility studies [75, 77–80].

Moreover, high-throughput screening (HTS) accelerates the assessment of many formulations, simplifying development processes, while advanced techniques such as proteomics provide comprehensive information about protein expression, isoforms, and interactions. In the process of drug discovery, HTS has played a vital role in rapidly identifying possible candidates [81–83]. Proteins and peptides are assessed to ensure their stability, safety, and excipient performance through the integration of physical, chemical, and biological methods. These techniques not only help in developing better formulations but also illustrate how proteins and peptides could propel pharmaceutical advancements in the future [84–87].

7.3.4 Quality Control Aspects of Proteins/Peptides Using Orthogonal Analytical Techniques

Quality control integrates methods like MS, CD, FTIR, HPLC, and DLS to verify the purity, stability, and functional integrity of proteins [70, 88–94]. Cryo-EM is utilized for examining large protein complexes, while computational methods like machine learning can forecast protein stability, reducing development expenses. These processes guarantee that proteins and peptides fulfill the strict criteria for medical use. The analysis of monoclonal antibodies using HRMS, for instance, is essential for ensuring batch consistency and detecting potential variations that may jeopardize safety or effectiveness. Cryo-EM has been crucial in studying viral proteins, contributing to the development of vaccines like those for SARS-CoV-2. DLS plays a vital role in creating nanoparticle-driven drug delivery systems as it guarantees the correct distribution of particle sizes. The stability of protein-based excipients is progressively forecasted using machine learning, reducing the necessity for extensive trial-and-error experiments [95, 96].

The quality control process ensures that proteins and peptides meet the stringent standards of pharmaceutical uses by employing these advanced methods, providing reliable and effective excipients for drug delivery systems.

7.4 Role of Natural Proteins/Peptides in Pharmaceutical Formulations

Natural proteins and peptides are increasingly being used as excipients in pharmaceutical formulations because of their sustainability, biodegradability, and biocompatibility. They decrease toxicity and immunogenicity while enhancing drug stability, solubility, bioavailability, and controlled release. In formulations including nanoparticles, for instance, albumin enhances bioavailability and solubility. These excipients are used to provide drugs orally, injectably, and topically. Controlled release is guaranteed, pharmaceuticals are strengthened, and their solubility is improved. Casein-based nanoparticles are being studied for the targeted administration of cancer treatments, and gelatin stabilizes proteins in capsules and microspheres. Their natural origin offers a safer alternative to synthetic excipients, driving pharmaceutical innovation [2, 87, 97–100].

7.4.1 Albumin: The Cornerstone of Parenteral Drug Formulations

Human serum albumin (HSA) is a crucial natural protein in pharmaceutical formulations, especially for injectable (parenteral) medication formulations. The main plasma protein, albumin, improves therapy results by increasing bioavailability, stabilizing hydrophobic medications, and acting as a vehicle for focused drug delivery. For example, albumin is used to stabilize monoclonal antibodies and vaccines, preventing aggregation and denaturation [101–104].

Albumin stabilizes hydrophobic drugs by forming complexes that improve solubility, allowing more efficient delivery. This is particularly important for biologic drugs that are prone to instability. Albumin's role in stabilizing monoclonal antibodies and vaccines ensures improved shelf-life and efficacy. In addition to stabilization, albumin serves as a carrier protein, enabling targeted drug delivery by binding to drugs and facilitating their transport across biological barriers. This reduces overall negative effects by increasing the precision of drug delivery to specific tissues [105–111].

Albumin-based formulations are being created to deliver anticancer drugs precisely. By eliminating the possibility of contamination from blood-derived sources, recombinant albumin—which is produced using biotechnological techniques—offers increased safety. It also provides enhanced consistency and scalability, making it a reliable excipient in large-scale pharmaceutical production. Nowadays, recombinant albumin is frequently used in medicinal formulations such as albumin-bound paclitaxel (ABRAXANE®) [112–114].

7.4.2 Gelatin and Casein: Versatile Excipients for Oral and Liquid Formulations

The pharmaceutical industry utilizes gelatin widely for tablets, hard capsules, and soft gel capsules. Its ability to form a gel makes it ideal for regulated drug delivery, especially for sensitive medications that need protection from degradation. Active pharmaceutical ingredients (APIs) are enclosed in gelatin, shielding them from the acidic conditions and digestive enzymes in the stomach. For example, enteric-coated capsule formulations often incorporate gelatin to help prevent the early degradation of APIs like vitamins or hormones. Moreover, it enhances the physical properties and release profiles of APIs in oral dosage forms by functioning as an emulsifier, stabilizer, and binder. For drugs requiring extended release, this regulated release provides enduring therapeutic advantages [115–119].

Casein, a protein obtained from milk, is highly regarded for its unique amphiphilic characteristics that allow it to stabilize emulsions of water and oil. This feature aids in providing medications with low solubility in injectable formats and is particularly beneficial for producing parenteral suspensions and emulsions. To enhance solubility and bioavailability, casein is employed in the formulation of injectable emulsions for drugs like paclitaxel. Moreover, casein forms micelles that enhance the body's distribution of lipophilic (fat-soluble) drugs by boosting their solubility and absorption. Furthermore, due to its micellar structure, casein serves effectively as an excipient in sustained-release formulations, facilitating long-acting formulations and controlled drug release over time [120–126].

7.4.3 Plant-Derived Proteins: A New Frontier in Pharmaceutical Formulations

Proteins sourced from plants are emerging as promising alternatives to animal-based excipients due to rising ethical and sustainability issues. These proteins derived from rice, quinoa, peas, and soy offer benefits such as reduced immunogenicity, sustainability, and environmental impact. Oral tablets and capsules are more effective when they enhance the solubility, stability, and bioavailability of poorly soluble medications. For example, soy protein isolates help stabilize hydrophobic drugs and improve their release profiles. Furthermore, in comparison to proteins sourced from animals, plant proteins are less likely to cause allergies and provoke fewer ethical concerns [127–132].

Furthermore, plant proteins can be chemically or genetically modified to improve their solubility, stability, and targeted drug delivery, among other functional characteristics. These changes can increase the therapeutic efficacy and drug absorption, especially in the gastrointestinal system [133–135].

7.4.4 Functional Applications of Natural Proteins/Peptides in Drug Delivery

Natural peptides and proteins significantly enhance pharmacokinetics and pharmacodynamics in drug delivery systems. Pharmaceutical researchers can develop excipients that meet the specific needs of different drugs by altering their properties, which improves both effectiveness and safety [133–135]. Gelatin and albumin are particularly important for stabilizing sensitive small-molecule drugs and biological medications. Gelatin protects medications in capsules from the breakdown caused by stomach acid, preserving drug integrity and therapeutic effectiveness, whereas albumin stabilizes monoclonal antibodies and biologics [136, 137]. Moreover, controlled-release formulations utilize natural proteins such as casein and gelatin to ensure a gradual, prolonged release of medication, which improves patient adherence and reduces side effects. Gelatin, for example, is used in extended-release capsules to maintain therapeutic drug levels for an extended period [138–141]. Moreover, albumin plays a crucial role in targeted drug delivery, particularly in cancer therapies, because it links medications to specific tissues and reduces overall toxicity. In addition, it is feasible to accurately deliver oral medications by modifying proteins from plants to enhance targeted delivery in the gastrointestinal system [25, 138, 142].

7.4.5 Biodegradability and Safety of Natural Proteins/Peptides

Biodegradable, safe, and biocompatible, natural proteins and peptides are increasingly favored in pharmaceutical formulations. As they originate from natural sources, these excipients offer numerous advantages compared to synthetic ones, including reduced toxicity risks and enhanced environmental sustainability [143–145]. Biodegradability refers to a material's capacity to break down naturally into nontoxic substances. Natural peptides and proteins are biodegradable, implying that once they have served their role in pharmaceutical formulations, the body metabolizes and assimilates them. This trait reduces the risk of toxicity, along with several other key benefits. Natural proteins and peptides are digested and eliminated by the body, unlike synthetic excipients, which reduces the risk of long-term toxicity and negative effects, particularly when medications are used chronically or for prolonged durations [146–148].

Biodegradable excipients also lessen environmental effects, as their degradation products are typically safer and less detrimental than those of synthetic options, which can endure and lead to environmental contamination [149, 150]. Moreover, since natural proteins and peptides are found naturally in the body, they are less prone to trigger immune reactions, thereby minimizing the risk of allergies, inflammation, or more serious side effects.

For instance, albumin, a protein found naturally, is extensively utilized in injectable products with limited immune responses because of its biocompatibility [151, 152]. When natural proteins and peptides are broken down, they are metabolized into amino acids and small peptides, which cells harmlessly use for different physiological functions, ensuring that these excipients do not lead to long-term damage or produce toxic metabolites, in contrast to certain synthetic materials that may generate harmful by-products during degradation [146, 153, 154].

Safety is a major concern when using excipients, particularly in biological formulations where the danger of immunological responses or toxicity is increased. The natural origin of proteins and peptides has a substantial impact on their safety profile. Natural proteins, such as albumin and gelatin, are already present in the human body; thus, they are less likely to cause allergic reactions than synthetic excipients or foreign compounds. Furthermore, plant-derived proteins, which are gaining popularity, are less allergenic than animal-derived proteins, lowering the likelihood of hypersensitivity reactions, which can cause serious side effects. The native amino acid sequences found in proteins such as albumin and casein are recognized and accepted by the body, resulting in improved integration and compatibility [155, 156].

This biocompatibility minimizes the chances of immune reactions, promoting safer interactions with the body's cells and tissues, especially in intravenous or subcutaneous drug formulations [80, 157]. The decomposition of natural proteins and peptides generates safe by-products like free amino acids, integral to the body's regular metabolic functions. These breakdown products are utilized in cellular processes or removed via natural routes, reinforcing the safety of natural proteins in pharmaceutical formulations [158, 159]. Moreover, the natural proteins' biodegradability guarantees that they do not build up in the body, reducing side effects. This is particularly crucial in long-acting or sustained-release medications, where excipients must be absorbed or eliminated safely without leading to adverse reactions over time. The capacity of natural proteins to decompose and be absorbed without building up in the body is an essential safety benefit in these formulations [160, 161].

7.4.6 Challenges Safety Considerations in the Use of Natural Proteins/Peptides as Excipients

Despite their biodegradability and safety, natural proteins and peptides have significant obstacles in maintaining a consistent safety profile across various formulations. One of the most significant issues is source variability. Natural proteins and peptides obtained from biological sources can differ based on their origin, extraction method, and processing conditions.

For example, albumin extracted from human blood can vary depending on the donor population and processing methods, affecting the excipient's purity and consistency. This diversity can have an impact on the safety and effectiveness of pharmacological formulations. To overcome this, extensive quality control techniques are required to ensure consistent manufacturing while maintaining the desirable features of these excipients [162, 163].

An additional obstacle is the possibility of pollutants. Proteins derived from animals or plants may contain contaminants such as infections, allergies, or pollutants. For example, animal-derived gelatin may be contaminated with prions or other diseases, posing health hazards. Recombinant proteins, such as recombinant human albumin, are frequently utilized to reduce these concerns by eliminating the requirement for animal-derived goods. However, preserving the integrity of these excipients through thorough testing and purification methods is critical to their safety [164, 165].

Immunogenicity also presents a challenge, even though natural proteins are often less immunogenic than synthesized compounds. Some people may still produce immunological reactions, particularly to animal-derived proteins. Patients who are allergic to cow's milk, for example, may have an immunological reaction to casein, a milk protein. Similarly, the use of plant-based proteins, such as soy or rice, may cause allergic reactions in certain people. This diversity in immunogenicity emphasizes the importance of diligent patient monitoring and tailored treatment strategies [89, 166].

Ultimately, incorporating natural proteins and peptides into pharmaceutical products poses considerable regulatory compliance difficulties. These excipients need to adhere to strict regulatory standards set by entities such as the FDA and European Medicines Agency (EMA). For instance, recombinant albumin utilized in injectable products must furnish comprehensive safety and quality data, including the protein's source, possible contaminants, and manufacturing processes, before its use in pharmaceutical items. As the utilization of natural proteins and peptides increases, regulatory frameworks need to evolve to address new safety issues [167–170].

7.5 Modifications of Natural Proteins/Peptides as Excipients

Natural proteins and peptides can be modified to enhance their effectiveness as excipients in pharmaceutical formulations, thanks to their versatility. These modifications aim to enhance stability, solubility, bioavailability, and effectiveness in drug delivery systems. Altering natural proteins and peptides allows for the development of excipient properties that meet the distinct needs of different therapeutic uses. Adjustments can be implemented through structural, chemical, or enzymatic modifications, along with genetic engineering methods [89, 171].

7.5.1 Structural Modifications

Structural modifications change the physical shape or configuration of a protein or peptide. These modifications can enhance the stability, solubility, and release characteristics of the drug formulation [171].

- *Cross-linking*: Cross-linking forms covalent bonds between protein molecules, enhancing structural stability and resistance to breakdown. Cross-linked proteins may be utilized in controlled-release formulations to delay the release of the active pharmaceutical ingredient (API). For instance, gelatin cross-linking is frequently used to enhance the mechanical strength and stability of hydrogels and capsules [14, 172–174].
- *Conjugation*: The incorporation of proteins or peptides with polymers, lipids, or targeting ligands enhances the efficacy of drug delivery. This modification could

enhance drug loading, boost solubility, and facilitate targeted delivery of therapeutic agents. As an illustration, linking albumin with pharmaceutical compounds can enhance the stability of the medication and facilitate targeted delivery to specific tissues [175, 176].

7.5.2 Chemical Modifications

Chemical alterations change the chemical structure of proteins and peptides, improving their functioning. These changes can enhance features such as solubility, stability, and resistance to enzymatic degradation [177, 178].

- *Acylation and alkylation*: Adding hydrophobic groups to proteins improves their solubility in lipophilic conditions, making them amenable for medicinal formulations. These changes can improve the protein's interaction with lipophilic medicines or delivery methods, increasing bioavailability [179, 180].
- *Glycosylation*: Glycosylation enhances protein stability, solubility, and resistance to enzymatic destruction. This alteration is especially beneficial for proteins used in parenteral formulations since it can extend their half-life and improve therapeutic efficacy. Biopharmaceutical formulations frequently include recombinant glycosylated proteins to lengthen their half-lives and decrease immunological recognition [181–183].

7.5.3 Genetic Engineering and Recombinant Technology

Advancements in biotechnology have enabled the genetic modification of proteins and peptides to improve their excipient properties. Recombinant DNA technology enables the creation of altered proteins possessing specific traits that do not normally exist [184].

Recombinant Proteins Recombinant technology addresses worries regarding the unpredictability and contamination linked to animal-derived proteins by enabling the regulated production of proteins like albumin, collagen, and insulin. In pharmaceutical applications, these proteins can be modified to enhance stability, and scalability, and minimize immunogenicity [185].

Protein Engineering Protein engineering involves modifying a protein's amino acid sequence to improve its functional properties. This may entail modifications to enhance solubility, reduce degradation, or boost binding affinity. Protein engineering, for example, can be employed to enhance the stability of casein or albumin, thereby improving their effectiveness as excipients in various drug formulations [186, 187].

7.5.4 Enzymatic Modifications

Enzymatic modifications are regulated changes made to the structure or characteristics of endogenous proteins and peptides using particular enzymes. This may be a useful method for improving natural excipients' performance without compromising their biocompatibility [188].

- *Proteolysis*: Controlled enzymatic cleavage of proteins can yield smaller peptides or fragments that have specific biological activity. These smaller peptides are ideal for use in cutting-edge drug delivery systems, such as peptide-based nanoparticles for targeted treatment, due to their potentially superior solubility, bioactivity, or target selectivity [1, 188].
- *Peptide synthesis*: Enzymes can also be utilized to create custom peptides with specific sequences designed to improve medication binding, stability, or release rates. This approach facilitates the development of excipients with targeted characteristics to satisfy particular pharmacological needs [189].

7.5.5 Hybrid and Composite Systems

Natural proteins and peptides can be combined with other excipients, such as lipids, synthetic polymers, or other biopolymers, to form hybrid or composite systems that maximize the advantages of each constituent. The purpose of these systems is to improve the excipient's overall performance in medication formulations [190].

- *Polymer-protein conjugates*: Excipients that offer a controlled release of medications can be created by combining proteins with biodegradable synthetic polymers, such as poly(lactic-co-glycolic acid) (PLGA).
- These conjugates are appropriate for a variety of drug delivery methods because they provide improved stability, solubility, and drug-encapsulating properties [191–193].
- *Protein-lipid systems*: By adding proteins to lipid-based delivery systems like liposomes or nanostructured lipid carriers (NLCs), poorly water-soluble medications can become more soluble and bioavailable. These technologies can enhance the targeted delivery of sensitive medications and prevent their degradation [194, 195].

Modifying natural proteins and peptides as excipients is a practical way to boost their efficacy in medication formulations. Their stability, solubility, and activity can be enhanced by a variety of methods, such as structural, chemical, genetic, and enzymatic modifications. These changes can help solve the limitations of native proteins and peptides by making them more compatible with various drug delivery platforms. It is anticipated that as biotechnological advancements continue, the possibility of further improvements in the design and application of modified natural proteins and peptides in drug formulations will increase, leading to more effective and patient-friendly pharmaceutical [145].

7.6 Challenges in Using Proteins/Peptides as Excipients

Peptides and proteins have a lot of potential as excipients in pharmaceutical formulations because of their functional flexibility, biocompatibility, and biodegradability. To guarantee effective commercialization, a number of issues related to their use in pharmaceutical products must be resolved. Stability is a major obstacle [196, 197]. Environmental factors may lead proteins and peptides to lose their structural integrity and function. Due to the heat-induced denaturation that affects proteins such as albumin and gelatin, compromising their functionality, thermal instability is a significant issue [198, 199].

An additional instance is casein, which is more stable in acidic environments but tends to aggregate more at elevated pH levels [200–203]. Additionally, the function of proteins may be altered by oxidative and hydrolytic degradation, especially in aqueous formulations, leading to a reduction in therapeutic effectiveness [159, 204].

Immunogenicity is another significant issue, which is particularly problematic for animal-derived proteins. Although some proteins, such as human serum albumin, are generally well tolerated, proteins from cows or pigs may activate immunological responses that result in inflammation or allergic reactions. For example, gelatin made from the collagen of cows or pigs has been connected to hypersensitivity reactions, particularly in vaccines. Additionally, during extraction or processing, proteins may get contaminated with endotoxins, which raise questions regarding immunological response and safety [205–207].

Expenses and scalability present additional challenges in the production of protein-based excipients. Extracting protein from natural sources such as casein or bovine gelatin entails expensive and labor-intensive methods. Ensuring consistency and yield is difficult because of the complexity and susceptibility to errors of the purification techniques used in large-scale production, like ion exchange chromatography and ultrafiltration [208–210].

Proteins and peptides provide significant regulatory and quality control issues when added to medication compositions. Proteins' efficacy as excipients may be impacted by batch variation brought on by variables like origin, diet, and processing conditions. Furthermore, endotoxin, host-cell proteins, or heavy metal contamination during manufacture can jeopardize safety and necessitate stringent testing to satisfy regulatory requirements [211–213].

Proteins and peptides function as excipients and have a major impact on the therapeutic result because of their pharmacological interactions and bioavailability. Proteins like casein and gelatin, for instance, can change the viscosity and rheological characteristics of formulations, which may have an effect on how quickly active pharmaceutical ingredients (APIs) are released. Extended-release versions may benefit even though injectable products may face difficulties. Proteins can also interact with APIs, reducing bioavailability or causing side effects, highlighting the need for compatibility studies [214–216].

Protein-based excipients have a tendency to deteriorate over time, raising concerns about their longevity and storage. Stability requires careful control over storage parameters like temperature, humidity, and light exposure. Although lyophilization is frequently used to preserve proteins, it can be expensive and occasionally fails to stop deterioration. Furthermore, protein aggregation during reconstitution may become possible [51, 198].

Excellent excipients are proteins and peptides because of their natural source and compatibility with biological systems. However, there are several obstacles to their use in medication formulations, such as difficulty with bioavailability, scalability, stability, immunogenic reactions, and regulatory barriers. Addressing these challenges through further research and optimization, as well as advances in manufacturing processes and regulatory harmonization, will be crucial for the successful commercialization of these promising natural excipients.

7.7 Future Prospects of Protein and Peptide Excipients

The future for protein and peptide excipients in pharmaceutical formulations is promising, driven by advancements in material science, biotechnology, and the growing understanding of biopolymers in drug delivery.

Proteins and peptides are expected to play a crucial role in shaping the future of pharmaceutical formulations because of the increasing demand for more effective, patient-friendly, and sustainable pharmaceutical products [217, 218].

The design and optimization of proteins have been greatly improved by developments in protein engineering, leading to improved stability, decreased immunogenicity, and customized activity for use as excipients. Proteins that are resistant to oxidation, heat denaturation, and enzymatic degradation can now be produced because of methods like recombinant DNA technology and protein redesign. This enhances the feasibility of utilizing proteins like albumin, gelatin, and casein in topical, parenteral, and oral products. Moreover, engineered proteins could be incorporated into advanced delivery systems such as hydrogels and nanocarriers, enabling improved targeting and more regulated drug release for enhanced patient results [161, 219, 220].

Biotechnology advancements are probably going to be aided by the increasing need for pharmaceutical excipients from renewable and sustainable sources. Proteins obtained from plants and microorganisms are starting to compete with those generated from animals. For instance, recombinant proteins like albumin and gelatin can be produced economically through microbial fermentation, providing a more scalable and ecologically friendly option. Protein-based excipients will become more economical and ecologically conscious as the pharmaceutical industry adopts green chemistry and sustainability more and more [221–223].

Protein and peptide excipients have enormous potential in the field of customized medicine. These resources can be tailored to each patient's unique requirements, including those with metabolic disorders or allergies. To enhance treatment effectiveness and minimize side effects, tailored formulations could include precisely dosed proteins within drug delivery systems. Additionally, patient adherence can be improved by incorporating protein-based excipients to create user-friendly dosage forms, like chewable tablets or oral dispersible systems, particularly for elderly or pediatric groups [224, 225].

More studies are also being conducted on protein and peptide excipients for sophisticated drug delivery systems. These systems, such as hydrogels, liposomes, and nanoparticles, offer sustained, targeted, and regulated drug release, enhancing therapeutic outcomes—especially for complex or biological treatments. To guarantee the secure transport of delicate biologics, proteins, and peptides are capable of stabilizing nanoparticles, enhancing bioavailability, and functioning as carriers for gene therapies and RNA-based treatments [226–228]. New regulatory advancements will be crucial as the application of protein and peptide excipients increases.

To ensure the effectiveness, safety, and cleanliness of these natural excipients, additional regulations are likely to be established. The integration of recombinant and altered proteins as excipients into pharmaceutical formulations will also be facilitated by more straightforward approval processes. Due to these regulatory enhancements, batch-to-batch variations will be reduced, ensuring consistent product quality [198, 229].

Additional advancements in quality control and characterization are also anticipated. Increased knowledge of protein structure, stability, and interactions with therapeutic components will result from developments in analytical methods including mass spectrometry, cryo-electron microscopy, and surface plasmon resonance. Better quality control will guarantee batch uniformity and raise the bar for drug development [92, 230].

Nanotechnology, when integrated with protein and peptide excipients, can enhance the solubility, stability, and bioavailability of poorly soluble medications. The compatibility of protein-based excipients with biological systems renders them suitable for use in nanocarriers, improving the in vivo safety of the carriers. Protein engineering can also enable more targeted drug delivery by binding to nanoparticles; this is particularly helpful in the treatment of cancer, as targeted administration can increase efficacy and decrease side effects [89, 231, 232].

Protein and peptide excipients in pharmaceutical formulations have a bright future ahead of them. As long as protein engineering, biotechnology, nanotechnology, and regulatory frameworks continue to advance, these natural excipients will remain essential in revolutionizing drug delivery methods, improving patient compliance, and paving the way for more sustainable and customized medications. If current challenges are addressed and these advancements are embraced, protein and peptide excipients will be essential in the development of next-generation therapeutic medications [233, 234].

Acknowledgment The authors expand their gratitude to all their respective academic institutions for support and encouragement. The authors also wish to acknowledge the use of advanced digital tools. Specifically, we utilized ChatGPT, QuillBot, and Grammarly for paraphrasing, refining, sentence structure, improving grammar, language precision, and spelling accuracy. These tools significantly contributed to enhancing the clarity, coherence, and overall quality of the text.

While these tools provided valuable assistance, the authors take full responsibility for the interpretation and presentation of the content within this monograph.

Competing Interests The authors have no conflicts of interest to declare that are relevant to the content of this chapter.

References

1. Purohit K, Reddy N, Sunna A. Exploring the potential of bioactive peptides: from natural sources to therapeutics. Int J Mol Sci. 2024;25(3):1391.
2. Bruno BJ, Miller GD, Lim CS. Basics and recent advances in peptide and protein drug delivery. Ther Deliv. 2013;4(11):1443–67.
3. Jain KK. An overview of drug delivery systems, Methods in molecular biology, vol. 2059; 2020. p. 1–54.
4. Brytan W, Padrela L. Structural modifications for the conversion of proteins and peptides into stable dried powder formulations: a review. J Drug Deliv Sci Technol. 2023;89:104992.
5. Cortes J, Saura C. Nanoparticle albumin-bound (nab™)-paclitaxel: improving efficacy and tolerability by targeted drug delivery in metastatic breast cancer. Eur J Cancer, Suppl. 2010;8(1):1–10.
6. Pehrsson M, Mortensen JH, Manon-Jensen T, Bay-Jensen AC, Karsdal MA, Davies MJ. Enzymatic cross-linking of collagens in organ fibrosis–resolution and assessment. Expert Rev Mol Diagn. 2021;21(10):1049–64.
7. Liao Y, Jiang X. Chemo-selective modification of cysteine residue: synthesis and application in the discovery of potential drug candidates. Explor Drug Sci. 2024;2:540–54.
8. Jiang H, Chen W, Wang J, Zhang R. Selective N-terminal modification of peptides and proteins: recent progresses and applications. Chinese Chem Lett. 2022;33(1):80–8.
9. Spicer CD, Davis BG. Selective chemical protein modification. Nat Commun. 2014;5:1–14.
10. Zhou G, Jin X, Zhu P, Yao J, Zhang Y, Teng L, et al. Human serum albumin nanoparticles as a novel delivery system for cabazitaxel. Anticancer Res. 2016;36(4):1649–56.
11. Shibin VP, Nair BR. Formulation and characterization of quercetin-loaded gelatin microspheres for drug delivery and their evaluation. Polym Bull [Internet]. 2024;81(16):14975–97. Available from:. https://doi.org/10.1007/s00289-024-05401-y.
12. Bin SS, Idrus RBH, Hwei NM. Gelatin microsphere for cartilage tissue engineering: current and future strategies. Polymers (Basel). 2020;12(10):1–16.
13. Mashabela LT, Maboa MM, Miya NF, Ajayi TO, Chasara RS, Milne M, et al. A comprehensive review of cross-linked gels as vehicles for drug delivery to treat central nervous system disorders. Gels. 2022;8(9):1–46.
14. Jayachandran B, Parvin TN, Alam MM, Chanda K, MM B. Insights on chemical crosslinking strategies for proteins. Molecules. 2022;27(23):8124.
15. Alavarse AC, Frachini ECG, da Silva RLCG, Lima VH, Shavandi A, Petri DFS. Crosslinkers for polysaccharides and proteins: synthesis conditions, mechanisms, and crosslinking efficiency, a review. Int J Biol Macromol. 2022;202(January):558–96.
16. Balthasar S, Michaelis K, Dinauer N, Von Briesen H, Kreuter J, Langer K. Preparation and characterisation of antibody modified gelatin nanoparticles as drug carrier system for uptake in lymphocytes. Biomaterials. 2005;26(15):2723–32.
17. Najjari Z, Sadri F, Varshosaz J. Smart stimuli-responsive drug delivery systems in spotlight of COVID-19. Asian J Pharm Sci. 2023;18(6):100873.
18. Liu D, Yang F, Xiong F, Gu N. The smart drug delivery system and its clinical potential. Theranostics. 2016;6(9):1306–23.

19. Sun Q, Yang Z, Qi X. Design and application of hybrid polymer-protein systems in cancer therapy. Polymers (Basel). 2023;15(9):2219.
20. Wu J, Sahoo JK, Li Y, Xu Q, Kaplan DL. Challenges in delivering therapeutic peptides and proteins: a silk-based solution. J Control Release. 2022;345:176–89.
21. Verma S, Goand UK, Husain A, Katekar RA, Garg R, Gayen JR. Challenges of peptide and protein drug delivery by oral route: current strategies to improve the bioavailability. Drug Dev Res. 2021;82(7):927–44.
22. Zhu Q, Chen Z, Paul PK, Lu Y, Wu W, Qi J. Oral delivery of proteins and peptides: challenges, status quo and future perspectives. Acta Pharm Sin B. 2021;11(8):2416–48.
23. Pereira AJ, de Campos LJ, Huihua Xing MCS. Peptide-based therapeutics: challenges and solutions. Med Chem Res. 2024;33:1275–80.
24. Mukherjee B, Karmakar S, Hossain C, Bhattacharya S. Peptides, proteins and peptide/protein-polymer conjugates as drug delivery system. Protein Pept Lett. 2014;21(11):1121–8.
25. Hong S, Choi DW, Kim HN, Park CG, Lee W, Park HH. Protein-based nanoparticles as drug delivery systems. Pharmaceutics. 2020;12(7):1–28.
26. Jain KK. Current status and future prospects of drug delivery systems, Methods in molecular biology, vol. 1141; 2014. p. 1–56.
27. Zhang Y, Zhang H, Ghosh D, Williams RO. Just how prevalent are peptide therapeutic products? A critical review. Int J Pharm. 2020;587:1–47.
28. Wang L, Wang N, Zhang W, Cheng X, Yan Z, Shao G, et al. Therapeutic peptides: current applications and future directions. Signal Transduct Target Ther. 2022;7(1):48.
29. Ghaly G, Tallima H, Dabbish E, Eldin NB, El-rahman MKA, Ibrahim MAA, et al. Anticancer peptides: status and future prospects. Molecules. 2023;28:1148.
30. Guo S, Wang J, Wang Q, Wang J, Qin S, Li W. Advances in peptide-based drug delivery systems. Heliyon [Internet]. 2024;10(4):e26009. Available from:. https://doi.org/10.1016/j.heliyon.2024.e26009.
31. Wu W, Kong X, Zhang C, Hua Y, Chen Y. Improving the stability of wheat gliadin nanoparticles – effect of gum Arabic addition. Food Hydrocoll [Internet]. 2018;80:78–87. Available from:. https://doi.org/10.1016/j.foodhyd.2018.01.042.
32. Ezpeleta I, Arangoa MA, Irache JM, Stainmesse S, Chabenat C, Popineau Y, et al. Preparation of Ulex europaeus lectin-gliadin nanoparticle conjugates and their interaction with gastrointestinal mucus. Int J Pharm. 1999;191(1):25–32.
33. Joye IJ, Nelis VA, McClements DJ. Gliadin-based nanoparticles: fabrication and stability of food-grade colloidal delivery systems. Food Hydrocoll [Internet]. 2015;44:86–93. Available from:. https://doi.org/10.1016/j.foodhyd.2014.09.008.
34. Weber C, Coester C, Kreuter J, Langer K. Desolvation process and surface characterisation of protein nanoparticles. Int J Pharm. 2000;194(1):91–102.
35. Wang H, Zou Q, Boerman OC, Nijhuis AWG, Jansen JA, Li Y, et al. Combined delivery of BMP-2 and bFGF from nanostructured colloidal gelatin gels and its effect on bone regeneration *in vivo*. J Control Release [Internet]. 2013;166(2):172–81. Available from:. https://doi.org/10.1016/j.jconrel.2012.12.015.
36. Cho YH, Jones OG. Assembled protein nanoparticles in food or nutrition applications [Internet], Advances in food and nutrition research, vol. 88. 1st ed. Elsevier Inc.; 2019. p. 47–84. Available from:. https://doi.org/10.1016/bs.afnr.2019.01.002.
37. Mirshahi T, Irache JM, Gueguen J, Orecchioni AM. Development of drug delivery systems from vegetal proteins: legumin nanoparticles. Drug Dev Ind Pharm. 1996;22(8):841–6.
38. Irache JM, Bergougnoux L, Ezpeleta I, Gueguen J, Orecchioni AM. Optimization and *in vitro* stability of legumin nanoparticles obtained by a coacervation method. Int J Pharm. 1995;126(1–2):103–9.
39. Ryu J, Kim H, Park HH, Lee HJ, Park JH, Rhee WJ, et al. Protein-stabilizing and cell-penetrating properties of α-helix domain of 30Kc19 protein. Biotechnol J. 2016;11(11):1443–51.
40. Park HJ, Kim EJ, Koo TY, Park TH. Purification of recombinant 30K protein produced in Escherichia coli and its anti-apoptotic effect in mammalian and insect cell systems. Enzyme Microb Technol. 2003;33(4):466–71.

41. Lee HJ, Park HH, Sohn Y, Ryu J, Park JH, Rhee WJ, et al. α-Galactosidase delivery using 30Kc19-human serum albumin nanoparticles for effective treatment of Fabry disease. Appl Microbiol Biotechnol [Internet]. 2016;100(24):10395–402. Available from:. https://doi.org/10.1007/s00253-016-7689-z.
42. Bricarello DA, Smilowitz JT, Zivkovic AM, German JB, Parikh AN. Reconstituted lipoprotein: a versatile class of biologically-inspired nanostructures. ACS Nano. 2011;5(1):42–57.
43. Davis ME, Chen Z, Shin DM. Nanoparticle therapeutics: an emerging treatment modality for cancer. Nat Rev Drug Discov. 2008;7(9):771–82.
44. Zheng G, Chen J, Li H, Glickson JD. Rerouting lipoprotein nanoparticles to selected alternate receptors for the targeted delivery of cancer diagnostic and therapeutic agents. Proc Natl Acad Sci U S A. 2005;102(49):17757–62.
45. Chasteen ND, Harrison PM. Mineralization in ferritin: an efficient means of iron storage. J Struct Biol. 1999;126(3):182–94.
46. Uchida M, Flenniken ML, Allen M, Willits DA, Crowley BE, Brumfield S, et al. Targeting of cancer cells with ferrimagnetic ferritin cage nanoparticles. J Am Chem Soc. 2006;128(51):16626–33.
47. Liang M, Fan K, Zhou M, Duan D, Zheng J, Yang D, et al. H-ferritin-nanocaged doxorubicin nanoparticles specifically target and kill tumors with a single-dose injection. Proc Natl Acad Sci U S A. 2014;111(41):14900–5.
48. Berillo D, Yeskendir A, Zharkinbekov Z, Raziyeva K, Saparov A. Peptide-based drug delivery systems. Medicina (B Aires). 2021;57:1209.
49. Keservani RK, Sharma AK, Jarouliya U. Protein and peptide in drug targeting and its therapeutic approach. Ars Pharm. 2015;56(3):165–77.
50. Haddadzadegan S, Dorkoosh F, Bernkop-Schnürch A. Oral delivery of therapeutic peptides and proteins: technology landscape of lipid-based nanocarriers. Adv Drug Deliv Rev [Internet]. 2022;182:114097Available from:. https://doi.org/10.1016/j.addr.2021.114097.
51. Rahban M, Ahmad F, Piatyszek MA, Haertlé T, Saso L, Saboury AA. Stabilization challenges and aggregation in protein-based therapeutics in the pharmaceutical industry. RSC Adv. 2023;13(51):35947–63.
52. Kumar M, Tomar M, Punia S, Dhakane-Lad J, Dhumal S, Changan S, et al. Plant-based proteins and their multifaceted industrial applications. LWT. 2022;154:112620.
53. Burns A, Olszowy P, Ciborowski P. 2 - Biomolecules. In: Ciborowski P, Silberring J, editors. Proteomic profiling and analytical chemistry. 2nd ed. Boston: Elsevier; 2016. p. 7–24.
54. Lorber B, Fischer F, Bailly M, Roy H, Kern D. Protein analysis by dynamic light scattering: methods and techniques for students. Biochem Mol Biol Educ. 2012;40(6):372–82.
55. Arzenšek D. Dynamic light scattering and application to proteins in solutions. Semin Univ Ljubljana. 2010; 1–18.
56. Behera A, Nayak AK, Mohapatra RK, Rabaan AA. Smart micro-and nanomaterials for drug delivery. Smart Micro-and Nanomater Drug Deliv. 2024;1:1–422
57. Shirvani H. Studying the physical state of phages using Dynamic Light Scattering (DLS). University of Jyväskylä; 2024.
58. Yao J, Li H, Lan J, Bao Y, Du X, Zhao Z, et al. Spectroscopic investigations on the interaction between nano plastic and catalase on molecular level. Sci Total Environ. 2023 Mar;10(863):160903.
59. Johnson C. Differential scanning calorimetry as a tool for protein folding and stability. Arch Biochem Biophys. 2012 Sept 27;531.
60. Kuril AK. Differential scanning calorimetry: a powerful and versatile tool for analyzing proteins and peptides. J Pharm Res Int. 2024;36(7):179–87.
61. Kunal Pal, Sarika Verma, Pallab Datta, Ananya Barui, S.A.R. Hashmi AKS. Advances in biomedical polymers and composites: materials and applications. 2023. 843 p.
62. Callies O, Daranas A. Application of isothermal titration calorimetry as a tool to study natural product interactions. Nat Prod Rep. 2016;33:881–904.
63. Upadhyay V, Lucas A, Patrick C, Mallela KMG. Isothermal titration calorimetry and surface plasmon resonance methods to probe protein-protein interactions. Methods. 2024;225:52–61.

64. Falconer RJ, Markelz AG. Terahertz spectroscopic analysis of peptides and proteins. J Infrared, Millimeter Terahertz Waves. 2012;33(10):973–88.
65. Dawson J. Determining the amino acid sequence of a protein. In: Introduction to biochemistry; 2021.
66. Desiderio DM. Mass spectrometry of peptides. Boca Raton: CRC Press; 2024. 440 p
67. Quiquampoix H, Linhares LA, Ramos CHI. Unlocking insights into folding, structure, and function of proteins through circular dichroism spectroscopy—a short review. Appl Biosci. 2023;2(4):639–55.
68. Perez-Iratxeta C, Andrade-Navarro MA. K2D2: estimation of protein secondary structure from circular dichroism spectra. BMC Struct Biol. 2008;8:25.
69. Rigoutsos I, Riek P, Graham RM, Novotny J. Sequence pattern descriptors for transmembrane structural details. 2010.
70. Hu Y, Cheng K, He L, Zhang X, Jiang B, Jiang L, et al. NMR-based methods for protein analysis. Anal Chem. 2021;93(4):1866–79.
71. Makyio H, Kato R. X-ray crystallography of sugar related proteins. Glycoscience: biology and medicine. 2014:175–82.
72. Rehm HJ, Reed G, Pühler A, Stadler PJW. Biotechnology: second, completely revised edition, vol. 1–12; 2008. p. 1–10194.
73. Sola RJ, Griebenow K. Effects of glycosylation on the stability of protein pharmaceuticals. J Pharm Sci. 2009;98(4):1223.
74. Behera BK. Biopharmaceuticals: challenges and opportunities. 1. | Boca Raton : CRC Press, 2021. |: 2020. 294 p.
75. Mire-Sluis A. Progress in the use of biological assays during the development of biotechnology products. Pharm Res. 2001;18:1239–46.
76. Paulie S, Perlmann P, Perlmann H. Enzyme linked immunosorbent assay. In: Cell biology: a laboratory handbook. StatPearls Publishing; 2023. p. 533–8.
77. Jarvi NL, Balu-Iyer SV. Immunogenicity challenges associated with subcutaneous delivery of therapeutic proteins. BioDrugs. 2021;35(2):125–46.
78. De Groot AS, Roberts BJ, Mattei A, Lelias S, Boyle C, Martin WD. Immunogenicity risk assessment of synthetic peptide drugs and their impurities. Drug Discov Today. 2023;28(10):103714.
79. Kaczmarek-Szczepańska B, Polkowska I, Małek M, Kluczyński J, Paździor-Czapula K, Wekwejt M, et al. The characterization of collagen-based scaffolds modified with phenolic acids for tissue engineering application. Sci Rep. 2023;13(1):1–12.
80. Varanko A, Saha S, Chilkoti A. Recent trends in protein and peptide-based biomaterials for advanced drug delivery. Adv Drug Deliv Rev. 2020;156:133.
81. Adnane M, de Almeida AM, Chapwanya A. Unveiling the power of proteomics in advancing tropical animal health and production. Trop Anim Health Prod. 2024;56(5):1–24.
82. Cui M, Cheng C, Zhang L. High-throughput proteomics: a methodological mini-review. Lab Investig. 2022;102(11):1170–81.
83. Mesri M. Advances in proteomic technologies and its contribution to the field of cancer. Adv Med. 2014;2014:238045.
84. Wang W, Ohtake S. Science and art of protein formulation development. Int J Pharm. 2019 Sept 10;568.
85. Bolje A, Gobec S. Analytical techniques for structural characterization of proteins in solid pharmaceutical forms: an overview. Pharmaceutics. 2021;13(4):534.
86. Bramham JE, Podmore A, Davies SA, Golovanov AP. Comprehensive assessment of protein and excipient stability in biopharmaceutical formulations using 1H NMR spectroscopy. ACS Pharmacol Transl Sci. 2021;4(1):288–95.
87. Bashir S, Fitaihi R, Abdelhakim HE. Advances in formulation and manufacturing strategies for the delivery of therapeutic proteins and peptides in orally disintegrating dosage forms. Eur J Pharm Sci. 2023;182:106374.

88. Deb PK, Al-Attraqchi O, Chandrasekaran B, Paradkar A, Tekade R. Protein/peptide drug delivery systems: practical considerations in pharmaceutical product development. In: Basic fundamentals of drug delivery; 2018. p. 651–84.
89. Wang L, Wang N, Zhang W, Cheng X, Yan Z, Shao G, et al. Therapeutic peptides: current applications and future directions. Signal Transduct Target Ther. 2022;7(1):1–27.
90. Bruque MG, Rodger A, Hoffmann SV, Jones NC, Aucamp J, Dafforn TR, et al. Analysis of the structure of 14 therapeutic antibodies using circular dichroism spectroscopy. Anal Chem. 2024;96(38):15151–9.
91. De Meutter J, Goormaghtigh E. Evaluation of protein secondary structure from FTIR spectra improved after partial deuteration. Eur Biophys J. 2021;50(3–4):613–28.
92. Deschamps E, Calabrese V, Schmitz I, Hubert-Roux M, Castagnos D, Afonso C. Advances in ultra-high-resolution mass spectrometry for pharmaceutical analysis. Molecules. 2023;28(5):2061.
93. Tamara S, Den Boer MA, Heck AJR. High-resolution native mass spectrometry. Chem Rev. 2021;122(8):7269.
94. Ma X. Recent advances in mass spectrometry-based structural elucidation techniques. Molecules. 2022;27(19):6466.
95. Benjin X, Ling L. Developments, applications, and prospects of cryo-electron microscopy. Protein Sci. 2020;29(4):872–82.
96. Chua EYD, Mendez JH, Rapp M, Ilca SL, Tan YZ, Maruthi K, et al. Better, faster, cheaper: recent advances in cryo–electron microscopy. Annu Rev Biochem. 2022;91:1.
97. Tsung TH, Tsai YC, Lee HP, Chen YH, Lu DW. Biodegradable polymer-based drug-delivery systems for ocular diseases. Int J Mol Sci. 2023;24(16):12976.
98. Kalpana Manivannan R, Sharma N, Kumar V, Jayaraj I, Vimal S, Umesh M. A comprehensive review on natural macromolecular biopolymers for biomedical applications: recent advancements, current challenges, and future outlooks. Carbohydr Polym Technol Appl. 2024;8:100536.
99. Bejenaru C, Radu A, Segneanu AE, Biţă A, Ciocîlteu MV, Mogoşanu GD, et al. Pharmaceutical applications of biomass polymers: review of current research and perspectives. Polymers (Basel). 2024;16(9):1182.
100. Giri NC, Giri NC. Protein and peptide drug delivery. In: Smart drug delivery. IntechOpen; 2022.
101. Shen X, Liu X, Li T, Chen Y, Chen Y, Wang P, et al. Recent advancements in serum albumin-based nanovehicles toward potential cancer diagnosis and therapy. Front Chem. 2021;9:746646.
102. Qu N, Song K, Ji Y, Liu M, Chen L, Lee RJ, et al. Albumin nanoparticle-based drug delivery systems. Int J Nanomedicine. 2024;19:6945.
103. Kopnova TY, Yakupova LR, Belogurova NG, Kudryashova EV. Human serum albumin grafted by monomeric and polymeric β-cyclodextrin as drug delivery system for levofloxacin with improved pharmacological properties. Futur Pharmacol. 2024;4(1):139–62.
104. Mishra V, Heath RJ. Structural and biochemical features of human serum albumin essential for eukaryotic cell culture. Int J Mol Sci. 2021;22(16):8411.
105. Khoder M, Abdelkader H, ElShaer A, Karam A, Najlah M, Alany RG. Efficient approach to enhance drug solubility by particle engineering of bovine serum albumin. Int J Pharm. 2016;515(1–2):740–8.
106. Naveenraj S, Anandan S. Binding of serum albumins with bioactive substances – Nanoparticles to drugs. J Photochem Photobiol C Photochem Rev. 2013;14(1):53–71.
107. Krishna RG, Monali RD, Ketki SJ, Milind JU. Strategies for improving hydrophobic drugs solubility and bioavailability. Int J Pharm Chem Anal. 2023;10(3):164–74.
108. Larsen MT, Kuhlmann M, Hvam ML, Howard KA. Albumin-based drug delivery: harnessing nature to cure disease. Mol Cell Ther. 2016;4(1):3.
109. Tincu CE, Andrițoiu CV, Popa M, Ochiuz L. Recent advancements and strategies for overcoming the blood–brain barrier using albumin-based drug delivery systems to treat brain cancer, with a focus on glioblastoma. Polymers (Basel). 2023;15(19):3969.

110. Mardikasari SA, Katona G, Csóka I. Serum albumin in nasal drug delivery systems: exploring the role and application. Pharmaceutics. 2024;16(10):1322.
111. Spada A, Emami J, Tuszynski JA, Lavasanifar A. The uniqueness of albumin as a carrier in nanodrug delivery. Mol Pharm. 2021;18(5):1862–94.
112. Wiedermann CJ. Recombinant versus plasma-derived human albumin: efficacy, safety, and future prospects in clinical medicine. Paracelsus Proc Exp Med. 2024;3(1):1–11.
113. Gupta V, Sengupta M, Prakash J, Tripathy BC. Production of recombinant pharmaceutical proteins. In: Basic and applied aspects of biotechnology; 2017. p. 77–101.
114. Tuan Giam Chuang V, Kragh-Hansen U, Otagiri M. Pharmaceutical strategies utilizing recombinant human serum albumin. Pharm Res. 2002;19(5):569–77.
115. Narang AS, Boddu SH. Excipient applications in formulation design and drug delivery. In: Excipient applications in formulation design and drug delivery; 2015. p. 1–681.
116. Bjorn VERGAUWENGuy VAN DE MOOTER. Protein based excipient for active pharmaceutical ingredients. 2017.
117. Gelatin as Pharmaceutical Excipient - Pharma Excipients [Internet]. [cited 2024 Dec 23]. Available from: https://www.pharmaexcipients.com/gelatin-as-excipient/
118. Naharros-Molinero A, Ángela Caballo-González M, De La Mata FJ, García-Gallego S, Naharros-Molinero A, De La Mata FJ, et al. Shell formulation in soft gelatin capsules: design and characterization. Adv Healthc Mater. 2024;13(1):2302250.
119. Gupta S, Kesarla R, Omri A. Formulation strategies to improve the bioavailability of poorly absorbed drugs with special emphasis on self-emulsifying systems. ISRN Pharm. 2013;2013:1–16.
120. Sadiq U, Gill H, Chandrapala J. Casein micelles as an emerging delivery system for bioactive food components. Foods. 2021;10(8):1965.
121. Casein | Definition, Properties, Manufacture, & Uses | Britannica [Internet]. [cited 2024 Dec 23]. Available from: https://www.britannica.com/science/casein
122. van de Langerijt TM, O'Mahony JA, Crowley SV. Structural, binding and functional properties of milk protein-polyphenol systems: a review. Molecules. 2023;28(5):2288.
123. Kolluru LP, Atre P, Rizvi SAA. Characterization and applications of colloidal systems as versatile drug delivery carriers for parenteral formulations. Pharmaceuticals. 2021;14(2):1–12.
124. Elzoghby A, El-Fotoh W, Elgindy N. Casein-based formulations as promising controlled release drug delivery systems. J Control Release. 2011;153:206–16.
125. Ranadheera S, Liyanaarachchi WS, Chandrapala J, Dissanayake M, Vasiljevic T. Utilizing unique properties of caseins and the casein micelle for delivery of sensitive food ingredients and bioactives. Trends Food Sci Technol. 2016;57:178.
126. Rajendran AT, Vadakkepushpakath AN. Natural food components as biocompatible carriers: a novel approach to glioblastoma drug delivery. Foods. 2024;13(17):2812.
127. Lokole PB, Byamungu GG, Mutwale PK, Ngombe NK, Mudogo CN, Krause RWM, et al. Plant-based nanoparticles targeting malaria management. Front Pharmacol. 2024;15:1440116.
128. Xiao X, Zou PR, Hu F, Zhu W, Wei ZJ. Updates on plant-based protein products as an alternative to animal protein: technology, properties, and their health benefits. Molecules. 2023;28(10):4016.
129. Fu Q, Zhao J, Rong S, Han Y, Liu F, Chu Q, et al. Research advances in plant protein-based products: protein sources, processing technology, and food applications. J Agric Food Chem. 2023;71(42):15429–44.
130. Dai S, Liao P, Wang Y, Tian T, Tong X, Lyu B, et al. Soy protein isolate-catechin non-covalent and covalent complexes: focus on structure, aggregation, stability and *in vitro* digestion characteristics. Food Hydrocoll. 2022;135:108108.
131. Zdraveva E, Gaurina Srček V, Kraljić K, Škevin D, Slivac I, Obranović M. Agro-industrial plant proteins in electrospun materials for biomedical application. Polymers (Basel). 2023;15(12):2684.
132. Nikbakht Nasrabadi M, Sedaghat Doost A, Mezzenga R. Modification approaches of plant-based proteins to improve their techno-functionality and use in food products. Food Hydrocoll. 2021;118:106789.

133. Nikbakht M, Doost A, Mezzenga R. Modification approaches of plant-based proteins to improve their techno-functionality and use in food products. Food Hydrocoll. 2021;118:106789.
134. Gomes A, Sobral PJDA. Plant protein-based delivery systems: an emerging approach for increasing the efficacy of lipophilic bioactive compounds. Molecules. 2021;27(1):60.
135. Kovačević J, Bechtold T, Pham T. Plant-based proteins and their modification and processing for vegan cheese production. Macromol. 2024;4(1):23–41.
136. Castañeda Ruiz AJ, Shetab Boushehri MA, Phan T, Carle S, Garidel P, Buske J, et al. Alternative excipients for protein stabilization in protein therapeutics: overcoming the limitations of polysorbates. Pharmaceutics. 2022;14(12):2575.
137. Akbarian M, Chen SH. Instability challenges and stabilization strategies of pharmaceutical proteins. Pharmaceutics. 2022;14(11):2533.
138. Sowjanya M, Debnath S, Lavanya P, Thejovathi R, Babu M. Polymers used in the designing of controlled drug delivery system. Res J Pharm Technol. 2017;10:903.
139. Msutu M, Elizabeth P, Consulting I, Description PP. Executive summary 1 215. 2004.
140. Adepu S, Ramakrishna S. Controlled drug delivery systems: current status and future directions. Molecules [Internet]. 2021;26(19):5905. Available from: https://pmc.ncbi.nlm.nih.gov/articles/PMC8512302/
141. Tapash K. Ghosh BRJ, editor. Theory and Practice of Contemporary Pharmaceutics. 2004. 564 p.
142. Wang H, Wu D, Wang P, Gao C, Teng H, Liu D, et al. Albumin nanoparticles and their folate modified counterparts for delivery of a lupine derivative to hepatocellular carcinoma. Biomed Pharmacother. 2023;167:115485.
143. Singh A. Biopolymers in drug delivery: a review. Pharmacologyonline. 2011;1:666–74.
144. Ciulla MG, Civera M, Sattin S, Kumar K. Nature-inspired and medicinally relevant short peptides. Open Explor. 2023;1(3):140–71.
145. Nugrahadi PP, Hinrichs WLJ, Frijlink HW, Schöneich C, Avanti C. Designing formulation strategies for enhanced stability of therapeutic peptides in aqueous solutions: a review. Pharmaceutics. 2023;15(3):935.
146. Peptides: Types, Applications, Benefits & Safety [Internet]. [cited 2024 Dec 23]. Available from: https://www.webmd.com/a-to-z-guides/what-are-peptides
147. Sairam AB, Sanmugam A, Pushparaj A, Mahesh Kumar G, Sundarapandian N, Balaji S, et al. Toxicity of polymeric nanodrugs as drug carriers. ACS Chem Heal Saf. 2023;30(5):236–50.
148. Ezike TC, Okpala US, Onoja UL, Nwike CP, Ezeako EC, Okpara OJ, et al. Advances in drug delivery systems, challenges and future directions. Heliyon. 2023;9(6):e17488.
149. Das A, Ringu T, Ghosh S, Pramanik N. A comprehensive review on recent advances in preparation, physicochemical characterization, and bioengineering applications of biopolymers. Polym Bull. 2023;80(7):7247–312.
150. Behrooznia Z, Nourmohammadi J. Polysaccharide-based materials as an eco-friendly alternative in biomedical, environmental, and food packaging. Giant. 2024;19:100301.
151. Dingman R, Balu-Iyer SV. Immunogenicity of protein pharmaceuticals. J Pharm Sci. 2019;108(5):1637–54.
152. Harris CT, Cohen S. Reducing immunogenicity by design: approaches to minimize immunogenicity of monoclonal antibodies. Biodrugs. 2024;38(2):205.
153. Ling ZN, Jiang YF, Ru JN, Lu JH, Ding B, Wu J. Amino acid metabolism in health and disease. Signal Transduct Target Ther. 2023;8(1):345.
154. Protein Metabolism | Anatomy and Physiology II [Internet]. [cited 2024 Dec 23]. Available from: https://courses.lumenlearning.com/suny-ap2/chapter/protein-metabolism/
155. Khan DA, Banerji A, Blumenthal KG, Phillips EJ, Solensky R, White AA, et al. Drug allergy: a 2022 practice parameter update. J Allergy Clin Immunol. 2022;150(6):1333–93.
156. Caballero ML, Krantz MS, Quirce S, Phillips EJ, Stone CA. Hidden dangers: recognizing excipients as potential causes of drug and vaccine hypersensitivity reactions. J Allergy Clin Immunol Pract. 2021;9(8):2968.

157. Amino Acid: Benefits & Food Sources [Internet]. [cited 2024 Dec 23]. Available from: https://my.clevelandclinic.org/health/articles/22243-amino-acids
158. Nie C, Zou Y, Liao S, Gao Q, Li Q. Peptides as carriers of active ingredients: a review. Curr Res Food Sci. 2023;7:100592.
159. Bellmaine S, Schnellbaecher A, Zimmer A. Reactivity and degradation products of tryptophan in solution and proteins. Free Radic Biol Med. 2020;160:696–718.
160. Adepu S, Ramakrishna S, Costa-Pinto R, Oliveira AL. Controlled drug delivery systems: current status and future directions. Molecules. 2021;26(19):5905.
161. Thang NH, Chien TB, Cuong DX. Polymer-based hydrogels applied in drug delivery: an overview. Gels. 2023;9(7):523.
162. Grácio M, Oliveira S, Lima A, Boavida FR. RuBisCO as a protein source for potential food applications: a review. Food Chem. 2023;419:135993.
163. Ketemepi HK, Bin AMA, Seelan JSS, Mohd Noor NQI. Extraction process and applications of mushroom-derived protein hydrolysate: a comprehensive review. Futur Foods. 2024;9:100359.
164. Svraka L, Ben AH, Johansen C. When recombinant proteins go wrong: the hidden pitfall of recombinant protein contamination. Cytokine. 2024;186:156830.
165. Herdiana Y, Sofian FF, Shamsuddin S, Rusdiana T. Towards halal pharmaceutical: exploring alternatives to animal-based ingredients. Heliyon. 2023;10(1):e23624.
166. Wang J, Zhou Y, Zhang H, Hu L, Liu J, Wang L, et al. Pathogenesis of allergic diseases and implications for therapeutic interventions. Signal Transduct Target Ther. 2023;8(1):1–30.
167. U.S. Food and Drug Administration. ANDAs for Certain Highly Purified Synthetic Peptide Drug Products That Refer to Listed Drugs of rDNA Origin – Guidance for Industry [Internet]. 2021. Available from: http://www.fda.gov/Drugs/GuidanceComplianceRegulatoryInformation/Guidances/default.htm
168. European Medicines Agency. Guideline on the Development and Manufacture of Synthetic Peptides [Internet]. Vol. 31. 2023. Available from: www.ema.europa.eu/contact
169. Current Good Manufacturing Practice (CGMP) Regulations | FDA [Internet]. [cited 2024 Dec 23]. Available from: https://www.fda.gov/drugs/pharmaceutical-quality-resources/current-good-manufacturing-practice-cgmp-regulations
170. Impact Story: Developing the Tools to Evaluate Complex Drug Products: Peptides | FDA [Internet]. [cited 2024 Dec 23]. Available from: https://www.fda.gov/drugs/regulatory-science-action/impact-story-developing-tools-evaluate-complex-drug-products-peptides
171. What is the production process of modified peptides? – Abyntek Biopharma.
172. Gerrard J. Protein-protein cross-linking in food: Methods, consequences, applications. Trends Food Sci Technol. 2002;13:391–9.
173. Fuentes-Lemus E, Hägglund P, López-Alarcón C, Davies MJ. Oxidative crosslinking of peptides and proteins: mechanisms of formation, detection, characterization and quantification. Molecules. 2022;27(1):15.
174. Rebezov M, Oboturova N, Statsenko E, Bachukin V, Katkova E, Khayrullin M, et al. Crosslinking methods for improving the properties of soy-protein based films for meat packaging: a review. Potravin Slovak J Food Sci. 2023;17:635–48.
175. Dissanayake T, Sun X, Abbey L, Bandara N. Recent advances in lipid-protein conjugate-based delivery systems in nutraceutical, drug, and gene delivery. Food Hydrocoll Heal. 2022;2:100054.
176. He R, Finan B, Mayer JP, DiMarchi RD. Peptide conjugates with small molecules designed to enhance efficacy and safety. Molecules [Internet]. 2019;24(10):1855. [cited 2024 Dec 23]. Available from: https://pmc.ncbi.nlm.nih.gov/articles/PMC6572008/
177. Naowarojna N, Cheng R, Lopez J, Wong C, Qiao L, Liu P. Chemical modifications of proteins and their applications in metalloenzyme studies. Synth Syst Biotechnol. 2021;6(1):32.
178. Sakamoto S, Hamachi I. Recent progress in chemical modification of proteins. Anal Sci. 2019;35(1):5–27.
179. Shang S, Liu J, Hua F. Protein acylation: mechanisms, biological functions and therapeutic targets. Signal Transduct Target Ther. 2022;7(1):1–30.

180. Protein Alkylation: Exploring Techniques and Applications – Creative Proteomics [Internet]. [cited 2024 Dec 23]. Available from: https://www.creative-proteomics.com/resource/protein-alkylation-biochemistry-proteomics.htm
181. Dammen-Brower K, Epler P, Zhu S, Bernstein ZJ, Stabach PR, Braddock DT, et al. Strategies for glycoengineering therapeutic proteins. Front Chem [Internet]. 2022;10:863118. Available from: www.frontiersin.org
182. Andreu A, Ćorović M, Garcia-Sanz C, Santos AS, Milivojević A, Ortega-Nieto C, et al. Enzymatic glycosylation strategies in the production of bioactive compounds. Catalysts. 2023;13(10):1359.
183. Dammen-Brower K, Epler P, Zhu S, Bernstein ZJ, Stabach PR, Braddock DT, et al. Strategies for glycoengineering therapeutic proteins. Front Chem. 2022;10:863118.
184. Khan S, Ullah MW, Siddique R, Nabi G, Manan S, Yousaf M, et al. Role of recombinant DNA technology to improve life. Int J Genomics. 2016;2016:2405954.
185. Jayakrishnan A, Wan Rosli WR, Tahir ARM, Razak FSA, Kee PE, Ng HS, et al. Evolving paradigms of recombinant protein production in pharmaceutical industry: a rigorous review. Science. 2024;6(1):9.
186. Sumida KH, Núñez-Franco R, Kalvet I, Pellock SJ, Wicky BIM, Milles LF, et al. Improving protein expression, stability, and function with proteinMPNN. J Am Chem Soc. 2024;146(3):2054–61.
187. Ebrahimi SB, Samanta D. Engineering protein-based therapeutics through structural and chemical design. Nat Commun. 2023;14(1):1–11.
188. Whitaker JR. Enzymatic modification of proteins applicable to foods. In: Food proteins; 1977. p. 95–155.
189. Akbarian M, Khani A, Eghbalpour S, Uversky VN. Bioactive peptides: synthesis, sources, applications, and proposed mechanisms of action. Int J Mol Sci. 2022;23(3):1445.
190. Opriş O, Mormile C, Lung I, Stegarescu A, Soran ML, Soran A. An overview of biopolymers for drug delivery applications. Appl Sci. 2024;14(4):1383.
191. Lee PW, Pokorski JK. Poly(lactic-co-glycolic acid) devices: production and applications for sustained protein delivery. Wiley Interdiscip Rev Nanomed Nanobiotechnol. 2018;10(5):e1516.
192. Lagreca E, Onesto V, Di Natale C, La Manna S, Netti PA, Vecchione R. Recent advances in the formulation of PLGA microparticles for controlled drug delivery. Prog Biomater. 2020;9(4):153–74.
193. Borandeh S, van Bochove B, Teotia A, Seppälä J. Polymeric drug delivery systems by additive manufacturing. Adv Drug Deliv Rev. 2021;173:349–73.
194. Corey RA, Stansfeld PJ, Sansom MSP. The energetics of protein–lipid interactions as viewed by molecular simulations. Biochem Soc Trans. 2019;48(1):25.
195. Mall J, Naseem N, Haider MF, Rahman MA, Khan S, Siddiqui SN. Nanostructured lipid carriers as a drug delivery system: a comprehensive review with therapeutic applications. Intell Pharm. 2024 Sept 23.
196. Zapadka KL, Becher FJ, Gomes dos Santos AL, Jackson SE. Factors affecting the physical stability (aggregation) of peptide therapeutics. Interface. Focus. 2017;7(6):20170030.
197. Maher S, Brayden DJ. Formulation strategies to improve the efficacy of intestinal permeation enhancers. Adv Drug Deliv Rev. 2021;177:113925.
198. Shi M, McHugh KJ. Strategies for overcoming protein and peptide instability in biodegradable drug delivery systems. Adv Drug Deliv Rev. 2023;199:114904.
199. Ma Y, Chen F. Plant protein heat-induced gels: formation mechanisms and regulatory strategies. Coatings. 2023;13(11):1899.
200. Post AE, Arnold B, Weiss J, Hinrichs J. Effect of temperature and pH on the solubility of caseins: environmental influences on the dissociation of αS- and β-casein. J Dairy Sci. 2012;95:1603–16.
201. Mediwaththe A, Huppertz T, Chandrapala J, Vasiljevic T. Effect of pH and shear on heat-induced changes in milk protein concentrate suspensions. Foods. 2024;13(10):1517.

202. Zhao X, Fu Y, Song C. Proteins. In: Essentials of food chemistry. Singapore: Springer; 2021. p. 49–121.
203. Talley K, Alexov E. On the pH-optimum of activity and stability of proteins. Proteins. 2010;78(12):2699.
204. Kehm R, Baldensperger T, Raupbach J, Höhn A. Protein oxidation – formation mechanisms, detection and relevance as biomarkers in human diseases. Redox Biol. 2021;42:101901.
205. Products TP. Guidance for industry: immunogenicity assessment for therapeutic protein products [excerpts], vol. 32. European Medicines Agency; 2017.
206. Immunogenicity of Protein-based Therapeutics | FDA [Internet]. [cited 2024 Dec 23]. Available from: https://www.fda.gov/vaccines-blood-biologics/science-research-biologics/immunogenicity-protein-based-therapeutics
207. Real-Fernandez F, Errante F, Di SA, Papini AM, Rovero P. Therapeutic proteins immunogenicity: a peptide point of view. Open Explor. 2023;1(5):377–87.
208. Rossino G, Marchese E, Galli G, Verde F, Finizio M, Serra M, et al. Peptides as therapeutic agents: challenges and opportunities in the green transition era. Molecules. 2023;28(20):7165.
209. Farooq S, Ahmad MI, Zheng S, Ali U, Li Y, Shixiu C, et al. A review on marine collagen: sources, extraction methods, colloids properties, and food applications. Collagen Leather. 2024;6(1):1–27.
210. Liu Y, Aimutis WR, Drake MA. Dairy, plant, and novel proteins: scientific and technological aspects. Foods. 2024;13(7):1010.
211. Wu L. Regulatory considerations for peptide therapeutics. In: RSC drug discovery series. Royal Society of Chemistry; 2019. p. 1–30.
212. Rýglová Š, Braun M, Suchý T, Hříbal M, Žaloudková M, Vištějnová L. The investigation of batch-to-batch variabilities in the composition of isolates from fish and mammalian species using different protocols. Food Res Int. 2023;169:112798.
213. Wang W, Ignatius AA, Thakkar SV. Impact of residual impurities and contaminants on protein stability. J Pharm Sci. 2014;103(5):1315–30.
214. Renukuntla J, Vadlapudi AD, Patel A, Boddu SHS, Mitra AK. Approaches for enhancing oral bioavailability of peptides and proteins. Int J Pharm. 2013;447:75.
215. Budai L, Budai M, Fülöpné Pápay ZE, Vilimi Z, Antal I. Rheological considerations of pharmaceutical formulations: focus on viscoelasticity. Gels. 2023;9(6):469.
216. Haddadzadegan S, Dorkoosh F, Bernkop-Schnürch A. Oral delivery of therapeutic peptides and proteins: technology landscape of lipid-based nanocarriers. Adv Drug Deliv Rev. 2022;182:114097.
217. Umeyor CE, Shelke V, Pol A, Kolekar P, Jadhav S, Tiwari N, et al. Biomimetic microneedles: exploring the recent advances on a microfabricated system for precision delivery of drugs, peptides, and proteins. Futur J Pharm Sci. 2023;9(1):1–34.
218. Jabeen N, Atif | Muhammad, Atif M, Chemistry M. Polysaccharides based biopolymers for biomedical applications: a review. Polym Adv Technol. 2024;35(1):e6203.
219. Vermonden T, Censi R, Hennink WE. Hydrogels for protein delivery. Chem Rev. 2012;112(5):2853–88.
220. Nassar N, Kasapis S. Fundamental advances in hydrogels for the development of the next generation of smart delivery systems as biopharmaceuticals. Int J Pharm. 2023;633:122634.
221. Bhatia SK, Bhatia S, Inda-Webb ME, Kourmentza K, Moon TS, Singh V, et al. Biotechnology for sustainable materials: innovating today for a greener tomorrow. Biotechnol Sustain Mater. 2024;1(1):1–2.
222. Gaur S, Kaur M, Kalra R, Rene ER, Goel M. Application of microbial resources in biorefineries: current trend and future prospects. Heliyon. 2024;10(8):e28615.
223. Ezeorba TPC, Okeke ES, Mayel MH, Nwuche CO, Ezike TC. Recent advances in biotechnological valorization of agro-food wastes (AFW): optimizing integrated approaches for sustainable biorefinery and circular bioeconomy. Bioresour Technol Reports. 2024;26:101823.
224. Marques L, Costa B, Pereira M, Silva A, Santos J, Saldanha L, et al. Advancing precision medicine: a review of innovative in silico approaches for drug development, clinical pharmacology and personalized healthcare. Pharmaceutics. 2024;16(3):332.

225. Su J, Yang L, Sun Z, Zhan X. Personalized drug therapy: innovative concept guided with proteoformics. Mol Cell Proteomics. 2024;23(3):100737.
226. van den Berg AIS, Yun CO, Schiffelers RM, Hennink WE. Polymeric delivery systems for nucleic acid therapeutics: approaching the clinic. J Control Release. 2021;331:121–41.
227. Al Fayez N, Nassar MS, Alshehri AA, Alnefaie MK, Almughem FA, Alshehri BY, et al. Recent advancement in mRNA vaccine development and applications. Pharmaceutics. 2023;15(7):1972.
228. Guo S, Wang J, Wang Q, Wang J, Qin S, Li W. Advances in peptide-based drug delivery systems. Heliyon. 2024;10(4):e26009.
229. Boddu SHS, Acharya D, Hala V, Jani H, Pande S, Patel C, et al. An update on strategies to deliver protein and peptide drugs to the eye. ACS Omega. 2023;8(39):35470–98.
230. WHO. Guidelines for the production and quality control of monoclonal antibodies and related products intended for medicinal use Replacement of Annex 3 of WHO Technical Report Series, No. 822. 2022.
231. Saif A, Anjum L, Faisal Z, Akram N, Shah YA, Irfan R, et al. Recent advances in protein-based nanoparticles and their applications in the delivery of bioactive compounds. Int J Food Prop. 2023;26(2):2881–95.
232. Petrovic S, Bita B, Barbinta-Patrascu ME. Nanoformulations in pharmaceutical and biomedical applications: green perspectives. Int J Mol Sci. 2024;25(11):5842.
233. Vora LK, Gholap AD, Jetha K, Thakur RRS, Solanki HK, Chavda VP. Artificial intelligence in pharmaceutical technology and drug delivery design. Pharmaceutics. 2023;15(7):1916.
234. Puccetti M, Pariano M, Schoubben A, Giovagnoli S, Ricci M. Biologics, theranostics, and personalized medicine in drug delivery systems. Pharmacol Res. 2024;201:107086.

Harnessing Natural Source Lipids for Advanced Drug Delivery Systems

8

Dishank Purandare, Arpana Patil, Abhishek Polshettiwar, Pratiksha Chavan, and Satish Polshettiwar

Abbreviation

AI	Artificial intelligence
ALA	Alpha-linolenic acid
API	Active pharmaceutical ingredients
Aqs	Aqueous
BBB	Blood-brain barrier
BHT	Butylated hydroxytoluene
CNS	Central nervous system
CRs	Controlled release
CVDs	Cardiovascular disease
DDS	Drug delivery systems
DHA	Docosahexaenoic acid
DNA	Deoxyribonucleic acid
EMA	European Medicines Agency
EPA	Eicosapentaenoic acid
EPR	Enhanced permeability and retention
FDA	Food and Drug Administration
GIT	Gastrointestinal tract
GLA	Gamma-linolenic acid
GMP	Good manufacturing practices
HDL	High-density lipoprotein

D. Purandare · A. Polshettiwar · S. Polshettiwar (✉)
School of Health Sciences and Technology, Dr Vishwanath Karad, MIT World Peace University, Pune, Maharashtra, India
e-mail: satish.polshettiwar@mitwpu.edu.in

A. Patil · P. Chavan
STES's Smt. Kashibai Navale College of Pharmacy, Pune, Maharashtra, India

J. S et al. (eds.), *Innovative Pharmaceutical Excipients: Natural Sources*,
https://doi.org/10.1007/978-981-96-7959-1_8

HPH	High-pressure homogenization
LBDDS	Lipid-based drug delivery systems
LDL	Low-density lipoprotein
LNPs	Lipid nanoparticles
MCTs	Medium-chain triglycerides
ML	Machine learning
MMPs	Matrix metalloproteinases
mRNA	Messenger RNA
MUFAs	Monounsaturated fats
NLCs	Nanostructured lipid carriers
NPs	Nanoparticles
O/W	Oil-in-water
PAA	Polyacrylic acid
PAT	Process analytical technologies
PD	Parkinson's disease
PEG	Polyethylene glycol
PLA	Polylactic acid
PLGA	Poly(lactic-co-glycolic acid)
PNIPAAm	Poly(N-isopropylacrylamide)
PUFAs	Polyunsaturated fatty acids
QC	Quality control
R&D	Research and development
SEDDS	Self-emulsifying drug delivery systems
siRNA	Small Interfering RNA
SLNs	Solid lipid nanoparticles
SRs	Sustained drug release
S-SEDDS	Solid self-emulsifying drug delivery systems
TDD	Targeted drug delivery
W/O	Water-in-oil

8.1 Introduction

Drug delivery systems (DDS) are critical for optimizing the therapeutic potential of pharmaceutical agents by enabling controlled release (CR), targeted distribution, and improved bioavailability [1]. Conventional drug delivery methods, such as oral tablets and injections, often face significant limitations, including poor solubility, rapid clearance, and non-specific distribution, which can result in suboptimal therapeutic effects and potential side effects [2]. To address these challenges, advanced DDS technologies have been developed to enhance drug stability, prolong therapeutic effects, and ensure precise drug localization at the target site [3]. Among these, lipid-based drug delivery systems (LBDDS) have garnered significant interest due to their unique ability to encapsulate both hydrophilic and hydrophobic drugs, facilitating improved solubility, stability, and controlled release [4]. Lipids play a crucial role in DDS, particularly for lipophilic drugs, by enhancing their bioavailability, stability, and biocompatibility. Their amphiphilic nature allows for favorable interactions with biological membranes, improving drug absorption while minimizing irritation and immune responses. Furthermore, lipid formulations, such as emulsions, micelles, liposomes, and solid lipid nanoparticle (SLNs), offer versatile platforms for achieving systemic and targeted drug delivery (TDD), bypassing first-pass metabolism through lymphatic absorption and providing sustained release (SR) to maintain steady therapeutic stages with reduced dosing frequency. This is mainly advantageous for managing chronic conditions where steady drug exposure is essential for efficacy and compliance. Moreover, lipid-based carriers provide a protective environment for sensitive molecules, such as peptides and biologics, shielding them from enzymatic degradation, oxidation, and environmental stressors, thereby preserving their therapeutic activity [5]. A key consideration in LBDDS is the selection of lipid excipients, which can be derived from natural or synthetic sources, each offering distinct advantages. Natural lipids, such as phospholipids from soybeans or egg yolk, are highly biocompatible and structurally similar to biological membranes, making them suitable for applications requiring prolonged circulation and minimal immunogenicity. However, their susceptibility to oxidation and batch-to-batch variability pose challenges in maintaining consistent formulation quality. In contrast, synthetic lipids are engineered to provide enhanced stability, specific regulators over particle size, and improved drug release profiles, making them ideal for use like in gene therapy and mRNA vaccine delivery, as seen in the success of COVID-19 vaccines. Despite these advantages, synthetic lipids can be associated with higher production costs and potential toxicity at high concentrations, necessitating careful formulation optimization. As research in lipid-based drug delivery continues to advance, emerging trends such as lipid hybrid systems, personalized medicine, and stimuli-responsive lipid carriers are driving the development of more targeted and efficient therapeutic solutions. The integration of lipid-based systems with nanotechnology, artificial intelligence (AI), and bioinformatics holds the potential to revolutionize drug delivery by enabling meticulousness medicine approaches and improving treatment outcomes across a wide range of medical conditions. The ongoing innovation in LBDDS underscores their importance in modern pharmaceutical science, offering promising avenues to address current drug delivery challenges and enhance patient care [6–9].

8.2 Natural Lipids

Natural lipids are essential biomolecules with diverse physiological roles. Primarily hydrophobic, some lipids exhibit amphipathic properties, enabling the formation of lipid bilayers in cell membranes. These bilayers maintain structural integrity, regulate cellular transport, and support processes like cell signaling and endocytosis. Lipids such as phospholipids, glycolipids, and cholesterol contribute to membrane fluidity, stability, and flexibility, with cholesterol playing a key role in preventing excessive rigidity or fluidity. Triglycerides serve as the body's primary energy storage, providing a dense energy source during caloric deficits. Lipids also aid in the absorption of fat-soluble vitamins (A, D, E, and K), vital for immune response and blood clotting. Unsaturated fatty acids, like omega-3 and omega-6, support heart health and reduce inflammation. Additionally, lipids promote neurological health and skin protection, and are pivotal in drug delivery systems, enhancing therapeutic efficacy and bioavailability. Their versatility underscores their critical role in health and innovation [10, 11].

8.2.1 Biocompatibility and Biodegradability

Natural lipids are highly appreciated for their biocompatibility and biodegradability, rendering them ideal for numerous biomedical and pharmaceutical applications. Biocompatibility refers to a material's potential to interact with biological systems without triggering toxicity, immune rejection, or inflammation. Since natural lipids are key components of cell membranes in all living organisms, they are well-tolerated by the body and have minimal adverse reactions. This is particularly advantageous in drug delivery systems, where lipid-based carriers like liposomes and lipid nanoparticles (LNPs) as shown in Fig. 8.1 safely transport therapeutic agents to targeted sites without triggering immune responses or cytotoxicity. Biodegradability is another critical property of natural lipids. These lipids break down naturally into non-toxic compounds through enzymatic processes in the body, minimizing the risk of accumulation and long-term toxicity.

For example, phospholipids and triglycerides are broken down by enzymes such as lipases, releasing fatty acids and glycerol that can be used for energy or other biological

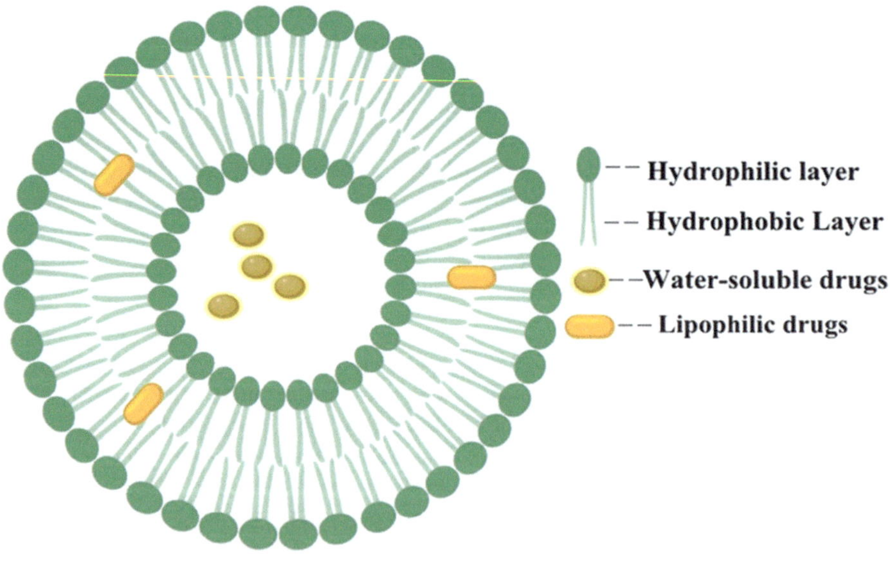

Fig. 8.1 Liposome-based DDS

functions. This natural metabolic process ensures that lipid-based carriers are safely cleared from the body after delivering their therapeutic payload, which is particularly important in drug delivery and tissue engineering applications. Biodegradable lipids are also valuable in medical devices and implants, as they degrade over time, eliminating the need for surgical removal. The biocompatibility and biodegradability of natural lipids make them essential in the development of LBDDS, gene therapies, and medical implants. These properties enable controlled, targeted delivery of drugs, proteins, or genetic material while ensuring safety and minimizing environmental impact. Natural lipids continue to play a crucial role in advancing medical technologies, improving therapeutic efficacy, and enhancing patient outcomes [12].

8.2.2 Encapsulation of Hydrophilic and Hydrophobic Drugs

Natural lipids are highly effective at encapsulating both hydrophilic and hydrophobic drugs, making them valuable in drug delivery systems. Due to their amphipathic nature, containing both hydrophilic and hydrophobic components, lipids are well-suited to enhance the solubility, stability, and bioavailability of a wide range of therapeutic agents. Hydrophobic drugs have low water solubility, which makes their delivery in the body difficult. Lipid-based delivery systems like liposomes, solid lipid nanoparticles (SLNs), and micelles help overcome this challenge by encapsulating the drugs in lipid environments, enhancing their solubility, stability, and bioavailability. For example, liposomes can encapsulate hydrophobic drugs like doxorubicin, improving solubility and targeting to cancer cells, while SLNs can solubilize compounds like curcumin, boosting their bioavailability and therapeutic potential [13]. Although hydrophilic drugs are water-soluble, they may face challenges such as rapid elimination or poor membrane permeability. Natural lipids can effectively encapsulate these drugs to enhance their stability and bioavailability. Liposomes can encapsulate hydrophilic drugs in their aqueous (Aqs) core, as seen with amphotericin B, an antifungal drug, that benefits from reduced toxicity and improved solubility in liposomal formulations. Moreover, LNPs are widely used for delivering sensitive hydrophilic molecules, such as mRNA in COVID-19 vaccines, protecting the mRNA and enabling efficient delivery to cells [14]. Lipid-based systems offer several advantages for both types of drugs. They improve solubility, protect drugs from degradation, and enhance bioavailability, ensuring efficient absorption in the body. These systems also allow for targeted delivery, enhancing therapeutic effectiveness while reducing toxic effects. Furthermore, lipid formulations enable controlled, sustained release, improving patient compliance by reducing dosing frequency. Overall, natural lipids play a critical role in advancing drug delivery technologies, offering safer, more effective treatments for both hydrophilic and hydrophobic drugs [15].

8.2.3 Lipid Solubility and Drug Stability

Natural lipids are crucial in enhancing both the solubility and stability of drugs, significantly improving their bioavailability and therapeutic efficacy. Due to their amphipathic nature, lipids contain both hydrophilic and hydrophobic components, allowing them to interact with a wide range of drugs, particularly those with poor solubility or stability. For hydrophobic drugs, lipids improve solubility by incorporating them into systems like liposomes, SLNs, and micelles. Liposomes, for example, encapsulate hydrophobic drugs like doxorubicin, improving solubility, reducing toxicity, and enhancing targeted delivery to cancer cells. Similarly, SLNs provide a stable matrix for drugs like curcumin, boosting their bioavailability. Natural lipids also protect drugs from degradation caused by environmental factors such as light, oxygen, and enzymes. Drugs like amphotericin B and mRNA are sensitive to degradation, and lipid-based systems help shield them, preserving their stability. For instance, mRNA vaccines use lipid nanoparticles to protect fragile mRNA, ensuring efficient delivery and stable storage. Additionally, lipid formulations enable controlled and sustained drug release, maintaining therapeutic levels over time and improving the overall therapeutic outcome. This is especially useful for drugs with short half-lives or those prone to rapid metabolism [16].

8.2.4 Controlled and Sustained Drug Release

Natural lipids are highly effective in enabling CR and SR, a key strategy to improve therapeutic efficacy and patient compliance. Lipid-based systems such as liposomes, SLNs, and LNPs can be engineered to release their drug payload gradually, maintaining optimal drug concentrations over extended periods. This controlled release reduces dosing frequency, enhances drug effectiveness, and minimizes side effects. Liposomes, for example, can be designed to degrade slowly in response to physiological conditions like pH or temperature, allowing for a steady release of drugs, as seen in cancer treatments like doxorubicin. SLNs, made of solid lipids, offer stability and protection to encapsulated drugs, releasing them slowly through tailored lipid composition and particle size. This is beneficial for drugs like curcumin, which requires improved bioavailability and sustained release. LNPs, used in mRNA vaccines, also offer controlled release properties. These nanoparticles protect and release mRNA gradually, ensuring it reaches target cells efficiently. The sustained release capability of lipid-based systems improves patient adherence by reducing the need for frequent dosing, ensuring stable drug concentrations, and reducing the risk of toxicity from peak levels. Additionally, these systems can be designed to target specific tissues or organs, enhancing drug efficacy and minimizing side effects [17, 18].

8.3 Sources of Natural Lipids

8.3.1 Plant-Based Lipids

Plant-based lipids, derived from natural sources such as seeds, fruits, and nuts, offer a treasure of health profits. They are rich in essential fatty acids, antioxidants, and vitamins. Monounsaturated fatty acids (MUFA), found in olive oil and

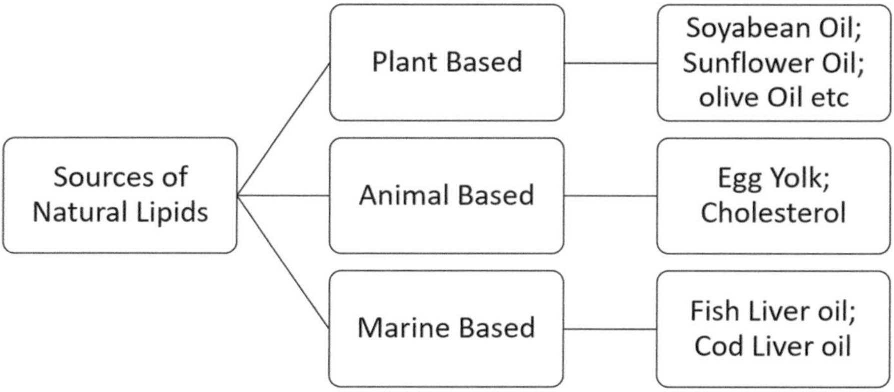

Fig. 8.2 Sources of natural lipids

avocado oil, help reduce bad cholesterol and boost good cholesterol levels. Polyunsaturated fatty acids (PUFA), found in flaxseed oil, hemp seed oil, and sunflower oil, reduce inflammation, support brain function, and improve cardiovascular health. Additionally, plant-based lipids aid in nutrient absorption, particularly of fat-soluble vitamins. Some oils, like coconut oil, have antimicrobial properties, making them useful in natural personal care. By incorporating plant-based lipids into your diet and skincare routine, you can enjoy these benefits and contribute to a healthier, more sustainable lifestyle [19]. Various sources are shown in Fig. 8.2.

8.3.1.1 Soybean Oil

Soybean oil is a widely consumed vegetable oil valued for its high value of PUFAs, primarily omega-6 fatty acids like linoleic acid. These compounds can help reduce LDL cholesterol and support cardiovascular health. However, excessive intake without balancing with omega-3-rich foods may lead to inflammation. Soybean oil also contains a moderate amount of MUFAs and small quantities of omega-3 fatty acids, such as alpha-linolenic acid (ALA). Its high smoke point (~450 °F) makes it suitable for frying, baking, and salad dressings, but mindful consumption is advised for optimal health [20].

8.3.1.2 Sunflower Oil

Sunflower oil is notable for its high vitamin E content, which acts as an antioxidant, and its rich polyunsaturated fat profile, particularly omega-6 fatty acids. It is also a source of monounsaturated fats, especially in high-oleic varieties. Known for lowering low-density lipoprotein (LDL) and increasing high-density lipoprotein (HDL) cholesterol, sunflower oil is commonly used in frying, sautéing, and salad dressings due to its neutral taste and high smoke point (~440 °F or 227 °C). However, balancing its omega-6 intake with omega-3-rich foods is recommended to mitigate potential inflammatory effects [21].

8.3.1.3 Coconut Oil

Rich in medium-chain triglycerides (MCTs), coconut oil provides quick energy and supports weight management. Its primary saturated fats, such as lauric acid, may raise HDL cholesterol. Despite potential benefits, its high saturated fat content has raised concerns regarding long-term cardiovascular health. With a moderate smoke point (~350 °F), it is ideal for high-heat cooking and baking when used in moderation [22].

8.3.1.4 Olive Oil

Extra virgin olive oil is praised for its monounsaturated fat content, mainly oleic acid, and antioxidants like phenolic compounds that promote heart health. A staple in the Mediterranean diet, it supports weight management and reduces inflammation. Its moderate smoke point (~375 °F) makes it best for sautéing, drizzling, and low-heat cooking offering significant health benefits when consumed daily in moderation [23].

8.3.1.5 Flaxseed Oil

It is rich in omega-3 fatty acids like ALA is known for reducing inflammation, supporting heart and cognitive health, and balancing hormones due to its phytoestrogenic lignans. Similarly, avocado oil is abundant in monounsaturated fats and antioxidants, promotes cardiovascular and eye health, excelling in both raw and high-heat cooking. Canola oil offers a balanced omega-6 to omega-3 ratio and supports heart health, while palm oil, despite concerns over saturated fats and deforestation, contributes antioxidants and flexibility in formulations. Peanut oil provides heart-friendly fats and antioxidants but suits only non-allergic users. Lastly, hemp seed oil, with its perfect omega ratio and gamma-linolenic acid (GLA), is ideal for raw dishes aiding hormonal and skin health [24].

8.3.2 Animal-Based Lipids

Animal-based lipids, derived from animal products like meat, eggs, dairy, and fish play essential roles in the body by supporting energy storage, brain function, cell integrity, and hormone production [25]. These lipids include triglycerides, which provide long-term energy and insulation; phospholipids such as lecithin in egg yolks, which are crucial for cell membrane structure and fat digestion; cholesterol, as shown in Fig. 8.3, is found in egg yolk and organ meats, which supports hormone synthesis, cell membrane stability, and bile acid production; fatty acids, which can be either saturated or unsaturated and provide energy while supporting the absorption of fat-soluble vitamins; sphingolipids is found in nerve tissue and crucial for nerve health and cell signaling; and omega-3 and omega-6 fatty acids, which are essential for regulating inflammation, supporting heart health, and maintaining brain function [25–29]. While these lipids are necessary for overall health, excessive consumption of saturated fats and cholesterol, particularly from red meat and full-fat dairy, can raise cholesterol levels and increase the risk of heart disease.

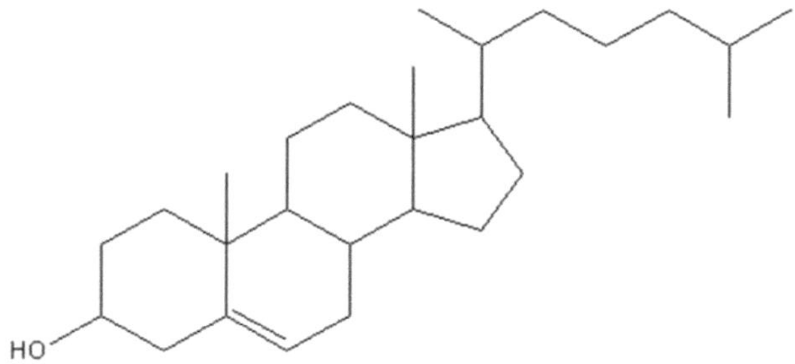

Fig. 8.3 Structure of cholesterol

A balanced intake, combining both animal-based and plant-based lipids, is vital for optimal health, supporting everything from immune function to mental well-being. Egg yolk is a rich source of essential lipids, including phospholipids such as lecithin, cholesterol, and triglycerides, which contain both saturated and unsaturated fatty acids [30]. Lecithin plays a crucial role in maintaining cell membrane integrity, supporting fat digestion and absorption, and serving as a primary source of choline—a nutrient vital for brain function, cognitive health, and liver function. Choline, a precursor to the neurotransmitter acetylcholine, contributes to memory and muscle control. Despite egg yolk's high cholesterol content, research suggests that it has minimal impact on blood cholesterol levels for most individuals when consumed in moderation. Cholesterol itself is a vital lipid necessary for cell membrane stability, hormone synthesis, and bile acid production, with the liver playing a key role in its regulation and distribution throughout the body. Beyond its biological functions, cholesterol is increasingly utilized in drug delivery systems, including liposomes, micelles, and nanoparticles, where it enhances formulation stability by preventing phospholipid phase transitions. This improves the delivery of poorly water-soluble drugs, offering benefits such as biocompatibility, membrane stabilization, and controlled drug release. Cholesterol-based carriers are widely used in pharmaceuticals due to their cost-effectiveness, biodegradability, and ability to accommodate both hydrophilic and lipophilic drugs with high loading efficiency [31]. Various chemical bonds, such as ester and hydrazone, regulate drug release, allowing for enzymatic or pH-sensitive delivery. Additionally, cholesterol prevents aggregation in aqueous environments, improving serum stability and enhancing drug absorption while minimizing gastrointestinal irritation and the first-pass effect [30–33].

8.3.3 Marine Lipids

Marine lipids, sourced from fish, krill, and microalgae, are rich in essential omega-3 fatty acids, particularly eicosapentaenoic acid (EPA) and docosahexaenoic acid (DHA), which provide numerous health benefits [32]. These bioactive lipids contribute to cardiovascular health by lowering triglycerides, regulating blood pressure, and enhancing circulation, while also playing a crucial role in reducing inflammation, making them beneficial for managing conditions such as arthritis, asthma, and inflammatory bowel disease. DHA is especially important for brain function, supporting cognitive health and potentially reducing the risk of neurodegenerative disorders like Alzheimer's disease [33]. Additionally, marine lipids support eye health by lowering the risk of age-related macular degeneration and dry eye syndrome. Fish oil, one of the most well-known sources, is obtained from fatty fish like mackerel, sardines, and salmon, providing a rich supply of omega-3s, along with vitamins A and D, which further promote immune function, bone health, and vision [34]. Algal oil, a plant-based alternative derived from microalgae, offers a sustainable and vegan-friendly source of DHA, with lower EPA content but free from contaminants like mercury. It supports brain development, heart health, and immune function while providing an ethical option for those avoiding fish-based products [35]. Other marine oils, such as cod liver, squid, and seal oils, are known for their bioavailability and high nutrient content, with cod liver oil being a notable source of vitamins A and D [36]. While the use of marine mammal oils has declined due to ethical concerns, sustainable sources of marine lipids continue to provide significant health advantages.

8.4 Lipid-Based Drug Delivery Systems (LBDDS)

LBDDS has appeared as a promising strategy to vanquish the limitations linked with conventional DDS. These systems utilize lipids to improve therapeutic agent's solubility, stability, and bioavailability. The unique structural and physicochemical properties of lipids enable their application in various forms, including liposomes, SLNs, NLCs, and SEDDS [37, 38]. These systems have distinct characteristics and advantages, making them appealing for diverse therapeutic applications.

8.4.1 Liposomes

Liposomes are biocompatible vesicles with phosphor-lipid bilayers that encapsulate an Aqs core, enabling them to carry both hydrophilic drugs in the core and hydrophobic drugs within the bilayers [38]. The mechanism of action of liposomes involves their interaction with biological membranes, facilitating drug release through fusion or endocytosis. Surface modification of liposomes with targeting ligands can enhance site-specific drug delivery, thereby minimizing systemic toxicity and improving therapeutic efficacy [39]. Liposomal formulations have been widely explored in cancer therapy, where they provide controlled drug release and reduced off-target effects. Furthermore, liposomes are extensively utilized in vaccine delivery to enhance antigen stability and immune response [40, 41]. Despite their numerous advantages, liposomal formulations face challenges related to stability, short circulation time, and production costs. Efforts to overcome these limitations include the incorporation of stabilizers such as cholesterol, surface PEGylation to extend circulation half-life, and the advance of stimuli-responsive liposomes that release drugs in response to environmental triggers.

8.4.2 Solid Lipid Nanoparticles (SLNs)

SLNs represent the first generation of lipid-based nanocarriers, characterized by a solid lipid matrix that remains in the solid state at physiological temperatures [42]. These NPs are made of biocompatible and biodegradable lipids like glyceryl behenate and stearic acid, stabilized with surfactants like polysorbates or poloxamers. The solid lipid core provides enhanced stability, protects the encapsulated drug from degradation, and enables controlled release. SLNs function by encapsulating drugs within the lipid matrix, which gradually releases the therapeutic agents through diffusion and lipid degradation [43]. This characteristic makes them suitable for conditions requiring SRs, in chronic disease managing. Their small particle size allows for improved bioavailability, enhanced permeation through biological barriers, and potential for targeting specific tissues. The clinical application of SLNs spans various therapeutic areas, including dermatology, where they improve drug penetration through the skin, and neurology, where they facilitate drug transport across the blood-brain barrier [44]. However, SLNs are associated with boundaries like low drug loading capacity and possible polymorphic transitions, which may lead to drug expulsion during storage. Advances in formulation techniques, including the use of lipid mixtures and surfactant combinations, aim to address these challenges and improve SLN performance.

8.4.3 Nanostructured Lipid Carriers (NLCs)

NLCs improve upon SLNs by combining solid and liquid lipids, creating a flexible matrix that enhances drug loading and prevents expulsion [45]. The disrupted crystalline structure increases space for drugs, providing better stability and a versatile release profile, making NLCs ideal for controlled transdermal and topical applications. Additionally, NLCs enhance drug absorption through the lymphatic system, bypassing first-pass metabolism and improving systemic bioavailability. The application of NLCs extends to various therapeutic areas, including cancer treatment, where they enable targeted delivery of chemotherapeutic agents with reduced side effects. In neurological disorders, NLCs facilitate the delivery of neuroprotective drugs with enhanced brain uptake [46]. Furthermore, NLC-based formulations have demonstrated potential in wound healing and inflammatory disorders by providing localized and sustained drug release. Despite their advantages, the complexity of NLC formulations poses challenges in large-scale production and quality control. Optimization of lipid selection, production parameters, and stabilization techniques is crucial to achieving consistent and reproducible formulations suitable for clinical applications.

8.4.4 Self-Emulsifying Drug Delivery Systems (SEDDS)

SEDDS are isotropic blends of lipids, surfactants, and co-solvents that spontaneously form fine oil-in-water (O/W) emulsions upon exposure to gastrointestinal fluids [47]. These systems have gained attention for their ability to improve the oral bioavailability of hydrophobic drugs by facilitating their dissolution and absorption in the gastrointestinal tract (GIT). The mechanism of action of SEDDS involves the dispersion of lipid droplets in the Aqueous setting of the GIT, creating a large surface area for drug absorption. This process enhances the solubilization of hydrophobic drugs, thereby improving their bioavailability. SEDDS formulations are particularly advantageous for drugs with low solubility as well as low permeability, as they bypass the dissolution step and promote direct absorption through the intestinal epithelium [48]. Applications of SEDDS include the delivery of lipophilic drugs such as cyclosporine, where they provide consistent drug absorption and reduced variability in plasma concentrations. Additionally, SEDDS have been explored for the delivery of nutraceuticals and poorly soluble natural compounds, enhancing their therapeutic potential. Challenges associated with SEDDS include the physical instability of emulsions during storage, potential drug precipitation upon dilution, and the need for specialized excipients to ensure formulation stability. Advancements in formulation strategies, such as the incorporation of S-SEDDS, aim to address these issues and improve the practicality of these systems for commercial applications.

LBDDS signifies a substantial development in pharmaceutical technology, donating solutions to vanquish the challenges of orthodox preparations. Liposomes, SLNs, NLCs, and SEDDS each provide a unique edge in terms of stability, bioavailability, and controlled drug release. While these systems have shown promising results in various therapeutic areas, additional R&D is essential to enhance their formulations, scale up production, and address existing limitations. With continuous advancements in lipid nanotechnology, LBDDS is poised to play a crucial role in the future of personalized and precision medicine.

8.5 Drug Encapsulation and Release

Efficient drug encapsulation and controlled release are critical aspects of DDS, ensuring therapeutic efficacy while reducing side effects. Various encapsulation techniques have been developed to protect APIs from degradation, enhance their bioavailability, and provide controlled release profiles [49]. Additionally, stimuli-responsive DDS has emerged as an innovative approach to achieving targeted and controlled drug delivery. This section explores the key drug encapsulation techniques and the mechanisms by which drugs are released from their carriers under specific physiological conditions.

8.5.1 Encapsulation Techniques

Drug encapsulation involves trapping therapeutic agents within a carrier matrix to enhance their stability, bioavailability, and CRs. Various techniques are employed based on drug properties, desired release profiles, and administration routes [50].

Common methods include solvent evaporation, nanoprecipitation, high-pressure homogenization, and emulsification [51]. In solvent evaporation, the drug and carrier material are dissolved in an organic solvent, emulsified in an aqueous phase, and then the solvent is removed, forming nanoparticles. This method is effective for lipophilic drugs, offering high encapsulation efficiency and tunable particle size and release properties. Nanoprecipitation is another widely used method, involving the rapid mixing of an organic solvent containing the drug and carrier material with an aqueous phase. The solvent diffuses into the aqueous medium, leading to the development of NPs [52]. This technique is particularly advantageous for hydrophobic drugs, providing narrow size distribution and high encapsulation efficiency. High-pressure homogenization is a mechanical method used to encapsulate drugs within lipid-based carriers such as SLNs and NLCs. The drug-loaded lipid phase is subjected to high shear forces under high pressure, resulting in the formation of nano-sized particles with uniform dispersion. This technique, suitable for both hydrophilic and hydrophobic drugs, ensures high stability of the encapsulated drug. The emulsification technique involves the formation of O/W or W/O (water-in-oil) emulsions to encapsulate the drug within lipid or polymeric carriers. Upon solvent removal or cooling, nanoparticles or microparticles are formed, entrapping the drug within the core [53]. This method is commonly used for controlled and sustained release formulations, particularly in transdermal and oral delivery applications. Each encapsulation technique suggests an exceptional edge in terms of drug loading, particle stability, and release kinetics, making them suitable for specific therapeutic applications. The selection of the encapsulation method depends on aspects like drug solubility, stability, and the desired release profile.

8.5.2 Drug Release Mechanisms

Once encapsulated, the delivery of the drug from the system plays a crucial role in determining its therapeutic effect. Drug release can occur through several mechanisms, including diffusion, matrix degradation, swelling, and osmotic pressure [54]. These mechanisms can be tailored to achieve immediate, sustained, or targeted release profiles based on the formulation design. Diffusion-based release occurs when the encapsulated drug gradually moves from the core of the carrier to the external environment through concentration gradients. This mechanism is commonly observed in polymeric and lipid-based nanoparticles, where the drug is slowly released over time, ensuring sustained therapeutic effects. The rate of diffusion is influenced by the porosity and hydrophobicity of the carrier material, with smaller particle sizes offering faster release rates due to increased surface area. Matrix degradation is another important release mechanism in biodegradable carriers, where the breakdown of the carrier material leads to the gradual release of the encapsulated drug [55]. This is particularly relevant in polymeric systems, where carriers such as PLA and PLGA degrade over time in response to physiological conditions, releasing the drug in a controlled manner. The degradation rate can be modified by adjusting polymer composition, molecular weight, and crystallinity.

Swelling-induced drug release occurs when hydrophilic carrier materials absorb biological fluids, leading to matrix expansion and subsequent diffusion of the drug. This mechanism is frequently used in hydrogel-based drug delivery systems, where the degree of swelling can be tailored to achieve specific release profiles in response to environmental conditions. Osmotic pressure-driven release mechanisms utilize the movement of water into the carrier system, leading to the generation of internal pressure that facilitates drug expulsion [56]. This mechanism is commonly employed in oral drug delivery systems, where osmotic pumps are designed to provide a sustained release of therapeutic agents over extended periods.

8.5.3 pH- and Temperature-Sensitive DDS

These DDS leverage physiological and external stimuli to achieve targeted and controlled drug release. pH-responsive systems, often using materials like PAA, chitosan, or Eudragit polymers, are designed to remain stable at neutral pH but release drugs in acidic environments, such as tumor microenvironments or the stomach, or in the alkaline conditions of the intestines, improving drug efficacy and minimizing systemic toxicity [57]. Similarly, temperature-sensitive systems employ thermos-responsive polymers like PNIPAAm or liposomes, which release drugs upon heating, often used in cancer therapies where localized hyperthermia enhances drug delivery to solid tumors [58]. These systems provide innovative solutions for precision medicine, reducing side effects while optimizing therapeutic outcomes.

8.5.4 Enzyme-Sensitive DDS

These systems take advantage of specific enzymes overexpressed in pathological conditions to trigger the release of drugs [59]. These systems are mainly relevant for applications in cancer, infectious diseases, and inflammatory disorders. Enzyme-responsive carriers are designed to degrade or undergo structural modifications in response to enzymatic activity, leading to targeted drug release. For instance, matrix metalloproteinases (MMPs), which are overexpressed in tumor tissues, can degrade specific peptide linkages in the carrier material, facilitating drug release at the tumor site [60]. In bacterial infections, enzyme-sensitive formulations can be designed to respond to bacterial enzymes such as lipases and proteases, enabling localized antibiotic release and reducing the risk of antimicrobial resistance. The development of advanced drug encapsulation and release strategies has significantly improved the efficacy and safety of therapeutic agents. Encapsulation techniques such as solvent evaporation, nanoprecipitation, and high-pressure homogenization provide tailored solutions for various drug delivery needs. Controlled drug release mechanisms, including diffusion, degradation, and swelling enable precise therapeutic effects, while stimuli-responsive systems offer site-specific drug delivery with minimal side effects.

8.6 Applications of LBDDS

LBDDS are gaining attention in pharmaceutical research for their ability to improve drug solubility, stability, and bioavailability. Their biocompatibility and interaction with biological membranes make them suitable for various therapeutic approaches. The versatility of lipid-based carriers, like liposomes, SLNs, NLCs, and SEDDS, has enabled their use in various drug delivery strategies. Some of the most notable applications include cancer therapy, vaccine delivery, treatment of neurological disorders, ophthalmic drug delivery, and oral and parenteral administration.

8.6.1 Cancer Therapy

LBDDS have revolutionized cancer therapy by addressing critical challenges such as poor solubility of chemotherapeutic agents, systemic toxicity, and the need for targeted drug delivery. Lipid nanoparticles offer a promising approach to delivering anticancer drugs by enhancing their circulation time, reducing off-target effects, and improving drug accumulation in tumor tissues through enhanced permeability and retention (EPR) effects [61]. Liposomes have been greatly studied for the delivery of chemotherapeutic medications such as doxorubicin and paclitaxel. Liposomal formulations provide a protective environment for the drug, reducing its degradation and enhancing its selective uptake by tumor cells [62]. Pegylated liposomes, such as Doxil, have demonstrated prolonged circulation times and increased tumor accumulation while minimizing cardiotoxicity. NLCs and SLNs have also shown significant potential in cancer therapy by providing controlled release profiles and targeted delivery. These systems can be loaded with targeting compounds such as folic acid or monoclonal antibodies to facilitate site-precise drug accumulation. Additionally, lipid-based carriers enable co-delivery of multiple therapeutic agents, including chemotherapeutics and immunomodulators, enhancing combination therapy approaches.

8.6.2 Vaccine Delivery (e.g., mRNA Vaccines)

Lipid-based delivery systems have played a crucial role in the recent success of mRNA-based vaccines [63]. LNPs have emerged as a critical technology for the delivery of mRNA vaccines, providing stability, protection, and efficient cellular uptake. LNPs facilitate the delivery of fragile mRNA molecules by encapsulating them within lipid bilayers, protecting them from enzymatic degradation and enhancing their intracellular uptake via endocytosis. The lipid components, including ionizable lipids, cholesterol, and PEGylated lipids, aid in the formation of stable nanoparticles that efficiently fuse with the target cell membranes, allowing for the release of mRNA into the cytoplasm [17]. The success of mRNA vaccines such as Moderna and Pfizer-BioNTech highlights the immense potential of lipid-based delivery systems in vaccine technology. Beyond COVID-19, LNPs are being explored for vaccines targeting infectious diseases such as influenza, HIV, and malaria, as well as therapeutic vaccines for cancer immunotherapy.

8.6.3 Neurological Disorders (Crossing the BBB)

The BBB poses a significant challenge for drug delivery to the central nervous system (CNS), limiting the effectiveness of many therapeutic agents for neurological disorders. LBDDS offer a potent method to overcome this barrier by enhancing drug permeability and facilitating targeted delivery to the brain [64]. Liposomes, SLNs, and NLCs have shown potential in delivering neuroprotective and neurotherapeutic agents for circumstances such as Alzheimer's disease, Parkinson's disease (PD), and brain tumors. NLCs, with their ability to bypass first-pass metabolism and provide sustained release, have been used for intranasal drug delivery to achieve direct nose-to-brain transport, effectively bypassing the BBB. This approach has shown promise in improving drug bioavailability and reducing systemic side effects in the treatment of neurodegenerative diseases.

8.6.4 Ophthalmic Drug Delivery

Ophthalmic drug delivery presents unique challenges due to the anatomical and physiological barriers of the eye, such as the corneal epithelium and tear turnover [65]. LBDDS offers a potent solution by enhancing drug retention time, penetration, and bioavailability. Liposomes have been researched for the delivery of an ophthalmic based delivery system, providing sustained release and improved drug distribution across ocular tissues [66]. Their expertise in encapsulating both type of drugs makes them suitable for treating numerous ocular dysfunctions, including glaucoma, dry eye disease, and ocular infections. SLNs and NLCs have also shown promise in ophthalmic applications by offering controlled drug release and improved corneal permeability. These systems can be formulated as eye drops, gels, or inserts, providing prolonged therapeutic effects with reduced dosing frequency. The incorporation of mucoadhesive polymers in lipid-based formulations further enhances their retention on the ocular surface, improving drug absorption and therapeutic efficacy.

8.6.5 Oral and Parenteral Delivery Systems

LBDDS have been widely utilized during oral and parenteral administration to improve the bioavailability of hydrophobic drugs. The gastrointestinal tract poses significant challenges to drug absorption due to enzymatic degradation and low solubility of certain drugs, which can be addressed using lipid-based carriers. SEDDS has been widely used in oral formulations to improve the solubility and absorption of hydrophobic drugs. These systems form fine O/W emulsions upon interaction with gastrointestinal fluids, facilitating drug dissolution and improving systemic bioavailability. SEDDS has been positively used in the delivery of drugs such as cyclosporine, improving therapeutic outcomes. Parenteral formulations, such as lipid emulsions and liposomes, are commonly used for intravenous drug delivery, providing a means of delivering poorly water-soluble drugs directly into the systemic circulation. Lipid emulsions are widely used for the delivery of anesthetic agents such as propofol, offering rapid onset of action and improved patient safety. Liposomal formulations for parenteral administration have also gained attention in targeted drug delivery for cancer and infectious diseases, offering controlled release and reduced systemic toxicity. Advances in lipid formulation technology, such as PEGylation and ligand functionalization, have further enhanced the potential of lipid-based carriers in systemic delivery of drugs [67].

LBDDS have demonstrated substantial potential in various therapeutic applications, ranging from cancer therapy to vaccine delivery and neurological disorders. Their capability to enhance drug solubility, stability, and targeted drug delivery (TDD) makes them valuable tools in modern pharmaceutical formulations, leading to improved treatment outcomes and patient compliance across multiple medical fields.

8.7 Technological Advances in Lipid Formulations

Technological advancements in lipid formulations have greatly improved their potential in drug delivery by addressing critical challenges related to formulation efficiency, stability, and scalability. Various advanced formulation techniques have been developed to produce lipid-based carriers with particular influence over particle size, morphology, and drug encapsulation, ensuring enhanced therapeutic efficacy [68]. High-pressure homogenization (HPH) is one of the utmost widely used techniques, which involves the application of intense pressure to force lipid dispersions through narrow orifices, resulting in nanoscale particles with uniform distribution. This method is particularly beneficial for industrial production due to its reproducibility and potential to achieve high encapsulation efficiency. Microfluidics, on the other hand, is an emerging technique that enables the precise manipulation of fluids at the microscale, facilitating the development of lipid nanoparticles with exceptional size control and composition. This technique offers advantages such as reduced solvent consumption, scalability, and rapid formulation optimization, making it ideal for personalized medicine and high-throughput screening applications [69]. Encapsulation efficiency and drug loading capacity are crucial parameters in LBDDS, as they directly influence the therapeutic performance and dosage requirements. Recent advancements in formulation strategies, such as the use of lipid blends, surfactant combinations, and co-solvents, have enabled the development of lipid carriers with improved drug loading capacity while maintaining stability and controlled release properties. The choice of lipid excipients, including solid and liquid lipids, plays a substantial part in determining the encapsulation efficiency of hydrophilic and lipophilic drugs. SLNs and NLCs have shown promising results in achieving high encapsulation efficiencies by providing a stable matrix for drug entrapment, minimizing premature release, and protecting sensitive bioactive compounds from degradation. Additionally, surface modifications with polymers, targeting ligands, and PEG have further enhanced the drug retention and circulation time of lipid-based carriers, thus reducing side effects [70]. Maintaining the stability of lipid formulations throughout their shelf life is a critical aspect of their successful application in pharmaceuticals. Lipid-based systems are prone to degradation due to aspects such as oxidation, hydrolysis, and aggregation, which can compromise their therapeutic potential. To tackle these limitations, several stabilization strategies have been developed, including the incorporation of antioxidants, such as tocopherols and butylated hydroxytoluene (BHT), to prevent oxidative degradation of lipid components. Additionally, freeze-drying and spray-drying techniques have been employed to convert lipid nanoparticles into dry powder forms, enhancing their stability and ease of handling. Scaling up lipid-based drug delivery systems for commercial production poses significant challenges, as the transition from laboratory-scale to industrial-scale manufacturing requires the optimization of multiple parameters to ensure batch-to-batch consistency, regulatory compliance, and cost-effectiveness [71]. Advances in large-scale production technologies, such as continuous manufacturing, supercritical fluid technology, and advanced emulsification methods, have facilitated the efficient production of lipid-based formulations with high throughput and reproducibility. Regulatory agencies, including the US FDA and the EMA, have developed procedures for the quality, safety, and efficacy of lipid-based therapeutics, necessitating stringent quality control measures during scale-up. Furthermore, process analytical technologies (PAT) have been integrated into production workflows to monitor critical quality attributes in real time, ensuring compliance with GMP and regulatory requirements.

Overall, the continuous evolution of lipid formulation technologies has paved pioneer for the progress of highly efficient, stable, and scalable DDS. These advancements are driving the commercialization of lipid-based therapeutics across various therapeutic areas, including oncology, infectious diseases, and neurological disorders, offering improved treatment outcomes and patient compliance. With ongoing research and innovation, lipid formulations are anticipated to play a substantial role in the imminent drug delivery, enabling the development of targeted, personalized, and more effective therapeutic interventions.

8.8 Challenges and Limitations

LBDDS offer numerous advantages in enhancing drug solubility, bioavailability, and targeted delivery; however, several challenges and limitations hinder their extensive acceptance. Prime challenges are lipid oxidation and stability issues, as lipids, particularly unsaturated fatty acids, are prone to oxidative degradation when exposed to oxygen, light, and heat [72]. Oxidation can compromise the stability of the formulation, causing the development of reactive derivatives that may affect the efficacy and safety of the encapsulated drug. Hydrolytic degradation is another concern, as it can result in the breakdown of lipid molecules into free fatty acids and glycerol, potentially altering the physiochemical characters of the drug delivery system [73]. To mitigate these issues, the inclusion of antioxidants, such as tocopherols and synthetic stabilizers, as well as the use of protective packaging, is often employed; however, achieving long-term stability remains a significant challenge. Additionally, manufacturing and scalability challenges pose significant barriers to the commercial production of LBDDS. This requires cautious optimization of process parameters, such as temperature, mixing conditions, and particle size uniformity, to ensure batch-to-batch consistency. Techniques such as high-pressure homogenization and microfluidics offer precise control over formulation parameters, but their scalability can be complex and resource-intensive. The requirement for specialized equipment and expertise further complicates the manufacturing process, increasing production costs and limiting accessibility for smaller pharmaceutical companies. Regulatory approval and safety concerns also present significant hurdles in the development and commercialization of lipid-based formulations. Agencies, such as the US FDA and the EMA, impose stringent guidelines on the quality, safety, and efficacy of these formulations. Safety concerns, including the potential for lipid accumulation, immune responses, and variability in patient response, must also be thoroughly addressed to gain regulatory approval. Furthermore, cost implications and market viability present challenges that impact the widespread adoption of LBDDS. The high cost of raw materials, specialized processing equipment, and stringent QC measures contribute to the overall expense of developing and producing lipid-based formulations. These cost considerations can limit the affordability and accessibility of LBDDS, particularly in low and middle income regions where resources are constrained. Additionally, market competition from established drug delivery technologies, such as polymer-based systems, poses a challenge in positioning lipid-based formulations as a preferred option. Contempt these limitations, current research efforts are aimed at improving formulation stability, optimizing manufacturing processes, and reducing costs to improve the feasibility and market acceptance of LBDDS. Addressing these limitations through innovative approaches and strategic collaborations will be crucial in attaching the potential of lipid-based carriers in modern medicine.

8.9 Future Trends in Lipid-Based Drug Delivery

The future of LBDDS is set to witness remarkable advancements driven by innovations in lipid nanoparticles, the development of lipid hybrid systems for combination therapies, the rise of personalized medicine, and emerging applications in immunotherapy and gene therapy. Innovations in lipid nanoparticles are paving the way for more effective drug delivery platforms with improved stability, and targeted release capabilities. Advances in lipid engineering have led to the design of sophisticated nanostructures, such as SLNs and NLCs, which offer controlled drug release, enhanced encapsulation efficiency, and improved pharmacokinetic profiles [74]. Furthermore, the integration of stimuli-responsive components into lipid nanoparticles is being explored to achieve site-specific drug unleash activated by environmental influences such as temperature and enzymatic activity, thus minimizing off-target consequences and refining treatment results. Another promising trend in LBDDs is the advance of lipid hybrid systems for combination therapies. These systems combine lipids with polymers, inorganic nanoparticles, or biomolecules to create multifunctional carriers capable of co-delivering multiple therapeutic agents. Lipid-polymer hybrid nanoparticles, for example, offer the structural integrity of polymers with the biocompatibility and biodegradability of lipids, enhancing drug stability and controlled release properties. Such hybrid systems are particularly beneficial for combination therapies in cancer treatment, where co-delivery of chemotherapeutic agents and targeted inhibitors can enhance synergistic effects and overcome drug resistance. As the field of precision medicine continues to evolve, personalized lipid-based drug delivery systems are gaining traction. Advances in nanotechnology and bioinformatics enable the customization of lipid carriers for targeted drug delivery based on individual genetic and physiological profiles. Functionalizing lipid nanoparticles with ligands like antibodies, peptides, or aptamers allows precise targeting of specific tissues or receptors, improving drug accumulation while reducing systemic exposure. This approach holds significant promise in the treating of chronic diseases such as cancer, cardiovascular diseases (CVDs), and neurodegenerative conditions. Additionally, the application of lipid-based systems in immunotherapy and gene therapy is rapidly emerging as a groundbreaking area of research. Lipid nanoparticles have shown immense potential as delivery vehicles for nucleic acid-based therapies, including mRNA, siRNA, and DNA, by protecting these fragile molecules from enzymatic degradation and facilitating efficient cellular uptake. The success of lipid nanoparticle-based COVID-19 mRNA vaccines has demonstrated the feasibility of using lipid carriers for gene therapy applications, paving the way for further advancements in treating genetic disorders and cancer. Immunotherapeutic applications are also gaining attention, with lipid-based formulations being developed to deliver immune-modulating mediators, such as cytokines and checkpoint inhibitors, to improve the response of immunity against diseases. As research in lipid-based drug delivery continues to progress, the integration of AI and machine learning (ML) in formulation design as well as optimization is expected to accelerate the progress of more operative and personalized treatments, further expanding the potential of lipid-based carriers in modern medicine.

8.10 Conclusion

Natural source lipids have emerged as a promising platform for advanced drug delivery systems. Lipid-based carriers, such as liposomes, SLNs, NLCs, and SEDDS, offer significant advantages in improving drug solubility, stability, and targeted delivery. These systems have demonstrated potential across various therapeutic applications, including cancer treatment, vaccine delivery, neurological disorders, and ophthalmic care. Despite the numerous benefits, challenges such as lipid oxidation, scalability issues, and regulatory hurdles must be addressed to facilitate widespread adoption. Ongoing research efforts are focused on optimizing formulation strategies, integrating advanced technologies such as nanotechnology and artificial intelligence, and developing personalized lipid-based therapies. Future advancements in lipid-based drug delivery systems are poised to revolutionize treatment modalities, enhancing therapeutic efficacy while minimizing adverse effects. With continued innovation, natural lipids hold immense potential in transforming drug delivery and refining patient results across a broad spectrum of medical uses.

Acknowledgment The authors express their gratitude toward the School of Health Sciences and Technology MIT WPU and STES's Smt. Kashibai Navale College of Pharmacy, Kondhwa, Pune, for offering facilities needed to complete this chapter.

References

1. Chen Q, Yang Z, Liu H, et al. Novel drug delivery systems: an important direction for drug innovation research and development. Pharmaceutics. 2024;16(5):674–710. https://doi.org/10.3390/pharmaceutics16050674.
2. Wen H, Jung H, Li X. Drug delivery approaches in addressing clinical pharmacology-related issues: opportunities and challenges. AAPS J. 2015;17(6):1327–40. https://doi.org/10.1208/s12248-015-9814-9.
3. Ezike TC, Okpala US, Onoja UL, et al. Advances in drug delivery systems, challenges and future directions. Heliyon. 2023;9(6):e17488. https://doi.org/10.1016/j.heliyon.2023.e17488.
4. Adepu S, Ramakrishna S. Controlled drug delivery systems: current status and future directions. Molecules. 2021;26(19):5905–50. https://doi.org/10.3390/molecules26195905.
5. Puri A, Loomis K, Smith B, et al. Lipid-based nanoparticles as pharmaceutical drug carriers: from concepts to clinic. Crit Rev Ther Drug Carrier Syst. 2009;26(6):523–80.
6. Wilson B, Geetha KM. Lipid nanoparticles in the development of mRNA vaccines for COVID-19. J Drug Deliv Sci Technol. 2022;74:103553. https://doi.org/10.1016/j.jddst.2022.103553.
7. Tenchov R, Bird R, Curtze AE, et al. Lipid nanoparticles – from liposomes to mRNA vaccine delivery, a landscape of research diversity and advancement. ACS Nano. 2021;15(11):16982–7015. https://doi.org/10.1021/acsnano.1c04996.
8. Swetha K, Kotla NG, Tunki L, et al. Recent advances in the lipid nanoparticle-mediated delivery of mRNA vaccines. Vaccines (Basel). 2023;11(3):658–76. https://doi.org/10.3390/vaccines11030658.
9. van Hoogevest P, Wendel A. The use of natural and synthetic phospholipids as pharmaceutical excipients. Eur J Lipid Sci Technol. 2014;116(9):1088–107. https://doi.org/10.1002/ejlt.201400219.
10. Dobreva M, Stefanov S, Andonova V. Natural lipids as structural components of solid lipid nanoparticles and nanostructured lipid carriers for topical delivery. Curr Pharm Des. 2020;26(36):4524–35.
11. Alfutaimani AS, Alharbi NK, Alahmari AS, et al. Exploring the landscape of lipid nanoparticles (LNPs): a comprehensive review of LNPs types and biological sources of lipids. Int J Pharm X. 2024;8:100305. https://doi.org/10.1016/j.ijpx.2024.100305.
12. Han X, Alameh MG, Xu Y, et al. Optimization of the activity and biodegradability of ionizable lipids for mRNA delivery via directed chemical evolution. Nat Biomed Eng. 2024;8(11):1412–24. https://doi.org/10.1038/s41551-024-01267-7.
13. Amiri S, Pashizeh F, Moeinabadi-Bidgoli K, et al. Co-encapsulation of hydrophilic and hydrophobic drugs into niosomal nanocarrier for enhanced breast cancer therapy: in silico and in vitro studies. Environ Res. 2023;239(Pt 2):117292. https://doi.org/10.1016/j.envres.2023.117292.

14. Hou X, Zaks T, Langer R, et al. Lipid nanoparticles for mRNA delivery. Nat Rev Mater. 2021;6(12):1078–94. https://doi.org/10.1038/s41578-021-00358-0.
15. Fitriani EW, Avanti C, Rosana Y, et al. Nanostructured lipid carriers: a prospective dermal drug delivery system for natural active ingredients. Pharmacia. 2024;71:1–15. https://doi.org/10.3897/pharmacia.71.e115849.
16. Patel P, Garala K, Singh S, et al. Lipid-based nanoparticles in delivering bioactive compounds for improving therapeutic efficacy. Pharmaceuticals (Basel). 2024;17(3):329–53. https://doi.org/10.3390/ph17030329.
17. Hald Albertsen C, Kulkarni JA, Witzigmann D, et al. The role of lipid components in lipid nanoparticles for vaccines and gene therapy. Adv Drug Deliv Rev. 2022;188:114416. https://doi.org/10.1016/j.addr.2022.114416.
18. Ashfaq R, Rasul A, Asghar S, et al. Lipid nanoparticles: an effective tool to improve the bioavailability of nutraceuticals. Int J Mol Sci. 2023;24(21):15764. https://doi.org/10.3390/ijms242115764.
19. Napier JA, Haslam RP, Beaudoin F, et al. Understanding and manipulating plant lipid composition: metabolic engineering leads the way. Curr Opin Plant Biol. 2014;19(100):68–75. https://doi.org/10.1016/j.pbi.2014.04.001.
20. Qin P, Wang T, Luo Y. A review on plant-based proteins from soybean: health benefits and soy product development. J Agric Food Res. 2022;7:100265. https://doi.org/10.1016/j.jafr.2021.100265.
21. Ehsandoost E, Eskandari MH, Keramat M, et al. Antioxidant activity and mechanism of action of phycocyanin in bulk sunflower oil and respective oil-in-water emulsion. Curr Res Food Sci. 2025;10:100981. https://doi.org/10.1016/j.crfs.2025.100981.
22. Cui J, Yang Z, Xu Y, et al. Lipidomics insight on differences in lipid profiles and phytosterol compositions of coconut oils extracted by classical and green solvents. Food Res Int. 2023;174(part 2):113653. https://doi.org/10.1016/j.foodres.2023.113653.
23. Bandara N, Chalamaiah M. Bioactives from agricultural processing by-products. Academic Press; 2019. p. 472–80. https://doi.org/10.1016/B978-0-08-100596-5.22408-6.
24. Fang X, Ge X-Y, Qin Z, et al. Nanoencapsulation of flaxseed oil: fabrication, characterization, and storage stability. LWT. 2024;214:117113. https://doi.org/10.1016/j.lwt.2024.117113.
25. Givens I. Animal nutrition and lipids in animal products and their contribution to human intake and health. Nutrients. 2009;1(1):71–82. https://doi.org/10.3390/nu1010071.
26. Prates JAM. The role of meat lipids in nutrition and health: balancing benefits and risks. Nutrients. 2025;17(2):350–65. https://doi.org/10.3390/nu17020350.
27. Zhao F, Li R, Liu Y, et al. Perspectives on lecithin from egg yolk: extraction, physicochemical properties, modification, and applications. Front Nutr. 2023;9:1082671. https://doi.org/10.3389/fnut.2022.1082671.
28. Kuang H, Yang F, Zhang Y, et al. The impact of egg nutrient composition and its consumption on cholesterol homeostasis. Cholesterol. 2018;2018:6303810. https://doi.org/10.1155/2018/6303810.
29. Leal AF, Suarez DA, Echeverri-Peña OY, et al. Sphingolipids and their role in health and disease in the central nervous system. Adv Biol Regul. 2022;85:100900. https://doi.org/10.1016/j.jbior.2022.100900.
30. Réhault-Godbert S, Guyot N, Nys Y. The golden egg: nutritional value, bioactivities, and emerging benefits for human health. Nutrients. 2019;11(3):684–710. https://doi.org/10.3390/nu11030684.
31. Ruwizhi N, Aderibigbe BA. The efficacy of cholesterol-based carriers in drug delivery. Molecules. 2020;25(18):4330–70. https://doi.org/10.3390/molecules25184330.
32. Burri L, Hoem N, Banni S, et al. Marine omega-3 phospholipids: metabolism and biological activities. Int J Mol Sci. 2012;13(11):15401–19. https://doi.org/10.3390/ijms131115401.
33. Lauritzen L, Brambilla P, Mazzocchi A, et al. DHA effects in brain development and function. Nutrients. 2016;8(1):6–23. https://doi.org/10.3390/nu8010006.
34. Das P, Dutta A, Panchali T, et al. Advances in therapeutic applications of fish oil: a review. Meas Food. 2024;13:100142. https://doi.org/10.1016/j.meafoo.2024.100142.

35. Santigosa E, Brambilla F, Milanese L. Microalgae oil as an effective alternative source of EPA and DHA for gilthead seabream (*Sparus aurata*) aquaculture. Animals (Basel). 2021;11(4):971–88. https://doi.org/10.3390/ani11040971.
36. McCullough PJ, Lehrer DS. Vitamin D, cod liver oil, sunshine, and phototherapy: safe, effective and forgotten tools for treating and curing tuberculosis infections – a comprehensive review. J Steroid Biochem Mol Biol. 2018;177:21–9. https://doi.org/10.1016/j.jsbmb.2017.07.027.
37. Preeti, Sambhakar S, Malik R, et al. Lipid horizons: recent advances and future prospects in LBDDS for oral administration of antihypertensive agents. Int J Hypertens. 2024;2024:2430147. https://doi.org/10.1155/2024/2430147.
38. Jaiswal P, Gidwani B, Vyas A. Nanostructured lipid carriers and their current application in targeted drug delivery. Artif Cells Nanomed Biotechnol. 2016;44(1):27–40. https://doi.org/10.3109/21691401.2014.909822.
39. Liu P, Chen G, Zhang J. A review of liposomes as a drug delivery system: current status of approved products, regulatory environments, and future perspectives. Molecules. 2022;27(4):1372–95. https://doi.org/10.3390/molecules27041372.
40. Gu Z, Da Silva CG, Van der Maaden K, et al. Liposome-based drug delivery systems in cancer immunotherapy. Pharmaceutics. 2020;12(11):1054–79. https://doi.org/10.3390/pharmaceutics12111054.
41. Tretiakova DS, Vodovozova EL. Liposomes as adjuvants and vaccine delivery systems. Biochem (Mosc) Suppl Ser A Membr Cell Biol. 2022;16(1):1–20. https://doi.org/10.1134/S1990747822020076.
42. Satapathy MK, Yen TL, Jan JS, et al. Solid lipid nanoparticles (SLNs): an advanced drug delivery system targeting brain through BBB. Pharmaceutics. 2021;13(8):1183–219. https://doi.org/10.3390/pharmaceutics13081183.
43. Akanda M, Mithu SH, Douroumis D. Solid lipid nanoparticles: an effective lipid-based technology for cancer treatment. J Drug Deliv Sci Technol. 2023;86:104709. https://doi.org/10.1016/j.jddst.2023.104709.
44. Geszke-Moritz M, Moritz. Solid lipid nanoparticles as attractive drug vehicles: composition, properties and therapeutic strategies. Mater Sci Eng C. 2016;68:982–94. https://doi.org/10.1016/j.msec.2016.05.119.
45. Naseri N, Valizadeh H, Zakeri-Milani P. Solid lipid nanoparticles and nanostructured lipid carriers: structure, preparation and application. Adv Pharm Bull. 2015;5(3):305–13. https://doi.org/10.15171/apb.2015.043.
46. Khan AA, Mudassir J, Mohtar N, et al. Advanced drug delivery to the lymphatic system: lipid-based nanoformulations. Int J Nanomedicine. 2013;8:2733–44. https://doi.org/10.2147/IJN.S41521.
47. Gursoy RN, Benita S. Self-emulsifying drug delivery systems (SEDDS) for improved oral delivery of lipophilic drugs. Biomed Pharmacother. 2004;58(3):173–82. https://doi.org/10.1016/j.biopha.2004.02.001.
48. Uttreja P, Karnik I, Youssef AAA, et al. Self-emulsifying drug delivery systems (SEDDS): transition from liquid to solid – a comprehensive review of formulation, characterization, applications, and future trends. Pharmaceutics. 2025;17(1):63–96. https://doi.org/10.3390/pharmaceutics17010063.
49. Alu'datt MH, Alrosan M, Gammoh S, et al. Encapsulation-based technologies for bioactive compounds and their application in the food industry: a roadmap for food-derived functional and health-promoting ingredients. Food Biosci. 2022;50(part A):101971. https://doi.org/10.1016/j.fbio.2022.101971.
50. Lammari N, Tarhini M, Miladi K, et al. Encapsulation methods of active molecules for drug delivery. Academic Press; 2021. p. 289–306. https://doi.org/10.1016/B978-0-12-819838-4.00008-0.
51. Sneha K, Kumar A. Nanoemulsions: techniques for the preparation and the recent advances in their food applications. Innovative Food Sci Emerg Technol. 2022;76:102914. https://doi.org/10.1016/j.ifset.2021.102914.

52. Rivas CJM, Tarhini M, Badri W, et al. Nanoprecipitation process: from encapsulation to drug delivery. Int J Pharm. 2017;532(1):66–81. https://doi.org/10.1016/j.ijpharm.2017.08.064.
53. Duong VA, Nguyen TT, Maeng HJ. Preparation of solid lipid nanoparticles and nanostructured lipid carriers for drug delivery and the effects of preparation parameters of solvent injection method. Molecules. 2020;25(20):4781–817. https://doi.org/10.3390/molecules25204781.
54. Bruschi ML. Strategies to modify the drug release from pharmaceutical systems. Woodhead; 2015. p. 37–62. https://doi.org/10.1016/B978-0-08-100092-2.00004-7.
55. Lee JH, Yeo Y. Controlled drug release from pharmaceutical nanocarriers. Chem Eng Sci. 2015;125:75–84. https://doi.org/10.1016/j.ces.2014.08.046.
56. Bruhn BR, Schroeder TB, Li S, et al. Osmosis-based pressure generation: dynamics and application. PLoS One. 2014;9(3):e91350. https://doi.org/10.1371/journal.pone.0091350.
57. Ding H, Tan P, Fu S, et al. Preparation and application of pH-responsive drug delivery systems. J Control Release. 2022;348:206–38. https://doi.org/10.1016/j.jconrel.2022.05.056.
58. Abuwatfa WH, Awad NS, Pitt WG, et al. Thermosensitive polymers and thermo-responsive liposomal drug delivery systems. Polymers (Basel). 2022;14(5):925–51. https://doi.org/10.3390/polym14050925.
59. Lee Y, Thompson DH. Stimuli-responsive liposomes for drug delivery. Wiley Interdiscip Rev Nanomed Nanobiotechnol. 2017;9(5):e1450. https://doi.org/10.1002/wnan.1450.
60. Yao Q, Kou L, Tu Y, et al. MMP-responsive 'smart' drug delivery and tumor targeting. Trends Pharmacol Sci. 2018;39(8):766–81. https://doi.org/10.1016/j.tips.2018.06.003.
61. Yaghmur A, Østergaard J, Mu H. Lipid nanoparticles for targeted delivery of anticancer therapeutics: recent advances in development of siRNA and lipoprotein-mimicking nanocarriers. Adv Drug Deliv Rev. 2023;203:115136. https://doi.org/10.1016/j.addr.2023.115136.
62. Fulton MD, Najahi-Missaoui W. Liposomes in cancer therapy: how did we start and where are we now. Int J Mol Sci. 2023;24(7):6615–38. https://doi.org/10.3390/ijms24076615.
63. Vasu M, Yadav P, Satyanarayana Y, et al. Golden threads of lipid nanoparticles (LNPs) contributing to COVID-19 mRNA vaccine: a review. BioNanoScience. 2024;14:3430–9. https://doi.org/10.1007/s12668-024-01589-5.
64. Kesharwani R, Jaiswal P, Patel DK, et al. Lipid-based drug delivery system (LBDDS): an emerging paradigm to enhance oral bioavailability of poorly soluble drugs. Biomed Mater Devices. 2023;1:648–63. https://doi.org/10.1007/s44174-022-00041-0.
65. Chandel A, Kandav G. Insights into ocular therapeutics: a comprehensive review of anatomy, barriers, diseases and nanoscale formulations for targeted drug delivery. J Drug Deliv Sci Technol. 2024;97:105785. https://doi.org/10.1016/j.jddst.2024.105785.
66. Mishra GP, Bagui M, Tamboli V, et al. Recent applications of liposomes in ophthalmic drug delivery. J Drug Deliv. 2011;2011:863734. https://doi.org/10.1155/2011/863734.
67. Kalepu S, Manthina M, Padavala V. Oral lipid-based drug delivery systems – an overview. Acta Pharm Sin B. 2013;3(6):361–72. https://doi.org/10.1016/j.apsb.2013.10.001.
68. Porter CJ, Williams HD, Trevaskis NL. Recent advances in lipid-based formulation technology. Pharm Res. 2013;30(12):2971–5. https://doi.org/10.1007/s11095-013-1229-7.
69. Maged A, Abdelbaset R, Mahmoud AA, et al. Merits and advances of microfluidics in the pharmaceutical field: design technologies and future prospects. Drug Deliv. 2022;29(1):1549–70. https://doi.org/10.1080/10717544.2022.2069878.
70. Priya S, Desai VM, Singhvi G. Surface modification of lipid-based nanocarriers: a potential approach to enhance targeted drug delivery. ACS Omega. 2022;8(1):74–86. https://doi.org/10.1021/acsomega.2c05976.
71. Gazzi R, Gelli R, Eugster R, et al. Transitioning the production of lipidic mesophase-based delivery systems from lab-scale to robust industrial manufacturing following a risk-based quality by design approach augmented by artificial intelligence. J Colloid Interface Sci. 2025;678(part C):595–607. https://doi.org/10.1016/j.jcis.2024.09.099.
72. Musakhanian J, Rodier JD, Dave M. Oxidative stability in lipid formulations: a review of the mechanisms, drivers, and inhibitors of oxidation. AAPS PharmSciTech. 2022;23(5):151. https://doi.org/10.1208/s12249-022-02282-0.

73. Woodard LN, Grunlan MA. Hydrolytic degradation and erosion of polyester biomaterials. ACS Macro Lett. 2018;7(8):976–82. https://doi.org/10.1021/acsmacrolett.8b00424.
74. Tapeinos C, Battaglini M, Ciofani G. Advances in the design of solid lipid nanoparticles and nanostructured lipid carriers for targeting brain diseases. J Control Release. 2017;264:306–32. https://doi.org/10.1016/j.jconrel.2017.08.033.

Gums and Mucilages: Versatile Natural Polymers

9

Poonam Taru, Dhanashree Walunj, Sabeena Sayd, and Rupesh Saindane

Abbreviations

MAE	Microwave-Assisted Extraction
UAE	Ultrasound-Assisted Extraction
SFE	Supercritical Fluid Extraction
EAE	Enzyme-Assisted Extraction
PHWE	Pressurized Hot Water Extraction
API	Active Pharmaceutical Ingredients
HPMC	Hydroxypropyl Methylcellulose
PEG	Polyethylene Glycol

P. Taru (✉)
School of Pharmacy, Vishwakarma University,
Kondhwa Pune, Maharashtra, India
e-mail: Poonam.taru@vupune.ac.in

D. Walunj
Department of Pharmaceutical Chemistry, PES Modern College of Pharmacy,
Pune, Maharashtra, India

S. Sayd
Department of Pharmacognosy, School of Pharmacy, Vishwakarma University,
Pune, Maharashtra, India
e-mail: sabeena.syed@vupune.ac.in

R. Saindane
Lotus Pharmaceutical Co, Ltd, Nantou, Taiwan

J. S et al. (eds.), *Innovative Pharmaceutical Excipients: Natural Sources*,
https://doi.org/10.1007/978-981-96-7959-1_9

9.1 Introduction

Gums and mucilages are natural polymers derived primarily from plants, though some originate from microbial or marine sources. These substances exhibit excellent biocompatibility, biodegradability, and non-toxic properties, making them indispensable in pharmaceutical, cosmetic, and food industries. While both are hydrophilic carbohydrates, gums are typically exudates from plants, and mucilages are intracellular polysaccharides produced in specific plant tissues [7].

9.1.1 Definition of Gums and Mucilages

9.1.1.1 Gums

These are natural polysaccharides that are soluble in water or form a viscous solution. They are commonly obtained as plant exudates, seed extracts, or from microbial sources. Gums are extensively used in food, pharmaceuticals, and industrial applications due to their gelling, emulsifying, and stabilizing properties, for example, gum arabic, guar gum, and xanthan gum.

9.1.1.2 Mucilages

These are complex polysaccharides stored in plant cells, particularly in seeds, roots, and barks. Unlike gums, mucilages are usually insoluble in alcohol and exhibit gel-like swelling properties when in contact with water. They function as water reservoirs in plants and are widely utilized in pharmaceuticals for their soothing and hydrating effects [3, 31]. Difference between gum and mucilage is mentioned in Table 9.1.

9.1.2 Classification of Gums and Mucilages (Refer Fig. 9.1)

9.1.2.1 Based on Source

Plant-Derived

Exudate gums: Gum arabic, gum tragacanth.
Seed gums: Guar gum, locust bean gum.
Extracted mucilages: Aloe vera, psyllium.

Microbial-Derived

Xanthan gum, gellan gum.

Table 9.1 Difference between gums and mucilages [7]

	Gums	Mucilages
Origin	Plant exudates (pathological)	Intracellular or seeds (physiological)
Function	Adhesive, protective	Water retention, dispersal
Solubility	Soluble in water/form viscous solution	Swell or partially dissolve
Examples	Gum arabic, xanthan gum	Aloe vera, psyllium mucilage

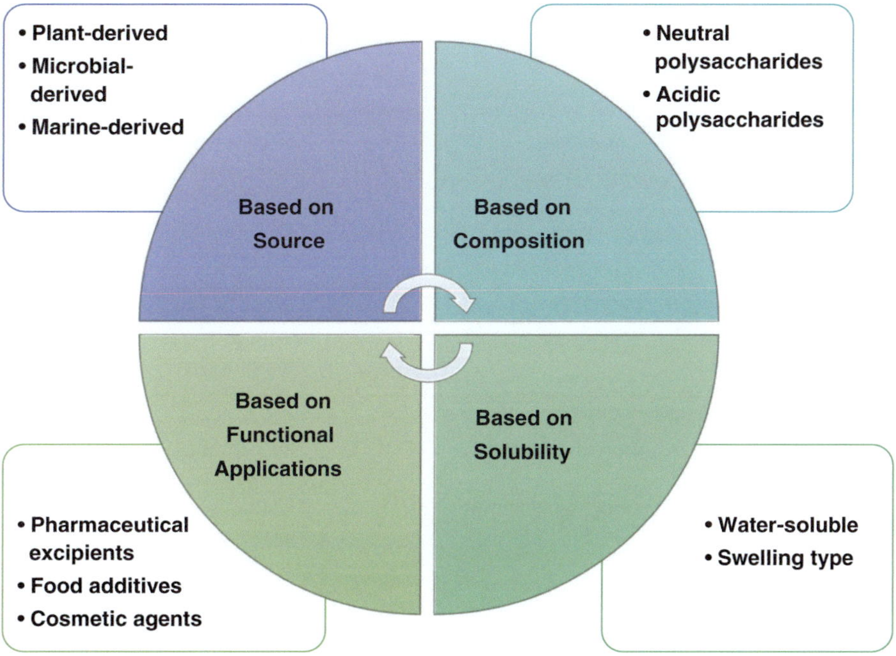

Fig. 9.1 Classification of gums and mucilages

Marine-Derived
Alginates, carrageenan.

9.1.2.2 Based on Composition

Neutral polysaccharides: Guar gum, locust bean gum.
Acidic polysaccharides: Gum arabic, alginate (contain uronic acids).

9.1.2.3 Based on Solubility

Water-soluble: Gum arabic, xanthan gum.
Swelling type: Psyllium mucilage, tragacanth.

9.1.2.4 Based on Functional Applications

Pharmaceutical excipients: Controlled-release agents (guar gum).
Food additives: Thickeners (carrageenan), stabilizers (xanthan gum).
Cosmetic agents: Moisturizers (aloe vera mucilage) [3].

9.1.3 Chemical Composition and Properties

9.1.3.1 Chemical Structure

Gums and mucilages are high-molecular-weight polysaccharides composed of repeating monosaccharide units such as glucose, mannose, galactose, arabinose,

Table 9.2 Comparative analysis of common natural polymers

Polymer	Source	Monosaccharides	Key functional groups	Properties	References
Gum arabic	Acacia tree exudate	Arabinose, galactose	-OH, -COOH	High solubility, emulsifying, stabilizing	[7]
Guar gum	Guar seeds	Mannose, galactose	-OH	High viscosity, thickening	[19]
Alginate	Brown seaweed	Mannuronic, guluronic acids	-COOH	Gel-forming, biocompatible	[1]
Carrageenan	Red algae	Galactose	-SO3H	Gelling, thickening, stabilizing	[23]
Psyllium mucilage	Plantain seeds	Xylose, arabinose, galacturonic acid	-OH, -COOH	Swelling, bulk-forming laxative	[18]
Aloe vera mucilage	Aloe leaves	Glucose, mannose	-OH	Hydrating, soothing	[31]

xylose, and rhamnose. These sugars are linked through glycosidic bonds, forming linear or branched structures that determine their functional properties. The presence of uronic acids (e.g., glucuronic and galacturonic acid) enhances their hydrophilicity and gel-forming ability. Functional groups like hydroxyl (-OH) contribute to their water solubility, while carboxyl (-COOH) and sulfate (-SO_3H) groups influence ionic interactions and gelation (e.g., in carrageenan). Structural variations impact their behavior—highly branched polymers like guar gum increase viscosity, whereas linear polymers like alginate exhibit superior gel-forming properties (refer to Table 9.2). These characteristics make gums and mucilages versatile excipients in pharmaceutical applications, supporting functions such as binding, thickening, stabilizing, and controlled drug release. Their natural origin, biocompatibility, and biodegradability further enhance their suitability for sustainable drug formulations [25, 32].

9.1.3.2 Physical Properties

Gums and mucilages exhibit strong hydrophilicity due to their polysaccharide structure, allowing them to absorb water and form gels, solutions, or dispersions. This property makes them highly valuable as thickeners, emulsifiers, and stabilizers in pharmaceutical formulations. Their viscosity is influenced by molecular weight, concentration, and degree of branching, with higher molecular weight and increased branching leading to greater viscosity, a crucial factor in controlled drug release systems [29]. The gel-forming ability of these biopolymers, as seen in alginate and carrageenan, depends on their interaction with divalent cations or specific environmental conditions such as pH and temperature. This property is widely exploited in wound dressings, encapsulation technologies, and sustained-release formulations. Certain gums and mucilages, such as psyllium husk, exhibit significant swelling, which enhances their application in bulk-forming laxatives and mucoadhesive drug delivery systems by facilitating prolonged retention and localized drug action. Thermal stability is another important characteristic, as most gums and mucilages can withstand specific temperatures without degradation, making them suitable for pharmaceutical processes involving heat and sterilization. Additionally, their stability across a broad pH range, as observed in xanthan and guar gum, enhances their versatility in acidic and basic formulations. Compatibility with other excipients and active pharmaceutical ingredients (APIs) is another advantage, as these natural polymers blend well without causing precipitation or degradation, ensuring the stability of complex formulations. Their hydroxyl groups contribute to swelling and emulsification, while the overall thermal stability of these biopolymers depends on their molecular structure and source [6].

9.1.3.3 Biological Compatibility

Non-toxic and metabolized easily by living systems. Some gums exhibit prebiotic activity, supporting gut microbiota.

9.1.4 Importance of Natural Polymers in Pharmaceuticals

Natural polymers have gained significant attention in pharmaceutical sciences due to their unique properties, which make them ideal for diverse applications. Their importance stems from their availability, versatility, biocompatibility, and sustainable nature.

9.1.4.1 Biodegradability

Natural polymers degrade into non-toxic byproducts, reducing environmental impact and enhancing patient safety. Examples include alginate and chitosan.

9.1.4.2 Biocompatibility

Their non-toxic and non-irritating nature ensures minimal immune response when used in drug formulations. For instance, hyaluronic acid is widely used in injectable formulations [3].

9.1.4.3 Versatile Applications

Drug Delivery Systems

Natural polymers play a crucial role in advanced drug delivery systems, particularly in controlled and sustained release formulations. Polymers such as guar gum and xanthan gum are widely utilized to regulate drug release, ensuring prolonged therapeutic effects. Additionally, mucoadhesive drug delivery systems incorporate polymers like chitosan to enhance drug bioavailability by improving adhesion to mucosal surfaces. These natural excipients contribute to the development of more effective and patient-friendly pharmaceutical formulations [17].

Tissue Engineering
Natural polymers like collagen and gelatin are used to create scaffolds for tissue regeneration.

Wound Healing
Polymers such as aloe vera and pectin are incorporated into hydrogels and dressings for promoting wound healing.

Regulatory Acceptance
Natural polymers are widely accepted by regulatory authorities, as they align with current trends in developing safer and eco-friendly pharmaceuticals.

Cost-Effectiveness
Compared to synthetic polymers, natural polymers are often more economical due to abundant availability.

9.1.5 Overview of Natural Sources of Polymers

Natural polymers are derived from diverse biological sources, offering a wide range of functionalities suited for pharmaceutical and industrial applications. These sources include plants, animals, microorganisms, and marine organisms, each contributing unique properties and applications.

9.1.5.1 Plant-Based Sources

Plants serve as the primary source of natural polymers due to their abundance and structural diversity. Plant-derived polymers include gums, such as gum arabic, guar gum, and locust bean gum, which function as stabilizers, emulsifiers, and binding agents in pharmaceutical formulations. Mucilages, obtained from seeds (e.g., psyllium), leaves (e.g., *Aloe vera*), and roots (e.g., *Mimosa pudica*), exhibit hydrophilic properties, forming gels or viscous solutions that enhance drug delivery systems. Pectins, primarily extracted from citrus peels and apple pomace, are widely utilized as gelling agents in both pharmaceutical and food applications. Additionally, starches derived from maize, wheat, and potatoes serve as essential excipients, acting as fillers and disintegrants in tablet formulations [3, 11, 30].

9.1.5.2 Animal-Based Sources

Natural polymers derived from animals are widely utilized in pharmaceuticals and biomedical applications due to their biocompatibility and unique bioactive properties. Collagen and gelatin, obtained from animal skin, bones, and connective tissues, play a crucial role in wound healing, tissue engineering, and as encapsulating agents in drug formulations. Chitosan, extracted from the exoskeleton of crustaceans such as shrimp and crab, exhibits antimicrobial properties and is extensively employed in drug delivery systems and wound care. Hyaluronic acid, naturally present in connective tissues, is commonly used in injectable formulations for joint disorders and cosmetic applications, owing to its hydrating and viscoelastic properties [12, 14].

9.1.5.3 Microbial Sources

Microorganisms produce high-quality natural polymers through fermentation, enabling controlled and scalable production for pharmaceutical and biomedical applications. Xanthan gum, synthesized by *Xanthomonas campestris*, serves as an effective thickening and stabilizing agent in various formulations. Gellan gum, produced by *Sphingomonas elodea*, functions as a gelling agent in pharmaceutical and food products due to its robust gel-forming properties. Dextran, a polysaccharide derived from *Leuconostoc* species, is widely utilized in blood plasma expanders and hydrogel formulations, highlighting its significance in medical and biotechnological applications [24].

9.1.5.4 Marine Sources

Marine organisms are a valuable source of polysaccharides with distinct functional properties, making them essential in pharmaceutical and biomedical applications. Alginates, extracted from brown seaweed such as Laminaria, are widely used in wound dressings, controlled drug release systems, and dental impressions due to their biocompatibility and gel-forming abilities. Carrageenan, derived from red algae, serves as a stabilizer and thickener in pharmaceutical formulations and food products. Similarly, agar, also obtained from red algae, is extensively utilized in microbial culture media, laxatives, and tablet formulations, owing to its gelling and stabilizing properties [4, 9].

9.2 Extraction and Purification Methods of Gum and Mucilage

9.2.1 Traditional and Modern Extraction Techniques

The extraction and purification of gums and mucilages have evolved significantly over time, transitioning from traditional methods to advanced modern techniques to improve efficiency, yield, and purity.

9.2.1.1 Traditional Techniques

Traditional extraction methods for gums and mucilages primarily involve water or solvent-based maceration, soaking, and precipitation, typically conducted at ambient or slightly elevated temperatures to prevent degradation of bioactive components. The process begins with maceration or soaking, where plant materials are immersed in water or appropriate solvents to facilitate the release of gums or mucilages [32].

This is followed by filtration to remove insoluble residues, ensuring a purified extract. The extracted biopolymer is then precipitated using alcohol or other suitable

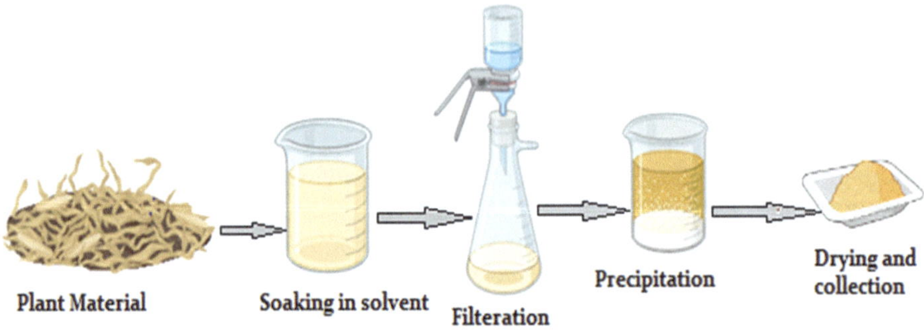

Fig. 9.2 Traditional method of extraction of gum and mucilages

agents, separated, and subjected to drying and pulverization to obtain a fine powder for further applications, as shown in Fig. 9.2. While these methods are cost-effective and simple, they are time-consuming and may yield products with variable purity, necessitating further refinement for consistency in pharmaceutical formulations [21].

9.2.1.2 Modern Techniques

Modern extraction techniques have significantly improved the efficiency, scalability, and quality of extracted gums and mucilages. Key advancements include:

Microwave-Assisted Extraction (MAE)

This technique uses microwave energy to heat the solvent and plant matrix, thereby accelerating the release of gum or mucilage. MAE is rapid, energy-efficient, and often results in higher purity extracts.

Ultrasound-Assisted Extraction (UAE)

By applying ultrasonic waves, UAE disrupts plant cell walls, facilitating the release of desired compounds. This method reduces extraction time and the amount of solvent required.

Supercritical Fluid Extraction (SFE)

SFE utilizes supercritical fluids, such as carbon dioxide, to extract gums and mucilages with minimal thermal degradation, making it particularly suitable for temperature-sensitive materials.

Enzyme-Assisted Extraction (EAE)

Specific enzymes are used in EAE to hydrolyze plant cell walls, enhancing the release of mucilage. This eco-friendly method results in high-quality, purer extracts.

Pressurized Hot Water Extraction (PHWE)

PHWE employs pressurized hot water as a solvent, offering an environmentally friendly solution for extracting hydrophilic components like gums and mucilages efficiently [13].

9.2.2 Purification Processes for Pharmaceutical-Grade Gums and Mucilages

The purification of gums and mucilages to pharmaceutical-grade standards is essential to remove impurities, enhance functional properties, and meet regulatory requirements. Below are the key purification processes:

9.2.2.1 Precipitation Methods

Precipitation is a widely used technique for purifying gums and mucilages, particularly in the pharmaceutical industry.

Alcohol Precipitation

Alcohols, such as ethanol or isopropanol, are added to the aqueous solution containing the gum or mucilage to induce precipitation. This process separates the desired polymer from impurities like proteins, tannins, and pigments [28].

Salt-Induced Precipitation

Certain salts are used to precipitate gums or mucilages based on their ionic interactions. This method enhances selectivity and purity.

pH Adjustment

The pH of the solution is altered to promote precipitation while maintaining the structural integrity of the polymer.

9.2.2.2 Filtration and Centrifugation

Physical separation techniques, such as filtration and centrifugation, are employed to remove particulate matter and insoluble impurities from gums and mucilages. Filtration, using membrane or vacuum systems, effectively separates solid contaminants, with fine filters ensuring the removal of even micro-level particulates. Centrifugation, particularly at high speeds, is used to separate insoluble impurities and enhance the purity of the gum or mucilage, making it especially useful for processing viscous solutions. These methods improve the overall quality and consistency of the extracted material [20].

9.2.2.3 Dialysis and Ultrafiltration

Advanced separation techniques like dialysis and ultrafiltration are used to refine gums and mucilages to pharmaceutical-grade purity. Dialysis and ultrafiltration are both effective methods for purifying gums and mucilages. Dialysis uses semipermeable membranes to remove low-molecular-weight impurities, salts, and solvents while retaining the high-molecular-weight gum or mucilage. Ultrafiltration, on the other hand, utilizes pressure-driven membranes with specific pore sizes to filter out impurities, ensuring that the polymer is retained. This method is particularly efficient for large-scale purification, minimizing product loss and ensuring the purity of the final extract. Both techniques play a crucial role in enhancing the quality and consistency of the extracted material [28].

9.2.2.4 Chromatographic Techniques

Chromatography techniques are employed to achieve the highest level of purity, particularly for pharmaceutical applications. Ion-exchange chromatography separates components based on their ionic charge, effectively removing charged impurities such as proteins or ionic contaminants. Size-exclusion chromatography utilizes a gel matrix to separate molecules according to size, ensuring the removal of smaller impurities. Affinity chromatography, on the other hand, employs specific binding interactions to selectively purify gums or mucilages with high specificity. These methods are essential for obtaining highly purified extracts suitable for use in sensitive pharmaceutical formulations [22].

9.3 Pharmaceutical Applications

9.3.1 Role as Excipients in Drug Formulations

Gums and mucilages play a critical role as excipients in pharmaceutical formulations due to their natural origin, biocompatibility, and multifunctionality. They are widely used for their physicochemical properties and ability to enhance the stability, efficacy, and patient acceptability of drug products [16].

9.3.1.1 Binding Agents

Gums and mucilages are used as binders in tablet formulations, providing cohesion to powdered materials and ensuring mechanical strength of the final product. Examples: Acacia gum, Mimosa mucilage, and guar gum.

9.3.1.2 Disintegrants

Facilitate the rapid breakdown of tablets upon ingestion, improving drug release and bioavailability. Example: Plantago ovata mucilage.

9.3.1.3 Suspending Agents

Stabilize suspensions by preventing sedimentation of dispersed particles, ensuring uniform dosing. Example: Xanthan gum and gum tragacanth.

9.3.1.4 Thickening Agents

Enhance the viscosity of liquid and semi-solid formulations, improving texture and stability. Example: Locust bean gum and carrageenan.

9.3.1.5 Controlled Drug Release

Their hydrophilic nature allows them to form gels or matrices that modulate drug release, enabling sustained or delayed-release formulations. Examples: Sodium alginate, pectin, and guar gum.

9.3.1.6 Film-Forming Agents

Used in coating tablets and capsules to mask taste, protect from environmental factors, and modify drug release. Example: Cellulose derivatives such as hydroxypropyl methylcellulose (HPMC).

9.3.1.7 Emulsifying Agents

Stabilize emulsions by reducing surface tension between oil and water phases. Example: Acacia gum in oil-in-water emulsions.

9.3.1.8 Bioadhesive Agents

Enhance the adhesion of dosage forms like buccal tablets or patches to mucosal surfaces, improving localized drug delivery. Example: Gum karaya and mucilage from Hibiscus species. [8, 15].

9.3.2 Binding, Thickening, Emulsifying, and Gelling Properties

Gums and mucilages possess diverse functional properties, making them highly valuable in pharmaceutical applications. Their natural origin and multifunctional capabilities in formulations offer significant advantages [27].

9.3.2.1 Binding Properties

Gums and mucilages function as binding agents by forming cohesive films around drug particles, enhancing the mechanical integrity of tablets. They are widely used in tablet formulations to improve compatibility and minimize friability. Natural binders, such as gum acacia, Mimosa mucilage, and guar gum, are commonly employed due to their superior compatibility with active pharmaceutical ingredients (APIs) compared to synthetic alternatives.

9.3.2.2 Thickening Properties

Gums and mucilages, due to their hydrophilic nature, absorb water and enhance the viscosity of formulations. This property makes them essential in syrups, suspensions, and topical gels for improving consistency. Common examples include xanthan gum and locust bean gum. By providing uniform texture, these natural agents enhance the appearance of formulations and improve patient acceptability. [2].

9.3.2.3 Emulsifying Properties

Gums and mucilages act as natural stabilizers in emulsions by forming protective layers around dispersed droplets, preventing coalescence. They are commonly used in oil-in-water emulsions for both oral and topical applications. Examples include gum tragacanth and acacia gum. As biodegradable emulsifiers, they offer an eco-friendly alternative to synthetic surfactants, reducing dependency on artificial additives.

9.3.2.4 Gelling Properties

Gums and mucilages form gels through hydration and molecular entanglement, creating a three-dimensional network. This property makes them valuable in controlled drug release systems, wound dressings, and hydrogel formulations. Examples include sodium alginate and pectin. Their ability to regulate drug release and provide excellent biocompatibility makes them ideal for sensitive pharmaceutical applications [10].

9.3.3 Comparative Efficacy with Synthetic Polymers

Gums and mucilages demonstrate comparable or superior efficacy to synthetic polymers in various pharmaceutical applications, offering distinct advantages and some limitations.

9.3.4 Advantages Over Synthetic Polymers

9.3.4.1 Biocompatibility and Safety

Natural polymers are generally non-toxic and biocompatible, minimizing the risk of adverse effects. In contrast, synthetic polymers such as polyvinyl alcohol (PVA) and polyethylene glycol (PEG) may present potential toxicity concerns in certain cases.

9.3.4.2 Renewability and Biodegradability

Gums and mucilages are renewable and biodegradable, making them eco-friendly alternatives to synthetic polymers.

9.3.4.3 Cost-Effectiveness

Natural sources are often more economical, especially when locally available.

9.3.4.4 Multi-functionality

Gums and mucilages often combine binding, thickening, emulsifying, and gelling properties, reducing the need for multiple excipients.

9.4 Challenges in Pharmaceutical Use

9.4.1 Variability in Quality Due to Source and Environmental Factors

One of the significant challenges in utilizing gums and mucilages in pharmaceutical applications is their variability in quality, which arises from differences in the plant species, geographical location, and environmental conditions during growth. Factors such as soil composition, climate, harvesting season, and post-harvest processing significantly influence the physicochemical properties of gums and mucilages, including viscosity, molecular weight, and chemical composition. This variability can result in batch-to-batch inconsistencies, affecting the reproducibility of pharmaceutical formulations. For example, differences in the degree of polymerization or impurity content may alter the gel-forming or binding capacity of a gum, leading to unpredictable performance in drug delivery systems.

Moreover, the presence of microbial contamination or residual solvents used during extraction processes can further compromise their quality and safety, requiring rigorous quality control measures. Standardization of these natural polymers is often difficult, as it necessitates a robust understanding of their chemical composition and interaction with other excipients under different conditions [26].

9.4.2 Stability and Solubility Issues

Stability and solubility challenges present another hurdle for the pharmaceutical use of gums and mucilages. Due to their hydrophilic nature, many gums and mucilages are prone to swelling and degradation when exposed to moisture or varying pH conditions. This characteristic can affect the shelf life of formulations containing these polymers, especially in humid climates. Additionally, the solubility of some gums and mucilages may be limited, requiring modification to enhance their functionality in certain applications [5]. Poor solubility can hinder their ability to uniformly disperse in aqueous or non-aqueous systems, impacting their effectiveness as emulsifiers, stabilizers, or drug carriers. Temperature sensitivity is another issue; some gums may lose their structural integrity at higher temperatures, making them unsuitable for heat-processed formulations. Oxidative degradation is also a concern, particularly for those with unsaturated or reactive functional groups, which may lead to a loss of functional properties over time. To address these issues, advanced techniques such as chemical or enzymatic modification, cross-linking, or blending with compatible polymers are being explored to enhance the stability and solubility profiles of these natural materials. Furthermore, incorporating proper packaging solutions and stabilizers in formulations can mitigate environmental effects and improve their performance [21].

9.4.3 Strategies for Overcoming Variability

9.4.3.1 Controlled Harvesting and Processing

One of the most effective strategies to minimize variability in the quality of gums and mucilages is the establishment of controlled harvesting and processing practices. By standardizing the conditions under which plant materials are collected, dried, and processed, pharmaceutical manufacturers can reduce the impact of environmental factors on the final product. For instance, optimizing the time of harvest based on the plant's growth cycle can ensure the highest yield and consistent quality of gums and mucilages. Harvesting at a specific season or growth stage when the plant's active constituents are most concentrated can help maintain uniformity. Additionally, standardizing post-harvest handling, including cleaning, drying, and grinding processes, is crucial to prevent degradation or contamination that might affect the physicochemical properties. Drying methods, such as freeze-drying or air-drying under controlled temperatures, can help preserve the stability of gums and mucilages by preventing excessive moisture content, which may lead to microbial contamination or degradation. Controlled extraction processes can also be optimized to ensure the purity and consistency of the extracted gum or mucilage, minimizing the risk of impurities that may affect the quality of the final product.

9.4.3.2 Chemical Characterization and Quality Control

Chemical characterization is essential to ensure the consistency and quality of gums and mucilages used in pharmaceutical formulations. Techniques such as chromatography, spectroscopy (FTIR, UV-VIS, NMR), and mass spectrometry are commonly employed to analyze the chemical composition, molecular weight distribution, functional groups, and purity of the polymers. A robust quality control system is vital for identifying batch-to-batch variations and ensuring that the natural polymers meet specific pharmacopoeial standards. Quality control measures should focus on assessing key parameters like viscosity, solubility, gelling properties, moisture content, and microbial contamination. The use of standardized testing procedures can minimize the effect of natural variations, ensuring that the gum or mucilage is suitable for pharmaceutical use. Regular chemical and microbial testing, along with proper documentation and traceability, are essential to maintain the integrity and safety of natural polymers. By implementing a rigorous quality assurance system, manufacturers can mitigate risks and improve the reproducibility of formulations containing gums and mucilages.

9.4.3.3 Blending with Other Excipients

Blending gums and mucilages with other excipients enhances stability, solubility, and viscosity while minimizing variability due to environmental factors. Combining them with hydrophilic polymers like HPMC or PEG improves drug release, while natural excipient blends optimize binding, gelling, and emulsifying properties. These blends also modify mechanical properties in dosage forms, ensuring better patient compliance and therapeutic efficacy.

9.5 Recent Advances in Gums and Mucilages

9.5.1 Chemical Modifications for Improved Performance

Chemical modifications enhance the properties of gums and mucilages, making them more suitable for advanced pharmaceutical applications. Derivatization, such as carboxymethylation and sulfation, improves solubility, viscosity, and bioadhesion. Cross-linking strengthens mechanical stability and enables controlled drug release. Graft copolymerization combines natural and synthetic benefits, creating superabsorbent and thermoresponsive systems. Functionalization with ligands like folic acid or antibodies facilitates targeted drug delivery. Nanostructure formation enhances drug loading and bioavailability through nanoparticles and nanogels. Stabilization techniques, including acetylation and esterification, improve emulsifying and suspending properties, ensuring formulation consistency.

9.5.2 Bioadhesive Properties for Targeted Drug Delivery

Gums and mucilages play a crucial role in targeted drug delivery due to their bioadhesive properties, enabling prolonged retention and localized drug release. Their high molecular weight and polysaccharide structures facilitate interactions with mucins, enhancing drug bioavailability. These polymers are widely used in oral (mucoadhesive tablets), buccal and sublingual (films for sustained drug release), vaginal and rectal (bioadhesive gels and suppositories), and nasal (bioadhesive nanoparticles and gels) drug delivery systems. Their ability to reduce dosing frequency, minimize systemic side effects, and enable site-specific treatment makes them highly valuable in pharmaceutical formulations.

9.5.3 Use in Novel Formulations: Hydrogels, Films and Nanoparticles

The versatility of gums and mucilages enables their use in advanced drug delivery systems like hydrogels, films, and nanoparticles, enhancing pharmaceutical applications. Hydrogels, formed due to their hydrophilic nature, are used for controlled drug release, wound healing, and tissue engineering. Films offer flexibility and biodegradability, making them ideal for buccal, transdermal, and rapid-dissolving applications, as well as tablet coatings. Nanoparticles improve drug stability, solubility, and targeted delivery, facilitating drug encapsulation, site-specific therapy, and sustained release. These innovations enhance drug efficacy, patient compliance, and treatment outcomes [6].

9.6 Sustainability and Biocompatibility

Gums and mucilages are natural polymers with significant advantages in sustainability and biocompatibility, making them ideal candidates for pharmaceutical applications. Their eco-friendly nature, biodegradability, and non-toxic characteristics align with the principles of green chemistry, offering superior alternatives to synthetic polymers.

9.6.1 Eco-friendly Nature of Natural Polymers

Gums and mucilages are sustainable biopolymers derived from renewable plant sources, including seeds, exudates, and mucilaginous tissues, ensuring a continuous and eco-friendly supply. Examples include gum arabic from *Acacia* species and mucilage from *Mimosa pudica*. Their extraction and processing have a significantly lower environmental impact than synthetic polymers, as they require minimal energy input and do not generate toxic byproducts. Furthermore, their bioprocessing can be achieved through green methodologies, such as aqueous extraction, eliminating the need for hazardous chemicals and promoting environmentally responsible pharmaceutical applications. [22].

9.6.2 Biodegradability and Non-toxic Characteristics

Gums and mucilages are environmentally friendly biopolymers due to their biodegradability, biocompatibility, and safe disposal. They readily decompose through microbial or enzymatic action, minimizing environmental pollution. For instance, gum arabic and guar gum fully degrade in soil, while mucilage-based films and hydrogels break down into harmless by-products. Their non-toxic and biocompatible nature ensures safety in pharmaceutical applications, exhibiting low immunogenicity and minimal risk of adverse reactions. Unlike synthetic polymers, these natural materials do not accumulate in ecosystems, enabling safe and sustainable disposal without long-term environmental impact. [30].

9.6.3 Benefits Over Synthetic Polymers in Green Chemistry

Gums and mucilages serve as sustainable alternatives to petroleum-based synthetic polymers, offering biodegradability and alignment with green chemistry principles. Unlike synthetic counterparts, their production and degradation do not release hazardous byproducts, promoting ecological balance. Additionally, their energy-efficient extraction and processing result in lower energy consumption and reduced greenhouse gas emissions. These natural polymers also demonstrate functional superiority, exhibiting bioadhesion, swelling, and gel formation—properties that are challenging to replicate synthetically. Their inherent structural complexity and multifunctionality minimize the need for extensive chemical modification, making them ideal for environmentally responsible pharmaceutical applications.

9.7 Pharmaceutical Applications

9.7.1 Drug Delivery Systems

Gums like guar gum and xanthan gum are used as release retardants in controlled drug delivery. Mucilages from *Mimosa pudica* or *Plantago ovata* (psyllium) are exploited for their gelling ability to prepare sustained-release formulations.

9.7.2 Applications in Food and Cosmetics

In the food industry, gums are utilized as thickeners, emulsifiers, and stabilizers in products such as sauces, ice creams, and baked goods. Mucilages, like psyllium, function as dietary fibers that support cholesterol reduction and promote gut health. Guar gum, in particular, enhances the shelf life of frozen desserts by preventing ice crystal formation. In cosmetics and personal care, gums and mucilages play a key role in formulations like lotions and gels, improving viscosity, texture, and hydration properties. Mucilages derived from aloe vera and *Mimosa pudica* are especially popular in soothing skincare products.

9.7.3 Environmental and Industrial Applications

Gums and mucilages provide eco-friendly solutions by serving as sustainable alternatives to synthetic polymers in industries such as textiles and paper, while their biodegradability minimizes environmental impact. Additionally, their potential in bioremediation has been explored, particularly in the development of gum-based hydrogels for pollutant adsorption in water purification, contributing to environmental sustainability [21].

9.8 Challenges and Future Directions

Despite their versatile applications, gums and mucilages face challenges, such as:

9.8.1 Standardization

Variability in chemical composition due to environmental and processing factors. Difficulty in scaling up the production of certain plant-based mucilages.

9.8.2 Technological Enhancements

Advancing the extraction, purification, and modification techniques to optimize their functionality. Developing composite formulations with synergistic effects.

9.8.3 Future Scope

The potential of gums and mucilages in pharmaceutical applications is expanding due to their natural origin, biodegradability, and biocompatibility. Advancements in polymer modification techniques are expected to enhance their physicochemical stability, solubility, and functionality, thereby improving their applicability in advanced drug delivery systems. The development of innovative formulations, including hydrogels, nano-based carriers, and mucoadhesive systems, will significantly broaden their therapeutic utility. Ensuring batch-to-batch consistency through rigorous standardization and quality control measures remains a critical challenge that requires systematic investigation. Furthermore, the integration of natural and synthetic polymers may lead to hybrid excipients with optimized performance characteristics. The widespread adoption of these natural polymers will be influenced by regulatory approvals and large-scale industrial feasibility. Research into plant-derived mucilages for controlled and targeted drug delivery holds promise for improving pharmacokinetic profiles and therapeutic efficacy. Additionally, their emerging applications in wound healing, tissue engineering, and nutraceutical formulations highlight their multifunctional potential. The incorporation of computational modeling and artificial intelligence to predict polymer behavior and optimize formulations may accelerate innovation in this domain. Future studies should also focus on developing environmentally sustainable and economically viable extraction methodologies to facilitate large-scale production and commercial utilization.

References

1. Alalor CA, Avbunudiogba JA, Augustine K. Isolation and characterization of mucilage obtained from *Colocasia esculenta*. Int J Pharm Bio Sci. 2014;4(1):25–9.
2. Alonso-Sande M, Teijeiro-Osorio D, Remunán-López C, Alonso MJ. Glucomannan, a promising polysaccharide for biopharmaceutical purposes. Eur J Pharm Biopharm. 2009;72(2):453–62.
3. Amiri MS, Mohammadzadeh V, Yazdi MET, Barani M, Rahdar A, Kyzas GZ. Plant-based gums and mucilages applications in pharmacology and nanomedicine: a review. Molecules. 2021;26(6):1770. https://doi.org/10.3390/molecules26061770.
4. Augst AD, Kong HJ, Mooney DJ. Alginate hydrogels as biomaterials. Macromol Biosci. 2006;6:623–33.
5. Beikzadeh S, Khezerlou A, Jafari SM, Pilevar Z, Mortazavian AM. Seed mucilages as functional ingredients for biodegradable films and edible coatings in the food industry. Adv Colloid Interf Sci. 2020;280:102164.
6. Choudhary A, Bains A, Sridhar K, Dhull SB, Goksen G, Sharma M, Chawla P. Recent advances in modifications of exudate gums: functional properties and applications. Int J Biol Macromol. 2024;271(Pt 2):132688. https://doi.org/10.1016/j.ijbiomac.2024.132688.
7. Choudhary PD, Pawar HA. Recently investigated natural gums and mucilages as pharmaceutical excipients: an overview. J Pharm (Cairo). 2014;2014:204849. https://doi.org/10.1155/2014/204849.
8. Chowdhury M, Sengupta A, Datta L, Chatterjee S. Role of mucilage as pharmaceutical additives and cytoprotective agents. J Innov Pharm Biol Sci. 2017;4:46–52.
9. d'Ayala GG, Malinconico M, Laurienzo P. Marine-derived polysaccharides for biomedical applications: chemical modification approaches. Molecules. 2008;13(9):2069–106. https://doi.org/10.3390/molecules13092069.
10. de Oliveira Filho JG, Lira MM, de Sousa TL, Campos SB, Lemes AC, Egea MB. Plant-based mucilage with healing and anti-inflammatory actions for topical application: a review. Food Hydrocoll Health. 2021;1:100012.
11. Divekar VB, Mohan K. Isolation and characterization of mucilage from *Lepidium sativum* Linn seeds. Int J Pharm Res Dev. 2010;2:1–5.
12. Edward MR. Modern pharmaceutics. Marcel Dekker; 2002.
13. Eskilsson CS, Björklund E. Analytical-scale microwave-assisted extraction. J Chromatogr A. 2000;902(1):227–50. https://doi.org/10.1016/s0021-9673(00)00921-3.
14. Evans WC. Trease and Evans' pharmacognosy. 14th ed. Harcourt Brace and Co; 1996.
15. Hamman H, Steenekamp J, Hamman J. Use of natural gums and mucilages as pharmaceutical excipients. Curr Pharm Des. 2015;21(33):4775–97. https://doi.org/10.2174/1381612821666150820100524.

16. Hezerlou A, Zolfaghari H, Banihashemi SA, Forghani S, Ehsani A. Plant gums as functional compounds for edible films and coatings in the food industry: a review. Polym Adv Technol. 2021;32:2306–26.
17. Jadav M, Pooja D, Adams DJ, Kulhari H. Advances in xanthan gum-based systems for the delivery of therapeutic agents. Pharmaceutics. 2023;15(2):402. https://doi.org/10.3390/pharmaceutics15020402.
18. Kirtikar KR, Basu BD. Indian medicinal plants. Periodical Expert's Book Agency; 1991.
19. Kokate CK, Purohit AP, Gokhale SB. A textbook of pharmacognosy. 21st ed. Nirali Prakashan; 2003.
20. Mahato N, Sinha M, Sharma K, Koteswararao R, Cho MH. Modern extraction and purification techniques for obtaining high-purity food-grade bioactive compounds and value-added co-products from citrus wastes. Food Secur. 2019;8(11):523. https://doi.org/10.3390/foods8110523.
21. Malviya R, Tyagi V, Singh D. Techniques of mucilage and gum modification and their effect on hydrophilicity and drug release. Curr Drug Deliv. 2020;14(3):214–22. https://doi.org/10.2174/1872211314666201204160641.
22. Marcondes Sari MH, Ferreira LM, Cruz L. The use of natural gums to produce nano-based hydrogels and films for topical application. Int J Pharm. 2022;626:122166. https://doi.org/10.1016/j.ijpharm.2022.122166.
23. Mondal S, Rahaman ST. An overview on isolation and characterization of mucilage from various species related to various plant families. Saudi J Med Pharm Sci. 2018;4(2):284–8.
24. Oluwatosin AO. Assessment of *Albizia zyga* gum as a binding agent in tablet formulations. Acta Pharma. 2005;55:263–76.
25. Patel MM, Chauhan GM, Patel LD. Mucilage of *Lepidium sativum* Linn. (Asario) and *Ocimum canum* Sims. (Bavchi) as emulgents. Int J Hosp Pharm. 1987;24:200–2.
26. Selvasekaran P, Chidambaram R. Food-grade aerogels obtained from polysaccharides, proteins, and seed mucilages: role as a carrier matrix of functional food ingredients. Trends Food Sci Technol. 2021;112:455–70.
27. Sharma DR, Sharma A, Kaundal A, Rai PK. Herbal gums and mucilage as excipients for pharmaceutical products. Res J Pharmacogn Phytochem. 2016;8(3):145–52.
28. Singh M, Raorane CJ, Alka, Shastri D, Raj V, Kim S-C, Tuteja M. Recent progress on modified gum Katira polysaccharides and their various potential applications. Polymers. 2022;14(17):3648. https://doi.org/10.3390/polym14173648.
29. Stephen AM, Churms SC. Food polysaccharides and their applications. Taylor & Francis; 2006.
30. Sunil M, Biswal PK, Mishra B, Sahoo S. Isolation, characterization, and pharmaceutical evaluation of the mucilage from *Polyalthia suberosa* leaves. Int J PharmTech Res. 2010;2(2):1455–9.
31. Taru P, Bhavya E. Mucilage-containing herbs used in traditional systems: an overview. GIS Sci J. 2022;9(4):1384–94.
32. Tosif MM, Najda A, Bains A, Kaushik R, Dhull SB, Chawla P, Walasek Janusz M. A comprehensive review on plant-derived mucilage: characterization, functional properties, applications, and its utilization for nanocarrier fabrication. Polymers (Basel). 2021;13(7):1066. https://doi.org/10.3390/polym13071066.

Natural Excipients in Solid Dosage Forms

10

Ria N. Patel, Tejal M. Keriwala, Vishwesh V. Upasani, Pranjal Gujarathi, Meghraj Suryawanshi, and Kuldeep Vinchurkar

10.1 Introduction to Natural Excipients

10.1.1 Definition and Importance

Natural excipients are plant-derived substances used in pharmaceutical formulations alongside active ingredients [1]. They offer several advantages over synthetic alternatives, including non-toxicity, cost-effectiveness, and widespread availability [2–4]. These excipients serve various functions in drug delivery systems, such as binders, diluents, lubricants, disintegrants, and coating agents [1]. Many natural excipients are hydrophilic and gel-forming, making them suitable for controlled

R. N. Patel
Department of Pharmaceutics, Sigma Institute of Pharmacy, Sigma University, Vadodara, Gujarat, India

T. M. Keriwala
Department of Quality Assurance, Vidhyadeep Institute of Pharmacy, Vidhyadeep University, Surat, Gujarat, India

V. V. Upasani
Department of Pharmaceutics, H.R. Patel Institute of Pharmaceutical Education and Research, Dhule, Maharashtra, India

P. Gujarathi (✉)
Department of Pharmacology, Vidhyadeep Institute of Pharmacy, Vidhyadeep University, Surat, Gujarat, India

Research Scholar, Bhagwan Mahavir Center for Advanced Research, Bhagwan Mahavir College of Pharmacy, Bhagwan Mahavir University, Surat, Gujarat, India

M. Suryawanshi · K. Vinchurkar
Department of Pharmaceutics, Sandip Institute of Pharmaceutical Sciences (SIPS), Affiliated to Savitribai Phule Pune University (SPPU, Pune), Nashik, Maharashtra, India

J. S et al. (eds.), *Innovative Pharmaceutical Excipients: Natural Sources*, https://doi.org/10.1007/978-981-96-7959-1_10

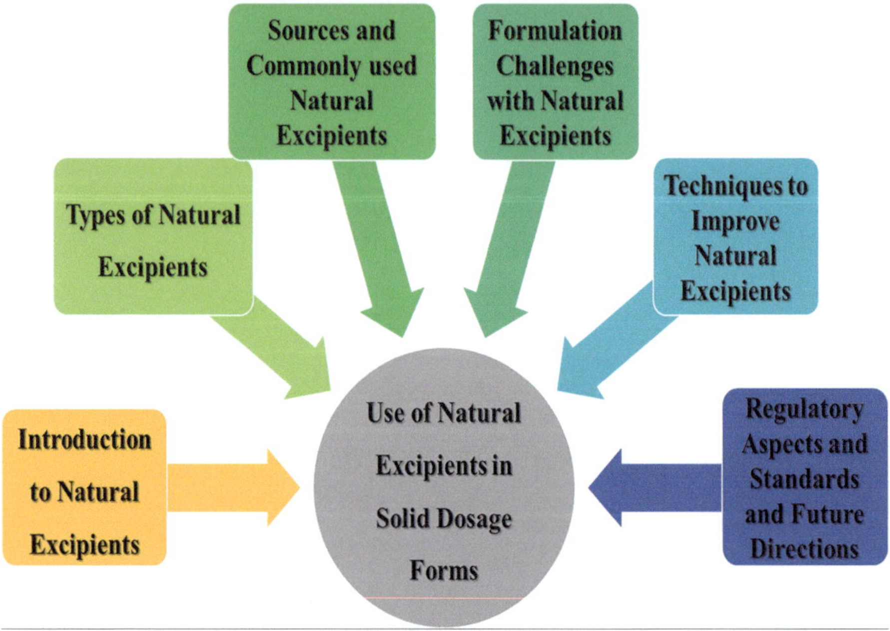

Fig. 10.1 Diagram of natural excipients in pharmaceutical applications: introductions, challenges, and future prospects

drug delivery systems [4]. The pharmaceutical industry's growing interest in natural polymers has led to their increased use in formulations [4]. Natural excipients are particularly valuable in addressing challenges in drug development, such as bioavailability, permeability, and poor solubility [2, 3]. As the concept of excipients evolves from inert vehicles to essential formulation components, the global excipient market is expected to grow rapidly, driven by emerging trends in the pharmaceutical industry [2, 3]. Other applications of natural excipients in solid dosage form are mentioned in Fig. 10.1.

10.1.2 Advantages Over Synthetic Excipients

Natural excipients offer several advantages over synthetic alternatives in pharmaceutical formulations. They are nontoxic, less expensive, and readily available [2, 3, 5, 6]. The concept of excipients has evolved from inert fillers to essential components that can influence drug absorption and release [2, 3]. Natural excipients are particularly valuable in novel drug delivery systems, addressing challenges such as bioavailability, permeability, and solubility [2, 3]. They are compatible with herbal formulations and can improve product shelf life [6]. Examples of natural excipients include carrageenan, thaumatin, and myrobalan [2, 3]. The global market for excipients is expected to grow rapidly, driven by emerging trends in the pharmaceutical industry [2, 3]. As the demand for plant-based and natural products increases, natural excipients are becoming increasingly important in pharmaceutical formulations [6].

10.2 Types of Natural Excipients

10.2.1 Binders

Natural excipients, derived from plant sources, are increasingly preferred in pharmaceutical formulations due to their safety, biocompatibility, and cost-effectiveness compared to synthetic alternatives [1, 7]. These excipients serve various functions, including binders, diluents, lubricants, disintegrants, and coating agents [1, 4]. Common natural excipients include gums, mucilages, and polysaccharides such as agar, starch, alginate, and cellulose [8]. Cellulose, the most abundant natural polymer, has diverse applications in pharmaceuticals, food, and chemical engineering [8]. Natural excipients are particularly suitable for controlled drug delivery systems and can be easily modified to meet specific formulation requirements [4, 7]. Additionally, they offer nutritional benefits and are associated with fewer regulatory issues compared to synthetic excipients [4, 7]. The use of agro-waste as a source of natural excipients presents an opportunity to address disposal concerns in developing countries [8].

10.2.2 Disintegrants

Natural disintegrants are gaining prominence in pharmaceutical formulations due to their safety, cost-effectiveness, and biocompatibility compared to synthetic alternatives [1, 9]. These plant-derived substances, including starches, gums, and mucilages, promote tablet breakup and enhance drug release [10]. Notable examples include silicified *Oryza sativa* Starch, *Plantago ovata* mucilage, and gum karaya, which have shown promising results in improving tablet disintegration and drug dissolution [11]. Other effective natural disintegrants include *Linum usitatissimum* mucilage, Ispaghula mucilage with banana powder, and *Lepidium sativum* seed mucilage [11]. Fenugreek gum has demonstrated super dissolving properties and anti-inflammatory effects when combined with diclofenac sodium [11]. The growing interest in natural disintegrants stems from their ability to enhance moisture penetration, disperse tablet matrices, and promote rapid drug release [10].

10.2.3 Fillers/Diluents

Natural excipients derived from plants are gaining prominence in pharmaceutical formulations due to their advantages over synthetic alternatives. These excipients are nontoxic, cost-effective, and readily available [1, 7]. They can serve various functions in drug formulations, including as binders, diluents, lubricants, disintegrants, and coating agents [1, 7]. Plant-derived gums and mucilages, such as carrageenan, agar, and gum acacia, are particularly versatile and can be easily modified to meet specific formulation needs [1, 7]. Natural excipients are biocompatible and can provide additional nutritional benefits to the dosage form [4, 7]. Their use in pharmaceutical applications is associated with fewer regulatory issues compared to synthetic excipients [7]. As a result, the pharmaceutical industry is increasingly inclined toward incorporating these natural excipients in both conventional and novel drug delivery systems [4, 7].

10.2.4 Lubricants

Natural excipients, derived from plant sources, are gaining prominence in pharmaceutical formulations due to their nontoxic nature, cost-effectiveness, and widespread availability [1, 7]. These excipients, primarily polymers of natural origin, can serve various functions in drug formulations, including as diluents, binders, disintegrants, and lubricants [1, 7]. Plant-derived gums and mucilages, such as carrageenan, agar, and gum acacia, meet many pharmaceutical requirements and offer stability with fewer regulatory concerns compared to synthetic alternatives [7, 12]. Natural excipients are biocompatible and can provide additional nutritional benefits to the dosage form [7]. Their adaptability allows for easy modification to meet specific formulation needs, making them potent and economical vehicles for delivering active pharmaceutical ingredients [7, 12]. As a result, pharmaceutical industries are increasingly inclined toward using these natural excipients in formulation development [4, 12].

10.2.5 Coating Agents

Natural polymers and plant-derived excipients have gained significant attention in drug delivery systems due to their non-toxicity, stability, renewability, and cost-effectiveness [13–15]. These biopolymers, including gums and mucilages, serve multiple functions in pharmaceutical formulations, such as coating materials, emulsifiers, and disintegrants [16]. They can influence drug release, absorption, and bioavailability, making them crucial components in novel dosage forms [14]. Plant polysaccharides, in particular, have shown promise as matrix formers in modified release formulations and even exhibit environmental-responsive gelation properties [14]. The versatility of natural excipients allows for their use in immediate, controlled, or sustained release formulations, depending on their aqueous solubility [15]. As research in excipient development continues to grow, natural polymers are increasingly being investigated for their potential to optimize drug delivery systems [13].

10.3 Sources of Natural Excipients

10.3.1 Plant-Based Excipients

Plant-based excipients have gained significant attention in pharmaceutical applications due to their biocompatibility, affordability, and sustainability [17]. These natural polymers, including gums and mucilages, offer advantages over synthetic alternatives in various drug delivery systems [18]. Plant-derived polysaccharides are extensively investigated for use in solid oral dosage forms and as matrix formers in modified release formulations [14]. Recent research has explored their potential in topical drug delivery, demonstrating effectiveness in delivering drugs through the skin [19]. Plant-based excipients can be chemically modified to suit specific needs and some exhibit environmental-responsive gelation characteristics, allowing for controlled drug release [14]. Despite challenges, the pharmaceutical industry is increasingly interested in developing innovative drug delivery systems using these natural excipients, driven by their non-toxicity, renewability, and potential to improve drug delivery efficiency [17, 19]. Table 10.1 explained about numerous plant-based polymers and its applications.

10.3.2 Animal-Derived Excipients

Animal-derived excipients play a significant role in pharmaceutical formulations, offering advantages like biocompatibility and low cost [20]. However, their use is limited by religious beliefs, potential toxicity, and the development of synthetic alternatives [20]. The US Food and Drug Administration (FDA) has established guidelines for evaluating new excipients, including those of animal origin [21]. In clinical applications, animal-derived surfactants have shown varying efficacy in treating respiratory distress syndrome in preterm infants, with porcine surfactants demonstrating some advantages over bovine surfactants [22]. Despite their widespread use in scientific experimentation, concerns about reproducibility and ethical issues have led to increased interest in nonanimal alternatives [23]. The transition to these alternatives is encouraged by regulations like the EU Directive 2010/63 and Organisation for Economic Co-operation and Development (OECD) guidelines, but challenges remain in increasing awareness and acceptance of nonanimal ingredients in research and pharmaceutical development [23].

10.3.3 Marine-Derived Excipients

Marine-derived compounds have gained significant attention in pharmaceutical and cosmetic industries due to their unique properties and biocompatibility. Marine polysaccharides are abundant, cost-effective, and have diverse applications as

Table 10.1 Plant-based natural excipients and its applications

Sr. no.	Material	Function	Descriptions
1	Cellulose	Thickener, gelling agent	Enhances stability and texture of formulations
2	Starch	Thickener, gelling agent	Provides rheological properties and influences drug release
3	Guar gum	Thickener, stabilizer	Improves consistency and feel of gels and creams
4	Carrageenan	Thickener, stabilizer, gelling agent	Increases viscosity and stability
5	Xanthan gum	Thickener, stabilizer	Enhances consistency and stability
6	Agar-agar	Gelling agent	Improves texture and structure of formulations
7	Olive oil	Emollients	Enhances skin hydration
8	Coconut oil	Emollients, thickener	Provides antimicrobial, anti-inflammatory, and moisturizing properties
9	Bees wax	Emollients, thickener	Enhances skin hydration and product stability
10	Shea butter	Emollients	Offers moisturizing and anti-inflammatory properties
11	Chamomile extract	Active agent	Provides anti-inflammatory and soothing effects
12	Aloe vera extract	Active agent	Offers soothing and wound-healing properties
13	Green tea extract	Active agent	Provides antiaging and anti-inflammatory effects
14	Calendula extract	Active agent	Offers anti-inflammatory and wound-healing properties
15	Myrrh resin	Healing agent	Provides antibacterial and anti-inflammatory properties
16	Frankincense resin	Fragrance, aromatic	Provides anti-inflammatory and antioxidant properties
17	Lavender oil	Fragrance, aromatic	Offers calming and antibacterial effects

excipients in pharmaceutical formulations [20]. The marine pharmaceutical pipeline has delivered FDA-approved drugs and continues to supply novel compounds for clinical trials [24]. Marine-derived molecules serve as active ingredients in cosmeceuticals, offering benefits such as skin hydration, UV protection, and antiaging properties [25]. Marine omega-3 phospholipids are particularly valuable for brain health, as they can cross the blood-brain barrier, unlike their triglyceride counterparts [26]. However, the use of marine-derived excipients faces challenges due to religious beliefs and potential toxicity issues [20]. Despite these limitations, the marine pharmaceutical and cosmeceutical pipelines remain active and promising for future developments.

10.4 Functional Properties of Natural Excipients

10.4.1 Physicochemical Characteristics

Natural excipients are gaining popularity in pharmaceutical formulations due to their biocompatibility, low toxicity, and cost-effectiveness compared to synthetic alternatives [1, 19]. These plant-derived substances, including gums and mucilages, offer various functional properties such as binding, disintegration, and lubrication in solid dosage forms [4]. Natural excipients can be easily modified to meet specific formulation needs and may provide additional nutritional benefits [1]. Their physicochemical characteristics are crucial for determining their suitability in different formulations. Common methods for analyzing solid excipients include flow properties, compressibility index, and angle of repose, while instrumental techniques such as Fourier Transform Infrared Spectroscopy (FTIR), Nuclear Magnetic Resonance (NMR), Scanning Electron Microscope (SEM), Particle size analysis (PSA), X-ray diffraction (XRD), and Differential Scanning Calorimetry (DSC) are used for more detailed characterization [27]. The versatility of natural excipients makes them suitable for use in both conventional and novel drug delivery systems, contributing to the overall quality and performance of pharmaceutical products [4, 19].

10.4.2 Biocompatibility and Safety

Recent studies have focused on the biocompatibility and safety of various biomaterials used in medical applications. Assad and Jackson (2019) reviewed the evaluation process for orthopedic biomaterials, emphasizing the importance of preclinical safety and efficacy testing [28]. Barman (2015) explored the biocompatibility of Zinc oxide (ZnO) nanoparticles, noting their potential in cancer therapy but highlighting the need for specific tumor cell targeting [29]. Ramot et al. (2016) examined Polylactic acid (PLA) and its copolymers, acknowledging their general biocompatibility but also reporting potential side effects and foreign body reactions [30]. Earlier work by Kyriacos et al. (1996) focused on PLA/PGA copolymers, discussing their suitability for orthopedic applications and potential for reducing osteopenia. However, they also noted challenges in sterilization and reported issues with cell proliferation. Despite some concerns, these materials show promise in various medical fields, particularly in soft tissue applications and osteochondral repair [31].

10.5 Role of Natural Excipients in Solid Dosage Forms

10.5.1 Tablets

Natural excipients are increasingly favored in pharmaceutical formulations due to their nontoxic, biodegradable, and cost-effective properties compared to synthetic alternatives [4, 32]. These plant-derived substances, such as gums and mucilages, serve multiple functions in drug delivery systems, including as binders, diluents, and disintegrants [4]. They play crucial roles in novel dosage forms, influencing drug release rates and absorption [2, 33]. Natural excipients are particularly valuable in nanoformulations, with examples including the use of gelatin and chitosan in developing encapsulated nanoparticulate [32]. The pharmaceutical industry's shift toward natural excipients is driven by their biocompatibility, wide availability, and potential to address challenges in drug development such as poor solubility and bioavailability [15, 33]. This trend aligns with the growing consumer preference for natural products, perceived as safer alternatives to synthetic excipients [33].

Case Study Title Formulation Development of Patient Friendly Dosage Form: All in One Natural Excipient as Binder, Diluent and Disintegrant [6].

Abstract Summary A natural excipient can be used as a binder, diluent, and disintegrant in oral disintegrating and immediate release tablet formulations.

Main Findings The natural excipient can be used as a binder, diluent, and disintegrant in oral disintegrating and immediate release tablet formulations, performing similarly to a synthetic superdisintegrant. The natural excipient is biocompatible, cost-effective, and can provide nutritional benefits.

Limitations No limitations addressed.

10.5.2 Capsules

Natural excipients derived from plants are gaining popularity in pharmaceutical formulations due to their nontoxic nature, low cost, and wide availability [3, 7]. These excipients, including gums, mucilages, and other plant-derived substances, can serve multiple functions such as diluents, binders, disintegrants, and lubricants in various dosage forms [13]. They offer advantages over synthetic counterparts, including biocompatibility, nutritional benefits, and fewer regulatory issues [4]. Natural excipients like carrageenan, thaumatin, agar, and gum acacia comply with pharmaceutical requirements and can be easily modified to meet specific formulation needs [3, 7]. The pharmaceutical industry's growing interest in natural excipients is driven by their potential to improve drug delivery systems, address challenges such as bioavailability and solubility, and meet the evolving demands of the market [3, 13].

Case Study Title Study of the Influence of the Excipients Used for the Synthesis of Microspheres Loaded with Quercetin: Their Characterization and Antimicrobial Activity [34].

Abstract Summary Chitosan-sodium alginate microspheres loaded with quercetin showed high encapsulation efficiency, release capacity, and antimicrobial activity.

Main Findings The complex coacervation method used for the encapsulation of quercetin in microspheres was efficient and offered the highest entrapment efficiency. Microspheres with lower polymer concentrations had a more wrinkled and rougher surface, while those with higher polymer concentrations had a smoother and more spherical appearance. Increasing the quercetin concentration positively influenced the entrapment efficiency, swelling index, and in vitro drug release.

Limitations The microsphere system may need further development to be suitable for oral administration. The antimicrobial effects of the quercetin-loaded microspheres need to be further evaluated compared to conventional antibiotics. There may be room for further optimization of the microsphere formulation parameters beyond the best performing samples (P1, P2, and P3).

10.6 Examples of Commonly Used Natural Excipients

10.6.1 Starch and Modified Starch

Recent advances in starch and modified starch applications as pharmaceutical excipients have expanded their potential in drug formulation. Chemical modifications, including esterification, etherification, and crosslinking, can enhance starch's hydrophobicity and improve its functionality for various applications [35]. These modifications address limitations of native starch, such as high digestibility and swelling index, making it more suitable for pharmaceutical use [36]. Modified starches have been employed in diverse drug delivery systems, including microparticles, nanoparticles, hydrogels, and scaffolds [37]. Dialdehyde starches show promise for direct drug attachment, while carboxymethyl starches enable pH-sensitive drug encapsulation and release [37]. Starch nanocrystals can enhance hydrogel properties and control drug release [37]. Some modified starches, such as sodium starch glycolate and pregelatinized starch, are already FDA-approved for use in various drug delivery systems [38].

10.6.2 Cellulose Derivatives

Recent advances in cellulose derivatives as pharmaceutical excipients have expanded their applications in drug formulation. These biocompatible and biodegradable polymers are extensively used in various dosage forms, including extended-release matrices, osmotic drug delivery systems, and bioadhesives [39]. Cellulose derivatives have gained popularity in developing nanoparticulate-based drug delivery systems for oral health improvement, protein and peptide delivery, and colorectal cancer treatment [40]. Innovative cellulose-based materials, such as micro-cellulose and nanocellulose, have shown promise in sustained drug delivery applications by modifying drug solubility and gelling behavior [41]. Furthermore, cellulose derivative-based composites have been researched for various biomedical applications, including osseointegration, hemodialysis, biosensors, and tissue engineering scaffolds [42]. These advancements highlight the versatility and potential of cellulose derivatives in pharmaceutical formulations and biomedical applications.

10.6.3 Gum and Mucilage (e.g., Xanthan Gum, Guar Gum)

Recent research has highlighted innovative applications for mucilage compounds like xanthan gum and guar gum across various industries. Xanthan gum's versatility stems from its tunable structural properties, allowing its use in drug delivery systems, scaffolds for cells, and as excipients in tablets [43]. Guar gum's unique properties, including thickening, emulsifying, and film-forming abilities, have led to its application in hydraulic fracturing, explosives, and bioremediation [44]. Both xanthan and guar gums are being explored for use in tissue engineering and oil recovery, with ongoing research focusing on enhancing their physicochemical properties through modifications [45]. These natural polymers are increasingly preferred over synthetic materials due to their biocompatibility, biodegradability, and cost-effectiveness. Current research is also investigating the development of nanomaterials from these gums for various biotechnological applications [46].

10.6.4 Alginates and Pectins

Recent research has highlighted innovative applications for alginates and pectins in various fields. Alginates have shown promise in pharmaceutical and biomedical applications, including drug formulation, tissue cultures, and cell transplantation for treating diabetes and neurodegenerative diseases [47]. Both alginates and pectins have been explored as sustainable alternatives to traditional plastic packaging in the food industry, with their unique functional properties allowing for the development of biofilms with specific physical-mechanical characteristics [48]. Alginates have also demonstrated potential in 3D bioprinting, drug delivery systems, and environmental applications such as sorption [49]. Their biocompatibility and ability to be modified make them particularly suitable for medical uses. These studies emphasize the versatility of alginates and pectins, showcasing their potential to address environmental and health needs through innovative technological applications.

10.7 Formulation Challenges with Natural Excipients

10.7.1 Compatibility Issues

Natural excipients are gaining attention in pharmaceutical formulations due to their nontoxic, cost-effective, and readily available nature compared to synthetic alternatives [2, 15]. These excipients have evolved from inert vehicles to essential components in drug delivery systems, addressing challenges such as bioavailability, permeability, and solubility [2, 15]. Plant-derived gums and mucilages are particularly promising, offering hydrophilic and gel-forming properties suitable for controlled drug delivery [4]. The development of new natural excipients is driven by safety and biodegradability concerns associated with synthetic options [50]. Natural excipients can serve as binders, diluents, and disintegrants in various dosage forms, including oral disintegrating tablets and immediate-release formulations [4]. As the pharmaceutical industry trends toward refining the physical structure of active pharmaceutical ingredients, the global excipient market is expected to grow rapidly [2, 15].

10.7.2 Stability Concerns

Natural polymers, such as gelatin, collagen, and guar gum, are increasingly used as excipients in pharmaceutical formulations due to their biocompatibility, biodegradability, and cost-effectiveness [51]. However, their application is accompanied by several stability concerns, including susceptibility to microbial contamination, which can compromise the safety and efficacy of the pharmaceutical product [51]. Additionally, batch-to-batch variability arising from differences in source and processing can lead to inconsistencies in physicochemical properties, affecting formulation quality [51]. Environmental factors influence the hydration behavior of these polymers, resulting in uncontrolled hydration rates that impact drug delivery performance. Moreover, exposure to heat and light can cause thermal and photodegradation, altering molecular weight and mechanical properties, potentially reducing formulation stability. Furthermore, natural polymers may interact with other excipients, leading to instability or reduced drug efficacy. Addressing these stability challenges is essential for the effective utilization of natural polymers in pharmaceutical formulation [51].

10.7.3 Variability in Composition

Excipient variability poses significant challenges in pharmaceutical development, affecting drug product functionality and performance [52, 53]. Variability can arise from raw material sources, manufacturing processes, and interactions with active ingredients [53]. Understanding excipient critical material attributes is crucial for implementing quality by design approaches and predicting biopharmaceutical behavior [54]. Advanced analytical techniques and statistical modeling are employed to characterize excipients and their impact on drug products [53]. Utilizing quantitative certificate of analysis data offers a comprehensive, cost-effective method to assess excipient lot-to-lot variability, enabling developers to examine manufacturing site variations, quantify prior experience domains, and rationally select excipient lots [55]. The International Pharmaceutical Excipients Council has developed guidelines for proper excipient selection and evaluation [52]. A thorough understanding of excipient properties, product interactions, and gastrointestinal tract effects on drug release is recommended for optimal formulation development [56].

10.8 Techniques to Improve Natural Excipients

10.8.1 Chemical Modification

Chemical modification of natural excipients, particularly starches, has emerged as a crucial strategy in pharmaceutical formulation development. Common modification techniques include oxidation, esterification, and etherification, which enhance properties like solubility, hydrophobicity, and thermoplasticity [54]. These modifications improve the functionality of excipients, allowing for better drug delivery, controlled release, and enhanced bioavailability [57]. Natural polymers are increasingly favored as excipients due to their non-toxicity, stability, and cost-effectiveness [4]. Plant-derived gums and mucilages, being hydrophilic and gel-forming, are particularly useful in controlled drug delivery systems [58]. The growing interest in natural excipients has led to the development of coprocessed excipients, which combine multiple functionalities in a single component [57]. These advancements in excipient technology are driving the creation of more effective and reliable pharmaceutical products.

10.8.2 Nanoformulation

Natural excipients are gaining prominence in pharmaceutical formulations due to their nontoxic, biodegradable, and cost-effective properties compared to synthetic alternatives [6, 15]. These excipients play crucial roles in novel drug delivery systems, particularly in nanoformulations, where they can influence drug release and absorption [32]. Nanoformulations using natural excipients have shown potential in improving the efficacy and stability of various products, including antiaging cosmetics [59]. Examples of natural excipients in nanoformulations include the substitution of synthetic lipids with natural and semisynthetic alternatives in lipid-based nanodrug delivery systems, and the use of gelatin and chitosan in encapsulated nanoparticulates [32]. The growing interest in natural products has led to increased research and development in this field, with global excipient markets expected to expand rapidly in response to emerging pharmaceutical industry trends [15].

Nanoformulation techniques have emerged as pivotal strategies to enhance the functional properties of natural excipients in pharmaceutical formulations. By reducing particle sizes to the nanometer scale, these methods improve solubility, stability, bioavailability, and controlled drug release profiles. Natural excipients, when modified using nanotechnology, exhibit superior physicochemical properties that make them more effective in drug delivery systems. Below is an overview of key nanoformulation approaches applied to natural excipients, supported by relevant references.

10.8.2.1 Nanoemulsions

Nanoemulsions are submicron-sized emulsions that enhance the delivery of hydrophobic drugs by improving their solubility and bioavailability. They consist of oil, water, and surfactants, forming a stable dispersion that enables efficient drug encapsulation. Natural excipients, such as triglyceride oils, can serve as the oil phase in nanoemulsions, facilitating the incorporation of lipophilic drugs. These nanoemulsions exhibit excellent kinetic stability, preventing phase separation and improving drug absorption. High-energy methods like high-pressure homogenization and ultrasonication are commonly employed to produce stable nanoemulsions [15].

10.8.2.2 Nanosuspensions

Nanosuspensions consist of pure drug nanoparticles stabilized by surfactants or polymers, enabling improved solubility and dissolution rates of poorly water-soluble drugs. Natural polymers like chitosan have been effectively utilized to stabilize these suspensions, enhancing their pharmaceutical applicability. Unlike traditional solubility-enhancing techniques, nanosuspensions allow high drug loading without requiring chemical modifications. Various methods such as high-pressure homogenization and media milling are frequently used to prepare nanosuspensions, ensuring uniform particle size and enhanced bioavailability [60].

10.8.2.3 Nanoparticles

Nanoparticles, particularly those derived from natural polymers, have gained substantial attention in drug delivery applications. Among them, chitosan nanoparticles are extensively studied due to their biocompatibility, biodegradability, and mucoadhesive properties. These nanoparticles can be synthesized through various techniques, including ionic gelation and emulsion solvent diffusion, which enable precise control over drug release and targeted delivery. The small particle size and high surface area improve drug absorption and therapeutic efficacy.

10.8.2.4 Nanocellulose

Nanocellulose, obtained from natural cellulose sources, has emerged as an innovative material for pharmaceutical applications. It is widely utilized in stabilizing Pickering emulsions, a type of emulsion stabilized by solid particles instead of traditional surfactants. Nanocellulose-based Pickering emulsions offer enhanced stability, making them an excellent choice for delivering bioactive compounds. Their ability to form a protective layer around dispersed droplets prevents coalescence and ensures long-term stability. These properties make nanocellulose a valuable excipient in nanoformulations.

10.8.2.5 Aquasomes

Aquasomes represent a unique class of nanoparticulate carrier systems comprising a solid nanocrystalline core coated with a carbohydrate layer. This structure facilitates the delivery of bioactive molecules while preserving their biological activity. The carbohydrate coating acts as a stabilizing shield, protecting sensitive biomolecules such as proteins, peptides, and vaccines from degradation. Aquasomes offer significant advantages in targeted drug delivery, particularly for therapeutic agents requiring high structural integrity and controlled release.

10.8.3 Hybrid Excipient Development

Coprocessed excipients have emerged as a promising alternative to single-component excipients in pharmaceutical formulations. These hybrid excipients, created by combining two or more existing excipients at the subparticle level, offer enhanced functionality and improved performance [61, 62]. They address the limitations of traditional excipients and provide cost-effective solutions for modern tablet manufacturing processes [61]. The development of new excipients requires demonstration of safety and absence of adverse reactions [63]. Proper selection of excipients is crucial in parenteral formulation development, particularly for lyophilized drug products, to achieve desired stability and efficacy [64]. Coprocessed excipients offer multifunctional activities, including improved flow properties, compressibility, and dissolution profiles [61]. As the pharmaceutical industry continues to evolve, these hybrid excipients are becoming increasingly important in meeting the demands of formulation experts and improving overall product quality [61].

10.9 Regulatory Aspects and Standards

Pharmaceutical excipients play a crucial role in drug formulations, affecting quality, efficacy, and safety [65]. While there are no specific regulations for excipients, their use is governed by medicine regulations in Europe and the United States [65, 66]. Natural excipients are gaining popularity due to their nontoxic nature, stability, and lower regulatory issues compared to synthetic alternatives [4]. They offer various functional benefits, including modulating solubility, enhancing bioavailability, and improving stability [66]. The International Pharmaceutical Excipients Council (IPEC) was established in 1991 to address industry concerns and harmonize international standards [66]. The development of new natural excipients is driven by safety and biodegradability concerns associated with synthetic excipients [50]. These natural alternatives can simplify formulation development, reduce operational costs, and maintain product quality while meeting regulatory requirements [50].

10.9.1 Approval Requirements

Natural excipients are gaining attention in pharmaceutical formulations due to their nontoxic, cost-effective, and readily available nature [7]. They offer advantages over synthetic counterparts, including biocompatibility and potential nutritional benefits [7]. However, regulatory challenges exist for novel excipients, as they are subject to stringent evaluation similar to new active substances [67]. The European Union has specific regulations and guidelines for excipients, derived from medicine regulations [65]. These include Community directives, European Pharmacopoeia standards, and guidelines from the European Agency for the Evaluation of Medicinal

Products [65]. Despite the benefits of natural excipients, their use in approved medicinal products remains limited due to regulatory complexities and potential approval delays [67]. The development of new natural excipients is necessary to address safety and biodegradability concerns associated with synthetic excipients, potentially simplifying formulation development and reducing operational costs [50].

10.9.2 Quality Control Measures

Pharmaceutical excipients play a crucial role in drug formulations, enhancing production, stability, and patient acceptability [68]. Natural excipients, derived from plants, are gaining popularity due to their nontoxic nature, availability, and cost-effectiveness [4]. However, regulatory authorities and manufacturers now prioritize quality control of excipients, as adulteration has led to adverse effects in patients [69]. The development of new excipients requires demonstrating their safety and absence of toxicity [63]. Regulatory mechanisms for excipient approval include Generally recognised as safe (GRAS) determination, food additive petition approval, or inclusion in an approved New drug application (NDA) [63]. Excipients can impact drug absorption, distribution, metabolism, and elimination, making their selection crucial for new formulations [68]. Recent regulatory initiatives focus on excipients for pediatric formulations, emphasizing the need for careful consideration in pharmaceutical development [68]. Recent guidelines focus on excipient safety, with appropriately identified and pure excipients generally considered safe when used in the correct context and concentration [70]. Advanced analytical techniques like near-infrared and Raman spectroscopy offer advantages in quality control, enabling rapid and reliable identification of excipients [70]. Ultimately, patient safety remains the primary focus of all regulatory requirements and quality control measures [70].

10.10 Future Trends and Research Opportunities

Natural excipients are gaining prominence in pharmaceutical formulations due to their nontoxic, biodegradable, and cost-effective properties compared to synthetic alternatives [6, 50]. These excipients play crucial roles in drug delivery systems, influencing drug release, absorption, and overall product stability [71]. The pharmaceutical industry is witnessing a shift toward functionality-related characterization of excipients, with a focus on developing novel drug delivery systems to address challenges such as bioavailability and solubility [15]. Natural excipients are being explored for use in both conventional dosage forms and advanced drug delivery systems, offering potential improvements in formulation development and operational costs [50]. As the global excipient market continues to grow, there is an increasing need for research and development of new natural excipients to meet the evolving demands of the pharmaceutical industry [6, 50].

10.10.1 Sustainable Sourcing

Sustainable sourcing of natural excipients in solid dosage forms is increasingly prioritized in the pharmaceutical industry, reflecting a commitment to environmental stewardship and product quality. Derived from renewable plant sources, natural excipients offer advantages such as biocompatibility and reduced ecological impact compared to synthetic alternatives. Their superior compatibility with human tissues minimizes adverse reactions, while utilizing renewable resources aligns with green chemistry principles, reducing the carbon footprint of pharmaceutical manufacturing. However, challenges such as ensuring uniformity in excipient properties due to plant source variations and navigating differing global regulatory standards require meticulous documentation and adherence to quality guidelines. Industry initiatives, such as sustainability information packages developed by the International Pharmaceutical Excipients Council (IPEC), aim to promote responsible sourcing practices and enhance transparency in the excipient supply chain. With consumer demand and regulatory pressures driving the shift toward sustainability, advancements in green chemistry and a focus on environmental, social, and governance (ESG) criteria are expected to influence excipient selection and sourcing strategies. Despite challenges, integrating sustainably sourced natural excipients into solid dosage forms presents significant opportunities, requiring a collaborative approach among suppliers, manufacturers, and regulatory bodies to balance environmental responsibility with product efficacy and safety [19].

10.10.2 Innovations in Natural Excipients

Natural excipients are gaining prominence in pharmaceutical formulations due to their advantages over synthetic alternatives. These plant-derived substances are nontoxic, cost-effective, and readily available [2, 6, 7, 13, 15]. Natural excipients have evolved from being inert bulk-forming agents to essential components that can influence drug release and absorption [6, 7, 13]. Examples include carrageenan, thaumatin, lard, and storax [7, 13, 15]. These excipients can serve various functions in conventional and novel drug delivery systems, such as diluents, binders, disintegrants, and lubricants [13]. The growing interest in natural products has led to an increased demand for plant-based excipients in the pharmaceutical industry [3, 6]. Natural excipients offer potential solutions to challenges in drug development, including bioavailability, permeability, and solubility issues [2, 15].

Innovations in natural excipients have significantly advanced the development of solid dosage forms in drug delivery systems. Zein, a biocompatible and biodegradable protein derived from corn, has been utilized to create nanoparticles, microparticles, and films, enhancing controlled drug release and stability in oral formulations [72]. Polysaccharides such as pectin, starch, guar gum, and alginate have been employed to fabricate modified-release systems like matrix tablets and beads, owing to their safety and gel-forming properties [73]. Additionally, natural deep eutectic solvents (NADES) have emerged as effective agents for improving the solubility, stability, and delivery of poorly soluble drugs like curcumin, offering a green approach to drug formulation [74]. These advancements underscore the potential of natural excipients to enhance the efficacy and safety of pharmaceutical products.

10.11 Conclusion

The increasing emphasis on safer, sustainable, and efficient pharmaceutical formulations has positioned natural excipients as valuable alternatives to synthetic counterparts. Their multifunctional roles, including binding, disintegration, and drug release modulation, make them indispensable in drug formulation. Plant-derived excipients, particularly polysaccharides and mucilages, have gained significant attention due to their nontoxic nature, ease of modification, and wide availability. Meanwhile, marine and animal-derived excipients, despite certain limitations, contribute to specialized drug delivery applications. As the pharmaceutical industry moves toward natural and patient-friendly excipients, their role in improving bioavailability, stability, and drug delivery efficiency will continue to expand.

Despite their numerous advantages, natural excipients face challenges such as physicochemical variability, stability concerns, and regulatory complexities, which can impact their widespread adoption. To address these limitations, formulation techniques such as chemical modifications, nanoformulations, and hybrid excipient development have been explored, demonstrating promising potential in enhancing their functionality and stability. Additionally, sustainable sourcing and innovative research efforts are driving the development of novel natural excipients that align with industry and regulatory requirements.

Moving forward, a multidisciplinary approach involving pharmaceutical scientists, regulatory authorities, and industry stakeholders is essential for optimizing the use of natural excipients in drug delivery systems. Advances in formulation technologies and evolving regulatory frameworks will further facilitate their integration into modern pharmaceutical applications. With continuous research and technological innovations, natural excipients are poised to play a transformative role in pharmaceutical sciences, contributing to the development of safer, more effective, and environmentally sustainable drug formulations.

References

1. Kaur M, Kaur S, Pal Singh A, Pal Singh A. A review-types of natural excipients. Certified Journal | 391 Kaur et al World. J Pharm Pharm Sci. 2021; https://doi.org/10.17605/OSF.IO/S4WKG.
2. Jain A, Radiya P, Wadekar R, Limaye S, Pawar C. Natural excipients-an alternative to synthetic excipients: a comprehensive review keywords: active pharmaceutical ingredient efficacy natural excipients safety. Int J Pharm Med Res. 2014;2:123–7.
3. Kumar T, Kumar Gupta S, Prajapati K, Tripathi DK, Sharma V, Jain P. Natural excipients: a review. AJPLS. 2012;2(1):97–108.

4. Dharmendra S, Surendra JK, Sujata M, Shweta S. Natural excipients—a review. Int J Pharm Bio Arch. 2012;3:1028–34.
5. Sabne PS, Avalaskar AN, Jadhav R, Sainkar PS. Natural excipients. Res J Pharm, Biol Chem Sci. 2013;4:1346–54.
6. Selvadurai S, Divya S, Geetha A, Rohini K, Anbazhagan S. Herbal excipients-significance of substances of natural origin as excipients. Int J Pharmtech Res. 2020;13:242–52.
7. Singh P, Mahmood T, Shameem A, Bagga P, Ahmad N. A review on herbal excipients and their pharmaceutical applications. Sch Acad J Pharm. 2016;5:53–7.
8. Vijayalakshmi A, Sanjeet MH, Sanjay RS, Roshini M, Jayakumari S, Joyce JS. Herbal excipients from agro waste: a review. Ann Phytomed Int J. 2024; https://doi.org/10.54085/ap.2024.13.1.50.
9. Kankanamge SU, Neranja AGK, Sandarenu KDS. Mini-review on natural disintegrants in pharmaceutical formulations. J Complement Altern Med Res. 2021;14(2):52–6.
10. Desai PM, Liew CV, Heng PWS. Review of disintegrants and the disintegration phenomena. J Pharm Sci. 2016;105:2545–55.
11. Nabila N, Karo-Karo SU. Exploring natural disintegrant as alternative excipients sources for tablet formulation: a systematic review. Int J Sci Res Arch. 2024;13:3296–305.
12. Mulla Mam N, Bias S, Manasi M. Review on herbal excipients. Res J Pharmacogn Phytochem. 2023;15(2):118–24.
13. Singh AV. Biopolymers in drug delivery: a review. Pharmacology. 2011;1:666–74.
14. Beneke CE, Viljoen AM, Hamman JH. Polymeric plant-derived excipients in drug delivery. Molecules. 2009;14:2602–20.
15. Kumar S, Gupta SK. Natural polymers, gums and mucilages as excipients in drug delivery. Polim Med. 2012;42:191–7.
16. Kalasz H, Antal I. Drug excipients. Curr Med Chem. 2006;13:2535–63.
17. Pal RS, Pal Y, Wal A, Wal P. Current review on plant based pharmaceutical excipients. Open Med J. 2019;6:1–5.
18. Prajapati VD, Jani GK, Moradiya NG, Randeria NP. Pharmaceutical applications of various natural gums, mucilages and their modified forms. Carbohydr Polym. 2013;92:1685–99.
19. Nammas M. Systematic review of plant-based excipients in topical drug delivery. Ibnosina J Med Biomed Sci. 2024;16:162–8.
20. Sabalingam S, Jayasuriya WJAB, Banukie WJA, Jayasuriya N. Pharmaceutical excipients of marine and animal origin: a review. Science Signpost Publishing; 2019.
21. Koo OM, Varia SA. Case studies with new excipients: development, implementation and regulatory approval. Ther Deliv. 2011;2:949–56.
22. Singh N, Halliday HL, Stevens TP, Suresh G, Soll R, Rojas-Reyes MX. Comparison of animal-derived surfactants for the prevention and treatment of respiratory distress syndrome in preterm infants. Cochrane Database Syst Rev. 2015; https://doi.org/10.1002/14651858.CD010249.pub2.
23. Cassotta M, Bartnicka JJ, Pistollato F, Parvatam S, Weber T, D'Alessandro V, Bastos LF, Coecke S. A worldwide survey on the use of animal-derived materials and reagents in scientific experimentation. Eng Life Sci. 2022;22:564–83.
24. Mayer AMS, Glaser KB, Cuevas C, Jacobs RS, Kem W, Little RD, McIntosh JM, Newman DJ, Potts BC, Shuster DE. The odyssey of marine pharmaceuticals: a current pipeline perspective. Trends Pharmacol Sci. 2010;31:255–65.
25. Alves A, Sousa E, Kijjoa A, Pinto M. Marine-derived compounds with potential use as cosmeceuticals and nutricosmetics. Molecules. 2020;25:2536.
26. Ahmmed MK, Ahmmed F, Tian H (Sabrina), Carne A, Bekhit AE. Marine omega-3 (n-3) phospholipids: a comprehensive review of their properties, sources, bioavailability, and relation to brain health. Compr Rev Food Sci Food Saf. 2020;19:64–123.
27. Sopyan I, Santi NMWS, Berlian AV, Meilina NE, Fauza Q, Apriyandi RA. A review: pharmaceutical excipients of solid dosage forms and characterizations. Int J Res Pharm Sci. 2020;11:1472–80.

28. Assad M, Jackson N. Biocompatibility evaluation of orthopedic biomaterials and medical devices: a review of safety and efficacy models. In: Encyclopedia of biomedical engineering. Elsevier; 2019. p. 281–309.
29. Barman A. Review on biocompatibility of ZnO nano particles. Springer India; 2015. p. 343–52.
30. Ramot Y, Haim-Zada M, Domb AJ, Nyska A. Biocompatibility and safety of PLA and its copolymers. Adv Drug Deliv Rev. 2016;107:153–62.
31. Athanasiou K. Sterilization, toxicity, biocompatibility and clinical applications of polylactic acid/polyglycolic acid copolymers. Biomaterials. 1996;17:93–102.
32. Jirwankar P, Agrawal S, Shaikh F, Borse K. Natural excipients: role in nano drug delivery system. Curr Nanomater. 2023; https://doi.org/10.2174/0124054615248522231211105002.
33. Mittal S, Pawar S. Herbal excipients: an overview. Int J Pharmacognosy Pharm. 2019;1:05–9.
34. Frent O-D, Duda-Seiman DM, Vicas LG, et al. Study of the influence of the excipients used for the synthesis of microspheres loaded with quercetin: their characterization and antimicrobial activity. Coatings. 2023;13:1376.
35. Wang X, Huang L, Zhang C, Deng Y, Xie P, Liu L, Cheng J. Research advances in chemical modifications of starch for hydrophobicity and its applications: a review. Carbohydr Polym. 2020;240:116292.
36. Bhatt P, Kumar V, Goel R, Sharma SK, Kaushik S, Sharma S, Shrivastava A, Tesema M. [Retracted] Structural modifications and strategies for native starch for applications in advanced drug delivery. Biomed Res Int. 2022; https://doi.org/10.1155/2022/2188940.
37. Lemos PVF, Marcelino HR, Cardoso LG, de Souza CO, Druzian JI. Starch chemical modifications applied to drug delivery systems: from fundamentals to FDA-approved raw materials. Int J Biol Macromol. 2021;184:218–34.
38. Garcia MAVT, Garcia CF, Faraco AAG. Pharmaceutical and biomedical applications of native and modified starch: a review. Starch Stärke. 2020; https://doi.org/10.1002/star.201900270.
39. Shokri J, Adibki K. Application of cellulose and cellulose derivatives in pharmaceutical industries. In: Cellulose—medical, pharmaceutical and electronic applications; 2013. https://doi.org/10.5772/55178.
40. Hosny KM, Alkhalidi HM, Alharbi WS, Md S, Sindi AM, Ali SA, Bakhaidar RB, Almehmady AM, Alfayez E, Kurakula M. Recent trends in assessment of cellulose derivatives in designing novel and nanoparticulate-based drug delivery systems for improvement of oral health. Polymers (Basel). 2021;14:92.
41. Sun B, Zhang M, Shen J, He Z, Fatehi P, Ni Y. Applications of cellulose-based materials in sustained drug delivery systems. Curr Med Chem. 2019;26:2485–501.
42. Oprea M, Voicu SI. Recent advances in composites based on cellulose derivatives for biomedical applications. Carbohydr Polym. 2020;247:116683.
43. Petri DFS. Xanthan gum: a versatile biopolymer for biomedical and technological applications. J Appl Polym Sci. 2015; https://doi.org/10.1002/app.42035.
44. Thombare N, Jha U, Mishra S, Siddiqui MZ. Guar gum as a promising starting material for diverse applications: a review. Int J Biol Macromol. 2016;88:361–72.
45. Bhat IM, Wani SM, Mir SA, Masoodi FA. Advances in xanthan gum production, modifications and its applications. Biocatal Agric Biotechnol. 2022;42:102328.
46. Albuquerque PBS, Coelho LCBB, Teixeira JA, Carneiro-da-Cunha MG. Approaches in biotechnological applications of natural polymers. AIMS Mol Sci. 2016;3:386–425.
47. Szekalska M, Puciłowska A, Szymańska E, Ciosek P, Winnicka K. Alginate: current use and future perspectives in pharmaceutical and biomedical applications. Int J Polym Sci. 2016;2016:1–17.
48. Van Rooyen B, De Wit M, Van Niekerk J. Pectin and alginate functional biopolymers: factors influencing structural composition, functional characteristics and biofilm development. Coatings. 2024;14:987.
49. Wawszczak A, Kocki J, Kołodyńska D. Alginate as a sustainable and biodegradable material for medical and environmental applications—the case studies. J Biomed Mater Res B Appl Biomater. 2024; https://doi.org/10.1002/jbm.b.35475.

50. Bahadur S, Roy A, Chanda R, Choudhury A. Natural excipient development: need and future. Asian J Pharm Res. 2016;4(1):12–5.
51. Dilip Bhandare S, Kavade E, Surse S, Bhandare D. Natural polymers: as pharmaceutical excipients and their applications in different pharmaceutical formulations—a review. World J Pharm Res. 2015;4(6):626–44.
52. Dave VS, Saoji SD, Raut NA, Haware RV. Excipient variability and its impact on dosage form functionality. J Pharm Sci. 2015;104:906–15.
53. Peng Soh J, Liew C, Sia Heng P. Impact of excipient variability on drug product processing and performance. Curr Pharm Des. 2015;21:5890–9.
54. Adewale Adetunji O. Chemically modified starches as excipients in pharmaceutical dosage forms. In: Chemical properties of starch; 2020. https://doi.org/10.5772/intechopen.88210.
55. Kushner J. Utilizing quantitative certificate of analysis data to assess the amount of excipient lot-to-lot variability sampled during drug product development. Pharm Dev Technol. 2013;18:333–42.
56. Zarmpi P, Flanagan T, Meehan E, Mann J, Fotaki N. Biopharmaceutical aspects and implications of excipient variability in drug product performance. Eur J Pharm Biopharm. 2017;111:1–15.
57. Ingle VB, Potdar SS, Mhetre RL, Kulkarni NS, Dhole SN. Natural and modified excipients in novel drug delivery system: a review. Res J Pharm Dosage Forms Technol. 2021;13:147–52.
58. Rani Earle R, Usha Ayalasomayajula L, Loknadh G, Ram Konda Reddy KS, Ratna Kanth L. A review on natural polymers used in pharmaceutical dosage forms. Int J Soc Res Methodol. 2016;3(3):1–12.
59. Basudkar V, Gharat SA, Momin MM, Shringarpure M. A review of anti-aging nanoformulations: recent developments in excipients for nanocosmeceuticals and regulatory guidelines. Crit Rev Ther Drug Carrier Syst. 2022;39:45–97.
60. Patravale VB, Date AA, Kulkarni RM. Nanosuspensions: a promising drug delivery strategy. J Pharm Pharmacol. 2004;56:827–40.
61. Desai U, Chaudhari PD, Phatak AA. A review: coprocessed excipients—an alternative to novel chemical entities. Int J Pharm Sci Rev Res. 2012;12(2):93–105.
62. Singh A, Maheshwari S, Vishwakarma VK, Mohammed S. Co-processed excipients: advances and future trends. Intell Pharm. 2024;2:130–5.
63. Patel H, Shah V, Upadhyay U. ChemInform abstract: new pharmaceutical excipients in solid dosage forms. ChemInform. 2012; https://doi.org/10.1002/chin.201232266.
64. Pramanick S, Singodia D, Chandel V. Excipient selection in parenteral formulation development. Pharma Times. 2013;45(3):65–77.
65. Artiges A. Excipients: regulatory aspects. Therapie. 1999;54:15–9.
66. Kumar D, Dureja H. Pharmaceutical excipients: global regulatory issues. Indonesian J Pharm. 2013;24:215–21.
67. Kozarewicz P, Loftsson T. Novel excipients—regulatory challenges and perspectives—the EU insight. Int J Pharm. 2018;546:176–9.
68. Elder DP, Kuentz M, Holm R. Pharmaceutical excipients—quality, regulatory and biopharmaceutical considerations. Eur J Pharm Sci. 2016;87:88–99.
69. DeMerlis CC. Regulation of pharmaceutical excipients. In: Excipient toxicity and safety. CRC Press; 1999. p. 91–118.
70. Franjko A. Regulatorni zahtjevi i analitičke metode u kontroli kvalitete pomoćnih tvari u lijekovima. Završni specijalistički; 2019.
71. Palcsó B, Zelkó R. Different types, applications and limits of enabling excipients of pharmaceutical dosage forms. Drug Discov Today Technol. 2018;27:21–39.
72. Berardi A, Bisharat L, AlKhatib HS, Cespi M. Zein as a pharmaceutical excipient in oral solid dosage forms: state of the art and future perspectives. AAPS PharmSciTech. 2018;19:2009–22.
73. Li L, Zhang X, Gu X, Mao S. Applications of natural polymeric materials in solid oral modified-release dosage forms. Curr Pharm Des. 2015;21:5854–67.
74. Jeliński T, Przybyłek M, Cysewski P. Natural deep eutectic solvents as agents for improving solubility, stability and delivery of curcumin. Pharm Res. 2019;36:116.

Applications of Natural Polymers in Controlled-Release Systems

11

Akshata Yashwant Patne, Kazi Akramuddaula, Divyang Patel, Meghraj Suryawanshi, and Kuldeep Vinchurkar

11.1 Introduction

Controlled-release systems (CRS) have revolutionized the therapeutic delivery of drugs by allowing the sustained and targeted release of therapeutic compounds, thereby enhancing therapeutic effectiveness, minimizing side effects, and enhancing patient compliance. Unlike immediate-release formulations, which dissolve rapidly and cause variability in drug concentration, CRS is designed to maintain the controlled plasma drug levels over time. This controlled release ensures consistent plasma drug concentration, thereby reducing toxicity risks and ensuring prolonged pharmacological effects. These systems are particularly useful for drugs with a narrow therapeutic index (NTI) or those requiring frequent dosing.

A. Y. Patne
Graduate Programs, Taneja College of Pharmacy, Tampa, FL, USA

Center for Research and Education in Nanobioengineering, Department of Internal Medicine, Morsani College of Medicine, University of South Florida, Tampa, FL, USA

K. Akramuddaula
Center for Research and Education in Nanobioengineering, Department of Internal Medicine, Morsani College of Medicine, University of South Florida, Tampa, FL, USA

D. Patel
Department of Pharmaceutics, Institute of Pharmaceutical Sciences Faculty of Pharmacy, Parul University, Vadodara, Gujarat, India

M. Suryawanshi (✉) · K. Vinchurkar
Department of Pharmaceutics, Sandip Institute of Pharmaceutical Sciences (SIPS), Affiliated to Savitribai Phule Pune University (SPPU, Pune), Nashik, Maharashtra, India

J. S et al. (eds.), *Innovative Pharmaceutical Excipients: Natural Sources*,
https://doi.org/10.1007/978-981-96-7959-1_11

11.1.1 Historical Development of Controlled-Release Systems

The field of controlled-release drug delivery has experienced remarkable evolution in the past few decades. Initially, conventional drug delivery systems relied on immediate-release formulations, where drugs dissolved rapidly upon administration, leading to quick systemic absorption. However, this approach often resulted in erratic plasma drug levels—high peak concentrations could lead to toxicity, while trough levels resulted in subtherapeutic effects [2]. To overcome these challenges, sustained-release tablets and capsules were developed, incorporating polymer-based matrix formulations or coatings to slow down drug dissolution and diffusion rates.

The second generation of CRS emerged with the introduction of biodegradable and biocompatible polymeric materials, such as poly(lactic-co-glycolic acid) (PLGA), alginate, and chitosan [110] (Fig. 11.1). These materials allowed for the development of more advanced platforms such as microspheres, nanoparticles, and hydrogels, which provided greater control over drug release kinetics. A notable example is DepoFoam®, a lipid-based drug delivery system designed to encapsulate drugs in multilamellar vesicles for extended-release over days or weeks [114]. Such innovations were particularly useful in managing chronic conditions like cancer and pain, where consistent drug levels are crucial for maintaining therapeutic efficacy.

The latest advancements in CRS have contributed to the rise of innovative and precise drug delivery systems that incorporate nanotechnology and stimuli-responsive materials. These third-generation systems are designed to release drugs in reaction to physiological signals like pH changes, temperature shifts, or enzymatic activity, allowing for site-specific drug delivery [89]. This innovation minimizes systemic toxicity while enhancing drug accumulation at diseased sites. One example of this approach is Lupron Depot®, an injectable sustained-release

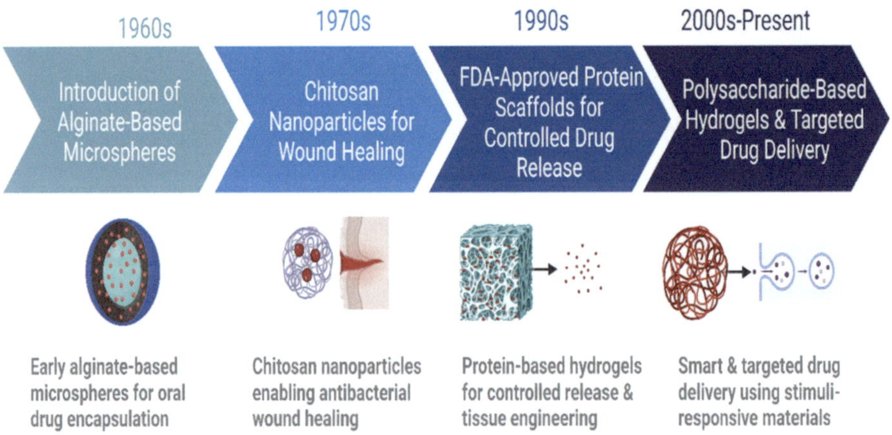

Fig. 11.1 Historical evolution of controlled-release systems

formulation of leuprolide acetate, which ensures continuous drug release for up to 3 months for the treatment of prostate cancer and endometriosis.

11.1.2 Market Trends and Economic Benefits

The controlled-release drug delivery market has experienced exponential growth due to advancements in precision medicine and materials science. According to market reports, the global CRS industry is projected to surpass $70 billion by 2027, driven by increasing demand for long-acting therapeutics and patient-centric formulations [60]. CRS technology reduces dosing frequency, improves medication adherence, and decreases hospitalization rates, leading to better health outcomes and cost savings for healthcare systems.

For example, extended-release formulations of antihypertensive medications have been shown to improve blood pressure regulation while reducing cardiovascular risks compared to immediate-release counterparts [167]. Similarly, osmotic-controlled-release oral delivery systems (OROS), such as nifedipine extended-release formulations, provide zero-order drug release kinetics, ensuring steady-state plasma concentrations with superior bioavailability.

Another significant economic and therapeutic advantage of CRS is site-specific drug targeting. By engineering polymeric carriers such as PLGA nanoparticles, drugs can accumulate at target tissues while avoiding systemic exposure, thus enhancing their therapeutic index [17]. This is particularly relevant in oncology, where doxorubicin-loaded nanoparticles have demonstrated superior tumor accumulation and lower cardiotoxicity than traditional chemotherapy formulations [69]. Personalized CRS formulations customized according to genetic profiles and disease conditions are under investigation, paving the way for patient-specific therapies. Additionally, the development of biodegradable and environmentally sustainable polymers aims to address concerns related to pharmaceutical waste and polymer accumulation in biological systems.

By optimizing drug release kinetics and targeting efficiency, controlled-release drug delivery continues to lead progressions in pharmaceutical innovation. As technological advancements drive the next generation of CRS, these advancements will continue to define the future of personalized medicine, ensuring safer, more effective, and patient-friendly therapeutic options.

11.1.3 Role of Natural Polymers in Controlled-Release Systems

Due to their inherent biocompatibility, biodegradability, and low toxicity, natural polymers are crucial in the development of controlled-release systems. Unlike synthetic polymers, which often require complex chemical processing and may pose long-term environmental hazards, natural polymers are derived from replenishable natural sources, making them a sustainable and eco-friendly alternative [146].

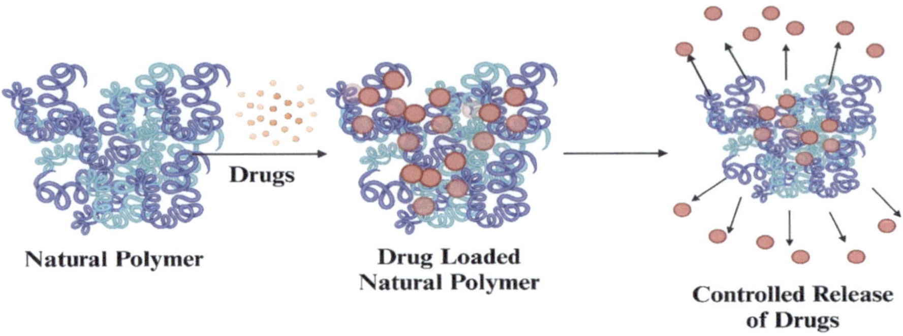

Fig. 11.2 Natural polymers for controlled release of drugs. (Created in https://BioRender.com)

Their widespread availability, reduced environmental footprint, and customizable chemical properties make them ideal for drug delivery uses. Furthermore, through chemical modifications, natural polymers can be adjusted to exhibit regulated degradation over time, enhanced mechanical strength, and tunable drug release kinetics, allowing their application across oral, parenteral, and transdermal delivery systems [125] (Fig. 11.2).

11.2 Classification of Polymers in Drug Delivery

11.2.1 Comparison of Natural and Synthetic Polymers in Drug Delivery

Natural and synthetic polymers are mainly distinguished by their source, degradation profile, toxicity, and sustainability. Natural polymers are readily metabolized via enzymatic or hydrolytic pathways, reducing concerns of bioaccumulation and toxic degradation by-products. In contrast, synthetic polymers offer greater control over structural properties but often lack predictable biodegradation and may induce immune responses or long-term toxicity [51] [196]. A detailed comparative analysis is provided below as shown in Table 11.1.

Despite the advantages of synthetic polymers in drug release control and mechanical strength, natural polymers offer a safer, more sustainable, and bioactive alternative, particularly for biomedical and pharmaceutical applications.

11.2.2 Historical Evolution of Natural Polymers in Drug Delivery

The application of natural polymers in medicine dates back centuries. Ancient civilizations used plant-derived materials, such as gums, resins, and animal-derived gelatin, for wound healing and medicinal formulations [211].

Table 11.1 Comparison of natural and synthetic polymers in drug delivery

Property	Natural polymers	Synthetic polymers
Biocompatibility	High, due to biological origin and structural similarity to extracellular matrix (ECM) proteins	May induce immune responses, cytotoxicity, or inflammatory reactions
Biodegradability	Enzymatically or hydrolytically degraded into nontoxic metabolites	Often nondegradable or slowly degraded, leading to potential accumulation in tissues
Toxicity	Minimal or none, safe metabolic by-products	May contain residual monomers or degradation products that are toxic
Sustainability	Renewable, eco-friendly, and biodegradable, reducing pharmaceutical waste	Derived from petrochemicals, contributing to environmental pollution
Mechanical strength	Variable can be enhanced through cross-linking or hybridization	Generally high, tunable via polymerization and copolymer modifications
Drug release control	Can be chemically modified for sustained or targeted release	More predictable and programmable but often requires complex chemical synthesis
Degradation half-life	Varies (days to months), depending on polymer type (e.g., alginate: ~5–14 days; chitosan: ~2–4 weeks).	Typically, longer (months to years) (e.g., PLGA: 3–6 months, Polycaprolactone (PCL): 2–3 years)

In the mid-twentieth century, the pharmaceutical industry recognized the potential of natural polymers, leading to the introduction of gelatin, alginate, and chitosan as pharmaceutical excipients for controlled drug release.

Key milestones in the evolution of natural polymers for CRS include:

- *1960s:* Introduction of alginate-based microspheres for oral drug encapsulation
- *1970s:* Development of chitosan nanoparticles for antibacterial wound healing applications
- *1990s:* FDA approval of protein-based scaffolds, expanding their role in tissue engineering and drug delivery
- *2000s–Present:* Emergence of polysaccharide-based hydrogels and hybrid natural-synthetic polymer systems, leading to targeted and stimuli-responsive drug delivery technologies.

One of the pioneering applications was the development of alginate-based insulin microcapsules, providing early examples of sustained drug release for diabetes therapy. Similarly, chitosan nanoparticles for anticancer drug delivery demonstrated improved tumor targeting and reduced systemic toxicity, highlighting their potential for precision medicine [20, 82]. With continuous advancements in biotechnology, natural polymers are increasingly being used in drug delivery as multifunctional biopolymer systems, combining targeting ligands, stimuli-responsiveness, and biomimetic properties for enhanced therapeutic outcomes.

11.2.3 Eco-friendly and Sustainable Controlled-Release Systems

The environmental impact of pharmaceutical manufacturing is an area of increasing concern, particularly regarding polymer-based drug delivery systems. Synthetic polymers, often derived from petroleum-based sources, contribute to plastic pollution, require complex disposal methods, and possess a high carbon footprint [173, 190].

Natural polymers, on the other hand, offer a sustainable solution due to their biodegradability, renewability, and eco-friendly degradation. The growing emphasis on green chemistry has accelerated the adoption of natural polymers for biodegradable drug formulations [4, 144].

Key sustainability benefits of natural polymers include:

- *Minimal Environmental Burden:* Biodegradable materials such as alginate and chitosan degrade into harmless by-products, reducing polymer accumulation in the environment [115].
- *Microbial Fermentation-Based Production:* The shift toward bacterial or fungal-based biopolymer synthesis (e.g., microbial alginate, recombinant gelatin) ensures scalability, batch consistency, and reduced energy-intensive processing [14].
- *Solvent-Free Fabrication Techniques:* Emerging nanotechnology approaches allow the manufacturing of natural polymer-based CRS using aqueous-based formulations, eliminating the need for toxic organic solvents [144].
- *Regulatory Alignment:* Global regulatory policies (US Food and Drug Administration [FDA], European Medicines Agency [EMA]) are increasingly favoring biodegradable polymeric drug carriers, driving the adoption of natural polymer-based formulations in commercial pharmaceuticals [141].
- *Eco-friendly CRS:* The demand for eco-friendly CRS is expected to rise significantly, as pharmaceutical companies shift toward biodegradable drug formulations that align with global sustainability goals [101].

11.2.4 Classification of Natural Polymers in Drug Delivery

Natural polymers play a crucial role in controlled-release drug delivery systems because of their capacity to improve drug stability, prolong release duration, and improve biocompatibility. These polymers can be broadly classified into polysaccharides, proteins, and other naturally derived materials. Each category offers distinct advantages, making them appropriate for specific drug delivery applications [75].

11.2.4.1 Polysaccharides

Polysaccharides, being highly hydrophilic and biodegradable, are widely employed in CRS due to their adjustable mechanical properties, simple chemical modification procedures, and excellent mucoadhesive characteristics. These polymers can be chemically modified to alter their molecular weight, cross-linking density, and degradation rate, providing accurate control over drug release patterns. Among polysaccharides, alginate, chitosan, and guar gum have been extensively studied for their controlled-release applications [76].

11.2.4.2 Alginate-Based Controlled-Release Systems

Alginate is a naturally derived anionic polysaccharide made up of β-D-mannuronic acid (M) and α-L-guluronic acid (G) that creates hydrogels via ionotropic gelation in the presence of divalent cations such as calcium (Ca^{2+}). The gelation process is mild and occurs at physiological conditions, making alginate highly suitable for encapsulating bioactive molecules, including proteins, peptides, and lives cells.

Drug-release profiles are heavily influenced by the molecular weight and M/G ratio of alginate. A higher guluronic acid content results in stronger gel networks, leading to slower drug diffusion and prolonged release. Studies on alginate-based microcapsules for insulin delivery have demonstrated sustained glycemic control in diabetic models, reducing the frequency of injections required. Clinical trials have explored alginate-based oral delivery systems for peptide drugs, such as insulin and calcitonin, with promising outcomes in improving bioavailability and patient adherence [68, 80, 168].

11.2.4.3 Chitosan-Based Mucoadhesive Drug Delivery

Chitosan, a polysaccharide derived from chitin, is widely employed in CRS due to its ability to adhere to mucosal membranes and facilitate drug absorption across epithelial barriers. The protonated amino groups on chitosan interact electrostatically with negatively charged mucosal surfaces, making it particularly effective for oral, nasal, and ophthalmic drug delivery systems [39, 133].

The viscosity and solubility of chitosan are determined by its degree of deacetylation and molecular weight, and drug-release behavior. High molecular weight chitosan forms viscous gels that provide extended drug release, while low molecular weight chitosan enhances drug permeation across epithelial barriers [122, 203].

Chitosan-derived nanoparticles have been developed for targeted anticancer drug delivery like doxorubicin, demonstrating improved tumor accumulation and reduced systemic toxicity. Furthermore, in vitro and in vivo investigations of chitosan hydrogels loaded with antibiotics have demonstrated their ability to provide controlled localized drug release, making them ideal for wound healing and infection control [208].

11.2.4.4 Guar Gum for Colon-Targeted Drug Delivery

Guar gum, a galactomannan polysaccharide, exhibits high viscosity and enzymatic degradability inside the colon, making it a perfect carrier for targeted drug delivery to the colon. Controlled degradation by colonic microflora ensures targeted drug release, offering a promising approach for treating inflammatory bowel diseases [181].

A clinical study exploring guar gum-based matrices as carriers for drug delivery of 5-aminosalicylic acid (5-ASA) demonstrated improved therapeutic efficacy in ulcerative colitis patients. The swelling and erosion behavior of guar gum in simulated gastrointestinal (GI) fluids have been extensively investigated to optimize release kinetics for colorectal cancer therapy and probiotic encapsulation [143].

11.2.5 Cross-Linking Strategies for Enhancing Polysaccharide Stability

Chemical modifications such as ionic cross-linking, covalent bonding, and graft polymerization are employed to enhance the stability of drug delivery systems based on polysaccharides.

- Ionic cross-linking (alginate-Ca^{2+}, chitosan-TPP (Sodium Tripolyphosphate)) improves mechanical strength and reduces burst release.
- Covalent modifications (PEGylation, acetylation) modulate hydrophilicity and degradation rates.
- Graft polymerization with synthetic polymers enhances mucoadhesive and pH-responsive properties [161, 195].

These cross-linking strategies have been successfully applied in alginate-chitosan polyelectrolyte complexes, demonstrating synergistic drug retention and prolonged bioavailability in gastrointestinal and pulmonary delivery systems [107, 185] (Fig. 11.3).

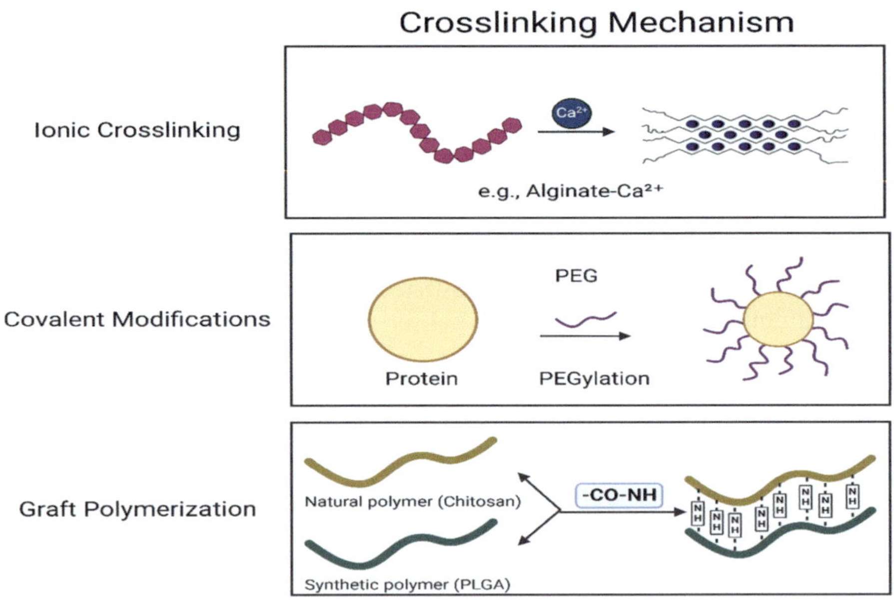

Fig. 11.3 Cross-linking strategies for enhancing polysaccharide stability

11.3 Proteins

Proteins exhibit excellent biocompatibility and bio-functionality because they resemble components of the extracellular matrix (ECM). They can be engineered into hydrogels, microspheres, and Scaffolds used in controlled drug delivery, tissue regeneration, and applications in regenerative medicine [24, 103].

11.3.1 Protein-Based Hydrogels for Tissue Engineering

Gelatin- and collagen-derived hydrogels replicate the hydration and elasticity of natural tissues, making them well-suited for wound healing, cartilage repair, and drug delivery. These hydrogels enable controlled drug release by regulating polymer degradation and diffusion [79, 147].

Mechanistically, protein-based hydrogels undergo enzymatic degradation via matrix metalloproteinases (MMPs), allowing enzyme-responsive drug release in diseased tissues. A study demonstrated that gelatin hydrogels loaded with vascular endothelial growth factor (VEGF) facilitated faster wound healing and angiogenesis in diabetic ulcer models [92, 97].

11.3.2 Advances in Recombinant Protein Technology

The advancement of recombinant DNA technology has made possible the large-scale production of bioengineered proteins that exhibit enhanced stability and reduced immunogenicity. Engineered recombinant collagen and silk fibroin scaffolds have been explored for long-acting drug-delivery implants. In ophthalmic delivery, recombinant gelatin nanoparticles encapsulating corticosteroids have been studied for age-related macular degeneration (AMD) treatment, showing sustained intraocular drug release over several months [21, 41].

11.4 Other Natural Polymers

Apart from polysaccharides and proteins, certain plant and animal-derived polymers such as shellac and natural rubber have been employed in specialized controlled-release applications.

Shellac for Gastro-Resistant Coatings

Shellac, a natural resin secreted by lac insects, has been widely used in sustained-release oral formulations. Its pH-sensitive solubility makes it an effective enteric coating agent, protecting drugs from gastric degradation while ensuring controlled release in the intestinal environment [56].

Recent research has explored shellac-based microencapsulation for probiotics, demonstrating enhanced bacterial viability in acidic conditions. This has paved the way for commercial probiotic formulations with improved gastrointestinal stability [40, 78].

11.4.1 Natural Rubber for Transdermal Patches

Natural rubber, though less common in drug delivery, has been investigated for its elastomeric properties in transdermal systems. Its flexibility and durability make it an excellent candidate for long-wear transdermal patches, such as those used in nicotine replacement therapy and chronic pain management [166].

Studies on natural rubber-latex membranes loaded with fentanyl have shown sustained analgesic effects for postoperative pain relief. Additionally, microporous rubber scaffolds are being researched for their effectiveness in bioactive wound dressings, offering continuous antiseptic release to prevent infection in chronic wounds [34, 46].

11.5 Mechanisms of Controlled Drug Release

Drug release from natural polymer-based systems is regulated by specific mechanisms that control the rate, duration, and location of drug availability. The mechanisms of diffusion-controlled, swelling-controlled, and biodegradation-controlled release each offer distinct benefits for various therapeutic applications. The selection of the release mechanism is influenced by factors like the physicochemical properties of the drug, the polymer matrix composition, and the desired pharmacokinetic profile. A thorough understanding of these mechanisms is crucial for developing optimized controlled-release systems that enhance drug efficacy and reduce adverse effects [12, 13].

11.5.1 Diffusion-Controlled Systems

In diffusion-controlled drug release, the drug migrates from the polymer matrix or a surrounding membrane at a rate dictated by Fick's laws of diffusion. The release kinetics depend on polymer porosity, cross-linking density, and molecular interactions between the drug and polymer. Diffusion-based systems can be classified into matrix systems and reservoir systems, each governed by different mathematical models that predict drug release profiles [58, 106].

11.5.1.1 Mathematical Models for Diffusion-Based Release

Different mathematical models have been created to characterize the release kinetics of drugs from diffusion-controlled polymer matrices:

1. *Higuchi Model*

   According to the Higuchi model, the release rate of drugs from a homogeneous matrix follows a square root of time function [102]:

$$Q = kt1/2Q = k\,t\,\wedge\{12\}$$

   where QQ represents the total amount of drug released, kk is a constant incorporating the drug diffusion coefficient, and tt is time. This model is widely used for hydrophilic and hydrophobic polymer matrices such as alginate and shellac-based systems.

2. *Korsmeyer-Peppas Model*

   This empirical model provides a more generalized equation for describing drug diffusion through polymeric systems:

$$Mt / M\infty = ktnM\_t / M\_\backslash\text{infty} = k\,t\wedge n$$

   where $Mt/M\infty M_t/M_\backslash$infty represents the fraction of drug released at time tt, kk is a rate constant, and nn is the diffusional exponent, which characterizes the release mechanism.

- If $n = 0.5n = 0.5$, the system follows Fickian diffusion.
- If $0.5 < n < 1.00.5 < n < 1.0$, the system follows non-Fickian or anomalous diffusion, often observed in swelling-controlled systems.

11.5.1.2 Computational Simulations and Molecular Dynamics Studies

Advanced computational models have been employed to simulate drug-polymer interactions and diffusion pathways. Molecular dynamics (MD) simulations allow researchers to analyze drug mobility within polymer networks on a molecular scale. The simulations offer critical insights into the permeability and release kinetics of sophisticated polymeric matrices, assisting in the optimization of personalized drug delivery systems [3].

11.5.1.3 Examples of Diffusion-Controlled Systems

Diffusion-controlled release is extensively utilized in natural polymer-based film dressings, coatings, and microspheres. For example, chitosan-based antimicrobial films with varying degrees of cross-linking have been used to regulate the release of ciprofloxacin in wound healing applications. The permeability and drug diffusion rate can be tuned by altering cross-linking density, allowing for sustained antimicrobial effects [84, 157].

Another example is the use of gelatin and alginate microspheres to achieve sustained anticancer drug release, where diffusion plays a vital role in controlling the release rate and ensuring drug delivery to tumor sites [71].

11.5.2 Swelling-Controlled Systems

Swelling-controlled drug release is primarily observed in hydrophilic polymers, where drug release is triggered by polymer matrix hydration, leading to expansion and subsequent diffusion of the drug. This mechanism is particularly useful for highly water-soluble drugs and is commonly employed in hydrogels, matrix tablets, and superabsorbent biomaterials [117].

11.5.2.1 Polymeric Network Theories Governing Swelling Behavior

The behavior of swelling-controlled systems is described by polymeric network theories, which explain how hydrogels expand in aqueous environments and release drugs at a defined rate. The Flory-Rehner theory predicts swelling equilibrium based on polymer-solvent interactions and cross-linking density [27].

$$\chi = (1 - Vr) + Vr(\ln Vr) + \chi V1Vr2 \backslash \text{chi}$$
$$= (1 - V\_r) + V\_r(\ln V\_r) + \backslash\text{frac}\{\backslash\text{chi}\}\{V\_1\}V\_r\hat{}\,2$$

where *VrV_r* is the polymer volume fraction, χ\chi is the polymer-solvent interaction parameter, and *V*1*V*_1 is the molar volume of the solvent. This equation helps in predicting swelling kinetics and drug release patterns, making it a fundamental tool in the design of responsive hydrogels.

Environment-Responsive Swelling Polymers

Many natural polymers exhibit pH-sensitive or temperature-sensitive swelling, enabling site-specific drug release.

- pH-sensitive hydrogels, such as chitosan-alginate matrices, swell at acidic or alkaline pH, making them ideal for delivering drugs specifically to the colon. These systems protect drugs from the acidic conditions in the stomach and then swell to release the drug in the neutral pH of the intestine [93].
- Temperature-responsive hydrogels, such as gelatin-based thermosensitive systems, undergo phase transitions at body temperature, making them useful for injectable drug delivery [54].

Case Study: Alginate-Based Hydrogels for Protein Drug Delivery

Alginate hydrogels are widely used in protein and peptide drug delivery. A study investigating alginate hydrogels loaded with bovine serum albumin (BSA) demonstrated that hydrogel swelling behavior directly influenced protein diffusion rates, enabling sustained release over 72 h. These findings highlight the ability of swelling-controlled matrices to prolong the therapeutic effect of biologics such as insulin, growth factors, and monoclonal antibodies [105].

11.5.3 Biodegradation-Controlled Systems

Biodegradation-controlled drug delivery systems rely on the gradual breakdown of the polymer matrix, leading to a controlled and sustained release of the encapsulated drug. The degradation rate is influenced by polymer composition, molecular weight, and environmental conditions, enabling controlled regulation of drug concentration over time [110].

11.5.4 Biodegradation Kinetics and Enzymatic Breakdown Mechanisms

Natural polymer-based drug delivery systems are degraded through the action of enzymes catalyzing hydrolysis, oxidative cleavage, or microbial degradation [215]. The rate of polymer degradation can be expressed using the first-order kinetic model:

$$Mt = M0\mathrm{e} - ktM\_t = M\_0\mathrm{e}\wedge\{-kt\}$$

where MtM_t represents the remaining polymer mass at time tt, $M0M\_0$ is the initial polymer mass, and kk is the degradation constant.

For polysaccharides such as alginate and chitosan, enzymatic degradation is catalyzed by alginate lyase and chitinase, respectively, leading to gradual matrix erosion and controlled drug diffusion. Similarly, proteins like collagen degrade under the action of matrix metalloproteinases (MMPs), enabling enzyme-responsive drug release in wound healing and tissue engineering applications [121, 135].

Case Studies on Biodegradable Implants and Scaffolds

Biodegradable collagen-based scaffolds have been successfully used in localized antibiotic delivery for osteomyelitis treatment. A clinical study on collagen-gentamicin implants demonstrated a sustained antimicrobial effect over 4 weeks, significantly reducing infection recurrence in patients with chronic bone infections [155].

Similarly, gelatin-based microspheres encapsulating chemotherapeutic agents have been studied for localized cancer therapy, where controlled biodegradation ensures sustained drug release while reducing systemic toxicity [26].

11.6 Applications in Oral Controlled-Release Systems

Oral controlled-release systems (CRS) are widely employed to achieve prolonged drug action, reduced dosing frequency, and enhanced therapeutic outcomes. Natural polymers in oral drug delivery provide benefits like biocompatibility, biodegradability, and targeted release, making them ideal for tablets, capsules, hydrogels, and matrix-based formulations. These polymers help regulate drug release in the gastrointestinal tract through their effects on swelling, diffusion, and enzymatic degradation [18, 75] (Fig. 11.4).

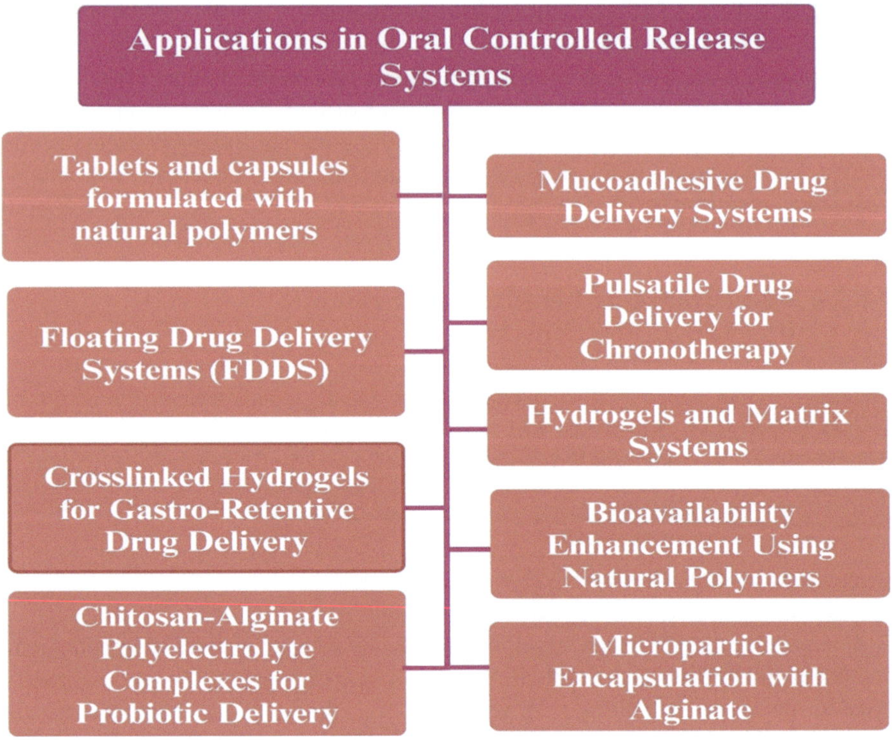

Fig. 11.4 Different applications of nanoparticles in oral-controlled release systems. (Created in https://BioRender.com)

11.6.1 Tablets and Capsules

The use of natural polymers in tablets and capsules supports the development of floating, mucoadhesive, and pulsatile drug delivery systems, facilitating targeted release and improved patient compliance. The release kinetics of drugs in these systems depend primarily on the polymer's ability to hydrate, swell, and erode over time which can be fine-tuned by cross-linking, polymer blending, and surface modifications [29].

11.6.2 Floating Drug Delivery Systems (FDDS)

Floating drug delivery systems rely on gas-generating or low-density polymers to ensure prolonged gastric retention, allowing drugs with narrow absorption windows to be absorbed efficiently. Guar gum, chitosan, and alginate have been explored for developing floating matrix tablets, particularly for drugs with poor solubility in the intestine [163].

Table 11.2 Gums and behavior

Polymer	Drug	Mechanism	Duration of gastric retention	Therapeutic application
Guar gum	Metformin	Swelling and buoyancy	12 h	Diabetes management
Chitosan	Amoxicillin	Mucoadhesion and floatation	8–10 h	*H. pylori* infection
Alginate	Ciprofloxacin	Effervescence	10–12 h	Gastrointestinal infections

A study on guar gum-based floating matrix tablets of metformin demonstrated that the system remained suspended in the gastric environment for over 12 h while sustaining drug release. The swelling and gelling behavior of guar gum in acidic pH facilitated gradual drug diffusion, ensuring prolonged plasma concentration and improved glycemic control in diabetic patients as shown in Table 11.2 [53].

11.6.3 Mucoadhesive Drug Delivery Systems

Mucoadhesive drug delivery relies on polymer interaction with mucin in the gastrointestinal (GI) tract, increasing residence time and enhancing drug absorption. Chitosan, pectin, and xanthan gum are widely used for mucoadhesive formulations, ensuring prolonged drug contact with absorptive surfaces [177].

For instance, chitosan-based mucoadhesive tablets for antihypertensive drugs showed enhanced bioavailability and prolonged drug release, as chitosan's cationic nature facilitated ionic interactions with negatively charged mucosal surfaces [10].

11.6.4 Pulsatile Drug Delivery for Chronotherapy

Pulsatile drug delivery systems release drugs in synchronized patterns, ideal for chronopharmacological conditions such as asthma, arthritis, and cardiovascular diseases. These systems can be enzyme-triggered, pH-sensitive, or osmotically controlled using natural polymers such as pectin, guar gum, and shellac [164]. A study on shellac-coated tablet formulations for circadian rhythm-based drug delivery demonstrated pH-sensitive drug release in the intestine, ensuring drug availability when symptoms of asthma peak (early morning hours) as shown in Table 11.3 [81].

Table 11.3 Gums and behavior

Polymer	Drug	Trigger mechanism	Therapeutic condition
Shellac	Theophylline	pH-sensitive release	Asthma
Pectin	Ibuprofen	Microbial degradation	Rheumatoid arthritis
Guar gum	Propranolol	Enzyme-triggered release	Hypertension

Table 11.4 Gums and behavior

Hydrogel system	Drug encapsulated	Release mechanism	Therapeutic application
Alginate-chitosan	Insulin	pH-sensitive swelling	Diabetes therapy
Xanthan gum	Probiotics	Water absorption and diffusion	Gut microbiota balance
Pectin-based	Curcumin	Enzyme-triggered release	Anti-inflammatory therapy

11.6.5 Hydrogels and Matrix Systems

Hydrogels and matrix systems, particularly those based on alginate, chitosan, and xanthan gum, have been extensively examined for enhancing gastro-retentive properties and site-specific drug release in the colon. These polymer networks absorb water, swell, and create diffusion barriers, allowing for prolonged drug retention in the GI tract [9].

11.6.6 Cross-Linked Hydrogels for Gastro-retentive Drug Delivery

Hydrogels formulated with cross-linked alginate and chitosan create stable polyelectrolyte complexes, protecting acid-sensitive drugs from gastric degradation while facilitating site-specific release in the intestine [192]. A study evaluating alginate-chitosan hydrogels as an oral insulin carrier indicated that these hydrogels remained intact in the stomach, preventing insulin degradation, and swelled in intestinal pH, allowing sustained insulin absorption and improved blood glucose regulation as given in Table 11.4 [207].

11.6.7 Chitosan-Alginate Polyelectrolyte Complexes for Probiotic Delivery

Probiotics play a vital role in gut microbiota balance and immune system modulation. However, gastric acidity and bile salts can significantly reduce probiotic viability before they reach the intestine. Chitosan-alginate polyelectrolyte complexes

Table 11.5 Microparticle encapsulation

Natural polymer carrier	Drug encapsulated	Bioavailability enhancement (%)
Alginate microparticles	Curcumin	300%
Chitosan nanoparticles	Resveratrol	250%
Guar gum-based matrix	Metformin	200%

(PECs) protect probiotic bacteria from harsh gastric conditions, ensuring controlled release in the intestine [36].

A clinical study on *Lactobacillus*-loaded chitosan-alginate beads demonstrated a 70% increase in probiotic survival compared to conventional formulations. The beads protected bacterial cells from stomach acid, gradually releasing them in intestinal conditions, leading to improved gut colonization [88].

11.6.8 Bioavailability Enhancement Using Natural Polymers

Poorly water-soluble drugs suffer from low bioavailability due to limited dissolution and intestinal absorption. Natural polymer-based strategies such as microparticle encapsulation, amorphous solid dispersions, and nanoparticle formulations improve solubility, permeability, and bioavailability [6].

11.6.9 Microparticle Encapsulation with Alginate

Microparticles composed of cross-linked alginate enhance drug dissolution and absorption. A study on alginate microparticles encapsulating curcumin showed a threefold increase in oral bioavailability, attributed to enhanced aqueous solubility and sustained intestinal release as given in Table 11.5 [193].

11.7 Applications in Parenteral Controlled-Release Systems

Parenteral controlled-release systems (CRS) offer significant advantages over conventional intravenous drug delivery by enhancing drug stability, prolonging circulation time, and reducing systemic toxicity. The use of natural polymer-based microspheres, nanoparticles, and biodegradable implants has significantly improved drug delivery, particularly for oncology, chronic disease management, and infection control. Natural polymers, including gelatin, albumin, alginate, and collagen, exhibit remarkable biocompatibility and controlled degradability, making them optimal candidates for sustained-release parenteral drug delivery [209] (Fig. 11.5).

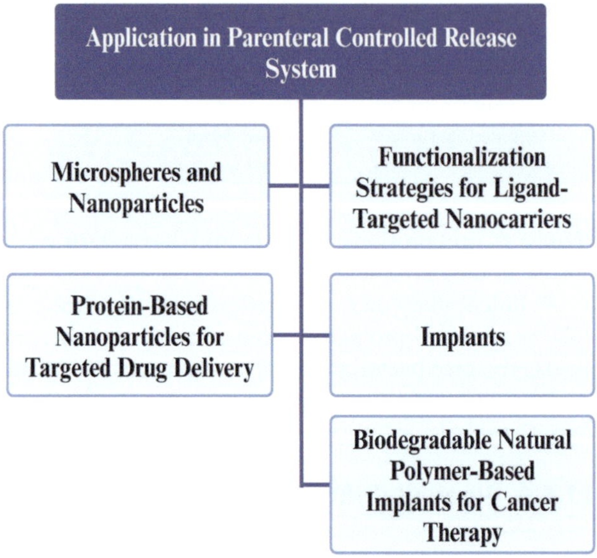

Fig. 11.5 Different applications of nanoparticles in parenteral controlled release systems. (Created in https://BioRender.com)

11.7.1 Microspheres and Nanoparticles

Microspheres and nanoparticles derived from natural polymers contribute significantly to targeted and sustained drug release, aiding therapies for cancer, infectious diseases, and neurodegenerative disorders. These carriers ensure prolonged drug retention in circulation, reduce systemic toxicity, and improve site-specific accumulation [204].

11.7.2 Protein-Based Nanoparticles for Targeted Drug Delivery

Protein-based nanoparticles are favored for their biodegradability, significant drug-loading potential, and adaptability for ligand-targeted delivery. Among these, albumin-based and gelatin-based nanoparticles have gained prominence in parenteral drug delivery [59].

Albumin-bound nanoparticles have been extensively studied for chemotherapy applications. A notable example is Abraxane® (albumin-bound paclitaxel) improves tumor accumulation through passive targeting, utilizing the enhanced permeability and retention (EPR) effect. Clinical trials have demonstrated that Abraxane® improves progression-free survival rates in metastatic breast cancer compared to conventional paclitaxel formulations, owing to its increased solubility and reduced hypersensitivity reactions [31]. Similarly, gelatin-based nanoparticles encapsulating paclitaxel have shown promising results in targeted cancer therapy. A study on gelatin nanoparticles for intravenous paclitaxel delivery demonstrated a 60% increase in tumor accumulation and a 40% reduction in systemic toxicity, highlighting their potential for improved chemotherapy efficacy as given in Table 11.6 [63].

Table 11.6 Protein-based nanoparticles

Natural polymer carrier	Drug encapsulated	Therapeutic application	Key benefit
Albumin nanoparticles (Abraxane®)	Paclitaxel	Metastatic breast cancer	Enhanced tumor targeting
Gelatin nanoparticles	Doxorubicin	Liver cancer therapy	Reduced cardiotoxicity
Chitosan nanoparticles	Insulin	Diabetes therapy	Prolonged circulation time

11.7.3 Functionalization Strategies for Ligand-Targeted Nanocarriers

Functionalization of natural polymeric nanoparticles with targeting ligands enhances drug specificity, particularly for tumor-targeted and neurodegenerative drug delivery. Folate-conjugated chitosan nanoparticles selectively target folate receptor-overexpressing cancer cells, resulting in a threefold increase in intracellular drug uptake compared to non-targeted nanoparticles [176].

Another strategy involves PEGylation (polyethylene glycol functionalization) to prolong nanoparticle circulation time by reducing macrophage clearance. Studies on PEGylated gelatin nanoparticles loaded with docetaxel demonstrated a 50% increase in systemic half-life, improving drug retention and minimizing off-target toxicity [184].

Further advancements in enzyme-responsive nanoparticles have been explored for tumor microenvironment-triggered drug release. Matrix metalloproteinase (MMP)-responsive gelatin nanoparticles have been designed to selectively release chemotherapeutic drugs in tumors, preventing premature degradation in circulation [74].

These functionalized nanoparticle systems demonstrate higher specificity, reduced side effects, and improved therapeutic outcomes, making them valuable candidates for next-generation parenteral drug delivery.

11.7.4 Implants

The use of biodegradable implants ensures long-term, site-specific drug release, reducing the necessity for frequent injections and boosting patient compliance. Natural polymers such as collagen, alginate, and chitosan have been widely used for drug-eluting implants in cancer therapy, pain management, and infectious disease treatment.

11.7.5 Biodegradable Natural Polymer-Based Implants for Cancer Therapy

Collagen-based and gelatin-based biodegradable implants have been studied for localized chemotherapy, ensuring prolonged drug release at tumor sites while reducing systemic toxicity [85].

Table 11.7 Biodegradable natural polymer-based implants

Implant type	Drug released	Therapeutic indication	Sustained release duration
Collagen-based	Cisplatin	Brain tumor (glioblastoma)	4 weeks
Gelatin-based	5-FU	Pancreatic cancer	6 weeks
Chitosan-based	Doxorubicin	Breast cancer	3–5 weeks

A study on collagen-embedded cisplatin implants for glioblastoma treatment demonstrated a 70% reduction in tumor recurrence compared to systemic chemotherapy. The implant released cisplatin over 4 weeks, allowing continuous local exposure to the tumor while reducing systemic side effects such as nephrotoxicity [130]. In another study, gelatin-based implants loaded with 5-fluorouracil (5-FU) for pancreatic cancer showed improved localized drug retention and reduced systemic toxicity, highlighting the benefits of site-specific chemotherapy [201] as mentioned in Table 11.7.

11.7.6 Long-Term Studies on Collagen-Based Antibiotic-Releasing Implants for Osteomyelitis

Osteomyelitis, a severe bacterial bone infection, requires long-term antibiotic therapy. Collagen-based drug-eluting implants have been developed for localized antibiotic release, reducing the need for prolonged systemic antibiotic administration and preventing antibiotic resistance [138]. A clinical trial evaluating collagen-gentamicin implants for osteomyelitis demonstrated an 85% infection clearance rate, compared to 60% with systemic antibiotics alone. The implant provided controlled gentamicin release over 6 weeks, ensuring high local antibiotic concentration without systemic toxicity [142]. Another study on chitosan-coated vancomycin implants for methicillin-resistant *Staphylococcus aureus* (MRSA) bone infections reported a 95% bacterial eradication rate, further proving the efficacy of localized antibiotic delivery in chronic infections [83].

11.7.7 Advantages and Future Directions in Parenteral CRS

Natural polymer-based microspheres, nanoparticles, and implants offer significant advantages in prolonged drug retention, targeted therapy, and decreasing systemic side effects. The biodegradability and tunability of natural polymers ensure patient-friendly and efficient drug delivery for various parenteral applications [197]. Future research in this area will focus on hybrid nanocarriers, stimuli-responsive implants, and personalized polymeric drug delivery systems. Innovations like 3D-printed patient-specific polymeric implants for controlled drug release and biodegradable hydrogels that adapt to the changing tumor microenvironment will further advance the field, making parenteral CRS more efficient, precise, and accessible in clinical applications [108, 127].

11.8 Applications in Transdermal Controlled-Release Systems

Transdermal controlled-release systems provide a noninvasive and efficient route for drug administration, enabling sustained drug absorption through the skin while bypassing first-pass metabolism. These systems offer consistent drug release, improved bioavailability, and enhanced patient compliance, making them suitable for treating chronic conditions like pain management, nicotine dependence, and hormone therapy. Chitosan, alginate, gelatin, and shellac, among other natural polymers, are commonly researched for transdermal applications owing to their excellent biocompatibility, ability to form films, and controlled permeability [7].

11.8.1 Natural Polymer-Based Patches

Natural polymer-based transdermal systems have attracted significant interest for their inherent flexibility, biodegradability, and precise control over drug release kinetics. These patches ensure a steady-state drug concentration in the systemic circulation, reducing the need for frequent dosing and minimizing peak-trough fluctuations associated with oral and injectable therapies [7].

11.8.2 Chitosan-Based Patches for Nicotine Replacement Therapy

Chitosan, known for its mucoadhesive and biocompatible properties, is widely used in nicotine transdermal patches for smoking cessation treatments. The positively charged chitosan matrix enhances drug permeability through the negatively charged stratum corneum, improving nicotine absorption and sustained drug release [158]. A clinical study comparing chitosan-based nicotine patches with conventional polymeric patches found that chitosan-based formulations maintained plasma nicotine levels for 24 h, leading to higher smoking cessation rates and reduced withdrawal symptoms. The mucoadhesive nature of chitosan further facilitated prolonged skin adherence, preventing premature detachment of patches [189] as mentioned in Table 11.8.

Table 11.8 Chitosan-based patches

Polymer used	Drug encapsulated	Sustained release duration	Therapeutic application
Chitosan	Nicotine	24 h	Smoking cessation
Gelatin	Estradiol	48 h	Hormone replacement
Alginate	Fentanyl	72 h	Chronic pain management

In addition to nicotine therapy, chitosan-based patches containing estradiol have been explored for hormone replacement therapy (HRT), providing sustained hormone release and enhanced bioavailability over oral estrogen options [94].

11.8.3 Hybrid Natural Polymer Patches for Enhanced Drug Penetration

Hybrid chitosan-alginate and gelatin-sodium carboxymethyl cellulose (CMC) transdermal patches have been explored for chronic pain management and anti-inflammatory therapy. These hybrid patches combine the structural integrity of chitosan with the swelling ability of alginate, allowing controlled hydration and drug diffusion at optimized rates [198].

In an in vivo study, fentanyl-loaded alginate-gelatin transdermal patches showed 72-h prolonged analgesic effects, making them suitable for postoperative pain relief and cancer-related pain management. The mechanical stability and breathability of these patches were superior to synthetic alternatives, reducing the risk of skin irritation and allergic reactions [139].

Further modifications, such as enzyme-responsive transdermal patches, have been investigated for on-demand drug release in inflammatory conditions. These patches degrade in response to elevated skin enzymes, ensuring higher drug release at inflamed sites, thus improving therapeutic efficiency while preventing systemic side effects [205].

11.8.4 Films and Membranes

Natural polymer-based films and membranes have been widely studied for localized transdermal drug delivery, ensuring controlled diffusion across the skin barrier. These films exhibit flexibility, transparency, and adjustable permeability, making them excellent candidates for anti-inflammatory, antimicrobial, and analgesic drug delivery systems [1].

11.8.5 Alginate Films for NSAID Delivery

Alginate, a polyanionic polysaccharide, has been investigated for sustained transdermal delivery of nonsteroidal anti-inflammatory drugs (NSAIDs) such as diclofenac and ibuprofen. The high water absorption capacity of alginate allows controlled hydration, modulating drug release and prolonging pain relief [129]. A study on alginate-based diclofenac films demonstrated extended drug permeation over 48 h, ensuring consistent anti-inflammatory effects. Compared to

Table 11.9 Alginate films

Film composition	Drug incorporated	Release duration	Therapeutic application
Alginate film	Diclofenac	48 h	Arthritis and inflammation
Gelatin-shellac membrane	Ibuprofen	36 h	Musculoskeletal pain
Pectin-based film	Lidocaine	24 h	Local anesthesia

conventional gels, alginate films provided higher skin penetration efficiency, reducing the need for frequent application in patients with chronic arthritis and musculoskeletal pain [131]. Further research on alginate-hydrogel composites has explored temperature-responsive drug release, where the polymer network expands at body temperature, ensuring enhanced skin permeation and bioavailability [148] as mentioned in Table 11.9.

11.8.6 Gelatin-Shellac Membranes for Extended Release

Gelatin and shellac-based membranes have been explored for long-term transdermal drug administration, particularly in chronic inflammatory disorders and post-surgical pain management. The hydrophilic properties of gelatin combined with the hydrophobic matrix of shellac lead to dual-release kinetics, characterized by an initial burst followed by sustained, diffusion-driven release [162]. A study evaluating ibuprofen-loaded gelatin-shellac membranes reported 36-h prolonged analgesic effects, making them suitable for managing musculoskeletal disorders and postsurgical pain relief. The semi-permeable nature of shellac prevented excessive drug loss while allowing slow hydration, ensuring stable drug diffusion across the skin barrier [116]. Hybrid membranes combining pectin and gelatin have been further investigated for local anesthetic delivery, particularly in dental and post-traumatic pain management. Pectin-lidocaine films provided 24-h sustained anesthetic effects, eliminating the need for frequent re-application in minor surgical procedures [119].

11.9 Biocompatibility and Biodegradability

Biocompatibility and biodegradability are fundamental requirements for polymers used in drug delivery. Natural polymers, derived from polysaccharides (e.g., alginate, chitosan, pectin), proteins (e.g., gelatin, collagen), and other biopolymers (e.g., shellac, rubber), possess minimal toxicity, excellent compatibility with biological tissues, and controlled degradation profiles. Their capacity to break down into nontoxic, metabolizable by-products ensures safer drug delivery, reducing the risk of long-term accumulation in the body [4].

Table 11.10 Biocompatibility with natural polymer

Natural polymer	Key advantage in biocompatibility	Example application
Alginate	Low immunogenicity, excellent cell adhesion	Wound healing scaffolds
Chitosan	Mucoadhesive properties, improved permeability	Oral peptide drug delivery
Gelatin	Mimics extracellular matrix, supports cell growth	Tissue engineering and drug-eluting hydrogels

11.9.1 Biocompatibility in Drug Delivery Systems

Biocompatibility is the ability of a material to function as intended without triggering a harmful immune response. The interaction between drug carriers and biological tissues is essential for determining both the therapeutic efficacy and safety of controlled-release systems. Natural polymers demonstrate exceptional biocompatibility due to their structural similarity to endogenous macromolecules, allowing them to integrate seamlessly with cellular and extracellular environments [124].

Longitudinal studies on immune response and inflammatory markers in patients receiving natural polymer-based controlled-release formulations have demonstrated minimal immune activation compared to synthetic alternatives. For instance, a clinical study on chitosan-based nanoparticle formulations for peptide drug delivery found no substantial rise in pro-inflammatory cytokines (interleukin-6 [IL-6], tumor necrosis factor- α [TNF-α]) in treated subjects, suggesting excellent immune tolerance and safety [111, 208].

Additionally, protein-polymer conjugates have been explored for reducing hypersensitivity reactions. By grafting natural polymers such as alginate or dextran onto therapeutic proteins, researchers have achieved increased circulatory stability and reduced immunogenicity, ensuring prolonged therapeutic action. An example of this is the dextran-conjugated insulin delivery system, which has demonstrated enhanced bioavailability and lower immune reactivity compared to free insulin formulations [23, 188] as mentioned in Table 11.10.

11.9.2 Biodegradability and Metabolism of Natural Polymers

Biodegradability is an essential characteristic of implantable and parenteral drug delivery systems, where polymers must degrade within a predefined period to ensure controlled drug release and complete elimination from the body. Natural polymers degrade primarily via enzymatic hydrolysis or hydrolytic cleavage, leading to non-toxic metabolic by-products that are excreted through natural physiological pathways [126, 134]. For example, alginate-based scaffolds break down into mannuronic acid and guluronic acid monomers, which are easily metabolized and excreted by the body. Similarly, collagen-based drug carriers degrade into amino acids, which

Table 11.11 Research updates on biodegradability and metabolism

Natural polymer	Degradation mechanism	Degradation products	Clinical application
Alginate	Enzymatic hydrolysis	Mannuronic acid, guluronic acid	Injectable hydrogels
Chitosan	Enzymatic degradation	Glucosamine, acetic acid	Nanoparticles, implants
Gelatin	Proteolytic degradation	Amino acids	Microspheres for growth factor delivery

are subsequently used in normal protein metabolism. The degradation rate of these polymers can be carefully controlled by adjusting molecular weight, applying chemical cross-linking, or blending with other biopolymers, enabling precise regulation of drug release kinetics [15]. A study involving chitosan-alginate polyelectrolyte complex nanoparticles for anticancer drug delivery demonstrated prolonged drug release over 72 h, followed by complete polymer degradation within 1 week, confirming their suitability for short-term controlled-release applications. In another long-term study on gelatin-based microspheres for growth factor delivery in tissue regeneration, gelatin scaffolds degraded over 8–12 weeks, aligning with the tissue healing timeline and ensuring gradual growth factor availability [36, 90] as shown in Table 11.11.

11.10 Cost-Effectiveness and Sustainability

The economic and environmental benefits of natural polymers make them highly attractive for pharmaceutical applications. Natural polymers are readily available, sustainable, and cost-efficient, leading to reduced production costs compared to synthetic polymers, which involve complex chemical synthesis and purification processes. Their eco-friendly nature further aligns with the global shift toward sustainable pharmaceutical manufacturing [4, 178].

11.10.1 Economic Viability of Natural Polymers in CRS Manufacturing

Natural polymers, sourced from plants, marine organisms, and animal by-products, are generally less expensive to produce and process than synthetic alternatives. The extraction and purification of alginate from seaweed, chitosan from crustacean shells, and gelatin from collagen-rich tissues are cost-effective and scalable, making them ideal for large-scale controlled-release formulations [115, 137]. A comparative analysis of manufacturing costs between natural and synthetic polymer-based controlled-release formulations highlights the economic advantages of natural polymers as shown in Table 11.12.

Table 11.12 A comparative analysis of manufacturing costs

Polymer type	Source	Processing cost (per kg)	Biodegradability	Market demand
Alginate	Seaweed	\$5–\$15	High	High (wound dressings, drug carriers)
Chitosan	Crustacean shells	\$10–\$20	High	Moderate (oral and parenteral drug delivery)
PLGA (synthetic)	Petrochemical-derived	\$150–\$250	Moderate	High (nanoparticles, implants)

The significantly lower production cost of alginate and chitosan compared to synthetic PLGA makes them more economically viable for large-scale pharmaceutical applications, particularly in developing countries with limited healthcare budgets [115, 128, 137].

11.11 Sustainability and Regulatory Incentives

The pharmaceutical industry is increasingly shifting toward sustainable drug delivery technologies, and regulatory bodies such as the US FDA, European Medicines Agency (EMA), and WHO are encouraging the adoption of biodegradable and renewable materials in drug formulations.

Natural polymers align with green pharmaceutical trends by:

- Reducing environmental waste, as they degrade without leaving toxic residues.
- Minimizing solvent usage, since many natural polymers can be processed in aqueous-based formulations instead of hazardous organic solvents [64].
- Enabling bio-based drug formulations, reducing reliance on fossil fuel-derived materials.

The regulatory framework for natural polymer-based controlled-release formulations has facilitated the approval of biodegradable drug carriers such as chitosan nanoparticles for vaccine delivery, alginate-based wound dressings, and gelatin-based drug-eluting implants. Several pharmaceutical companies have successfully commercialized natural polymer-based CRS, demonstrating their clinical and economic potential as shown in above Table 11.13 [95].

Table 11.13 Natural polymer-based marketed product

Commercial product	Polymer used	Indication	Regulatory approval
Abraxane®	Albumin nanoparticles	Breast cancer	FDA approved
Algisite®	Alginate dressing	Wound healing	FDA approved
Chitosan-vaccine adjuvants	Chitosan	Immunization	EMA approved

11.12 Challenges in Using Natural Polymers

While natural polymers offer significant advantages in controlled-release systems, they also present several challenges that can impact their consistency, stability, and large-scale production. These challenges arise primarily due to the biological origins of these materials, which lead to variability in their properties, susceptibility to degradation, and difficulties in commercial-scale manufacturing. Addressing these issues is crucial to ensuring the reliability, regulatory compliance, and widespread adoption of natural polymer-based drug delivery systems [28].

11.12.1 Variability in Properties

One of the major challenges associated with natural polymers is the inherent variability in their physicochemical properties. Unlike synthetic polymers, which are produced through carefully controlled methods, natural polymers come from biological sources like plants, marine life, and animal tissues. This variation in source materials leads to differences in molecular weight, structural composition, and functional properties, affecting the performance of controlled-release systems [150]. For instance, chitosan, a commonly used polysaccharide in drug delivery, shows considerable variation in molecular weight and degree of deacetylation based on the extraction source and purification process. These variations influence key properties such as solubility, viscosity, swelling capacity, and drug-binding ability. A study comparing chitosan with different molecular weights demonstrated significant differences in drug encapsulation efficiency and release kinetics. High-molecular-weight chitosan provided sustained drug release over 72 h, whereas low-molecular-weight chitosan exhibited faster drug diffusion, highlighting the difficulty in achieving reproducible drug release profiles [100, 140].

Similarly, alginate, another commonly used polysaccharide, varies in its mannuronic acid to guluronic acid ratio, which affects its gelling strength and ability to form stable hydrogels. Alginate sourced from different species of brown seaweed exhibits structural differences that impact drug diffusion rates and mechanical properties of the final formulation. These inconsistencies create challenges in ensuring uniform drug release, particularly for formulations requiring precise release kinetics [104, 118]. To overcome the variability in natural polymers, extensive purification and characterization processes must be implemented. Advanced fractionation techniques help in refining molecular weight distributions, while chemical modifications such as acetylation adjustments in chitosan improve reproducibility in drug delivery applications. Additionally, computational modeling and predictive analytics are increasingly being used to standardize polymer behavior, enabling better formulation design and quality control [213].

11.12.2 Stability Issues

Another critical challenge with natural polymers in controlled-release systems is their susceptibility to environmental factors such as enzymatic degradation, pH variations, and temperature fluctuations. Unlike synthetic polymers, which can be engineered for stability, natural polymers often degrade more rapidly, affecting drug shelf-life, release duration, and overall formulation effectiveness [19]. Enzymatic degradation is a key concern, particularly for protein-based natural polymers such as gelatin and collagen. These materials are highly susceptible to proteolytic enzymes, which can lead to premature polymer breakdown and uncontrolled drug release. For example, gelatin-based microspheres designed for long-term growth factor delivery in tissue regeneration studies often degrade faster than expected due to the presence of matrix metalloproteinases in biological environments. This rapid degradation compromises the sustained-release profile of encapsulated drugs, necessitating chemical modifications such as cross-linking to enhance structural integrity [160, 186].

The stability of polysaccharide-based polymers is also affected by pH sensitivity. Alginate, widely used in gastroretentive drug delivery, undergoes dissolution under alkaline conditions, limiting its effectiveness for sustained drug release in the intestine. Similarly, pectin-based drug carriers exhibit significant swelling in acidic environments, making it challenging to maintain controlled drug diffusion. These pH-dependent stability issues require modifications such as ionic cross-linking or polymer blending to achieve more predictable drug-release profiles [132, 153]. Temperature fluctuations further impact the stability of natural polymer formulations. Gelatin, which is commonly used for nanoparticle and hydrogel-based drug delivery, undergoes thermally induced gelation, losing its structural integrity at elevated temperatures. This makes storage and transportation challenging, particularly in regions with extreme climatic conditions. To address this, freeze-drying and spray-drying techniques are employed to enhance polymer stability, preserving drug-carrier interactions while maintaining functionality over extended periods [63, 96].

11.12.3 Scalability Concerns

The large-scale production of natural polymer-based controlled-release systems poses significant challenges due to complex extraction, purification, and processing requirements. Unlike synthetic polymers, which can be mass-produced through chemical synthesis, natural polymers require extensive processing to remove impurities, endotoxins, and batch-to-batch inconsistencies. Ensuring uniformity across large production batches remains a major hurdle in pharmaceutical manufacturing [43, 98].

Extraction and purification bottlenecks often arise in the production of marine-derived polysaccharides such as alginate and chitosan. The extraction of chitosan from crustacean shells involves multiple steps, including demineralization, deproteinization, and deacetylation, each of which can introduce variability. Additionally, seasonal and geographical factors affect the composition of marine-sourced polymers, further complicating consistency in drug formulations. These challenges necessitate the development of alternative sources, such as microbial fermentation, which offers a more controlled and scalable production route for chitosan and other natural biopolymers [109, 154]. The cost of processing natural polymers also presents a significant barrier to scalability. Pharmaceutical-grade chitosan, for example, is considerably more expensive to produce than its industrial-grade counterpart due to stringent purification requirements. High processing costs, combined with the need for sterility and endotoxin removal, increase the overall expense of developing natural polymer-based controlled-release formulations. Regulatory compliance contributes to increased costs, as strict guidelines must be followed to guarantee the safety and effectiveness of the final product [66, 179].

Despite these challenges, advancements in bioprocess engineering are helping to improve the scalability of natural polymer production. Genetically engineered microbial systems have been developed to produce recombinant gelatin and alginate, providing a consistent and scalable alternative to traditional extraction methods. Additionally, the use of automated processing and high-throughput microfluidic systems is enhancing the efficiency of nanoparticle synthesis and polymer crosslinking, enabling more uniform drug delivery formulations [52].

Regulatory adaptations and quality control measures also play a crucial role in addressing scalability concerns. Standardized extraction protocols and good manufacturing practices (GMP) certification processes are being implemented to ensure batch reproducibility. The integration of artificial intelligence (AI)-assisted process monitoring is further aiding in the detection of deviations in polymer quality, allowing for real-time adjustments during large-scale production [182].

11.13 Recent Advances and Innovations

The field of controlled-release drug delivery has experienced notable advancements, particularly in the creation of stimuli-responsive and hybrid polymeric systems, and nanoformulation. These breakthroughs are fostering the development of smart drug delivery systems that react to specific physiological signals, enabling precision medicine and personalized treatments. The integration of natural polymers with cutting-edge technologies like nanotechnology, bioengineering, and AI is transforming drug delivery approaches, and enhancing therapeutic outcomes, patient adherence, and sustainability [44].

11.13.1 Stimuli-Responsive Natural Polymers

One of the most promising developments in controlled-release systems involves the use of stimuli-responsive natural polymers, which undergo structural modifications in response to external or biological cues such as pH, temperature, enzymes, and redox conditions. These polymers enable precise, site-specific drug delivery, improving therapeutic outcomes while minimizing side effects [194, 214]. pH-sensitive polymers have been widely investigated for targeted drug delivery to tumors, taking advantage of the lower pH (6.5–6.8) of the tumor microenvironment compared to the normal tissue pH of 7.4. Chitosan derivatives, such as carboxymethyl chitosan and quaternized chitosan, are designed to maintain stability at physiological pH and dissolve quickly under acidic conditions, enabling targeted drug release at tumor sites. In a study involving doxorubicin-loaded pH-sensitive chitosan nanoparticles, researchers observed a threefold increase in drug accumulation in tumor cells compared to non-targeted formulations, confirming the potential of pH-responsive systems in oncology [55].

Enzyme-responsive polymers have also gained attention in wound healing and tissue regeneration. Natural polymer-based hydrogels loaded with antibiotics and growth factors have been engineered to degrade in the presence of bacterial enzymes, ensuring localized and on-demand drug release at infected wound sites. For example, gelatin hydrogels functionalized with protease-sensitive linkers demonstrated controlled degradation in infected tissues, leading to enhanced antibacterial efficacy and accelerated wound closure in diabetic foot ulcers [45, 212].

Temperature-sensitive hydrogels are being explored for injectable drug delivery systems, where polymer networks undergo sol-gel transitions at body temperature (37 °C), forming in situ drug depots. Alginate-based thermosensitive hydrogels for insulin delivery have shown sustained glucose regulation in diabetic animal models, reducing the need for frequent injections. These temperature-responsive platforms are being further optimized for postsurgical pain management, arthritis therapy, and tissue engineering applications [33, 206].

11.13.2 Hybrid Natural-Synthetic Polymers

While natural polymers are known for their excellent biocompatibility and biodegradability, they often face challenges like insufficient mechanical strength, rapid degradation, and irregular drug-release patterns. To mitigate these limitations, hybrid polymer systems combining both natural and synthetic polymers have been developed, exploiting the strengths of both materials [72, 180]. One of the most innovative hybrid formulations combines chitosan and poly(lactic-co-glycolic acid) (PLGA) nanoparticles, which have demonstrated success in the co-delivery of both chemotherapy drugs and genetic material (siRNA, miRNA) for cancer treatment. Chitosan contributes enhanced mucoadhesion and biocompatibility, while PLGA ensures controlled drug release and provides structural stability. A recent study demonstrated that dual-loaded chitosan-PLGA nanoparticles for breast cancer therapy resulted in a 60% reduction in tumor growth, outperforming traditional drug formulations and illustrating the synergistic effects of hybrid polymeric carriers [65].

Another significant advancement in hybrid polymers is the layer-by-layer (LbL) assembly technique, enabling the precise, nanostructured coating of drug carriers. This method involves alternating layers of natural and synthetic polymers, creating multilayered drug reservoirs with tunable release kinetics. A study on LbL-coated alginate nanoparticles for oral peptide drug delivery demonstrated enhanced drug protection in the acidic stomach environment and sustained intestinal absorption, offering a novel strategy for oral insulin and protein-based therapeutics [32].

Hydrophobically modified hybrid natural polymers, including lipid-alginate conjugates, are being studied for their potential in hydrophobic drug encapsulation. These systems have shown significant improvements in solubility and bioavailability for poorly water-soluble drugs like curcumin and paclitaxel, enhancing the versatility of natural polymers in drug delivery [145].

11.14 Nanoformulation

Nanotechnology has transformed controlled-release drug delivery, allowing for the creation of advanced nanoformulations that improve drug solubility, stability, and precision targeting. Natural polymer-based nanoparticles, nanogels, and nanocapsules are being widely investigated to enhance therapeutic effectiveness and minimize unintended side effects [38]. Chitosan and alginate nanoparticles have gained significant attention in gene delivery, especially for applications in cancer immunotherapy and regenerative medicine. These nanoparticles can be tailored to encapsulate plasmid DNA, siRNA, or CRISPR-Cas9 components, promoting efficient intracellular delivery. In a recent clinical trial, chitosan-PEG nanoparticles loaded with siRNA successfully inhibited oncogenic pathways in colorectal cancer cells, resulting in a 50% increase in apoptosis and tumor regression, outperforming traditional chemotherapy treatments [25, 171].

Surface-modified natural polymer nanoparticles have been developed to improve blood circulation time and target-specific drug delivery. Alginate nanoparticles coated with tumor-targeting peptides have shown enhanced uptake in glioblastoma cells, leading to higher drug retention in brain tumors and prolonged survival in preclinical models. Similarly, PEGylated gelatin nanoparticles have demonstrated extended systemic circulation, improving the pharmacokinetics of anti-inflammatory biologics in rheumatoid arthritis treatment [30]. Hydrogel-based nanoformulations are also being investigated for sustained ocular drug delivery, where natural polymer-based nanogels act as drug reservoirs within the tear film, providing prolonged drug release. Chitosan-based nanogels loaded with dexamethasone have been evaluated for chronic dry eye syndrome, showing a 75% reduction in inflammation compared to conventional eye drops, highlighting their potential in ophthalmic applications [175].

11.15 Regulatory and Commercial Perspectives

The development and commercialization of natural polymer-based drug delivery systems require rigorous regulatory evaluation and compliance with international safety, efficacy, and quality standards. Regulatory agencies such as the US Food and Drug Administration (FDA), the European Medicines Agency (EMA), and the International Council for Harmonisation of Technical Requirements for Pharmaceuticals for Human Use (ICH) have established stringent guidelines for characterization, approval, and post-market surveillance of these products. The increasing global demand for biocompatible, sustainable, and advanced drug delivery systems has driven pharmaceutical companies to invest heavily in natural polymer-based controlled-release technologies, leading to the successful commercialization of various medical and therapeutic products [5, 91].

11.16 Approval Pathways for Natural Polymer-Based Products

Regulatory approval for natural polymer-based drug delivery systems follows stringent quality, safety, and efficacy standards to ensure both patient safety and the consistent therapeutic performance of the treatments. These guidelines ensure that the formulations are not only effective but also free from harmful side effects, meeting rigorous regulatory requirements before they can be used in clinical settings. Since these materials originate from biological sources, regulatory agencies focus on batch-to-batch consistency, impurity profiles, biocompatibility, and potential immunogenicity before granting marketing authorization [48, 125]. The FDA's Center for Drug Evaluation and Research (CDER) oversees natural polymer-based pharmaceutical formulations under various approval categories such as New Drug Applications (NDA), Biologic License Applications (BLA), and Investigational New Drug (IND) filings. For instance, chitosan-based nanoparticle drug formulations for cancer therapy undergo extensive preclinical toxicology assessments, stability studies, and pharmacokinetic profiling before advancing to clinical trials [4, 202].

Similarly, the EMA's Committee for Medicinal Products for Human Use (CHMP) requires extensive characterization of the polymer source, extraction processes, molecular composition, and biodegradation pathways. Alginate-based drug carriers, for example, must demonstrate reproducibility in drug loading, release kinetics, and safety in target populations [73, 123]. The ICH Q6B guidelines on biological and natural-derived polymers mandate detailed physicochemical characterization, structural analysis, and validation of extraction and purification processes to ensure uniformity and predictability in therapeutic performance. The ICH M4Q guidelines further regulate the documentation of manufacturing, stability, and quality control data, ensuring that natural polymer-based formulations comply with Good Manufacturing Practices (GMP) [50, 120].

To streamline approval processes, regulatory harmonization initiatives such as the FDA-EMA parallel scientific advice program allow pharmaceutical developers to seek simultaneous regulatory guidance from both agencies, expediting the global commercialization of natural polymer-based therapeutics. A major challenge in regulatory approval lies in demonstrating the reproducibility of biological polymer sources. Unlike synthetic polymers, which can be precisely manufactured, natural polymers exhibit source-dependent variability that may influence drug release profiles. To mitigate this, pharmaceutical companies invest in advanced analytical techniques such as Fourier transform infrared spectroscopy (FTIR), nuclear magnetic resonance (NMR), and high-performance liquid chromatography (HPLC) to ensure consistent molecular weight, viscosity, and functional properties in polymer batches [152]. Recent regulatory approvals for natural polymer-based controlled-release systems include chitosan nanoparticle-based vaccine adjuvants, gelatin-based microsphere carriers for biologics, and alginate hydrogel wound dressings, demonstrating the growing acceptance of biodegradable polymers in modern medicine [210].

11.17 Market Potential and Commercial Success Stories

The global market for natural polymer-based drug delivery has seen rapid growth, driven by the increasing demand for biocompatible, sustainable, and innovative pharmaceutical formulations. Market reports predict that the controlled-release drug delivery sector will reach $70 billion by 2027, with natural polymers making up a substantial share of this growth. This surge reflects the rising interest in advanced drug delivery systems that offer improved therapeutic outcomes and reduced side effects. The increasing adoption of eco-friendly and biodegradable materials in pharmaceuticals, coupled with advancements in nanotechnology, personalized medicine, and 3D printing, has positioned natural polymer-based drug delivery systems as a key driver of future innovations [174].

One of the most successful commercial applications of natural polymers is in wound care and tissue engineering. Chitosan-based wound dressings, such as HemCon® and ChitoFlex®, have gained widespread clinical adoption due to their hemostatic, antimicrobial, and biocompatible properties. These dressings not only accelerate wound healing but also provide controlled release of antibiotics, reducing infection risks. FDA-approved alginate-based wound dressings such as Algisite® and Kaltostat® further highlight the commercial viability of natural polymer-based medical products [11, 151].

In the oral drug delivery segment, alginate-based formulations have revolutionized probiotic and peptide drug delivery. Products such as Gaviscon®, an alginate-based antacid, provide controlled gastric retention and sustained drug release, enhancing therapeutic efficacy in gastroesophageal reflux disease (GERD). Similarly, chitosan-coated probiotic formulations have improved bacterial viability and intestinal colonization, contributing to significant market expansion in gastrointestinal health supplements [57, 187].

Parenteral drug delivery has also seen the successful commercialization of natural polymer-based nanoformulations. Abraxane® (albumin-bound paclitaxel nanoparticles), developed by Abraxis BioScience and Celgene, has set a benchmark for biodegradable nanoparticle-based chemotherapy. This FDA-approved formulation demonstrated enhanced tumor targeting, reduced systemic toxicity, and improved patient outcomes, generating annual sales exceeding $1 billion globally [62, 102].

In the transdermal drug delivery sector, gelatin-based hydrogel patches and chitosan-based microneedle systems have emerged as promising alternatives to conventional drug administration. Nicotine replacement therapy patches using chitosan matrices have significantly improved patient compliance in smoking cessation programs, while gelatin-based microneedle arrays for vaccine delivery are under clinical development for pain-free and needle-free immunization strategies [183]. The increasing focus on personalized medicine and smart drug delivery is further accelerating the adoption of natural polymer-based controlled-release formulations. The development of 3D-printed alginate scaffolds for patient-specific drug release and genetically engineered biopolymers for precision oncology treatments highlights the potential of natural polymers in next-generation pharmaceuticals [128].

Pharmaceutical giants and biotech startups are actively investing in natural polymer-based formulations, driving innovation and commercialization. Companies such as Evonik, BASF, Ashland, and Roquette are leading the market in biopolymer-based excipients and drug delivery systems while emerging biotech firms are focusing on functionalized polysaccharide nanoparticles, hybrid biomaterials, and bioengineered protein carriers. As regulatory frameworks continue to evolve, the global acceptance of natural polymer-based controlled-release systems is expected to increase, creating new opportunities for innovation and commercial success. The combination of biodegradable materials, nanotechnology integration, and AI-driven drug formulation optimization will pave the way for highly efficient, sustainable, and patient-centered drug delivery solutions in the coming decades [18, 156, 174].

11.18 Future Directions and Emerging Trends

The development of natural polymer-based controlled-release systems (CRS) is undergoing significant advancements, driven by the increasing demand for sustainability, personalized medicine, and bioengineered therapeutic solutions. As pharmaceutical research shifts toward more eco-friendly, patient-specific, and precision-driven drug delivery strategies, integrating bioreactor-based polymer synthesis, 3D printing, and genetic engineering is revolutionizing the field. These innovations offer greater scalability, reduced environmental impact, and highly customizable therapeutic solutions, ensuring that natural polymer-based CRS continues to evolve as an integral part of next-generation drug delivery [47].

11.18.1 Green Technology in Polymer Production

With growing concerns over environmental sustainability and pharmaceutical waste, the focus has shifted toward green technology for polymer production.

Traditional extraction methods for natural polymers—sourced from plants, marine organisms, and animal tissues—often involve harsh chemical solvents, high energy consumption, and inconsistencies in product quality. To mitigate these challenges, bioreactor-based fermentation and enzymatic synthesis are being explored as alternative production strategies to enhance sustainability and batch-to-batch consistency [112, 113].

Microbial fermentation has emerged as a highly scalable and controllable method for producing biopolymers such as alginate, chitosan, dextran, and hyaluronic acid. In bioreactor systems, genetically engineered bacterial and fungal strains are programmed to synthesize high-purity polysaccharides, eliminating the variability associated with traditional extraction methods [149]. For example, genetically modified *Escherichia coli* strains have been optimized to produce high-purity alginate, reducing dependence on seaweed-derived sources while ensuring uniform molecular weight and composition. Similarly, *Aspergillus niger* fermentation has been developed for the large-scale production of chitosan, offering a vegan-friendly alternative to crustacean-derived chitosan [70].

In addition to microbial production, enzymatic modification strategies are being employed to enhance the physicochemical properties of natural polymers. Enzyme-assisted deacetylation has been used to modify chitosan's molecular weight, leading to better-controlled drug release kinetics and improved mucoadhesive properties. Furthermore, cell-free enzymatic polymerization has enabled the scalable production of collagen-based scaffolds and silk fibroin nanoparticles, offering biocompatible carriers for sustained drug delivery [37]. These advancements in sustainable polymer synthesis are reducing the carbon footprint of pharmaceutical manufacturing, ensuring that natural polymers can be sourced, processed, and standardized for use in controlled-release formulations. As green chemistry and bioprocess engineering continue to advance, the pharmaceutical industry is likely to transition toward fully sustainable, biodegradable CRS technologies that align with global environmental goals [86, 136].

11.18.2 Personalized Medicine Approaches

The future of drug delivery is moving toward personalized medicine, with treatments designed to suit individual genetic profiles, disease conditions, and pharmacokinetic responses. Natural polymers play a key role in the development of tailored CRS formulations, especially in areas like 3D-printed personalized oral tablets, individualized implants, and gene-based treatments [16].

11.18.3 3D Printing of Personalized CRS Tablets

The integration of 3D printing in pharmaceutical sciences has enabled the fabrication of customized oral controlled-release tablets using biodegradable natural polymers. By incorporating alginate, chitosan, and cellulose derivatives into bio-inks, researchers can precisely engineer drug-loaded matrices with tailored release kinetics [61]. A study on 3D-printed alginate scaffolds loaded with chemotherapy drugs demonstrated that drug release kinetics could be fine-tuned by modifying scaffold geometry, ensuring tumor-targeted therapy while reducing systemic toxicity. This approach is promising for oncology patients, where treatment regimens require adjustments based on tumor progression and genetic markers [35]. Additionally, multilayered 3D-printed CRS tablets have been developed, allowing for sequential drug release based on programmed dissolution profiles, enhancing treatment efficacy in chronic conditions such as diabetes and cardiovascular diseases [191].

11.18.4 Natural Polymer-Based Personalized Implants

3D printing is also revolutionizing the field of implantable CRS, where natural polymer-based hydrogels are being utilized to create patient-specific, biodegradable implants for localized drug release. For instance, chitosan-alginate composite implants have been successfully fabricated for targeted antibiotic delivery in bone infections, leading to improved bone regeneration and reduced recurrence of infection [77].

Personalized biopolymer implants offer precision-controlled drug release, particularly in postsurgical therapy and chronic disease management. Research on collagen-based drug-eluting implants has demonstrated long-term release of anti-inflammatory agents, reducing postoperative complications while promoting tissue healing. These implantable CRS technologies are expected to replace conventional systemic drug delivery methods, minimizing side effects and enhancing localized drug availability [159, 165].

11.18.5 Gene Therapy and CRISPR-Based Drug Delivery

Gene therapy is another rapidly evolving area where natural polymer-based CRS are proving instrumental in targeted genetic modifications. Advances in CRISPR-Cas9 gene editing have prompted the development of chitosan-DNA nanoparticles and alginate-coated viral vectors for efficient intracellular gene delivery [170, 172]. A study investigating chitosan-based CRISPR/Cas9 nanoparticle delivery for cystic fibrosis treatment demonstrated a 50% increase in gene correction efficiency, highlighting the potential of biopolymer-based carriers in nonviral gene therapy. The use of natural polymer nanoparticles offers several advantages over viral vectors, including lower immunogenicity, improved stability, and reduced risk of insertional mutagenesis [87].

Beyond CRISPR, biodegradable polymer-based siRNA delivery systems are being explored for cancer immunotherapy and neurodegenerative disease treatment. Alginate-based nanocarriers encapsulating therapeutic siRNA molecules have demonstrated significant tumor suppression in preclinical models of triple-negative breast cancer, highlighting the transformative potential of natural polymers in the future of genetic medicine [99].

11.19 Future Outlook: Toward Smarter, Sustainable CRS

The next generation of CRS technologies will likely focuses on the integration of smart biomaterials, artificial intelligence-driven formulation design, and self-assembling biopolymers. Natural polymers, owing to their biodegradability and functionality, will continue to be a major component of next-gen drug delivery platforms, particularly in oncology, regenerative medicine, and precision nanomedicine [49, 67].

Key emerging trends include:

- Microneedle-assisted natural polymer patches, improving the transdermal delivery of biologics such as insulin and vaccines [199].
- Bioengineered polysaccharide-lipid hybrid films, ensuring enhanced drug penetration through biological barriers [42].
- AI-driven CRS formulation models, enabling real-time optimization of polymer matrices for targeted therapy [169].
- 3D-printed personalized transdermal patches, tailored for patient-specific drug release in chronic disease management [200].

As the pharmaceutical industry continues to transition toward sustainable and highly efficient drug delivery models, natural polymer-based controlled-release systems (CRS) will continue to lead the way in medical innovation, providing safer, more effective, and environmentally sustainable therapies [8, 22].

11.20 Conclusion

Natural polymers have emerged as vital components in controlled-release systems (CRS), offering biocompatibility, biodegradability, and sustainability for efficient drug delivery. Their ability to regulate drug release through diffusion, swelling, and enzymatic degradation has enabled advancements in oral, parenteral, and transdermal applications. Recent innovations in nanotechnology and stimuli-responsive materials have further enhanced precision in drug targeting while minimizing side effects. However, challenges such as variability, mechanical instability, and scalability remain, necessitating continued research into hybrid polymer systems and computational modeling for optimized release kinetics. Regulatory advancements have facilitated commercialization, promoting the adoption of biodegradable drug carriers. Future developments, including 3D-printed drug delivery devices and bioengineered polymers, will further refine CRS, ensuring safer, more effective, and patient-friendly therapeutic solutions that align with the future of modern medicine.

References

1. Abbaspour M, Makhmalzadeh BS, Behjat R, Shoja S, Ahangari Z. Evaluation of the antimicrobial effect of chitosan/polyvinyl alcohol electrospun nanofibers containing mafenide acetate. Jundishapur J Microbiol. 2015;8(10):e24239. https://doi.org/10.5812/jjm.24239.
2. Adepu S, Ramakrishna S. Controlled drug delivery systems: current status and future directions. Molecules. 2021;26(19):5905. https://doi.org/10.3390/molecules26195905.
3. Afzal MAF, Lehmkemper K, Sobich E, Hughes TF, Giesen DJ, Zhang T, Krauter CM, Winget P, Degenhardt M, Kyeremateng SO, Browning A, Shelley JC. Molecular-level examination of amorphous solid dispersion dissolution. Mol Pharm. 2021;18(11):3999–4014. https://doi.org/10.1021/acs.molpharmaceut.1c00289.
4. Agarwal S, Singh S. A review: natural polymer loaded matrix tablets of antipyretic drug. Int J Pharm Sci Med. 2024;9:46–61. https://doi.org/10.47760/ijpsm.2024.v09i01.004.
5. Ahmed R. Ensuring quality medicine: a comprehensive overview of EMA and DGDA's history, structure and functions. J Chem Pharm Res. 2024;16:209. https://doi.org/10.20944/preprints202408.0031.v1.
6. Alalaiwe A, Carpinone PL, Alshahrani SM, Alsulays BB, Ansari MJ, Anwer M, Alshehri S, Alshetaili A. Influence of chitosan coating on the oral bioavailability of gold nanoparticles in rats. Saudi Pharm J. 2019;27(2):171–5. https://doi.org/10.1016/j.jsps.2018.09.011.
7. Alam A, Machale MU, Yadav RP, Sharma M, Patel AK. Role of transdermal drug delivery system. Asian J Pharm Res Dev. 2021;9(3):137–43. https://doi.org/10.22270/ajprd.v9i3.976.
8. Aldawood FK, Andar A, Desai S. A comprehensive review of microneedles: types, materials, processes, characterizations and applications. Polymers. 2021;13(16):2815. https://doi.org/10.3390/polym13162815.
9. Amidon S, Brown J, Dave VS. Colon-targeted oral drug delivery systems: design trends and approaches. AAPS PharmSciTech. 2015;16(4):731–41. https://doi.org/10.1208/s12249-015-0350-9.
10. Andersen T, Bleher S, Flaten GE, Tho I, Mattsson S, Škalko-Basnet N. Chitosan in mucoadhesive drug delivery: focus on local vaginal therapy. Mar Drugs. 2015;13(1):222–36. https://doi.org/10.3390/md13010222.
11. Archana D, Dutta PK, Dutta J. Chitosan: a potential therapeutic dressing material for wound healing. In: Chitin and chitosan for regenerative medicine. New Delhi: Springer India; 2015. p. 193–227. https://doi.org/10.1007/978-81-322-2511-9_8.
12. Armutcu C. The investigation of parameters affecting ibrutinib release from chitosan/ tripolyphosphate/carbon nanofiber composite microspheres. Turk J Chem. 2022;46(5):1632–41. https://doi.org/10.55730/1300-0527.3466.
13. Azeem MK, Islam A, Khan RU, Rasool A, Anees Ur Rehman Qureshi M, Rizwan M, Shuib RK, Rehman A, Sadiqa A. Guar gum/poly ethylene glycol/graphene oxide environmentally friendly hybrid hydrogels for controlled release of boron micronutrient. R Soc Open Sci. 2023;10(12):231157. https://doi.org/10.1098/rsos.231157.
14. Bachleitner S, Ata Ö, Mattanovich D. The potential of CO_2-based production cycles in biotechnology to fight the climate crisis. Nat Commun. 2023;14(1):6978. https://doi.org/10.1038/s41467-023-42790-6.
15. Baranwal J, Barse B, Fais A, Delogu G, Kumar A. Biopolymer: a sustainable material for food and medical applications. Polymers. 2022;14(5):983. https://doi.org/10.3390/polym14050983.
16. Beer N, Hegger I, Kaae S, Bruin MLD, Genina N, Alves TL, Hoebert J, Sporrong SK. Scenarios for 3D printing of personalized medicines—a case study. Explor Res Clin Soc Pharm. 2021;4:100073. https://doi.org/10.1016/j.rcsop.2021.100073.
17. Begines B, Ortiz T, Pérez-Aranda M, Martínez G, Merinero M, Argüelles-Arias F, Alcudia A. Polymeric nanoparticles for drug delivery: recent developments and future prospects. Nanomaterials (Basel). 2020;10(7):1403. https://doi.org/10.3390/nano10071403.

18. Bhatt P, Kumar V, Subramaniyan V, Nagarajan K, Sekar M, Chinni SV, Ramachawolran G. Plasma modification techniques for natural polymer-based drug delivery systems. Pharmaceutics. 2023;15(8):2066. https://doi.org/10.3390/pharmaceutics15082066.
19. Bhowmik A, Sultana R, Ahmed T, Sultana S. Design and development of atenolol matrix tablet employing natural and synthetic polymers. J Appl Pharm Sci. 2013;3(9):103–8. https://doi.org/10.7324/japs.2013.3919.
20. Bhumkar DR, Joshi HM, Sastry M, Pokharkar VB. Chitosan reduced gold nanoparticles as novel carriers for transmucosal delivery of insulin. Pharm Res. 2007;24:1415–26.
21. Bini E, Foo CWP, Huang J, Karageorgiou V, Kitchel B, Kaplan DL. RGD-functionalized bioengineered spider dragline silk biomaterial. Biomacromolecules. 2006;7(11):3139–45. https://doi.org/10.1021/bm0607877.
22. Bird D, Ravindra NM. Transdermal drug delivery and patches—an overview. Med Devices Sens. 2020;3(6):e10069. https://doi.org/10.1002/mds3.10069.
23. Briand VA, Kumar CV, Kasi RM. Protein-polymer conjugates. In: Encyclopedia of polymer science and technology; 2011. https://doi.org/10.1002/0471440264.pst325.
24. Brown BN, Badylak SF. Extracellular matrix as an inductive scaffold for functional tissue reconstruction. Transl Res. 2014;163(4):268–85. https://doi.org/10.1016/j.trsl.2013.11.003.
25. Cai L, Xu J, Yang Z, Tong R, Dong Z, Wang C, Leong KW. Engineered biomaterials for cancer immunotherapy. Medcomm. 2020;1(1):35–46. https://doi.org/10.1002/mco2.8.
26. Campbell R, Shim H, Choi JM, Park M, Byun E, Islam S, Song SH, Kim A. Implantable cisplatin synthesis microdevice for regional chemotherapy. Adv Healthc Mater. 2020;10(3):2001582. https://doi.org/10.1002/adhm.202001582.
27. Çaykara T, Bozkaya U, Kantoğlu Ö. Network structure and swelling behavior of poly(acrylamide/crotonic acid) hydrogels in aqueous salt solutions. J Polym Sci B Polym Phys. 2003;41(14):1656–64. https://doi.org/10.1002/polb.10500.
28. Chandakavathe BN, Kulkarni R, Rajmane AD, Mane SL, Waykule NJ. Development of colon targeted extended drug delivery system for mebendazole using grafted natural polymers. Indian J Pharm Educ Res. 2024;58(3s):s837–47. https://doi.org/10.5530/ijper.58.3s.85.
29. Chechare DD, Siddaiah M. Mucoadhesive drug delivery system: a review. Am J Pharmtech Res. 2018;8(6):16–22. https://doi.org/10.46624/ajptr.2018.v8.i6.002.
30. Chen C, Duan Z, Yuan Y, Li R, Pang L, Liang J, Xu X, Wang J. Peptide-22 and cyclic RGD functionalized liposomes for glioma targeting drug delivery overcoming BBB and BBTB. ACS Appl Mater Interfaces. 2017;9(7):5864–73. https://doi.org/10.1021/acsami.6b15831.
31. Chen N, Brachmann CB, Liu X, Pierce DW, Dey J, Kerwin W, Li Y, Zhou S, Hou S, Carleton M, Klinghoffer RA, Palmisano M, Chopra R. Albumin-bound nanoparticle (nab) paclitaxel exhibits enhanced paclitaxel tissue distribution and tumor penetration. Cancer Chemother Pharmacol. 2015;76(4):699–712. https://doi.org/10.1007/s00280-015-2833-5.
32. Chen N, Johnson MM, Collier MA, Gallovic MD, Bachelder EM, Ainslie KM. Tunable degradation of acetalated dextran microparticles enables controlled vaccine adjuvant and antigen delivery to modulate adaptive immune responses. J Control Release. 2018;273:147–59. https://doi.org/10.1016/j.jconrel.2018.01.027.
33. Chou HY, Weng CC, Lai JY, Lin SY, Tsai HC. Design of an interpenetrating polymeric network hydrogel made of calcium-alginate from a thermos-sensitive pluronic template as a thermal-ionic reversible wound dressing. Polymers. 2020;12(9):2138. https://doi.org/10.3390/polym12092138.
34. Choudhuri AH, Uppal R. A comparison between intravenous paracetamol plus fentanyl and intravenous fentanyl alone for postoperative analgesia during laparoscopic cholecystectomy. Anesth Essays Res. 2011;5(2):196. https://doi.org/10.4103/0259-1162.94777.
35. Dang W, Yi K, Ju E, Jin Y, Xu Y, Wang H, Chen W-C, Wang K, Wang Y, Tao Y, Li M. 3D printed bioceramic scaffolds as a universal therapeutic platform for synergistic therapy of osteosarcoma. ACS Appl Mater Interfaces. 2021;13(16):18488–99. https://doi.org/10.1021/acsami.1c00553.

36. Darwish WM, Bayoumi NA, Ebeid NH. Biocompatible mucoadhesive nanoparticles for brain targeting of ropinirole hydrochloride: formulations, radiolabeling and biodistribution. Biopolymers. 2022;113(6):e23489. https://doi.org/10.1002/bip.23489.
37. Deljavan Ghodrati A, ÇOmoĞLu T. MukoadezİF PolİmerlerİN Kolon Hedeflİ İlaç Taşiyici Sİstemlerde Kullanimi: Detayli Bİr İnceleme. Ank Univ Eczaci Fak Derg. 2024;48(2):5–5. https://doi.org/10.33483/jfpau.1400920.
38. Dergunov SA, Khabiyev AT, Shmakov SN, Kim MD, Ehterami N, Weiss MC, Birman VB, Pinkhassik E. Encapsulation of homogeneous catalysts in porous polymer nanocapsules produces fast-acting selective nanoreactors. ACS Nano. 2016;10(12):11397–406. https://doi.org/10.1021/acsnano.6b06735.
39. Desai N, Rana D, Salave S, Gupta R, Patel P, Karunakaran B, Sharma A, Giri J, Benival D, Kommineni N. Chitosan: a potential biopolymer in drug delivery and biomedical applications. Pharmaceutics. 2023;15(4):1313. https://doi.org/10.3390/pharmaceutics15041313.
40. Dhillon S, Renwick S, Berdy B, Dhillon AS. Pediatric probiotic culture survival study in acidic pH using an in vitro model. J Emerg Invest. 2022; https://doi.org/10.59720/22-099.
41. Dinjaski N, Kaplan DL. Recombinant protein blends: silk beyond natural design. Curr Opin Biotechnol. 2016;39:1–7. https://doi.org/10.1016/j.copbio.2015.11.002.
42. Dong W-J, Ye J, Wang W, Yang Y, Wang H, Sun T, Gao L, Liu Y. Self-assembled lecithin/chitosan nanoparticles based on phospholipid complex: a feasible strategy to improve entrapment efficiency and transdermal delivery of poorly lipophilic drug. Int J Nanomedicine. 2020;15:5629–43. https://doi.org/10.2147/ijn.s261162.
43. Ebbers HC, Fehrmann B, Ottosen M, Hvorslev N, Høier P, Hwang JW, Chung J, Lim HT, Lee S, Hong JH, Rezk MF. Batch-to-batch consistency of SB4 and SB2, etanercept and infliximab biosimilars. BioDrugs. 2020;34(2):225–33. https://doi.org/10.1007/s40259-019-00402-0.
44. Elmowafy M, Shalaby K, Elkomy MH, Alsaidan OA, Gomaa HAM, Abdelgawad MA, Mostafa EM. Polymeric nanoparticles for delivery of natural bioactive agents: recent advances and challenges. Polymers. 2023;15(5):1123. https://doi.org/10.3390/polym15051123.
45. Erkoc P, Seker-Polat F, Bagcı-Önder T, Kızılel S. Gelatin methacryloyl hydrogels in the absence of a crosslinker as 3D glioblastoma multiforme (GBM)-mimetic microenvironment. Macromol Biosci. 2018;18(3):1700369. https://doi.org/10.1002/mabi.201700369.
46. Ershad R, Hossain M, Alam MS, Asaduzzaman AKM. A comparative randomised clinical study between nebulised fentanyl and intravenous fentanyl for post operative pain relief. J Bangladesh Soc Anaesthesiol. 2016;27(1):27–30. https://doi.org/10.3329/jbsa.v27i1.28997.
47. Farin M, Maisha JT, Gibson I, Arafat MT. Additive manufacturing: a bespoke solution for drug delivery. Rapid Prototyp J. 2024;30(9):1717–36. https://doi.org/10.1108/rpj-09-2023-0326.
48. Farooq MA, Aquib M, Ghayas S, Bushra R, Khan DH, Parveen A, Wang B. Whey protein: a functional and promising material for drug delivery systems recent developments and future prospects. Polym Adv Technol. 2019;30(9):2183–91. https://doi.org/10.1002/pat.4676.
49. Feldman D. Polymers and polymer nanocomposites for cancer therapy. Appl Sci. 2019;9(18):3899. https://doi.org/10.3390/app9183899.
50. Filipović VV, Babić MM, Vuković JS, Vukomanović M, Rubert M, Hofmann S, Müller R, Tomić SL. Biodegradable hydrogel scaffolds based on 2-hydroxyethyl methacrylate, gelatin, poly(β-amino esters), and hydroxyapatite. Polymers. 2021;14(1):18. https://doi.org/10.3390/polym14010018.
51. Fischer KM, Scott TE, Browe DP, McGaughey TA, Wood C, Wolyniak MJ, Freeman JW. Hydrogels for skeletal muscle regeneration. Regen Eng Transl Med. 2021;7:353–61.
52. Fischer M, Gebhard F, Hammer TR, Zurek C, Meurer G, Marquardt C, Hoefer D. Microbial alginate dressings show improved binding capacity for pathophysiological factors in chronic wounds compared to commercial alginate dressings of marine origin. J Biomater Appl. 2017;31(9):1267–76. https://doi.org/10.1177/0885328217702173.
53. Fitriani L, Abdillah R, Ben ES. Formulation of metformin HCl floating tablet using HPC, HPMC K100M, and the combinations. J Sains Farm Klin. 2017;4(1):79. https://doi.org/10.29208/jsfk.2017.4.1.201.

54. Fundueanu G, Constantin M, Bucătariu S, Ascenzi P. Poly(*N*-isopropylacrylamide-co-*N*-isopropylmethacrylamide) thermo-responsive microgels as self-regulated drug delivery system. Macromol Chem Phys. 2016;217(22):2525–33. https://doi.org/10.1002/macp.201600324.
55. Gao C, Zheng H, Xing L, Shu M, Che S. Designable coordination bonding in mesopores as a pH-responsive release system. Chem Mater. 2010;22(19):5437–44. https://doi.org/10.1021/cm100667u.
56. Gately N, Kennedy JE. The development of a melt-extruded shellac carrier for the targeted delivery of probiotics to the colon. Pharmaceutics. 2017;9(4):38. https://doi.org/10.3390/pharmaceutics9040038.
57. Gheorghiţă R, Lobiuc A, Dimian M, Covaşă M. Alginate: from food industry to biomedical applications and management of metabolic disorders. Polymers. 2020;12(10):2417. https://doi.org/10.3390/polym12102417.
58. Ghormade JM, Yadav SK, Burakle PV, Raut AS. The role of matrix tablet in controlled release drug delivery system. GSC Biol Pharm Sci. 2023;23(1):220–5. https://doi.org/10.30574/gscbps.2023.23.1.0157.
59. Gil D, Frank-Kamenetskii A, Barry J, Reukov V, Xiang Y, Das A, Varma A, Kindy MS, Banik NL, Vertegel A. Albumin-assisted method allows assessment of release of hydrophobic drugs from nanocarriers. Biotechnol J. 2017;13(1):201700337. https://doi.org/10.1002/biot.201700337.
60. Goldmann N, Skalicky SE, Weinreb RN, Guedes RAP, Baudouin C, Zhang X, van Gestel A, Blumenthal EZ, Kaufman PL, Rothman R. Defining functional requirements for a patient-centric computerized glaucoma treatment and care ecosystem. J Med Artif Intell. 2023;6:3.
61. Goyanes Á, Chang H, Sedough D, Hatton GB, Wang J, Buanz ABM, Gaisford S, Basit AW. Fabrication of controlled-release budesonide tablets via desktop (FDM) 3D printing. Int J Pharm. 2015;496(2):414–20. https://doi.org/10.1016/j.ijpharm.2015.10.039.
62. Gradishar WJ, Tjulandin S, Davidson N, Shaw H, Desai N, Bhar P, Hawkins M, O'Shaughnessy J. Phase III trial of nanoparticle albumin-bound paclitaxel compared with polyethylated castor oil–based paclitaxel in women with breast cancer. J Clin Oncol. 2005;23(31):7794–803. https://doi.org/10.1200/jco.2005.04.937.
63. Gülsu A, Kıllı B, Alper M. Paclitaxel delivery by cationic gelatin nanoparticles. Chemistryselect. 2022;7(1):e202103495. https://doi.org/10.1002/slct.202103495.
64. Guo M, Wang G. Milk protein polymer and its application in environmentally safe adhesives. Polymers. 2016;8(9):324. https://doi.org/10.3390/polym8090324.
65. Hady WEAE, Mohamed EA, Soliman O, Sabbagh HMEL. In vitro-in vivo evaluation of chitosan-PLGA nanoparticles for potentiated gastric retention and anti-ulcer activity of diosmin. Int J Nanomed. 2019;14:7191–213. https://doi.org/10.2147/ijn.s213836.
66. Hahn T, Tafi E, Paul A, Salvia R, Falabella P, Zibek S. Current state of chitin purification and chitosan production from insects. J Chem Technol Biotechnol. 2020;95(11):2775–95. https://doi.org/10.1002/jctb.6533.
67. Hajebi S, Rabiee N, Bagherzadeh M, Ahmadi S, Rabiee M, Roghani-Mamaqani H, Tahriri M, Tayebi L, Hamblin MR. Stimulus-responsive polymeric nanogels as smart drug delivery systems. Acta Biomater. 2019;92:1–18. https://doi.org/10.1016/j.actbio.2019.05.018.
68. Hariyadi DM, Islam N. Current status of alginate in drug delivery. Adv Pharmacol Pharm Sci. 2020;2020:1–16. https://doi.org/10.1155/2020/8886095.
69. Hasanzadeh A, Hamblin MR, Kiani J, Noori H, Hardie JM, Karimi M, Shafiee H. Could artificial intelligence revolutionize the development of nanovectors for gene therapy and mRNA vaccines? Nano Today. 2022;47:101665. https://doi.org/10.1016/j.nantod.2022.101665.
70. Hay ID, Rehman ZU, Ghafoor A, Rehm BHA. Bacterial biosynthesis of alginates. J Chem Technol Biotechnol. 2010;85(6):752–9. https://doi.org/10.1002/jctb.2372.
71. He W, Zhang Y, Qu Y, Liu M, Li G, Pan L, Xu X, Shi G, Hao Q, Liu F, Gao Y. Research progress on hydrogel-based drug therapy in melanoma immunotherapy. BMB Rep. 2024;57(2):71–8. https://doi.org/10.5483/bmbrep.2023-0160.

72. He X, Kawazoe N, Chen G. Preparation of cylinder-shaped porous sponges of poly(L-lactic acid), poly(DL-lactic-co-glycolic acid), and poly(ε-caprolactone). Biomed Res Int. 2014;2014:1–8. https://doi.org/10.1155/2014/106082.
73. Huang KS, Yang CH, Kung CP, Grumezescu AM, Ker MD, Lin YS, Wang CY. Synthesis of uniform core–shell gelatin–alginate microparticles as intestine-released oral delivery drug carrier. Electrophoresis. 2013a;35(2–3):330–6. https://doi.org/10.1002/elps.201300194.
74. Huang S, Shao K, Liu Y, Kuang Y, Li J, An S, Guo Y, Ma H, Jiang C. Tumor-targeting and microenvironment-responsive smart nanoparticles for combination therapy of antiangiogenesis and apoptosis. ACS Nano. 2013b;7(3):2860–71. https://doi.org/10.1021/nn400548g.
75. Idrees H, Zaidi SZJ, Sabir A, Khan RU, Zhang X, Hassan S-U. A review of biodegradable natural polymer-based nanoparticles for drug delivery applications. Nanomaterials. 2020;10(10):1970.
76. Iglesias N, Galbis E, Valencia C, de Paz MV, Galbis JA. Reversible pH-sensitive chitosan-based hydrogels. Influence of dispersion composition on rheological properties and sustained drug delivery. Polymers. 2018;10(4):392. https://doi.org/10.3390/polym10040392.
77. Inzana JA, Trombetta RP, Schwarz EM, Kates SL, Awad HA. 3D printed bioceramics for dual antibiotic delivery to treat implant-associated bone. Infection. 2015;30:232–47. https://doi.org/10.22203/ecm.v030a16.
78. Iqbal U, Westfall S, Prakash S. Novel microencapsulated probiotic blend for use in metabolic syndrome: design and *in-vivo* analysis. Artif Cells Nanomed Biotechnol. 2018;46(sup3):116–24. https://doi.org/10.1080/21691401.2018.1489270.
79. Ivory-Cousins T, Nurzyńska A, Klimek K, Baines DK, Truszkiewicz W, Pałka K, Douglas T. Whey protein isolate/calcium silicate hydrogels for bone tissue engineering applications—preliminary in vitro evaluation. Materials. 2023;16(19):6484. https://doi.org/10.3390/ma16196484.
80. Jain D, Bar-Shalom D. Alginate drug delivery systems: application in context of pharmaceutical and biomedical research. Drug Dev Ind Pharm. 2014;40(12):1576–84. https://doi.org/10.3109/03639045.2014.917657.
81. Jain D, Raturi R, Jain V, Bansal P, Singh R. Recent technologies in pulsatile drug delivery systems. Biomatter. 2011;1(1):57–65. https://doi.org/10.4161/biom.1.1.17717.
82. Jhaveri J, Raichura Z, Khan T, Momin M, Omri A. Chitosan nanoparticles-insight into properties, functionalization and applications in drug delivery and theranostics. Molecules. 2021;26(2):272.
83. Jiang J, Li Y, Fang T, Zhou J, Li X, Wang Y, Dong J. Vancomycin-loaded nano-hydroxyapatite pellets to treat MRSA-induced chronic osteomyelitis with bone defect in rabbits. Inflamm Res. 2011;61(3):207–15. https://doi.org/10.1007/s00011-011-0402-x.
84. Johnson MD, Lloyd J, Tekkam S, Crooke SN, Witherden DA, Havran WL, Finn MG. Degradable hydrogels for the delivery of immune-modulatory proteins in the wound environment. ACS Appl Bio Mater. 2020;3(8):4779–88. https://doi.org/10.1021/acsabm.0c00301.
85. Kadiyala P, Li D, Núñez FM, Altshuler D, Doherty R, Kuai R, Yu M, Kamran N, Edwards M, Moon JJ, Löwenstein PR, Castro MG, Schwendeman A. High-density lipoprotein-mimicking nanodiscs for chemo-immunotherapy against glioblastoma multiforme. ACS Nano. 2019;13(2):1365–84. https://doi.org/10.1021/acsnano.8b06842.
86. Kar S, Sanderson H, Roy K, Benfenati E, Leszczyński J. Green chemistry in the synthesis of pharmaceuticals. Chem Rev. 2021;122(3):3637–710. https://doi.org/10.1021/acs.chemrev.1c00631.
87. Karayianni M, Sentoukas T, Skandalis A, Pippa N, Pispas S. Chitosan-based nanoparticles for nucleic acid delivery: technological aspects, applications, and future perspectives. Pharmaceutics. 2023;15(7):1849. https://doi.org/10.3390/pharmaceutics15071849.
88. Kareem RA, Razavi SH, Mousavi ZE. Effect of two types of wall matrix on viability of *Lactobacillus plantarum* under simulated gastrointestinal fluids. Research Square; 2023. https://doi.org/10.21203/rs.3.rs-3493871/v1.

89. Karimi M, Ghasemi A, Sahandi Zangabad P, Rahighi R, Moosavi Basri SM, Mirshekari H, Amiri M, Shafaei Pishabad Z, Aslani A, Bozorgomid M, Ghosh D, Beyzavi A, Vaseghi A, Aref AR, Haghani L, Bahrami S, Hamblin MR. Smart micro/nanoparticles in stimulus-responsive drug/gene delivery systems. Chem Soc Rev. 2016;45(5):1457–501. https://doi.org/10.1039/c5cs00798d.
90. Katuwavila NP, Perera ADLC, Samarakoon SR, Soysa P, Karunaratne V, Amaratunga GAJ, Karunaratne DN. Chitosan-alginate nanoparticle system efficiently delivers doxorubicin to McF-7 cells. J Nanomater. 2016;2016:1–12. https://doi.org/10.1155/2016/3178904.
91. Kavasidis I, Lallas E, Leligou HC, Oikonomidis G, Karydas DM, Gerogiannis VC, Karageorgos A. Deep transformers for computing and predicting ALCOA+ data integrity compliance in the pharmaceutical industry. Appl Sci. 2023;13(13):7616. https://doi.org/10.3390/app13137616.
92. Keshavarz M, Smith Q. Gelatin-mediated vascular self-assembly via a YAP-MMP signaling axis. Adv Funct Mater. 2024;34(37):2402360. https://doi.org/10.1002/adfm.202402360.
93. Kevadiya BD, Chen L, Zhang L, Thomas MB, Davé RN. Fenofibrate nanocrystal composite microparticles for intestine-specific oral drug delivery system. Pharmaceuticals. 2019;12(3):109. https://doi.org/10.3390/ph12030109.
94. Kim S-M, Kim SE, Lee DY, Choi D. Serum estradiol level according to dose and formulation of oral estrogens in postmenopausal women. Sci Rep. 2021;11(1):3585. https://doi.org/10.1038/s41598-021-81201-y.
95. Kim S. Recent trends and notable advances of alginate based nano-particles for effective biomedical materials: wound healing and drug delivery. Key Eng Mater. 2023;945:27–32. https://doi.org/10.4028/p-fsw5mj.
96. Kim S, Lee J. Effective polymeric dispersants for vacuum, convection and freeze drying of drug nanosuspensions. Int J Pharm. 2010;397(1–2):218–24. https://doi.org/10.1016/j.ijpharm.2010.07.010.
97. Kim SW, Kawai T, Wang D, Yang Y. Engineering a dual-layer chitosan–lactide hydrogel to create endothelial cell aggregate-induced microvascular networks in vitro and increase blood perfusion in vivo. ACS Appl Mater Interfaces. 2016;8(30):19245–55. https://doi.org/10.1021/acsami.6b04431.
98. Korchak JA, Wiest EF, Zubair AC. How do we assess batch-to-batch consistency between extracellular vesicle products? Transfusion. 2022;63(2):279–87. https://doi.org/10.1111/trf.17156.
99. Kozielski KL, Ruiz-Valls A, Tzeng SY, Guerrero-Cázares H, Rui Y, Li Y, Vaughan HJ, Gionet-Gonzales M, Vantucci CE, Kim J, Schiapparelli P, Al-kharboosh R, Quiñones-Hinojosa A, Green JJ. Cancer-selective nanoparticles for combinatorial siRNA delivery to primary human GBM in vitro and in vivo. Biomaterials. 2019;209:79–87. https://doi.org/10.1016/j.biomaterials.2019.04.020.
100. Küçükgülmez A, Çelik M, Yanar Y, Şen D, Polat H, Kadak AE. Physicochemical characterization of chitosan extracted from *Metapenaeus stebbingi* shells. Food Chem. 2011;126(3):1144–8. https://doi.org/10.1016/j.foodchem.2010.11.148.
101. Kumar A, Thakur VK, Nezhad HY, Lee KS. Prospects of sustainable polymers. Sci Rep. 2024;14(1):9430. https://doi.org/10.1038/s41598-024-59439-z.
102. Kwon IK, Lee SC, Han B, Park K. Analysis on the current status of targeted drug delivery to tumors. J Control Release. 2012;164(2):108–14. https://doi.org/10.1016/j.jconrel.2012.07.010.
103. Laurencin CT, Jiang T, Kumbar SG, Nair LS. Biologically active chitosan systems for tissue engineering and regenerative medicine. Curr Top Med Chem. 2008;8(4):354–64. https://doi.org/10.2174/156802608783790974.
104. Lee KY, Mooney D. Alginate: properties and biomedical applications. Prog Polym Sci. 2012;37(1):106–26. https://doi.org/10.1016/j.progpolymsci.2011.06.003.
105. Lin N, Gèze A, Wouessidjewe D, Huang J, Dufresne A. Biocompatible double-membrane hydrogels from cationic cellulose nanocrystals and anionic alginate as complexing drugs

codelivery. ACS Appl Mater Interfaces. 2016;8(11):6880–9. https://doi.org/10.1021/acsami.6b00555.
106. Liu JY, Xu M. An exact solution to the moving boundary problem with fractional anomalous diffusion in drug release devices. ZAMM Z fur Angew Math Mech. 2003;84(1):22–8. https://doi.org/10.1002/zamm.200410074.
107. Liu P, Zhao X. Facile preparation of well-defined near-monodisperse chitosan/sodium alginate polyelectrolyte complex nanoparticles (CS/SAL NPs) via ionotropic gelification: a suitable technique for drug delivery systems. Biotechnol J. 2013;8(7):847–54. https://doi.org/10.1002/biot.201300093.
108. Liu Y, Yang G, Baby T, Tengjisi, Chen D, Weitz DA, Zhao CX. Stable polymer nanoparticles with exceptionally high drug loading by sequential nanoprecipitation. Angew Chem. 2020;59(12):4720–8. https://doi.org/10.1002/anie.201913539.
109. Lukova P, Katsarov P. Contemporary aspects of designing marine polysaccharide microparticles as drug carriers for biomedical application. Pharmaceutics. 2023;15(8):2126. https://doi.org/10.3390/pharmaceutics15082126.
110. Makadia HK, Siegel SJ. Poly lactic-co-glycolic acid (PLGA) as biodegradable controlled drug delivery carrier. Polymers. 2011;3(3):1377–97. https://doi.org/10.3390/polym3031377.
111. Malek-Khatabi A, Tabandeh Z, Nouri A, Mozayan E, Sartorius R, Rahimi S, Jamaledin R. Long-term vaccine delivery and immunological responses using biodegradable polymer-based carriers. Acs Appl Bio Mater. 2022;5(11):5015–40. https://doi.org/10.1021/acsabm.2c00638.
112. Maniar D, Hohmann KF, Jiang Y, Woortman AJJ, Dijken JV, Loos K. Enzymatic polymerization of dimethyl 2,5-furandicarboxylate and heteroatom diamines. Acs Omega. 2018;3(6):7077–85. https://doi.org/10.1021/acsomega.8b01106.
113. Maniar D, Jiang Y, Woortman AJJ, Dijken JV, Loos K. Furan-based copolyesters from renewable resources: enzymatic synthesis and properties. ChemSusChem. 2019;12(5):990–9. https://doi.org/10.1002/cssc.201802867.
114. Mantripragada S. A lipid based depot (DepoFoam technology) for sustained release drug delivery. Prog Lipid Res. 2002;41(5):392–406. https://doi.org/10.1016/s0163-7827(02)00004-8.
115. Manyam SSD, Arumilli S, Pakalapati P, Sarella PNK. Formulation and evaluation of sustained release atorvastatin tablets using natural polymers, with a focus on okra gum. Int J Pharm Sci Med. 2023;8(10):11–22. https://doi.org/10.47760/ijpsm.2023.v08i10.003.
116. Mason L, Moore RA, Edwards J, Derry S, McQuay HJ. Topical NSAIDs for chronic musculoskeletal pain: systematic review and meta-analysis. BMC Musculoskelet Disord. 2004;5(1):28. https://doi.org/10.1186/1471-2474-5-28.
117. Matsliah L, Goder D, Giladi S, Zilberman M. In vitro characterization of novel multidrug-eluting soy protein wound dressings. J Biomater Appl. 2020;35(8):978–93. https://doi.org/10.1177/0885328220975178.
118. Mettler E, Zimmermann U, Hansen T, Ehrhart F, Zimmermann H, Weber M. Poloxamer 188 as a supplement to barium cross-linked ultra-high viscosity alginate for immunoisolation of transplanted islet cells. J Postgenomics Drug Biomarker Dev. 2015;05(04):1000157. https://doi.org/10.4172/2153-0769.1000157.
119. Mishra RK, Majeed ABA, Banthia AK. Development and characterization of pectin/gelatin hydrogel membranes for wound dressing. Int J Plast Technol. 2011;15(1):82–95. https://doi.org/10.1007/s12588-011-9016-y.
120. Mustafa NS, Omer MAA, Garlnabi M, Ismail H. Reviewing of general polymer types, properties and application in medical field. Int J Sci Res. 2016;5(8):212–21. https://doi.org/10.21275/art2016772.
121. Nagpal K, Singh SK, Mishra DN. Chitosan nanoparticles: a promising system in novel drug delivery. Chem Pharm Bull. 2010;58(11):1423–30. https://doi.org/10.1248/cpb.58.1423.
122. Nakayama R, Katsumata K, Niwa Y, Namiki N. Dependence of water-permeable chitosan membranes on chitosan molecular weight and alkali treatment. Membranes. 2020;10(11):351. https://doi.org/10.3390/membranes10110351.

123. Nam YS, Bae MS, Kim S, Noh I, Suh JKF, Lee KB, Kwon IK. Mechanism of albumin release from alginate and chitosan beads fabricated in dual layers. Macromol Res. 2011;19(5):476–82. https://doi.org/10.1007/s13233-011-0501-1.
124. Namasivayam KR. Untitled. Asian J Pharm. 2016;10(4) https://doi.org/10.22377/ajp.v10i04.891.
125. Ngwuluka NC, Ochekpe NA, Aruoma OI. Naturapolyceutics: the science of utilizing natural polymers for drug delivery. Polymers. 2014;6(5):1312–32. https://doi.org/10.3390/polym6051312.
126. Ogay V, Mun EA, Kudaibergen GK, Baidarbekov M, Kassymbek K, Zharkinbekov Z, Saparov A. Progress and prospects of polymer-based drug delivery systems for bone tissue regeneration. Polymers. 2020;12(12):2881. https://doi.org/10.3390/polym12122881.
127. Pandey AK, Dwivedi AK. Recent advancement in wound healing dressing material. Int J Res Pharm Sci. 2019;10(3):2572–7. https://doi.org/10.26452/ijrps.v10i3.1512.
128. Paul S, Das B, Sharma HK. A review on bio-polymers derived from animal sources with special reference to their potential applications. J Drug Deliv Ther. 2021;11(2):209–23. https://doi.org/10.22270/jddt.v11i2.4763.
129. Pedro SN, Mendes MSM, Neves BM, Almeida IF, Costa P, Correia-Sá I, Vilela C, Freire MG, Silvestre AJD, Freire CSR. Deep eutectic solvent formulations and alginate-based hydrogels as a new partnership for the transdermal administration of anti-inflammatory drugs. Pharmaceutics. 2022;14(4):827. https://doi.org/10.3390/pharmaceutics14040827.
130. Pérez JE, Fritzell S, Kopecký J, Visse E, Darabi A, Siesjö P. The effect of locally delivered cisplatin is dependent on an intact immune function in an experimental glioma model. Sci Rep. 2019;9(1):5632. https://doi.org/10.1038/s41598-019-42001-7.
131. Postolović KS, Antonijević MD, Ljujić B, Radenković S, Kovačević MM, Hiezl Z, Pavlović S, Radojević ID, Stanić Z. Curcumin and diclofenac therapeutic efficacy enhancement applying transdermal hydrogel polymer films, based on carrageenan, alginate and poloxamer. Polymers. 2022;14(19):4091. https://doi.org/10.3390/polym14194091.
132. Prabowo HA, Budianto E, Hudiyono S. Encapsulation of partially purified bromelain from pineapple cores in alginate-pectin beads as a targeted antiplatelet agent. Int J Appl Pharm. 2021;13(3):176–82. https://doi.org/10.22159/ijap.2021v13i3.41158.
133. Quiñones JP, Peniche H, Peniche C. Chitosan based self-assembled nanoparticles in drug delivery. Polymers. 2018;10(3):235. https://doi.org/10.3390/polym10030235.
134. Ramana BV, Vyshnavi K. A review article on natural polymers in novel drug delivery system. Int J Multidiscip Res. 2024;6(6) https://doi.org/10.36948/ijfmr.2024.v06i06.31051.
135. Rambhia KJ, Peter X. Controlled drug release for tissue engineering. J Control Release. 2015;219:119–28. https://doi.org/10.1016/j.jconrel.2015.08.049.
136. Ran J. The application and prospect of green chemistry in pharmaceutical field. Theor Nat Sci. 2024;58(1):55–9. https://doi.org/10.54254/2753-8818/58/20241335.
137. Rattanamanee A, Niamsup H, Srisombat L-O, Punyodom W, Watanesk R, Watanesk S. Role of chitosan on some physical properties and the urea controlled release of the silk fibroin/gelatin hydrogel. J Polym Environ. 2014;23(3):334–40. https://doi.org/10.1007/s10924-014-0703-6.
138. Riau AK, Mondal D, Aung TT, Murugan E, Chen L, Lwin NC, Zhou L, Beuerman RW, Liedberg B, Venkatraman SS, Mehta JS. Collagen-based artificial corneal scaffold with anti-infective capability for prevention of perioperative bacterial infections. ACS Biomater Sci Eng. 2015;1(12):1324–34. https://doi.org/10.1021/acsbiomaterials.5b00396.
139. Rk G. Transdermal drug delivery system: an emphasis on transdermal patches. Pharm Drug Regul Aff J. 2023;6(1):1–5. https://doi.org/10.23880/pdraj-16000147.
140. Rodrigues S, Dionísio M, López CR, Grenha A. Biocompatibility of chitosan carriers with application in drug delivery. J Funct Biomater. 2012;3(3):615–41. https://doi.org/10.3390/jfb3030615.
141. Rosenboom JG, Langer R, Traverso G. Bioplastics for a circular economy. Nat Rev Mater. 2022;7(2):117–37. https://doi.org/10.1038/s41578-021-00407-8.

142. Seman CNZC. In vitro elution and dissolution of tobramycin and gentamicin from calcium phosphate. Afr J Pharm Pharmacol. 2011;5(20):2283–91. https://doi.org/10.5897/ajpp11.591.
143. Sahoo BK, Pattajoshi SP, Pattajoshi S. A promising approach to provide appropriate colon target drug delivery systems of metronidazole: development and evaluation. J Biomed Pharm Res. 2020;9(3) https://doi.org/10.32553/jbpr.v9i3.760.
144. Samuel HS, Ekpan F-DM, Ori MO. Biodegradable, recyclable, and renewable polymers as alternatives to traditional petroleum-based plastics. Asian J Environ Res. 2024;1(3):152–65. https://doi.org/10.69930/ajer.v1i3.86.
145. Sasaki H, Sunagawa Y, Takahashi K, Imaizumi A, Fukuda H, Hashimoto T, Wada H, Katanasaka Y, Kakeya H, Fujita M, Hasegawa K, Morimoto T. Innovative preparation of curcumin for improved oral bioavailability. Biol Pharm Bull. 2011;34(5):660–5. https://doi.org/10.1248/bpb.34.660.
146. Satchanska G, Davidova S, Petrov PD. Natural and synthetic polymers for biomedical and environmental applications. Polymers (Basel). 2024;16(8):1159. https://doi.org/10.3390/polym16081159.
147. Schloss AC, Williams DM, Regan L. Protein-based hydrogels for tissue engineering. Adv Exp Med Biol. 2016;940:167–77. https://doi.org/10.1007/978-3-319-39196-0_8.
148. Serra E, Saubade F, Ligorio C, Whitehead KA, Sloan AJ, Williams DW, Hidalgo-Bastida A, Verran J, Malic S. Methylcellulose hydrogel with *Melissa officinalis* essential oil as a potential treatment for oral candidiasis. Microorganisms. 2020;8(2):215. https://doi.org/10.3390/microorganisms8020215.
149. Serra MC, Casas A, Toubarro D, Barros A, Teixeira JA. Microbial hyaluronic acid production: a review. Molecules. 2023;28(5):2084. https://doi.org/10.3390/molecules28052084.
150. Shabbir Ali A, Gul R, Usman H, Naseer F. Applications of natural polymers in regenerative medicine and tissue engineering. Pharmacol Commun. 2023;2(1):47–61. https://doi.org/10.55627/pharma.002.01.0299.
151. Shahzadi M, Hassan T, Saeed S. Application of natural polymers in wound dressings. Proc Pak Acad Sci B Life Environ Sci. 2022;59(2):19–23. https://doi.org/10.53560/ppasb(59-2)688.
152. Shamili G, Santhi G. Identification and characterization of bioactive compounds of leaves of Justicia Gendarussa Burm F. Int J Sci Res Biol Sci. 2019;6(1):145–53. https://doi.org/10.26438/ijsrbs/v6i1.145153.
153. Shi J, Liu X, Sun X, Cao S. Hybrid alginate beads with thermal-responsive gates for smart drug delivery. Polym Adv Technol. 2011;22(11):1539–46. https://doi.org/10.1002/pat.1677.
154. Silva TH, Alves A, Ferreira B, Oliveira JM, Reys LL, Ferreira R, Sousa RA, Silva SS, Mano JF, Reis RL. Materials of marine origin: a review on polymers and ceramics of biomedical interest. Int Mater Rev. 2012;57(5):276–306. https://doi.org/10.1179/1743280412y.0000000002.
155. Simpson J, Peacock JL, Maxwell-Armstrong C. Use of a gentamicin-impregnated collagen sheet (Collatamp®) following implantation of a sacral nerve stimulator for faecal incontinence. Color Dis. 2012;14(4):e200. https://doi.org/10.1111/j.1463-1318.2011.02886.x.
156. Singh R, Malviya R, Sharma PK. Extraction and characterization of tamarind seed polysaccharide as a pharmaceutical excipient. Pharm J. 2011;3(20):17–9. https://doi.org/10.5530/pj.2011.20.4.
157. Singh TRR, Woolfson AD, Donnelly RF. Investigation of solute permeation across hydrogels composed of poly(methyl vinyl ether-co-maleic acid) and poly(ethylene glycol). J Pharm Pharmacol. 2010;62(7):829–37. https://doi.org/10.1211/jpp.62.07.0003.
158. Smith SS, Keller PA, Kobinsky K, Baker TB, Fraser D, Bush T, Magnusson B, Zbikowski SM, McAfee T, Fiore MC. Enhancing tobacco quitline effectiveness: identifying a superior pharmacotherapy adjuvant. Nicotine Tob Res. 2012;15(3):718–28. https://doi.org/10.1093/ntr/nts186.
159. Solano AGR, Pereira ADF, Pinto FCH, Ferreira LGR, Barbosa LA, Fialho SL, Patrício PSDO, Silva-Cunha A, Silva GRD, Pianetti GA. Development and evaluation of sustained-release etoposide-loaded poly(ε-caprolactone) implants. AAPS PharmSciTech. 2013;14(2):890–900. https://doi.org/10.1208/s12249-013-9977-6.

160. Solorio L, Zwolinski C, Lund AW, Farrell MJ, Stegemann JP. Gelatin microspheres crosslinked with genipin for local delivery of growth factors. J Tissue Eng Regen Med. 2010;4(7):514–23. https://doi.org/10.1002/term.267.
161. Somo SI, Langert KA, Yang CY, Vaicik MK, Ibarra V, Appel AA, Akar B, Cheng MH, Brey EM. Synthesis and evaluation of dual crosslinked alginate microbeads. Acta Biomater. 2018;65:53–65. https://doi.org/10.1016/j.actbio.2017.10.046.
162. Song J, Odekerken JCE, Löwik DWPM, López-Pérez PM, Welting TJM, Yang F, Jansen JA, Leeuwenburgh SCG. Influence of the molecular weight and charge of antibiotics on their release kinetics from gelatin nanospheres. Macromol Biosci. 2015;15(7):901–11. https://doi.org/10.1002/mabi.201500005.
163. Sopyan I, Sriwidodo S, Wahyuningrum R, Norisca Aliza P. A review: floating drug delivery system as a tool to improve dissolution rate in gastric. Int J Appl Pharm. 2020;12(4):51–4. https://doi.org/10.22159/ijap.2020v12i4.38415.
164. Sowmya PS, Dp V, Nayek S. Pulsatile drug delivery system: a formulation approach for treatment of diseases. Int J Curr Pharm Res. 2020;12(3):16–21. https://doi.org/10.22159/ijcpr.2020v12i3.38328.
165. Sternberg K, Petersen S, Grabow N, Senz V, Schwabedissen HEMZ, Kroemer HK, Schmitz KP. Implant-associated local drug delivery systems based on biodegradable polymers: customized designs for different medical applications. Biomed Tech. 2013;58(5):417–27. https://doi.org/10.1515/bmt-2012-0049.
166. Suksaeree J, Waiprib R, Simchareon W, Phaechamud T, Pichayakorn W. Deproteinized natural rubber latex as matrix base of meloxicam transdermal patches. Key Eng Mater. 2022;914:3–8. https://doi.org/10.4028/p-591atj.
167. Sumarsono A, Sumarsono N, Das SR, Vaduganathan M, Agrawal D, Pandey A. Economic burden associated with extended-release vs immediate-release drug formulations among Medicare Part D and Medicaid beneficiaries. JAMA Netw Open. 2020;3(2):e200181.
168. Sun C. Application of alginate saline gel for drug delivery, cancer treatment and wound dressing. Appl Comput Eng. 2024;60(1):165–9. https://doi.org/10.54254/2755-2721/60/20240863.
169. Tang RC, Yang IH, Lin FH. Current role and potential of polymeric biomaterials in clinical obesity treatment. Biomacromolecules. 2023;24(8):3438–49. https://doi.org/10.1021/acs.biomac.3c00388.
170. Tang W, Liu J, Ding B. Nucleic acid nanostructure for delivery of CRISPR/Cas9-based gene editing system. Interdiscip Med. 2022;1(1):e20220014. https://doi.org/10.1002/inmd.20220014.
171. Tao Y, Hou X, Zuo F, Li X-X, Pang Y, Jiang G. Application of nanoparticle-based siRNA and CRISPR/Cas9 delivery systems in gene-targeted therapy. Nanomedicine. 2019;14(5):511–4. https://doi.org/10.2217/nnm-2018-0522.
172. Thevendran R, Solayappan M. Recognizing CRISPR as the new age disease-modifying drug: strategies to bioengineer CRISPR/Cas for direct in vivo delivery. Biotechnol J. 2023;18(9):e2300077. https://doi.org/10.1002/biot.202300077.
173. Tiwari AK, Gautam M, Maurya HK. Recent development of biodegradation techniques of polymer. Int J Res Granthaalayah. 2018;6(6):414–52.
174. Tong X, Pan W, Su T, Zhang M, Dong W, Qi X. Recent advances in natural polymer-based drug delivery systems. React Funct Polym. 2020;148:104501. https://doi.org/10.1016/j.reactfunctpolym.2020.104501.
175. Tsai C-H, Wang PY, Lin IC, Hu H, Liu GS, Tseng CL. Ocular drug delivery: role of degradable polymeric nanocarriers for ophthalmic application. Int J Mol Sci. 2018;19(9):2830. https://doi.org/10.3390/ijms19092830.
176. Ullah S, Azad AK, Nawaz A, Shah KU, Iqbal M, Albadrani GM, Al-Joufi FA, Sayed AA, Abdel-Daim MM. 5-fluorouracil-loaded folic-acid-fabricated chitosan nanoparticles for site-targeted drug delivery cargo. Polymers. 2022;14(10):2010. https://doi.org/10.3390/polym14102010.

177. Umadevi S, Thiruganesh R, Suresh S, Reddy KB. Formulation and evaluation of chitosan microspheres of aceclofenac for colon-targeted drug delivery. Biopharm Drug Dispos. 2010;31(7):407–27. https://doi.org/10.1002/bdd.722.
178. Untitled. Int J Pharm Sci Res. 2024;15(7) https://doi.org/10.13040/ijpsr.0975-8232.15(7).1886-96.
179. Valades-Aguilar BA, Rivera-González TI, Rangel-López R, Luna-Bárcenas G, Franco-Molina MÁ, Rodríguez-Padilla C, Zárate-Triviño DG. Influence of chitosan purification on the immunomodulator potential of chitosan nanoparticles on human monocytes. Polymers. 2024;16(23):3390. https://doi.org/10.3390/polym16233390.
180. Vasile C, Pamfil D, Stoleru E, Baican M. New developments in medical applications of hybrid hydrogels containing natural polymers. Molecules. 2020;25(7):1539. https://doi.org/10.3390/molecules25071539.
181. Vemula SK, Bontha VK. Colon targeted guar gum compression coated tablets of flurbiprofen: formulation, development, and pharmacokinetics. Biomed Res Int. 2013;2013:1–8. https://doi.org/10.1155/2013/287919.
182. Vora LK, Gholap AD, Jetha K, Singh TRR, Solanki HK, Chavda VP. Artificial intelligence in pharmaceutical technology and drug delivery design. Pharmaceutics. 2023;15(7):1916. https://doi.org/10.3390/pharmaceutics15071916.
183. Walker N, Parag V, Verbiest M, Laking G, Laugesen M, Bullen C. Nicotine patches used in combination with E-cigarettes (with and without nicotine) for smoking cessation: a pragmatic, randomised trial. Lancet Respir Med. 2020;8(1):54–64. https://doi.org/10.1016/s2213-2600(19)30269-3.
184. Walkey C, Olsen JB, Guo H, Emili A, Chan WCW. Nanoparticle size and surface chemistry determine serum protein adsorption and macrophage uptake. J Am Chem Soc. 2012;134(4):2139–47. https://doi.org/10.1021/ja2084338.
185. Wang AN, Wu L, Jia L, Li XL, Yu S. Alginate-chitosan microspheres for controlled release of tea polyphenol. Adv Mater Res. 2010;152–153:1726–9. https://doi.org/10.4028/www.scientific.net/amr.152-153.1726.
186. Wang P, Meng X, Wang R, Yang W, Yang L, Wang J, Wang DA, Fan C. Biomaterial scaffolds made of chemically cross-linked gelatin microsphere aggregates (C-GMSs) promote vascularized bone regeneration. Adv Healthc Mater. 2022a;11(13):e2102818. https://doi.org/10.1002/adhm.202102818.
187. Wang X, Gao S-K, Yun S, Zhang M, Peng L, Li Y, Zhou Y. Microencapsulating alginate-based polymers for probiotics delivery systems and their application. Pharmaceuticals. 2022b;15(5):644. https://doi.org/10.3390/ph15050644.
188. Wang Y, Wu C. Site-specific conjugation of polymers to proteins. Biomacromolecules. 2018;19(6):1804–25. https://doi.org/10.1021/acs.biomac.8b00248.
189. Ways TMM, Lau WM, Khutoryanskiy VV. Chitosan and its derivatives for application in mucoadhesive drug delivery systems. Polymers. 2018;10(3):267. https://doi.org/10.3390/polym10030267.
190. Webb HK, Arnott J, Crawford RJ, Ivanova EP. Plastic degradation and its environmental implications with special reference to poly (ethylene terephthalate). Polymers. 2012;5(1):1–18.
191. Wei H, Luo Y, Ma R, Li Y. Three-dimensional printing multi-drug delivery core/shell fiber systems with designed release capability. Pharmaceutics. 2023;15(9):2336. https://doi.org/10.3390/pharmaceutics15092336.
192. Wen H, He B, Wang H, Chen F, Li P, Cui M, Li Q, Pan W, Yang X. Structure-based gastro-retentive and controlled-release drug delivery with novel 3D printing. AAPS PharmSciTech. 2019;20(2):68. https://doi.org/10.1208/s12249-018-1237-3.
193. Wu H, Hou X, Branford-White C, Sun X, Tao L, Um-i-Zahra S, Zhu LM. Drug-loaded microparticles prepared by the one-step deposition of calcium carbonate/alginate onto cotton fabrics. J Appl Polym Sci. 2015;132(40):42618. https://doi.org/10.1002/app.42618.
194. Wu P, Gao J, Prasad P, Dutta K, Kanjilal P, Thayumanavan S. Influence of polymer structure and architecture on drug loading and redox-triggered release. Biomacromolecules. 2021;23(1):339–48. https://doi.org/10.1021/acs.biomac.1c01295.

195. Xin H, Brown HR, Naficy S, Spinks GM. Mechanical recoverability and damage process of ionic-covalent PAA m-alginate hybrid hydrogels. J Polym Sci B Polym Phys. 2015;54(1):53–63. https://doi.org/10.1002/polb.23899.
196. Xing Z-C, Han S-J, Shin Y-S, Kang I-K. Fabrication of biodegradable polyester nanocomposites by electrospinning for tissue engineering. J Nanomater. 2011;2011(1):929378.
197. Xu Y, Kim CS, Saylor DM, Koo D. Polymer degradation and drug delivery in PLGA-based drug–polymer applications: a review of experiments and theories. J Biomed Mater Res B Appl Biomater. 2016;105(6):1692–716. https://doi.org/10.1002/jbm.b.33648.
198. Yadav AV, Urade MN. Formulation and evaluation of chitosan based transdermal patches of lornoxicam for prolonged drug release and to study the effect of permeation enhancer. Indian J Pharm Educ Res. 2019;53(1):88–96. https://doi.org/10.5530/ijper.53.1.12.
199. Yang Y, Zhou R, Wang Y, Zhang Y, Yu J, Gu Z. Recent advances in oral and transdermal protein delivery systems. Angew Chem. 2022;135(10):e202214795. https://doi.org/10.1002/ange.202214795.
200. Yeung C, Chen S, King B, Lin H, King K, Akhtar F, Diaz G, Wang B, Zhu J, Sun W, Khademhosseini A, Emaminejad S. A 3d-printed microfluidic-enabled hollow microneedle architecture for transdermal drug delivery. Biomicrofluidics. 2019;13(6):064125. https://doi.org/10.1063/1.5127778.
201. Yuan W, Chen Y, Sun W. Hemostatic absorbable gelatin sponge loaded with 5-fluorouracil for treatment of tumors. Int J Nanomed. 2013;8:1499. https://doi.org/10.2147/ijn.s41462.
202. Zahariev N, Katsarov P, Lukova P, Pilicheva B. Novel fucoidan pharmaceutical formulations and their potential application in oncology—a review. Polymers. 2023;15(15):3242. https://doi.org/10.3390/polym15153242.
203. Zhang C, Li J. Preparation of a novel chitosan/glycerol/matrine drug-loading film and the study of its release behavior in vitro. Int J Biomed Mater Res. 2018;6(1):20. https://doi.org/10.11648/j.ijbmr.20180601.14.
204. Zhang H, Wu F, Li Y, Yang X, Huang J, Lv T, Zhang Y, Chen J, Chen H, Gao Y, Liu G, Jia L. Chitosan-based nanoparticles for improved anticancer efficacy and bioavailability of mifepristone. Beilstein J Nanotechnol. 2016;7:1861–70. https://doi.org/10.3762/bjnano.7.178.
205. Zhang S, Jin AT, Tang W, Zhang RY, Jing L, Zhou Y, Zhang H, Lennerz JK, Korzenik JR, Langer R, Traverso G. Thermo-responsive polymers targeting inflammation in murine colitis. bioRxiv. 2023a; https://doi.org/10.1101/2023.12.28.573545.
206. Zhang W, Wang B, Xiang G, Jiang T, Zhao X. Photodynamic alginate Zn-Mof thermosensitive hydrogel for accelerated healing of infected wounds. ACS Appl Mater Interfaces. 2023b;15(19):22830–42. https://doi.org/10.1021/acsami.2c23321.
207. Zhang Y, Wei W, Lv P, Wang L, Ma G. Preparation and evaluation of alginate–chitosan microspheres for oral delivery of insulin. Eur J Pharm Biopharm. 2011;77(1):11–9. https://doi.org/10.1016/j.ejpb.2010.09.016.
208. Zhao D, Yu S, Sun B, Gao S, Guo S, Zhao K. Biomedical applications of chitosan and its derivative nanoparticles. Polymers. 2018;10(4):462. https://doi.org/10.3390/polym10040462.
209. Zhao F, Shen G, Chen C, Xing R, Zou Q, Ma G, Yan X. Nanoengineering of stimuli-responsive protein-based biomimetic protocells as versatile drug delivery tools. Chem Eur J. 2014;20(23):6880–7. https://doi.org/10.1002/chem.201400348.
210. Zhao K, Gao Y, Hu G, Wang L, Cui S, Jin Z. N-2-hydroxypropyl trimethyl ammonium chloride chitosan as adjuvant enhances the immunogenicity of a VP2 subunit vaccine against porcine parvovirus infection in sows. Vaccine. 2021;9(9):1027. https://doi.org/10.3390/vaccines9091027.
211. Zhao L, Zhou Y, Zhang J, Liang H, Chen X, Tan H. Natural polymer-based hydrogels: from polymer to biomedical applications. Pharmaceutics. 2023;15(10):2514.
212. Zhou X, Ning X, Chen Y, Chang H, Lu D, Pei D, Geng Z, Zeng Z, Guo C, Huang J, Yu S, Guo H. Dual glucose/ROS-sensitive injectable adhesive self-healing hydrogel with photothermal antibacterial activity and modulation of macrophage polarization for infected diabetic wound healing. ACS Mater Lett. 2023a;5(12):3142–55. https://doi.org/10.1021/acsmaterialslett.3c00925.

213. Zhou Y, Bore S, Tao AR, Paesani F, Arya G. Many-body potential for simulating the self-assembly of polymer-grafted nanoparticles in a polymer matrix. npj Comput Mater. 2023b;9:224. https://doi.org/10.26434/chemrxiv-2023-8s90g.
214. Zhou Y, Wang S, Ying X, Wang Y, Geng P, Deng A, Yu Z. Doxorubicin-loaded redox-responsive micelles based on dextran and indomethacin for resistant breast cancer. Int J Nanomed. 2017;12:6153–68. https://doi.org/10.2147/ijn.s141229.
215. Zumstein M, Kohler HPE, McNeill K, Sander M. High-throughput analysis of enzymatic hydrolysis of biodegradable polyesters by monitoring cohydrolysis of a polyester-embedded fluorogenic probe. Environ Sci Technol. 2017;51(8):4358–67. https://doi.org/10.1021/acs.est.6b06060.

Natural Polymers in Oral Drug Delivery 12

Niharika Gokhale, Aakansha Meval, Dishant Gupta, Kuldeep Vinchurkar, and Meghraj Suryawanshi

12.1 Introduction

Oral drug delivery remains the most preferred and widely used route for administering therapeutic agents due to its convenience, cost-effectiveness, and high patient compliance [17]. This route is favored in both chronic and acute conditions, as it allows patients to self-administer medications without the need for healthcare professionals, reducing hospital visits and associated medical costs. As a result, approximately 60% of all marketed pharmaceutical formulations are oral dosage forms, encompassing tablets, capsules, solutions, suspensions, and emulsions [5].

The efficiency of oral drug delivery systems depends on multiple factors, including the physicochemical properties of the drug, such as solubility, permeability, stability, and release kinetics. These properties significantly influence drug absorption in the gastrointestinal (GI) tract and its subsequent bioavailability. The human GI tract presents numerous physiological challenges, including variations in pH, enzymatic degradation, and efflux transport mechanisms, which can impact drug stability and absorption. To address these challenges, various formulation strategies have been developed to enhance drug delivery via the oral route.

Among the most widely explored strategies are immediate-release, controlled-release, and targeted drug delivery systems [7]. Immediate-release formulations are

N. Gokhale · A. Meval
Faculty of Pharmacy, Medi-Caps University, Indore, Madhya Pradesh, India

D. Gupta
Swami Vivekanand College of Pharmacy, Indore, Madhya Pradesh, India

K. Vinchurkar (✉) · M. Suryawanshi
Department of Pharmaceutics, Sandip Institute of Pharmaceutical Sciences (SIPS), Affiliated to Savitribai Phule Pune University (SPPU, Pune), Nashik, Maharashtra, India

J. S et al. (eds.), *Innovative Pharmaceutical Excipients: Natural Sources*,
https://doi.org/10.1007/978-981-96-7959-1_12

designed for rapid drug dissolution and absorption, while controlled-release systems regulate drug release over a prolonged period, improving therapeutic outcomes by maintaining steady plasma drug concentrations. Targeted drug delivery systems, on the other hand, aim to direct drugs to specific sites within the GI tract, enhancing efficacy while minimizing side effects. These advanced delivery systems rely heavily on functional excipients, particularly polymers, which play a critical role in optimizing drug-release profiles, stability, and overall therapeutic performance.

Key Benefits of Oral Drug Delivery Systems

1. *Ease of Administration:* Oral drug delivery is non-invasive and does not require specialized training, making it the most patient-friendly route [12]. This is especially beneficial for pediatric, geriatric, and chronically ill patients who require long-term medication.
2. *Cost-Effectiveness:* Compared to parenteral and transdermal delivery systems, oral formulations have lower manufacturing and production costs. Large-scale production of tablets and capsules requires relatively simple processes, making them economically viable for pharmaceutical industries and healthcare providers [13].
3. *Flexibility in Dosage Forms:* Oral dosage forms can be customized to meet specific therapeutic needs. They allow modifications such as enteric coatings, sustained-release formulations, and multi-unit particulate drug delivery systems, enabling better control over drug release and absorption [3].
4. *Systemic Absorption:* The oral route facilitates effective systemic drug delivery through absorption in the gastrointestinal tract, primarily via passive diffusion or carrier-mediated transport [21, 32]. Drugs with favorable solubility and permeability properties can achieve high systemic bioavailability. Despite these advantages, oral drug delivery systems face challenges such as enzymatic degradation, poor solubility, and limited permeability, which necessitate the use of advanced materials like natural polymers to overcome these barriers [1].

12.1.1 Importance of Natural Polymers in Pharmaceutical Applications

Natural polymers have emerged as a vital class of materials in pharmaceutical sciences due to their biocompatibility, biodegradability, and functional versatility [9]. These polymers, derived from renewable sources such as plants, animals, and microorganisms, offer unique physicochemical properties that make them highly valuable in drug delivery and formulation development. Compared to synthetic polymers, natural polymers provide safety advantages, reduced toxicity, and environmental sustainability, making them a preferred choice for modern pharmaceutical applications.

The pharmaceutical industry continuously explores novel excipients to improve drug formulation properties, enhance drug bioavailability, and ensure patient compliance. Natural polymers play a crucial role in these advancements, offering excellent gelling, swelling, and film-forming abilities, which are essential for designing controlled-release and targeted drug delivery systems. Additionally, their mucoadhesive properties allow for enhanced drug retention at specific sites, leading to improved therapeutic efficacy.

A major advantage of using natural polymers in pharmaceuticals is their ability to interact with biological systems without causing significant adverse effects. For example, plant-derived polysaccharides such as chitosan and alginate have been widely used in drug delivery systems due to their biodegradability, non-toxicity, and bioadhesive properties. Similarly, protein-based polymers like gelatin and silk fibroin have found extensive applications in encapsulation and scaffold formulations due to their superior mechanical strength and biocompatibility. The incorporation of these polymers in drug formulations ensures prolonged drug action, targeted delivery, and enhanced stability of pharmaceutical compounds.

Furthermore, as the pharmaceutical industry moves toward sustainable and green technologies, natural polymers provide an eco-friendly alternative to synthetic excipients. The rising concern over synthetic polymer waste and its impact on the environment has driven researchers to focus on biodegradable and renewable materials for pharmaceutical applications. Natural polymers not only reduce environmental burden but also align with regulatory guidelines advocating for safer and more sustainable drug delivery approaches.

Thus, the use of natural polymers in pharmaceuticals represents an innovative and sustainable strategy for improving drug formulations while minimizing toxicity and environmental impact. Their diverse applications range from tablet coatings and controlled-release matrices to nanoparticle-based delivery systems, demonstrating their potential in advancing drug delivery technologies.

12.1.2 Functional Roles of Natural Polymers in Drug Delivery

Natural polymers exhibit multifunctional properties that enhance drug formulation and delivery. Their applications span across controlled drug release, bioadhesion, stabilization, targeted drug delivery, and sustainability, making them indispensable in modern pharmaceutical sciences (Table 12.1).

1. *Controlled Release of Drugs*
   One of the primary functions of natural polymers in drug delivery is their ability to modulate drug- release kinetics through matrix formation, swelling, and ero-

Table 12.1 Common natural polymers and their pharmaceutical applications

Polymer	Source	Key applications
Cellulose	Plant cell walls	Controlled-release formulations
Chitosan	Shellfish shells	Bioadhesive systems, nanoparticles
Alginate	Brown algae	Hydrogels, encapsulation
Gelatin	Animal collagen	Capsules, film-forming agent
Natural gums	Plant exudates	Thickening agents, bioadhesives

sion mechanisms [2]. Controlled-release formulations ensure that the drug is released at a predetermined rate, maintaining therapeutic drug levels in the bloodstream and reducing the frequency of dosing. Polysaccharides such as cellulose derivatives (e.g., hydroxypropyl methylcellulose, HPMC) and chitosan have been widely used to develop extended-release tablets and hydrogel-based drug delivery systems. These polymers create a viscous gel barrier around the drug, controlling its diffusion and release over an extended period.

2. *Bioadhesion and Mucoadhesion*
   Bioadhesion is another critical property of natural polymers that enhances drug retention at specific sites within the body. Mucoadhesive polymers, such as chitosan and alginate, interact with mucosal surfaces in the gastrointestinal tract, oral cavity, or vaginal mucosa, prolonging drug residence time and enhancing bioavailability [19]. This is particularly beneficial for drugs with low solubility or rapid clearance, ensuring sustained therapeutic effects. Mucoadhesive formulations have been extensively utilized in buccal, nasal, and gastrointestinal drug delivery systems.
3. *Stabilization of Sensitive Drugs*
   Many pharmaceutical compounds, especially peptides, proteins, and nucleic acids, are prone to degradation by enzymes, oxidation, or hydrolysis. Natural polymers serve as excellent stabilizing agents by providing a protective matrix that shields drugs from chemical and enzymatic degradation [4]. For instance, gelatin-based microcapsules and silk fibroin nanoparticles have been employed to encapsulate unstable drug molecules, enhancing their shelf life and therapeutic effectiveness. Additionally, polysaccharides such as pectin and carrageenan act as film-forming agents that protect sensitive drugs in harsh gastric environments.
4. *Targeted Drug Delivery*
   Natural polymers play an essential role in site-specific drug delivery, ensuring that therapeutic agents reach the intended site of action while minimizing systemic side effects). Polymeric nanoparticles, microspheres, and liposomes composed of natural biopolymers have been designed for tumor-targeting, colon-specific, and brain-targeted drug delivery. For example, chitosan-modified nanoparticles enhance drug transport across the blood-brain barrier (BBB), facilitating treatment of neurological disorders. Similarly, pH-sensitive polymers such as pectin and guar gum enable colon-targeted drug delivery, releasing drugs specifically in the large intestine.
5. *Sustainability and Environmental Benefits*
   The pharmaceutical industry's growing commitment to green chemistry and eco-friendly practices has increased the demand for biodegradable and renewable excipients [16, 45]. Unlike synthetic polymers, which contribute to non-biodegradable waste, natural polymers offer a sustainable alternative with minimal environmental impact. Biopolymers such as starch, alginate, and natural gums are not only biodegradable but also renewable, making them highly suitable for modern pharmaceutical formulations. The use of plant-derived polysaccharides aligns with the principles of circular economy and sustainable development, reducing dependency on fossil fuel-based synthetic polymers.

12.1.3 Advantages of Natural Polymers Over Synthetic Counterparts

Natural polymers have garnered significant attention in pharmaceutical applications due to their biodegradability, biocompatibility, and sustainable origin. Compared to synthetic polymers, natural polymers provide safer, environmentally friendly, and functionally versatile alternatives, particularly in oral drug delivery systems. Their ability to interact with biological systems without eliciting immunogenic or toxic responses makes them an ideal choice for drug formulation and biomedical applications.

The growing emphasis on green pharmaceutical technologies and sustainability has further reinforced the importance of naturally derived polymers. These biopolymers, obtained from plant, animal, microbial, and marine sources, exhibit a range of physicochemical properties that enable their use in controlled-release systems, bioadhesive formulations, nanoparticle-based delivery, and biodegradable scaffolds.

Natural polymers offer significant advantages over their synthetic counterparts, particularly in the context of oral drug delivery systems:

1. *Biocompatibility:* One of the most significant advantages of natural polymers is their excellent biocompatibility, meaning they are inherently safe for biological systems and do not elicit adverse reactions [14]. Unlike synthetic polymers, which may require additional processing to reduce toxicity, irritation, or immunogenicity, natural polymers are recognized by the body as biocompatible materials. For example, chitosan (derived from crustacean shells), alginate (from brown algae), and gelatin (from animal collagen) are widely used in drug formulations, wound healing, and tissue engineering due to their low immunogenicity and high tolerability. Their ability to form hydrogels, nanoparticles, and bioadhesive films further enhances their pharmaceutical applications.
2. *Biodegradability:* Natural polymers are biodegradable, meaning they break down into non-toxic by-products upon degradation, making them environmentally friendly [11]. This contrasts with synthetic polymers, which often persist in the environment, leading to pollution and disposal challenges. For instance, starch, pectin, and carrageenan degrade naturally into harmless oligosaccharides, while proteins like silk fibroin and gelatin are enzymatically broken down into amino acids that can be safely metabolized by the body. This biodegradability feature is especially beneficial in the development of temporary drug carriers, bioresorbable implants, and targeted delivery systems, where controlled degradation is a key requirement.
3. *Renewable Sources:* Unlike synthetic polymers, which are often petroleum-derived and non-renewable, natural polymers are obtained from sustainable and renewable sources [6]. The use of plant-derived polymers such as cellulose, guar gum, and arabinoxylan reduces dependence on fossil fuels, making these materials more environmentally sustainable. Microbial sources, including xanthan gum from *Xanthomonas campestris* and pullulan from *Aureobasidium pullulans*, provide high-yield, biodegradable polymers with consistent quality and bioac-

tivity. Similarly, marine-derived polymers like alginate and chitosan have gained popularity due to their abundance in nature and minimal ecological footprint. The ability to source natural polymers from agricultural and marine waste also aligns with the principles of green chemistry and circular economy, further reducing environmental impact.

4. *Versatility:* Natural polymers contain diverse functional groups (-OH, -COOH, -NH2, -SH) that enable chemical modifications and functionalization, allowing for tailored drug delivery applications [8, 28, 43, 49]. Unlike synthetic polymers, which often require complex synthesis processes to introduce specific functionalities, natural polymers can be easily modified to alter their solubility, mucoadhesion, mechanical strength, or drug-loading capacity.
   For example:
   - Chitosan can be chemically quaternized to improve solubility in physiological conditions.
   - Alginate beads can be ionically cross-linked with calcium to create pH-sensitive drug carriers.
   - Hyaluronic acid can be functionalized for targeted drug delivery in cancer therapy.

   This versatility allows for the design of responsive, stimuli-sensitive, and controlled-release formulations, enhancing therapeutic efficacy while minimizing side effects.
5. *Eco-friendliness:* The production and disposal of natural polymers are less harmful to the environment compared to synthetic polymers [18]. The extraction and processing of natural polymers typically require fewer harsh chemicals, generating less hazardous waste. In contrast, synthetic polymer synthesis often involves toxic solvents, catalysts, and by-products that pose significant environmental and health risks. Moreover, natural polymer-based pharmaceutical excipients are generally recognized as safe (GRAS) by regulatory authorities, reducing toxicological concerns associated with their use in oral, parenteral, and transdermal drug formulations.
   For instance:
   - Pectin-based drug formulations are widely accepted in food and pharmaceutical industries due to their biocompatibility and digestibility.
   - Cellulose derivatives such as hydroxypropyl methylcellulose (HPMC) and ethylcellulose are widely used as tablet coatings and controlled-release matrices, offering a safe and non-toxic alternative to synthetic polymers.

   By choosing natural polymers, the pharmaceutical industry can minimize exposure to toxic chemicals, ensuring safer formulations for both patients and the environment.

12.1.4 Challenges Addressed by Natural Polymers

1. *Enhanced Drug Stability:* One of the foremost challenges in drug formulation is ensuring the stability of active pharmaceutical ingredients (APIs) throughout the

manufacturing, storage, and administration processes. Many drugs, especially peptides, proteins, and highly sensitive small molecules, are prone to degradation due to enzymatic activity, oxidation, or hydrolysis in the gastrointestinal (GI) tract [10]. Natural polymers help stabilize and protect drugs by forming physical or chemical barriers that reduce drug exposure to enzymatic degradation, light, moisture, and temperature fluctuations [15].

For instance:

- Alginate and pectin-based hydrogels create a protective matrix that encapsulates proteins and peptides, shielding them from degradation.
- Chitosan nanoparticles can electrostatically interact with negatively charged APIs, preventing premature breakdown.
- Hyaluronic acid-based nanocarriers offer stability for biologics and gene therapies by reducing enzymatic degradation.

By employing natural polymer-based protective formulations, pharmaceutical scientists can extend drug shelf life, improve therapeutic efficacy, and enhance patient compliance.

2. *Improved Bioavailability:* One of the foremost challenges in drug formulation is ensuring the stability of active pharmaceutical ingredients (APIs) throughout the manufacturing, storage, and administration processes. Many drugs, especially peptides, proteins, and highly sensitive small molecules, are prone to degradation due to enzymatic activity, oxidation, or hydrolysis in the gastrointestinal (GI) tract [10]. Natural polymers help stabilize and protect drugs by forming physical or chemical barriers that reduce drug exposure to enzymatic degradation, light, moisture, and temperature fluctuations.

   For instance:

   - Alginate and pectin-based hydrogels create a protective matrix that encapsulates proteins and peptides, shielding them from degradation.
   - Chitosan nanoparticles can electrostatically interact with negatively charged APIs, preventing premature breakdown.

   Hyaluronic acid-based nanocarriers offer stability for biologics and gene therapies by reducing enzymatic degradation.

   By employing natural polymer-based protective formulations, pharmaceutical scientists can extend drug shelf life, improve therapeutic efficacy, and enhance patient compliance.

3. *Customizable Drug Release:* Controlled drug release is crucial for sustained therapeutic effects, reduced side effects, and improved patient compliance. Synthetic polymers have traditionally been used to design controlled-release formulations, but many pose toxicity, biocompatibility, or regulatory concerns. Natural polymers, on the other hand, offer tunable drug-release mechanisms based on their hydrophilic, hydrophobic, or pH-sensitive nature [20]. These biopolymers can be chemically modified or combined to create customized drug-release profiles such as:

   - Immediate release for rapid therapeutic action.
   - Sustained release for prolonged drug action.
   - Targeted release at specific sites within the GI tract or other tissues.

Examples of natural polymer-based controlled-release systems:

- *Hydrogel-Based Drug Delivery:* Alginate and gelatin hydrogels swell in aqueous environments, gradually releasing encapsulated drugs.
- *Mucoadhesive Delivery:* Chitosan and guar gum enhance residence time in the GI tract, improving local drug absorption.
- *pH-sensitive Release:* Pectin and carrageenan matrices degrade selectively in colonic pH, ensuring targeted colonic drug delivery.
- *Enzyme-Responsive Systems:* Polysaccharides like starch undergo enzymatic degradation, enabling controlled breakdown and release.

Such customizable drug-release systems allow for optimized pharmacokinetics, reduced toxicity, and better therapeutic efficacy, making natural polymers indispensable in modern drug delivery.

Natural polymers are indispensable in the development of oral drug delivery systems, offering a harmonious balance between efficacy, safety, and sustainability. Subsequent chapters will delve deeper into the classification, mechanisms, and innovative applications of these polymers in pharmaceutical sciences.

12.2 Classification of Natural Polymers

12.2.1 Polysaccharides

Polysaccharides are naturally occurring high-molecular-weight biopolymers composed of monosaccharide units linked by glycosidic bonds. These polymers exhibit excellent biocompatibility, biodegradability, and hydrophilicity, making them highly valuable in pharmaceutical formulations.

Examples: Cellulose, chitosan, alginate.

Key Properties

- *Biodegradability:* Naturally degraded by enzymatic or hydrolytic processes, making them eco-friendly.
- *Biocompatibility:* Safe for biomedical and pharmaceutical applications.
- *Hydrophilicity:* Capable of interacting with aqueous environments, ideal for hydrogel formulations.
- *Gelling and Thickening Abilities:* Useful in matrix systems for controlled drug release.
- *Mucoadhesive Properties:* Prolongs drug retention time in the gastrointestinal (GI) tract, improving bioavailability (Table 12.2).

Table 12.2 Categorized natural polymers with their key properties and applications

Category	Examples	Key properties	Applications
Polysaccharides	Cellulose, chitosan, alginate	Biodegradability, biocompatibility	Matrix tablets, hydrogels, nanoparticles
Proteins	Gelatin, silk fibroin	Structural strength, biocompatibility	Capsules, films, scaffold formulations
Lipids and others	Gum arabic, beeswax	Emulsification, stability	Bioadhesive tablets, sustained-release coatings

Applications

1. *Cellulose and Its Derivatives*

   Source and Structure: Derived from plant cell walls, cellulose is a linear polysaccharide composed of β-D-glucose units [22].

Applications:

- Used as a binder, disintegrant, and film-forming agent in tablets and coatings.
- Hydroxypropyl methylcellulose (HPMC) is commonly employed in sustained-release matrix tablets.
- Microcrystalline cellulose (MCC) serves as an excipient in solid dosage forms due to its excellent compressibility.

2. *Chitosan*

   Source and Structure: A cationic polysaccharide derived from chitin, primarily found in crustacean shells.

Applications:

- Nanoparticles and microspheres for targeted drug delivery.
- Mucoadhesive agent for buccal, nasal, and gastrointestinal drug delivery systems.
- Used in wound dressings and antimicrobial formulations due to its biocompatibility and antibacterial properties.

3. *Alginate*

   Source and Structure: Extracted from brown seaweed, alginate consists of mannuronic and guluronic acid residues.

Applications:

- Forms hydrogels for controlled drug release in oral and topical formulations.
- Utilized in matrix tablets for gastroretentive drug delivery.
- Employed in tissue engineering scaffolds and wound dressings due to its gel-forming ability in the presence of divalent cations like calcium.

Polysaccharides play a crucial role in modern drug delivery by providing enhanced stability, controlled release, and bioadhesion, making them indispensable excipients in pharmaceuticals.

12.2.2 Proteins

Proteins are natural polymers with unique structural and functional properties that make them ideal carriers for drug delivery. Their biodegradability, biocompatibility, and ability to interact with biological molecules allow for targeted and sustained drug delivery applications [23].

Key Properties of Proteins in Drug Delivery

Biodegradable: Naturally broken down into amino acids, avoiding toxic accumulation.

Biocompatible: Well tolerated by biological tissues.

Structural Versatility: Can form hydrogels, scaffolds, and nanoparticles for drug encapsulation.

Controlled-Release Capabilities: Can be chemically modified or cross-linked to alter drug-release profiles (Table 12.2).

12.2.2.1 Gelatin

Source and Structure:

Gelatin is derived from collagen, found in animal connective tissues. It is a protein composed of glycine, proline, and hydroxyproline.

Applications:

- Widely used in soft and hard capsules.
- Employed in microspheres and nanoparticles for sustained and targeted drug delivery.

12.2.2.2 Silk Fibroin

Source and Structure:

Silk fibroin is derived from the cocoons of the silkworm, *Bombyx mori*. It consists of repetitive amino acid sequences, primarily glycine, alanine, and serine.

Applications:

- Used in film and scaffold formulations for controlled drug release.
- Provides excellent biocompatibility and mechanical strength.

12.2.3 Lipids and Others

Natural lipids and related materials such as gums and waxes are gaining attention due to their functional versatility (Table 12.2).

12.2.3.1 Natural Gums

- *Examples:* Gum arabic, guar gum, and xanthan gum.
- *Applications:*
  - Act as emulsifying agents and stabilizers.
  - Used in the preparation of bioadhesive tablets and hydrogels [24].

12.2.3.2 Waxes

- *Examples:* Beeswax and carnauba wax.
- *Applications:*
  - Utilized in sustained-release formulations.
  - Provide a protective coating for tablets and capsules.

12.3 Mechanisms of Natural Polymers in Drug Delivery

Natural polymers play a crucial role in enhancing the efficacy and precision of drug delivery systems. Their unique physicochemical properties allow them to interact with biological systems effectively. Below, the primary mechanisms by which natural polymers facilitate drug delivery are discussed:

12.3.1 Controlled Drug Release

Natural polymers are extensively used to achieve controlled drug release, ensuring a consistent therapeutic effect over time.

- *Mechanism:* T6hese polymers form matrix systems or hydrogels that regulate the diffusion of drugs, enabling a sustained-release profile.
- *Examples:*
  - *Chitosan:* Forms hydrogels that swell in aqueous environments, controlling drug release [25].
  - *Cellulose Derivatives:* Hydroxypropyl methylcellulose (HPMC) is commonly used in extended-release formulations.
- *Applications:* Matrix tablets, implantable systems, and injectable depots utilize natural polymers for controlled drug delivery.

12.3.2 Bioadhesion and Mucoadhesion

Bioadhesion and mucoadhesion refer to the ability of a material to adhere to biological or mucosal surfaces, enhancing drug residence time at the site of action or absorption.

- *Mechanism:* Polymers interact with the mucosal layer through hydrogen bonding, electrostatic interactions, or van der Waals forces.
- *Examples:*
  - *Alginate:* Exhibits excellent mucoadhesive properties and is used in buccal and intestinal delivery systems.
  - *Guar Gum:* Enhances the bioavailability of poorly absorbed drugs by prolonging mucosal contact [26].
- *Applications:* Mucoadhesive tablets, films, and patches for oral, buccal, and nasal drug delivery.

Table 12.3 Mechanism of natural polymers in drug delivery

Mechanism	Polymer examples	Key applications
Controlled release	Chitosan, HPMC	Matrix tablets, injectable systems
Bioadhesion	Alginate, guar gum	Mucoadhesive tablets, films
pH-sensitive delivery	Pectin, carrageenan	Colon-targeted and gastric drug delivery

12.3.3 pH-Sensitive and Targeted Delivery

Natural polymers can respond to changes in pH, allowing targeted delivery to specific regions of the gastrointestinal tract (Table 12.3).

- *Mechanism:* Polymers swell, degrade, or release their payload in response to pH changes [27].
- *Examples:*
  - *Pectin:* Degrades in the colon's alkaline environment, enabling colon-targeted delivery.
  - *Carrageenan:* pH-sensitive properties make it suitable for targeted drug release in acidic environments like the stomach.
- *Applications:* Colon-specific delivery of drugs for treating inflammatory bowel disease and colorectal cancer.

12.4 Role of Natural Polymers in Formulation Development

Natural polymers play a pivotal role in the development of innovative pharmaceutical formulations. Their biocompatibility, tunable properties, and ability to enhance drug stability and bioavailability make them indispensable in various drug delivery systems. Below, we delve into their application in matrix tablets, hydrogels, microcapsules, and nanoformulations (Table 12.4).

12.4.1 Matrix Tablets and Granules

Matrix systems are widely used for controlled drug release, leveraging the properties of natural polymers to sustain and modulate the release profile of active pharmaceutical ingredients (APIs) [8, 28, 43, 49].

12.4.1.1 Mechanism

Natural polymers form a gel-like matrix upon contact with gastrointestinal fluids, which regulates drug diffusion and provides a sustained-release profile [29]. Polymers such as hydroxypropyl methylcellulose (HPMC) and guar gum are commonly employed in these formulations.

Table 12.4 Applications of natural polymers in formulations

Formulation type	Natural polymers used	Key features	Applications
Matrix tablets	HPMC, xanthan gum, guar gum	Sustained release, biocompatibility	Antihypertensives, antibiotics
Hydrogels	Alginate, pectin, gelatin	Swelling, controlled release	Wound healing, tissue regeneration
Microcapsules	Chitosan, alginate	Controlled release, drug protection	Protein delivery, nasal formulations
Nanoparticles	Chitosan, alginate	Targeted delivery, improved bioavailability	Cancer therapy, antiviral delivery
Nanogels and nanofibers	Gelatin, silk fibroin	High surface area, tunable drug release	Cancer therapy, wound healing

12.4.1.2 Applications

- *Hydrophilic Matrix Tablets:* Polysaccharides like xanthan gum and cellulose derivatives are utilized for prolonged drug release [30]. For example, HPMC-based tablets have shown effective control over drug release for antihypertensive medications.
- *Granules:* Chitosan-based granules have demonstrated potential in enhancing the solubility and bioavailability of poorly water-soluble drugs. These granules can be tailored for site-specific release [31, 42].

12.4.2 Hydrogels and Microcapsules

Natural polymers are the backbone of hydrogel and microcapsule formulations, which are used for a variety of drug delivery applications, including controlled release and targeted therapy [21, 32].

12.4.2.1 Hydrogels

Hydrogels are three-dimensional, cross-linked polymeric networks capable of absorbing and retaining significant amounts of water or biological fluids. Polymers such as alginate, pectin, and gelatin are commonly used in hydrogel formulations [33].

Advantages:

- Excellent biocompatibility.
- Tunable swelling behavior for controlled drug release.
- Effective in wound healing and tissue regeneration applications [34].

12.4.2.2 Microcapsules

Microcapsules are spherical particles with a polymeric shell encapsulating the drug, providing controlled release and protection from environmental degradation.

Applications:

- *Alginate Microcapsules:* Widely used for encapsulating sensitive drugs like proteins and peptides [35].
- *Chitosan-Based Microcapsules:* Provide mucoadhesion and enhanced permeability, particularly in nasal and buccal drug delivery [36].

12.4.3 Nanoformulations Using Natural Polymers

Nanoformulations have revolutionized drug delivery by improving drug solubility, stability, and targeted delivery. Natural polymers offer a versatile platform for creating nanoparticles, nanogels, and nanofibers [37].

12.4.3.1 Polymeric Nanoparticles

- *Chitosan Nanoparticles:* Widely researched for their ability to deliver anticancer and antiviral drugs effectively. They offer controlled release and cellular uptake [38].
- *Alginate Nanoparticles:* Utilized in oral insulin delivery due to their protective properties against the acidic gastric environment [39].

12.4.3.2 Nanogels and Nanofibers

- *Nanogels:* Hydrophilic networks formed using polymers like gelatin and hyaluronic acid. These are extensively used for targeted drug delivery in cancer therapy [40].
- *Nanofibers:* Electrospun fibers using polymers such as silk fibroin have shown promise in localized drug delivery and wound healing applications [41].

12.5 Case Studies and Applications

Natural polymers demonstrate versatile applications across various pharmaceutical formulations. Below are expanded case studies with five examples in each category, emphasizing the significant contributions of polysaccharides, natural gums, and proteins [42–44].

12.5.1 Polysaccharides in Sustained-Release Formulations

Polysaccharides are integral to sustained-release systems, controlling drug release and enhancing therapeutic efficacy.

1. *Alginate Beads for Metformin*
   Alginate beads with calcium ions show prolonged metformin release, ensuring better glycemic control in diabetic patients [45].

2. *Chitosan-HPMC Matrix for Ibuprofen*
   A combination of chitosan and HPMC in matrix tablets provided sustained release of ibuprofen over 12 h.
3. *Pectin-Based Microspheres for Diclofenac Sodium*
   Pectin microspheres extended the release of diclofenac, reducing dosing frequency in pain management [46].
4. *Xanthan Gum in Tablet Coatings*
   Xanthan gum-coated tablets delayed drug release, suitable for colon-targeted delivery [47].
5. *Cellulose Derivatives in Antihypertensive Drugs*
   HPMC was employed in sustained-release tablets for losartan potassium, improving patient compliance [48].

12.5.2 Natural Gums for Bioadhesive Tablets

Natural gums enhance bioadhesion and retention, leading to improved drug absorption.

1. *Guar Gum Tablets for Diclofenac Sodium*
   Guar gum demonstrated effective mucoadhesion and extended retention for better pain management.
2. *Gum Arabic in Chewable Vitamin Tablets*
   Provided stability, enhanced taste masking, and facilitated nutrient delivery
3. *Xanthan Gum in Buccal Tablets*
   Buccal tablets for propranolol HCl prepared with xanthan gum improved bioavailability and prolonged retention.
4. *Karaya Gum in Gastroretentive Systems*
   Used for gastroretentive floating tablets, ensuring better control over drug release in the stomach
5. *Locust Bean Gum for Drug Suspension Stabilization*
   Enhanced suspension stability and uniform drug distribution in liquid formulations [49–50].

12.5.3 Proteins for Encapsulation and Targeted Delivery

Proteins exhibit remarkable potential for encapsulation and precision targeting in drug delivery systems.

1. *Gelatin Nanoparticles for Doxorubicin*
   Improved drug stability and targeted tumor cell delivery in cancer therapy [51].
2. *Silk Fibroin for Curcumin Encapsulation*
   Controlled release of curcumin increased bioavailability and efficacy against inflammatory diseases [52].

3. *Whey Protein Nanoparticles for Insulin Delivery*
   Oral insulin-loaded whey protein nanoparticles enhanced bioavailability and reduced enzymatic degradation [53].
4. *Albumin Nanoparticles for Anticancer Drugs*
   Paclitaxel-loaded albumin nanoparticles (Abraxane®) improved drug solubility and efficacy in cancer treatment [54].
5. *Casein-Based Micelles for Hydrophobic Drugs*
   Hydrophobic drugs like docetaxel showed enhanced solubility and targeted delivery using casein micelles [55].

12.5.4 Variability in Sources and Composition

The natural origin of these polymers leads to batch-to-batch variability, affecting their performance and reliability.

1. *Source Dependency*
   Natural polymers like pectin and alginate are derived from plant or marine sources, making their composition dependent on environmental factors such as climate, soil conditions, and harvesting techniques. This variability affects the physicochemical properties critical for drug delivery applications.
2. *Complex Composition*
   Polymers such as gum arabic or guar gum have complex structures with varying molecular weights and functional groups. These inconsistencies complicate standardization and reproducibility in formulations.
3. *Seasonal and Regional Variability*
   Seasonal changes impact the quality of natural polymers, such as chitosan from crustacean shells, which can have fluctuating deacetylation degrees depending on the harvesting period.
4. *Impurities and Contamination*
   Natural polymers are prone to impurities such as residual proteins, salts, or microbial contamination, requiring extensive purification processes.
5. *Supply Chain Issue*s
   The sourcing of natural polymers like locust bean gum or karaya gum depends on specific geographical regions, making consistent supply vulnerable to political, economic, or environmental changes [56].

12.5.5 Stability and Scalability Issues

Natural polymers often face challenges in maintaining stability during formulation and large-scale production.

1. *Environmental Sensitivity*
   Polymers such as gelatin and xanthan gum are sensitive to temperature, pH, and moisture, leading to degradation or denaturation under unsuitable conditions.

2. *Shelf Life Concerns*
   Natural polymers may degrade over time due to oxidation or microbial activity, limiting their shelf life in drug formulations.
3. *Scalability Constraints*
   The production and processing of natural polymers like silk fibroin or casein are often labor-intensive and costly, making large-scale manufacturing difficult.
4. *Chemical Modification Requirements*
   Many natural polymers require chemical modifications to enhance their properties for drug delivery, but these modifications can compromise their stability and biocompatibility.
5. *Formulation Challenges*
   In nanoformulations or hydrogels, achieving uniform particle size or gel consistency using natural polymers can be challenging due to their inherent variability [57].

12.5.6 Regulatory and Quality Considerations

The regulatory framework for natural polymers poses challenges in meeting stringent quality standards for pharmaceutical applications.

1. *Lack of Standardization*
   Unlike synthetic polymers, natural polymers lack a standardized quality control framework, leading to inconsistencies in pharmaceutical formulations.
2. *Complex Regulatory Approval*
   Obtaining regulatory approval for formulations containing natural polymers like chitosan or alginate is challenging due to their variable composition and potential impurities.
3. *Quality Assurance Issues*
   Ensuring consistent quality across batches requires rigorous testing for purity, molecular weight, and functional properties, which can be time-consuming and expensive.
4. *Biocompatibility and Safety Testing*
   Extensive safety evaluations are required to confirm the absence of toxic contaminants in polymers such as guar gum or gum arabic.

Table 12.5 Regulatory challenges with their examples and impact

Challenge	Examples	Impact
Variability in sources	Seasonal variability in chitosan, impurities in gum arabic	Inconsistent drug performance
Stability issues	Temperature sensitivity of gelatin, shelf life issues in alginate formulations	Reduced product reliability
Regulatory challenges	Lack of standardization for chitosan, complex approval processes for natural polymer formulations	Delayed market entry and increased development costs

5. *Cost Implications*
   The need for stringent quality checks and purification processes increases the overall cost of formulations using natural polymers, affecting their commercial viability (Table 12.5).

12.6 Future Directions

The future of natural polymers in drug delivery and formulation science is promising, with emerging trends focusing on sustainability, the integration of artificial intelligence (AI), and the exploration of novel polymer sources (Table 12.6). These advancements aim to address current challenges while expanding the applications of natural polymers.

12.6.1 Sustainability and Eco-friendly Polymer Sourcing

A growing emphasis on environmental sustainability has led to the exploration of green and renewable sources for natural polymers.

1. *Biowaste Utilization*
   Agro-industrial waste such as fruit peels (pectin), fish scales (collagen), and shellfish by-products (chitosan) are being repurposed to extract natural polymers, reducing environmental waste.
2. *Renewable Plant Sources*
   Sustainable cultivation of plants like guar (guar gum), cassava (starch), and flax (cellulose) ensures a renewable supply of polymers with minimal environmental impact.
3. *Eco-friendly Extraction Techniques*
   Green chemistry approaches, including enzyme-assisted extraction and supercritical fluid extraction, are being adopted to reduce chemical waste during polymer isolation.
4. *Biodegradability and Circular Economy*
   Natural polymers such as xanthan gum and carrageenan are being integrated into formulations that degrade naturally, aligning with the principles of a circular economy.

Table 12.6 Future directions and its potential impact in polymer-based drug delivery system

Future direction	Examples	Potential impact
Sustainability in sourcing	Use of agro-industrial waste (pectin, chitosan)	Reduced environmental impact, cost-effective raw material sourcing
AI in formulation design	AI-driven optimization of chitosan-based nanocarriers	Improved efficiency in drug development
Exploring new polymers	Fucoidan from algae, mucilage from okra	Expanded applications and novel formulation possibilities

5. *Regenerative Agriculture*
   Regenerative farming practices are being explored to ensure the sustainable production of raw materials like alginate and gum arabic, improving soil health and carbon sequestration.

12.6.2 Role of Artificial Intelligence in Polymer-Based Formulation Design

AI-driven technologies are revolutionizing the design and optimization of polymer-based drug delivery systems.

1. *Polymer Property Prediction*
   Machine learning models are being used to predict the physicochemical properties of polymers, such as solubility, viscosity, and molecular weight, enabling tailored formulation design.
2. *Formulation Optimization*
   AI algorithms help identify the optimal combination of polymers, excipients, and active pharmaceutical ingredients for desired drug-release profiles.
3. *Simulation and Modeling*
   Computational tools simulate drug–polymer interactions and release kinetics, reducing the need for extensive experimental trials.
4. *High-Throughput Screening*
   AI enables the rapid screening of polymer libraries to identify candidates with superior drug delivery potential.
5. *Personalized Medicine Applications*
   AI is being leveraged to design polymer-based formulations tailored to individual patient needs, advancing the field of personalized drug delivery.

12.6.3 Exploring Underutilized Natural Polymers

The discovery and utilization of novel natural polymers hold immense potential for future applications in drug delivery.

1. *Marine-Derived Polymers*
   Polymers such as fucoidan (brown algae), ulvan (green algae), and sponge collagen are under investigation for their bioactive and drug-carrying properties.
2. *Uncommon Plant Polysaccharides*
   Polysaccharides like mucilage from okra and polysaccharides from cactus are being explored for their mucoadhesive and gelling properties.
3. *Animal-Derived Proteins*
   Less commonly utilized proteins like keratin (feathers) and sericin (silkworms) show promise as biocompatible and biodegradable excipients.

4. *Microbial Biopolymers*
   Polymers like bacterial cellulose and exopolysaccharides (e.g., gellan gum, dextran) offer unique structural and functional advantages in drug delivery.
5. *Fungal-Derived Polymers*
   Fungi-derived chitosan and glucans are emerging as sustainable and bioactive materials for innovative drug delivery solutions.

12.7 Conclusion

Natural polymers, including polysaccharides and proteins, have emerged as indispensable materials in pharmaceutical sciences due to their biocompatibility, biodegradability, and tunable physicochemical properties. Polysaccharides such as cellulose, chitosan, and alginate provide excellent mucoadhesion, gelling capabilities, and controlled-release properties, making them ideal for applications in matrix tablets, hydrogels, and nanoparticulate drug carriers. On the other hand, protein-based polymers like gelatin and silk fibroin offer structural versatility, mechanical resilience, and bioactive functionality, supporting their use in microspheres, scaffolds, and sustained drug delivery formulations.

The renewable and sustainable nature of these biopolymers makes them environmentally friendly alternatives to synthetic polymers, addressing concerns regarding ecotoxicity and long-term accumulation. Their functional adaptability, achieved through chemical modifications and formulation innovations, enables precise control over drug-release kinetics, bioavailability, and targeting efficiency.

Despite their numerous advantages, challenges such as batch variability, extraction complexities, and potential immunogenicity remain areas of active research. Advanced processing techniques, including nanotechnology, cross-linking strategies, and hybrid polymer formulations, continue to enhance their performance in modern drug delivery systems.

As pharmaceutical sciences shift toward more sustainable, patient-centric, and advanced drug delivery approaches, natural polymers will continue to play a pivotal role in the development of next-generation therapeutics. Subsequent chapters will further explore the classification, mechanisms, and emerging innovations surrounding these biomaterials, paving the way for enhanced pharmaceutical formulations and biomedical applications.

References

1. Ahmed Z, et al. Advances in oral drug delivery. J Pharm Sci. 2020;109(2):345–60.
2. Banik S, et al. Controlled drug release from natural polymers. Drug Deliv Today. 2021;12(3):245–60.
3. Chawla A, et al. Oral dosage form development. Curr Pharm Des. 2019;25(4):561–78.
4. Goyal R, et al. Stabilizing agents in oral formulations. J Drug Dev. 2019;10(5):125–38.
5. Gupta M, et al. Oral drug delivery technologies. Int J Pharm Sci. 2020;112(4):452–67.
6. Gupta R, Pathak P. Renewable polymers in pharmaceuticals. Sustain Chem. 2019;8(6):310–26.

7. Kumar A, Bhowmik D. Innovations in drug delivery. Pharma Times. 2019;45(7):24–31.
8. Kumar P, et al. Versatility of natural polymers. Adv Drug Formul. 2021a;18(2):215–29.
9. Mishra R, et al. Natural polymers in modern drug delivery. Biom J. 2018;9(3):198–213.
10. Mitra A, et al. Gastrointestinal stability of drugs. Drug Metab Rev. 2019;51(1):67–80.
11. Patel K, et al. Biodegradability of natural polymers. Policy Sci. 2018;23(4):145–60.
12. Patra S, et al. Advantages of oral drug delivery. Pharm Rev. 2021;16(9):489–505.
13. Rao R, et al. Cost-effective drug delivery systems. PharmaWorld. 2017;29(8):41–50.
14. Reddy V, et al. Biocompatibility of polymers. Mater Med. 2020;11(5):310–26.
15. Roy A, et al. Improving drug bioavailability. Ther Adv. 2020;14(6):201–13.
16. Sharma P, et al. Eco-friendly polymers. Green Chem Adv. 2021a;19(2):145–60.
17. Shah D, et al. Oral drug delivery: past and present. Pharm Horiz. 2018;22(7):125–37.
18. Shah S, et al. Eco-friendly pharmaceutical excipients. Drug Dev J. 2020;30(3):278–94.
19. Singh R, Kumar M. Bioadhesive drug delivery systems. Adv Pharm. 2020;15(4):299–312.
20. Verma N, et al. Customizing drug release profiles. Drug Release Sci. 2021;17(9):342–56.
21. Singh T, et al. Systemic absorption through oral routes. J Drug Sci. 2022a;14(8):412–25.
22. Kumar R, et al. Chitosan in drug delivery: advances and applications. J Pharm Sci. 2023a;112(3):124–35.
23. Smith J, et al. Alginate-based drug delivery systems: an overview. Int J Drug Deliv. 2022;14(5):567–78.
24. Jones D, et al. Functional natural gums in pharmaceutical formulations. Pharmac. 2021;
25. Patil S, et al. Advances in hydrogels for controlled drug release. Drug Deliv Res J. 2023;18(2):145–60.
26. Singh A, et al. Natural polymers in mucoadhesive drug delivery. J Pharm Innov. 2021;9(4):210–25.
27. Kumar R, et al. pH-responsive natural polymers: applications and challenges. Int J Pharm. 2023b;32(6):567–89.
28. Kumar S, et al. Matrix tablets for controlled drug delivery: a comprehensive review. Int J Pharm Sci. 2021b;13(4):245–58.
29. Patil P, et al. Advances in hydrophilic matrix systems for sustained drug release. J Control Release. 2020;320:310–28.
30. Deshmukh V, et al. Xanthan gum-based hydrophilic tablets for prolonged release of antihypertensive drugs. Pharm Innov J. 2019;8(3):27–33.
31. Rana R, et al. Chitosan granules: solubility enhancement and site-specific drug delivery. Curr Pharm Des. 2023a;29(2):178–86.
32. Singh A, et al. Hydrogels and their applications in drug delivery: current trends and future perspectives. Polym Med. 2022b;12(3):190–210.
33. Sharma D, et al. Biopolymeric hydrogels: advances in controlled drug delivery applications. Carbohydr Polym. 2020;245:116537.
34. Choudhary S, et al. Role of hydrogels in wound healing and tissue engineering. Mater Today Proc. 2018;5(10):21234–40.
35. Aggarwal A, et al. Alginate microcapsules for encapsulation of bioactive compounds: a review. Food Hydrocoll. 2021;120:106874.
36. Kumar A, et al. Chitosan microcapsules: mucoadhesive systems for nasal and buccal drug delivery. Eur J Pharm Biopharm. 2019;141:50–8.
37. Rajput S, et al. Natural polymeric nanoparticles for targeted drug delivery: current status and future challenges. J Drug Deliv Sci Technol. 2022;67:102849.
38. Jain P, et al. Chitosan-based nanoparticles: revolutionizing drug delivery for cancer therapy. Adv Drug Deliv Rev. 2023;196:114105.
39. Mitra S, et al. Alginate nanoparticles for oral insulin delivery: overcoming gastric degradation. J Biomed Mater Res A. 2020;108(12):2541–50.
40. Tiwari R, et al. Nanogels as a promising carrier for anticancer drug delivery. Colloids Surf B: Biointerfaces. 2021;206:111938.
41. Yadav S, et al. Electrospun nanofibers using silk fibroin for localized drug delivery applications. Mater Sci Eng C. 2018;91:295–309.

42. Rana P, et al. Advances in alginate-based drug delivery systems: applications and insights. Drug Dev Res. 2023b;14(2):245–58.
43. Kumar A, et al. Applications of chitosan in controlled drug delivery: a review. Pharm Innov J. 2021c;10(4):156–73.
44. Singh M, et al. Natural gums as excipients in bioadhesive drug delivery systems. Int J Pharm Sci. 2020;45(7):122–35.
45. Sharma S, et al. Functional properties of gum Arabic in chewable tablets. Drug Formul Res. 2021b;12(6):345–67.
46. Jain K, et al. Gelatin nanoparticles in cancer therapeutics: formulation and targeting potential. Oncol Pharm Sci. 2022;15(3):211–30.
47. Yadav S, et al. Silk fibroin-based encapsulation systems for drug delivery: a comprehensive review. J Pharm Biotechnol. 2019;8(4):98–115.
48. Sharma R, et al. Natural polymers in drug delivery: challenges and opportunities. J Drug Dev Sci. 2022;18(3):199–213.
49. Kumar P, et al. Variability and standardization challenges in natural polymer-based formulations. Pharm Innov. 2021d;15(4):276–88.
50. Yadav V, et al. Stability concerns in natural polymer systems: insights and solutions. Drug Formul Deliv Sci. 2020;11(2):89–103.
51. Patel S, et al. Regulatory aspects of natural polymers in pharmaceuticals: an overview. Int J Pharm Sci. 2019;45(6):123–36.
52. Singh K, et al. Ensuring quality in natural polymer-based drug delivery systems. Pharm Biotechnol J. 2023;14(1):78–91.
53. Smith J, et al. Sustainable approaches to natural polymer sourcing. Green Chem Pharm Appl. 2023;17(4):345–61.
54. Gupta R, et al. Artificial intelligence in polymer-based drug delivery systems. J Comput Pharm. 2022;12(3):212–27.
55. Zhao Y, et al. Marine-derived polymers for advanced drug delivery applications. Pharm Mar Biotechnol. 2021;9(2):123–39.
56. Patel S, et al. Emerging trends in underutilized natural polymers. Polysaccharide Adv Pharm. 2020;6(4):97–115.
57. Lee A, et al. Biopolymers in personalized medicine: a computational perspective. Front Drug Design Deliv. 2024;8(1):34–49.

Optimizing Bioavailability Through Natural Excipients: A Sustainable Approach to Drug Delivery

13

Kevinkumar Garala, Reena Ughreja, Rajan Ranpara, and Nirali Dholaria

13.1 Introduction

The effectiveness of a pharmacological agent depends on the ability of the formulation to deliver the active component to the desired site in the required amount and rate to bring about the desired pharmacological activity [99]. This feature of the dosage form is commonly known as biological availability, or more in a word, bioavailability. For the majority, a direct relationship between the plasma concentrations of drugs and their pharmacological response exists. Thus, bioavailability is described as both the rate and extent (amount) of absorption of the unchanged drug from its formulation.

Route of administration highly influences the bioavailability of a drug. Generally, in the following order and with few exceptions: *Parenteral > Oral > R ectal > Topical.* In the parenteral route, an intravenous injection of a drug automatically provides 100% bioavailability since the absorption process is circumvented. However, two valid reasons for most drugs being administered orally relate to stability and convenience. In such a case, the dose available to the patient, termed as the bioavailable dose, would be often less than the administered dose. The systemic availability, or simply put, availability, is the amount of drug that reaches systemic circulation, which relates to the degree of absorption. The bioavailable fraction, F is the fraction of the dosage administered that subsequently enters the systemic circulation [17].

K. Garala (✉) · R. Ughreja · R. Ranpara
School of Pharmaceutical Sciences, Atmiya University, Rajkot, Gujarat, India
e-mail: kevin.garala@atmiyauni.ac.in

N. Dholaria
Ernest Mario School of Pharmacy, Rutgers University, New Brunswick, NJ, USA

J. S et al. (eds.), *Innovative Pharmaceutical Excipients: Natural Sources*, https://doi.org/10.1007/978-981-96-7959-1_13

Absolute bioavailability represents the measurement of systemic availability of a drug when administered orally relative to its intravenous administration. The symbol F is used to express absolute bioavailability. The absolute bioavailability determines the inherent absorption properties of a drug from an extravascular site. *Relative or comparative bioavailability* is applied in calculating systemic availability of a drug following an oral administration relative to an oral standard formulation of the same drug. This may include an aqueous or nonaqueous solution or a suspension. It has the symbol Fr. Unlike absolute bioavailability, it measures the absorption property of a drug from its special formulation [36].

13.1.1 Role of Excipients in Drug Formulations

The Latin word excipere, meaning "to take to oneself," has been derived for the word "excipient." It would therefore be some sort of thing hosting the active ingredient in a drug product. As far as the medication preparation is concerned, the definition of an excipient would be regarded as the blend that is not the active agent. The European Pharmacopoeia defines an excipient as "any part, other than the active pharmaceutical ingredient, of a medicinal product or used in making it" [41].

Excipients are essentially the inactive major components which can be thought of as a carrier or base helping the active pharmaceutical ingredient (API) along. They play a key role in many properties that are fundamentally important to the product, including stability, biopharmaceutical characteristics, visual appeal, patient acceptability, and efficacy in manufacturing [53]. In most instances, a single excipient is combined with another within the composition of the pharmaceutical product to come up with the required properties of the product.

Pharmacologically active compounds, more often referred to as drugs or active pharmaceutical ingredients (APIs), are rarely administered directly in their pure form. Instead, they are formulated into carefully designed dosage forms which enhance their potency, make them more full-proof and safer, and in return, assist the patient in tolerating them in a clinical setting. These pharmaceutical dosage forms serve as a basis for uniform and accurate dosing, quality, efficacy, safety, stability, and increased patient acceptability and compliance [108]. At first, dosage forms had to be generated using pharmacologically inert substances, also known as excipients, introduced into the API for filling in volume to an acceptable dosing unit. However, advances in pharmaceutical technology have brought about the identification and formulation of excipients that contribute to specific functions beyond mere volume contribution including helping in the formulation of the dosage form and aiding in delivery of drugs from innovative dosage forms [103]. In fact, the functions of excipients in dosage forms relate to all aspects of the final product, including its manufacturability, the stability of the API, dose uniformity, and effective delivery of the API to the systemic circulation after administration as well as acceptable organoleptic properties for maximum compliance by the patient.

13.1.2 Natural Excipients: An Overview

The natural environment is full of diverse varieties of valuable elements that directly and indirectly benefit the living organisms by maintaining optimal health. Abundant resources of natural excipients include binders, diluents, sweeteners, colorants, preservatives, film formers, and their derivatives in the form of plants, animals, and minerals [68]. They have an inert and biodegradable characteristic with a major level of biocompatibility. They illustrate minimal toxicological effects and cost-effectiveness. Hence, pharmaceutical companies are much interested in the usage of these excipients for the formulation of innovative drug formulations, cosmetic products, and food items [95]. Few are mentioned in Table 13.1.

Several research studies reveal that the indiscriminate use of synthetic excipients in formulations is a big challenge concerned with their toxicity, incompatibility, unavailability, and high cost. Incorporation of natural excipients in the pharmaceutical field has placed a quest for novel techniques toward preparation of safer and more efficacious drugs for the patients [16].

There are structural complexities in mixtures and variances of productivity associated with natural origin chemicals, which may be influenced by geographical locations and season. The processes associated with this group of chemicals are slow and cost-prohibitive. Additionally, there are numerous sources of obtaining natural excipients, such as plants, animals, and sea living organisms among others. More importantly, all the diverse derivatives from natural excipients can also be prepared using similar modification methods, such as grafting and cross-linking. In the last years, research focus shifted toward components derived from plants, like polysaccharides, gums, and mucilages and volatile oils. Such plant-based polymers have scope in pharmaceutical preparations involving solid and liquid dosage forms, controlled and sustained drug delivery, osmotic drug delivery, microencapsulation, injectables, buccal films, and implants [78]. These ingredients are used as binders, disintegrants, diluents, emulsifiers, suspending agents, coloring agents, flavoring agents, gelling agents, bioadhesives, and viscosity modifiers.

The chemically modified natural excipients have been widely used for the last few decades to surmount the disadvantages of their native forms. They offer a putative role not only in the pharmaceutical domain but also in other areas like food, textile, and paper industries. Pharmaceutically, natural excipients are being preferred over synthetic to ensure the proper administration of pharmaceuticals to precise sites, as well as optimal times, is also critical. Additionally, these excipients

Table 13.1 Classification of natural excipients

Plant-based	Animal-based	Mineral-based	Marine-based
Corn, potato, rice starch	Gelatin	Silica	Alginates (from algae)
Cellulose, wood pulp	Lactose	Calcium salts	Seaweed carrageenans
Gums	Animal fat, stearic acid	Magnesium stearate	Chitin
Pectin	Cholesterol	Talc	

enhance bioavailability and stability, which improve patient compliance and ensures safety and efficacy in the delivery process [45]. Further work will be needed to elucidate the various characteristics and potential uses of these excipients in drug delivery systems to further optimize pharmaceutical formulations.

13.2 Types of Natural Excipients

13.2.1 Plant-Derived Excipients: Natural Resources for Improving Bioavailability

Plants have been used as sources of various active compounds for therapeutic purposes for many centuries. They show excellent properties, such as biocompatibility and biodegradability, with very low toxicity; hence, they are of great interest in the development of potent and green drug delivery systems [95]. The following review discusses how specific excipients derived from plants can enhance bioavailability:

Corn, potato, rice starch: Starch is a versatile, widely available polysaccharide widely used in drug delivery systems with diverse functions. As a binder, it ensures tablet integrity; its property as a disintegrant prompts rapid breakdown of tablets in the alimentary tract, hence drug release [96]. The second starch can be engineered into either nanoparticles or microspheres with drug molecules to carry these into a protective environment away from degradation, thus increasing absorption and bioavailability [83].

Cellulose, wood pulp: Another popular excipient is cellulose, a major structural constituent of plant cell walls. Similar to starch, it acts as both a binder and disintegrant (D. [77]). In addition, certain cellulose derivatives are very often used in sustained-release products: hydroxypropyl methylcellulose, for instance. It forms a kind of gel matrix with aqueous fluids, delaying release of the drug and lengthening the duration of therapeutic activity [92].

Gums: Plant-derived gums include guar gum and acacia gum, which are complex polysaccharides exhibiting high thickening and stabilizing properties [5]. As a binder in drug delivery, gums maintain tablet integrity, and as a suspending agent prevent settling in liquid formulations. Mucoadhesive properties can also be found in some gums such as guar gum, which adhere to the mucosal lining of the gastrointestinal tract and prolong the time a drug resides within it, thus enhancing its absorption [73].

Pectin: It is a naturally available cell wall polysaccharide from the fruits which is employed as gelling agent, thickening and stabilization agent for food, cosmetic as well as pharmaceutical drugs. It provides beneficial gelling properties within formulating the oral-controlled delivery systems for drugs in release control from matrices derived based on pectin wherein varied mechanisms control drug releases, for instance, through swelling erosion or diffusion mechanism, allowing controlled bioavailability and frequent administrations [7, 71]. By strategically adding such plant-based excipients in drug formulations, researches and pharmaceutical scientists could find the unique properties they leverage to optimize drug release profiles, enhance bioavailability, and ultimately improve the final therapeutic outcomes.

13.2.2 Marine-Based Excipients: Diving into Bioavailability Enhancement from the Ocean

Ocean makes up an immense diverse number of biotas, which host the largest source of biologically active compounds with promising drugs delivery. Marine-derived excipients gain enormous attention in the scientific world nowadays due to their compatibility towards biodegradable nature along with unique physicochemical properties that can boost bioavailability of multiple drugs [94].

Alginates (from algae): Alginates are linear polysaccharides isolated from brown seaweed, having mannuronic and guluronic acid in varied ratios. Due to their ability to gelate in the presence of divalent cations like calcium, alginates are applied as drug delivery agents [9]. Alginate-based hydrogels and microbeads may be used as drug carriers; drugs encapsulated within them can be released in controlled fashion and may protect sensitive drugs from gastrointestinal tract degradation. Release control might offer significant improvement of absorption as well as bioavailability of drugs because of potential mucoadhesive properties [59].

Seaweed carrageenans: Another sulfated polysaccharide obtained from red seaweed show varying gelling and thickening abilities according to their composition. Similar to alginates, carrageenans also can form gels; it is under investigation for controlled drug release applications [44]. Mucoadhesion characteristics of certain types including kappa-carrageenan contribute further to increase the resident time of the drug at the absorption site due to enhanced bioavailability [61].

Chitin: It is a linear polysaccharide, present extensively in the exoskeletons of crustaceans. Although chitin as a polymer is very sparingly soluble in solvents, as an endogenous biocompatible molecule in nature, it does gain importance [57]. Its deacylated derivative, chitosan, though a very sparingly known occurrence in nature, has widely been studied for purposes like drug delivery. It has mucoadhesive properties since at physiological pH it carries a positive charge and interacts with the negatively charged mucosal surfaces. Thus, mucoadhesion increases the contact time of drugs at the site of absorption, which in turn increases the absorption and bioavailability of drugs [80].

13.2.3 Animal-Based Excipients: Traditional Foundations for Drug Delivery

While plant-based and synthetic excipients have gained significant attraction, animal-derived materials have long been employed in pharmaceutical applications and can still provide several useful properties.

Gelatin: It is an excipient made from collagen, which exists in animal skin and bones. It is the most used excipient especially in capsule production. Its ability to form transparent, easily soluble gels is very useful for the encapsulation of drugs in a flavorless and odorless fashion. Gelatin capsules can easily pass the throat and dissolve quickly within the stomach, releasing their content for absorption [117].

Lactose: One of the main ingredients of whey is lactose, a disaccharide that is created as a byproduct of the dairy industry. Lactose has been a common filler in tablet manufacture for many years. Lactose is better suited as a tablet filler because of its excellent compressibility [14]. The pharmaceutical industry makes extensive use of the purest form of αlactose as an excipient. Lactose plays a variety of roles in dose formulations. It is utilized as a carrier for dry powder inhalation products, as a sweetener in liquid formulations, and as a filler, diluent, or bulking agent in tablets and capsules [85]. Because of its affordability, low hygroscopicity, low sweetness, compatibility with active substances and other excipients, bland taste, good chemical and physical stability, and adequate water solubility, lactose is primarily used in pharmaceuticals as an excipient. These characteristics make it the best pharmaceutical excipient for dose formulations; it is thought to be used in roughly 70% of these formulations [50].

Animal fat, stearic acid: It is a saturated fatty acid produced from animal fats among others. As a lubricant it is normally included in tablet formulations It keeps ingredients from sticking to manufacturing equipment when they are made. So, it enables production to occur as smooth and efficient. This contributes to tablet hardness, plus the time it will disintegrate [30].

Cholesterol: Despite all the adverse health effects generally attributed to cholesterol, it plays a vital role in pharmaceutical formulation. Specifically, it is a constituent in topical creams and ointments. Cholesterol, which increases drug penetration into the skin, is very useful as an excipient in direct drug delivery to the action site [114].

13.2.4 Mineral-Based Excipients: The Inorganic Foundations of Drug Delivery

Mineral-derived excipients are overshadowed by their organic counterparts in many instances, yet they are fundamental components in most drug formulations. Their inert nature, stability, and unique physicochemical properties make them invaluable for enhancing drug product manufacturing, stability, and performance [46].

Silica: It is available as colloidal silicon dioxide or magnesium trisilicate and used in pharmaceutical formulation as an established ingredient [35]. It acts as a glidant which helps prevent the agglomeration of powder particles and hence helps ensure smooth flow in tablet manufacturing. This leads to uniform filling of tablet dies, which consequently ensures uniform dosing and weight [20]. Silica contains pores and hence is an excellent desiccant; the pores absorb much moisture, making it useful in safeguarding the shelf-life of moisture-sensitive drugs against degradation [109]. Silica can adsorb drugs onto its surface, thereby modifying drug release profiles and improving bioavailability in some formulations [65].

Calcium Salts: Calcium salts, such as calcium carbonate, calcium phosphate, and calcium sulfate, are multipurpose mineral excipients with a wide range of uses. It is a good diluent which act as fillers, bulking up tablets and capsules to achieve practical sizes for processing and administration, especially for low-dose drugs [10]. They are useful buffering agents when present in an appropriate form in liquid formulations, hence resisting changes in pH, thus stabilizing and maintaining the solubility of sensitive drugs [53]. It may serve as a supplement of calcium. In this regard, calcium salts can contribute to nutritional value to the drug product [49].

Magnesium Stearate: Magnesium stearate is a salt of stearic acid and magnesium; it is a very common lubricant in tablet manufacturing [60]. As an anti-adherent, it prevents tablet ingredients sticking onto punches and dies in the process of compression, and prevents any abrasions during ejection from happening on tablets or equipment [111]. Although primarily a lubricant, magnesium stearate also improves powder flow, which further enhances the uniformity of tablet weight and content. Too much magnesium stearate can form a hydrophobic layer on the surface of tablets, which may delay drug dissolution and absorption. Its concentration needs to be optimized with care [6].

Talc: The hydrated magnesium silicate mineral talc has been employed traditionally as glidant and lubricant. It will improve powder flow and reduce friction during tablet compression [84]. Talc has recently become the subject of criticism for its possible asbestos fiber contamination that causes health issues [40]. It can be replaced with safer options of silicon dioxide and magnesium stearate.

13.3 Mechanisms of Natural Excipients for Improving Drug Bioavailability

Natural excipients play a crucial role in pharmaceutical formulations due to their biocompatibility, non-toxicity, and ability to enhance the bioavailability of therapeutic agents. The mechanisms by which these natural substances improve drug bioavailability are multifaceted, involving enhancements in solubility, permeability, stability, and release kinetics.

13.3.1 Increasing the Drug Solubility

Low solubility is the most major impediment to oral bioavailability for many drugs [58]. Among the naturally occurring excipients, those from plants-for example, gums, such as guar gum [89], and polysaccharides, such as starch [101] and pectin [63] share similar uses as solubilizers in their surfactant, co-solvent, or complexation properties. The bases of weakly soluble drugs can introduce hydrophilic excipients to hydrophilic complexes or micelles, which improve their rate of dissolution in the gastrointestinal tract. For instance, cyclodextrins are oligosaccharides naturally occurring [33]. They form inclusion complexes that contain lipophilic drugs, which greatly improve their aqueous solubility.

13.3.2 Permeability and Absorption Enhancement

Bio-enhancers and improving drug permeability are other key functions of natural excipients [81]. Modulators of lipid bilayer function or tight junction integrity enhance saponins, terpenoids, and natural oils that make drugs permeate better across biological membranes. These substances temporarily increase the fluidity of the membrane and decrease the efflux of drugs, thus allowing for a higher rate of absorption. Formulations using natural permeation enhancers could be useful in certain drugs that fall under the Biopharmaceutics Classification System classification as Class III or Class IV, with poor permeability. Aloe vera gel may be used as a potential buccal permeation enhancer at concentrations of 0.25–2% w/v for didanosine [76]. It was demonstrated in one of the study that chitosan might improve the buccal mucosal tissue's absorption of the transforming growth factor-β (TGF-β) [87]. Enhanced intestinal absorption of fexofenadine was seen by pretreatment of capsaicin [12]. Another study showed caraway's bioenhancing effect on the pharmacokinetics of rifampicin, isoniazid, and pyrazinamide. It could be explained by increased mucosal to serosal penetration and its impact on P-gp efflux [25].

13.3.3 Stabilization of Drug Molecules

The stability of the drug molecule will determine the period throughout the shelf life when the product will be effective. Natural polymers like alginate, carrageenan, and cellulose derivatives provide stabilization through the development of a protective matrix that will cover the API. This matrix protects the drug from hydrolysis-related degradation as well as the chemical instability related to oxidation. In addition, these excipients also help preserve the drug in its amorphous form, which is more soluble than the crystalline drug. Moisture-related degradation of nitrazepam was studied using microcrystalline cellulose, corn starch, mannitol, and saccharose. Degradation was correlated with the nitrogen adsorption energy of the excipients which was determined using the Brunauer–Emmett–Teller (BET) equation. Materials with higher adsorption energies caused less degradation [31]. Stability issues are also arising due to photodegradation. Thoma demonstrated that the natural food coloring curcumin can both efficiently stabilize the photolabile calcium antagonist nifedipine [104]. Apart from their use as solubility enhancer, cyclodextrins have ability to stabilize drug candidates [51].

13.3.4 Controlled and Sustained Release

Natural hydrogels and biopolymers, such as chitosan, xanthan gum, and agarose, have been quite widely exploited as excipients in controlled release drug delivery.

Swelling in response to biological fluids generates a gel matrix that controls the drug release during the process. It can be formulated for a prolonged release and maintain the drug level for a long time such that the therapeutic effect is sustained. This method improves not only the bioavailability but also the compliance of patients due to reduction in dosing frequency and side effects due to peak plasma concentration. In one of the study, sustained release of ferrous sulfate was achieved for 7 h by preparing gum arabic pellets [11]. Near zero order drug release was achieved by both xanthan and karaya gums, with the erosion mechanism predominating, particularly in karaya gum matrices [72]. Tragacanth, either by itself or in conjunction with other polymers, achieved appropriate release prolongation when utilized as the carrier in the construction of one- and three-layer matrices [91]. Clays and clay minerals (including smectites, kaolinite, and fibrous clay minerals) have been extensively applied for prolonged release, especially of basic drugs. Modified clay, sepiolite, that had been heated demonstrated regulated release of antifungal and antibacterial isothiocyanates for food preservation and hygienic kitchen protection [2].

13.4 Formulation Strategies Using Natural Excipients

Natural excipients have gained attention in pharmaceutical formulations due to their biocompatibility, multifunctionality, and environmental sustainability. These are materials derived from renewable resources, such as plants, marine organisms, and animals, which play a central role in designing advanced drug delivery systems and achieving the targeted therapeutic outcome. Key formulation strategies that exploit unique properties of natural excipients are discussed below.

13.4.1 Solid Dosage Forms

Natural excipients are an integral part in the preparation of solid dosages such as tablets, capsules, and powders. Multiparticulate systems, such as pellets, granules, and microspheres, are advanced drug delivery platforms offering controlled and targeted release profiles. Among such excipients, cellulose derivatives, starches, and gums are binders, disintegrants, and fillers, with a mechanism to give the required strength, stability, and ease of drug delivery. Microcrystalline cellulose is a frequently used natural excipient offering compressibility and uniformity in making tablets [105]. Starch derivatives such as pregelatinized starch improve disintegration and dissolution, thus affecting the bioavailability [86]. Gums, natural, particularly guar gum and xanthan gum, are being used as release-modifying agents in sustained-release formulations on account of their hydrophilic and gelling characters [70]. They find an advantage over synthetic versions owing to their compatibility with the APIs and biodegradable nature. Further enteric coating for protecting drugs from gastric fluids, with natural polymers providing acid resistance, is advanced application of natural polymers. Shellac as an important coating material for sustained release or colon targeting applications [39] is described in one of the study.

13.4.2 Liquid Dosage Forms

Natural excipients include mostly polysaccharides like gums and oils that substantially contribute to the stability viscosity, and palatability of liquid dosage forms: syrups, suspensions, and emulsions [8]. For suspensions, xanthan and carrageenan gums increase the viscosity and improves suspensions stability by making it homogeneous and consistent. Oils such as castor oil and sesame oil are used as solvent systems and emulsifying agents in oral and injectable delivery systems [54]. Plant-based surfactants, such as lecithin, are used in emulsions to enhance solubilization and drug delivery [115]; these natural excipients assure physicochemical stability, providing safety for the patient due to reduced potential toxicities related with synthetic counterparts.

13.4.3 Semi-solid and Topical Dosage Forms

Semi-solid formulations, such as creams, gels, and ointments, find natural excipients acting as a thickening agent, an emollient, or stabilizer. Aloe vera gel and chitosan are often used for their hydrating and healing properties in topical formulations [112]. Beeswax and lanolin are natural bases in ointments to enhance smoothness and permeability through the skin [75]. Some examples of polysaccharides that are used in gels due to their excellent viscosity and gel-forming properties are guar gum and xanthan gum. Alginates and carrageenans also offer bioadhesive properties, which further increase the retention of the drug at the application site [18]. These excipients possess inherent biocompatibility with minimal irritancy; hence, they are ideal for use in dermatological and transdermal delivery systems.

13.4.4 Advanced Drug Delivery Systems

Furthermore, chemical modification of natural excipients, for example, through crosslinking, now allows the tailoring of drug release profiles in solid dosage systems [66]. Insulin-loaded chitosan microspheres engineered by emulsion cross-linking method were found superior in terms of sustaining protein release over conventional insulin therapy [52]. Targeted drug delivery, with alginate beads used for gastric retention or intestinal delivery, was achieved in study by Amiri et al. [74]. Site-specific drug delivery using nanoparticles stabilized with natural polymers to enhance drug loading and release was achieved. Functions of hyaluronic acid nanoparticles for tumor-targeting drugs were studied for tumor-targeting purpose, enzyme-sensitive biodegradable modality, pH-sensitive component, reduction-sensitive component, and the gel backbone are discussed in one of the study [64]. Chitosan, with its mucoadhesive and permeation-enhancing properties, is widely used in nanoparticle formulations for targeted drug delivery [37]. Spheroids and organoids hold broad application prospects in regenerative medicine and disease therapy. One such study included OSTE+ (off-stoichiometry thiol-ene epoxy) polymer chip fabrication which enabled the in-chip release of spheroids from droplets, thereby supporting cultivation of tumor cell spheroids [79]. Lecithin, either from soy or egg, is a critical component in liposome formulations that enhances the solubility and bioavailability of poorly soluble drugs [22]. These excipients enable site-specific drug delivery and reduced systemic toxicity, making them invaluable for personalized medicine approaches.

13.5 Advantages of Natural Excipients Over Synthetic Ones

Natural excipients have various advantages that synthetic ones do not possess [34]. Increasingly, they are more sought after in pharmaceutical formulations. Most of them derive from renewable sources, such as plants [11, 72, 101], animals [3, 42], or marine organisms [3, 19, 85], and are anticipated to be biocompatible as well as biodegradable, thus more in line with the growing movement toward sustainable pharmaceutical practices.

An inherent advantage with natural excipients is that they have a profile for safety [21]. The toxicity and allergenicity of most of the natural excipients, such as alginates [107], chitosan [42], and pectins [67], are compared to their synthetic counterparts and are relatively low. Their biodegradability is good for reduced environmental impact and is thus favorable for applications requiring sustainability and regulatory compliance [15].

However, natural excipients tend to be multifunctional. For instance, alginate or chitosan can serve as gelling agents, bioadhesives, and drug release modifiers or exhibit bioactivities such as antimicrobial and anti-inflammatory activities [34]. Such multifunctionality enhances formulation efficiency and delivers a better therapeutic output. Such excipients of natural origin are well compatible with biological systems due to the similarity in their structure with endogenous molecules [55, 56]. Such properties support improved absorption, less immunogenicity and better acceptability by patients. Their wide versatility for the acceptance of various modifications, such as grafting or crosslinking, broadens their application to include advanced drug delivery systems. Synthetic excipients often are nonbiodegradable and synthesized through complex processes, which compromise the values of safety, scalability, and environ mental impact. Natural excipients, therefore, have inbuilt functionality combined with sustainability for the needs of the pharmaceutical world of today.

13.6 Challenges Using Natural Excipients

Although there are various natural excipients gaining vast attention in pharmaceutical formulations, what with their biocompatibility, safety features, and sustainability, they are not without challenges. Several problems crop up while using such bio-based materials in drug delivery systems, like compositional variability, microbial contamination susceptibility, regulatory challenges, limited functionality, among others.

13.6.1 Variability in Composition

One of the challenges with natural excipients is that they would be bound to vary in chemical composition. In contrast, synthetic excipients will always have limited variability from batch to batch, while natural sources include sources such as plants or marine organisms whose batch-to-batch variability arises from variable environmental conditions; cultivation practices; and harvest times [93]. Such variability affects the physicochemical properties of an excipient, including solubility, viscosity, and stability, and therefore the performance in drug formulations becomes unpredictable [21]. As such, quality consistency and reliable performance still represent an enormous headache for pharmaceutical manufacturers. Other factors such as agroclimatic conditions like light, temperature, soil quality, plant age, and harvest time may also affect the quality and consistency of natural excipients [62]. Sometimes, in order to achieve product specifications, batch blending is necessary due to a wider range of natural variation, which results in bigger composition changes [55]. Performing quality control checks for every raw ingredient utilized in the formulation becomes quite difficult. Nonetheless, Good and Manufacturing Practices (GMP) outlines a number of unavoidable procedures for raw material quality control, such as accurately identifying the right medicinal plants and putting in place adequate storage, cleaning, and other procedures.

13.6.2 Contamination with Microorganisms and Instability Problems

Natural excipients, especially plant and animal-based ones, are more prone to microbial contamination [93]. These materials have high moisture contents and are organic in nature [113]. Thus, these can harbor bacteria, fungi, or yeast, which may adversely affect the final product's safety and efficacy [98]. Sterilization or antimicrobial treatment is usually very severe and may severely degrade the properties of the excipients or alter their functional behavior [88]. Natural excipients may also degrade over time due to the presence of enzymes or inherent instability, therefore affecting shelf-life and quality of the drug product [69].

13.6.3 Challenges Regarding Regulatory and Standardization

Natural excipients and, in general, their regulatory landscape are complicated and not as well defined compared to synthetic excipients. The lack of an independent origin and, therefore the inherent variability of these materials can complicate the establishment of quality standards, as well as regulatory approval [32]. In many cases, severe characterization and safety data, including source, composition, potential allergens, and impurities, will be requested by regulatory agencies. There is not really standardized protocol for the assessment of natural excipients that can delay development process and complicate market entry.

13.6.4 Low Functional Capability

Though natural excipients have many advantages, their functional capabilities might be inferior as compared with their synthetic counterparts [26]. For example, they may not completely mimic the precise control in terms of drug-release profiles or specific physicochemical properties demanded by some other advanced drug delivery systems. Hence, their function could be negatively affected by environmental conditions including pH or temperature alterations, and they are, therefore, utilized less extensively in various pharmaceutical preparations [93]. Sometimes, these limitations can be overcome only by drastic formulation optimization and the introduction of extra stabilizers or modifiers [106].

13.7 Regulatory and Safety Considerations

The inclusion of natural excipients in pharmaceutical formulations carries with it a lot of promise, since these substances are biocompatible with potential applications leading towards better patient compliance. However, regulatory and safety issues concerning these materials must be dealt with properly to serve product quality, efficacy, and safety issues.

13.7.1 Regulatory Issues

Even though excipients are pharmacologically inert, they can be considered the real facilitators of medication product performance. Although it can be minimized, unintended variability in the excipients' characteristics may be inevitable. The source, the excipient-making procedure, or the dosage form manufacturing method could all be the cause of the variability. Such variability in their sources along with diverse origins makes it quite difficult to withstand the regulatory scrutiny [32]. Synthetic excipients are defined and standardized, but natural excipients exhibit significant compositional variation, dependent on the environment, cultivation conditions, and extraction methods. These natural excipients are therefore a challenge to the regulatory bodies like US Food and Drug Administration (FDA) or European Medicines Agency (EMA), which demand thorough characterization to assure not only the identity but also the purity and consistency of the excipient. Complete documentation with source information, manufacturing process, and stability profiles is required to clear the regulator's checks. Further complicating the approval process would be the lack of specific monographs for most natural excipients in pharmacopeias, which means protracted timelines and a higher cost for the pharmaceutical development groups.

The FDA has an excipient database called the Inactive Ingredient Database that lists the excipient's usage history, which users can consult. To ensure quality of excipients when more than one monograph exists for a specific excipient, all test procedures shall be followed and the results shall agree with those listed in such compendiums. Without monographs, the manufacturer would have to use some test procedures similar to those found in other monographs with findings being similar. Test procedures should always be validated and must meet current compendiums according to ICH Q2 Validation Guidance [38].

13.7.2 Safety and Toxicological Aspects

The safety evaluation is one of the critical aspects when there are plans to introduce natural excipients in drug formulations. These may have residual impurities, allergens, or potentially toxic constituents that can be introduced during extraction or processing [102]. If there are any adverse effects suspected, it is very important to have thorough toxicological evaluation, which should comprise in vitro and in vivo studies. Natural excipients are also likely to be contaminated by microbes due to their organic nature. Hence, more stringent sterilization methods have to be used. However, any conventional sterilization technique, such as autoclaving or irradiation, may alter the chemical structure or functional properties of the excipient to a level that makes it unsafe and ineffective. The challenge remains in balancing issues of microbial safety with the integrity of the excipient at the formulation stage.

13.7.3 Standardization and Quality Control

Quality consistency is the key to natural excipients for pharmaceuticals. Quality control is built using stringent mechanisms, such as standard operating procedures for testing and batch analysis to reduce variability while ensuring safety. Every regulatory agency has safety testing procedures that must be followed when a novel excipient is launched to the pharmaceutical market. Both agencies and specialists employ test methods, with a focus on those that have already been confirmed with trustworthy safety data and can serve as a substitute for preclinical in vivo experiments [1]. The excipient's safety test procedures should always adhere to Good Laboratory Practice and the Guiding Principles on the Use of Animals in Toxicology. GMP compliance ensures that formulations are reliable and reproducible, thus ensuring the patient's health while working toward approval [32].

13.8 Case Studies and Real-World Applications of Pharmaceutical Excipients

The number of pharmaceutical excipients currently available is huge and encompasses hundreds of substances. But among those proven to significantly enhance the therapeutic benefit of drugs, the majority are typically small molecules or macromolecules. Such include surfactants, lipids, oils, ion exchange resins, cyclodextrins hydrogels, and hydrophobic polymers [23, 28, 100].

13.8.1 Traditional and Nontraditional Applications of Lipid Excipients

Lipid-based excipients have been used traditionally and innovatively. Traditionally, they serve as a solvent for drugs, stabilizer, surfactant, antioxidant, and adjuvant in the formulation of vaccines. Not only these lipid excipients have been used as a typical component in formulation but also nonconventional, like membrane permeability enhancers, inhibitors of gut wall metabolism and efflux pumps, or even as atypical active pharmaceutical ingredients [13, 27, 90].

13.8.2 Polyionic Hydrocolloids and Innovative Applications

Excipients such as alginate and chitosan, traditionally employed as gelling agents, have been repurposed for the intestinal delivery of peptide and protein therapeutics [42]. Similarly, layered 2:1 phyllosilicates—commonly used as thickening agents—have been explored for their antimicrobial properties [48]. Surface-modified nanostructured aluminosilicates like montmorillonite have shown promise in reducing dietary cholesterol absorption in preclinical models [43].

13.8.3 Role of the Tocol Family in Pharmaceutical Applications

Compounds from the tocol family, including tocopherols, tocotrienols, and their derivatives (e.g., tocopherol succinate and α-tocopherol-polyethylene glycol 1000-succinate [TPGS]), demonstrate a wide range of applications. They are employed as drug delivery carriers, antioxidants, and therapeutic agents [27].

13.8.4 Advances in Methacrylic Polymers

Basic butylated methacrylic polymers have significantly enhanced the permeability of poorly absorbable substances [4, 47]. Additionally, Eudragit® polymers, traditionally used for enteric coating and controlled release, have been investigated for their efficacy in solid dispersions to improve solubilization and maintain the physical stability of poorly soluble drugs [82, 97].

13.8.5 Mineral-Based Excipients' Uses

The diverse applications of minerals as excipients include both conventional and unconventional uses. For instance, minerals are employed not only as fillers, binders, and lubricants but also as functional agents with significant implications for the drug's in vivo performance. Nontraditional roles include atypical pharmaceutical ingredients in over-the-counter medications [13, 27].

13.8.6 Excipients in Oral Drug Products and Dietary Supplements

The usage of excipients in oral drug products, as well as in dietary supplements, has been a theme in various conferences and workshops. Recently, in a symposium at the 2014 AAPS Annual Meeting in San Diego with a theme entitled Excipients as Atypical Actives in Nutraceuticals and Pharmaceuticals: Application and Development Considerations [29], presenters discussed the dietary role of excipients, excipients as inhibitors of the absorption of dietary cholesterol, challenges associated with the use of atypical actives in drug products, and regulatory perspectives on the nontraditional uses of excipients.

13.8.7 Regulatory Considerations About Osmotically Active Excipients

In their critical review, Chen et al. commented upon the pharmacokinetic implications of osmotically active excipients such as sorbitol, mannitol, D-xylose, and PEG 400 for the bioavailability and bioequivalence of a drug belonging to Biopharmaceutics Classification System (BCS) Class III, namely ranitidine, cimetidine, and amoxicillin. It has been demonstrated that knowledge of the dose-response relationships of these "active" excipients would optimize their application in drug formulations while offering scope for biowaivers without compromising the quality and efficacy of the product. The authors stress recognition of "active" excipients during formulation development for both generic and innovator pharmaceuticals. They further suggested further mechanistic studies to unveil the effect that such excipients may pose on the absorption, bioavailability, and bioequivalence of drugs [24].

A study by Wang et al. tested 20 excipients on the oral bioavailability of a potent opioid analgesic, nalbuphine, in humans and rats with in vivo studies [110]. The researchers applied PEG 400 and PEG 2000 as excipients, preservatives such as methyl and propyl parabens and sodium benzoate, and surfactants like Tween 20/40/80, Cremophor EL/RH40, Solutol HS-15, Pluronic F68, and sodium lauryl sulfate. An improvement in the oral bioavailability of nalbuphine was noted due to the association of the excipients' blocking capacity for UDP-glucuronosyltransferase 2B7, which appeared to be the major metabolic enzyme for nalbuphine, was proved with. Tween 20 and PEG 400, both alone and in combination, appeared to be the most active excipients.

There are several excipients that are used in various drug carriers such as micro- and nanoparticles and administered by the oral, parenteral, and other routes of administration that together with their primary role of modifying the pharmacokinetics and biodistribution of the drug exhibit bioactivities in the body by modulating cellular and physiological processes [116]. Such pharmaceutical excipients include hyaluronic acid, chitosan and chitosan derivatives, pluronics, polar lipids such as pegylated phospholipids and cationic lipids, cyclodextrins, and N-(2-hydroxylpropyl) methacrylamide copolymers.

13.9 Opportunities and Future Directions for Natural Excipients in Drug Delivery

Natural-origin excipients have remarkable properties that explain why they cannot be disregarded for the creation of more advanced drug delivery systems with controlled release. Often biocompatible, nontoxic, biodegradable, and responsive to various external stimuli, they appear to be excellent candidates for pharmaceutical and nutraceutical applications.

Marine-derived materials, including many types of polysaccharides such as alginates, are valuable sources of pharmaceutical formulation, with diverse chemical and biological properties for use as excipients or, in some instances, as active components in dietary supplements. Such versatility and functional diversity allow them to be applied to various drug delivery platforms and improve the performance of formulations along with the therapeutic outcome.

Clay minerals are widely used materials in drug products both as excipients and active agents. In recent years, and based on their high retention capacities as well as swelling and colloidal properties, clays have been proposed as very useful materials for modulating drug delivery. Clays are used to delay and/or target drug release or even improve drug dissolution. Finally, new strategies are reported for increasing drug stability and simultaneously modifying drug delivery patterns through the use of clay minerals.

Only such excipients that comply with FDA standards can be incorporated into medicines and foods. Many of the marine (alginate) and animal excipients (lactose and gelatin) are highly crucial in drug manufacture and synthetic substitutes have been discovered to replace them only very occasionally. We cannot exclude these natural excipients while developing novel dosage formulations.

13.10 Conclusion

The integration of natural-origin excipients into novel drug delivery systems offers unparalleled advantages in terms of safety, functionality, and sustainability. Their unique characteristics, combined with adherence to regulatory standards, make them essential components in the design of innovative dosage forms. Excluding these natural materials from pharmaceutical development would significantly limit the scope for advancing drug delivery technologies and improving therapeutic efficacy.

References

1. Abrantes CG, Duarte D, Reis CP. An overview of pharmaceutical excipients: safe or not safe? J Pharm Sci. 2016;105(7):2019–26. https://doi.org/10.1016/j.xphs.2016.03.019.
2. Aguzzi C, Cerezo P, Viseras C, Caramella C. Use of clays as drug delivery systems: possibilities and limitations. Appl Clay Sci. 2007;36(1–3):22–36. https://doi.org/10.1016/j.clay.2006.06.015.

3. Alamgir ANM. Bioactive compounds and pharmaceutical excipients derived from animals, marine organisms, microorganisms, minerals, synthesized compounds, and pharmaceutical drugs; 2018. p. 311–406. https://doi.org/10.1007/978-3-319-92387-1_4.
4. Alasino RV, Ausar SF, Bianco ID, Castagna LF, Contigiani M, Beltramo DM. Amphipathic and membrane-destabilizing properties of the cationic acrylate polymer Eudragit® E100. Macromol Biosci. 2005;5(3):207–13. https://doi.org/10.1002/mabi.200400168.
5. Amiri MS, Mohammadzadeh V, Yazdi MET, Barani M, Rahdar A, Kyzas GZ. Plant-based gums and mucilages applications in pharmacology and nanomedicine: a review. Molecules. 2021;26(6):1770. https://doi.org/10.3390/molecules26061770.
6. Ariyasu A, Hattori Y, Otsuka M. Delay effect of magnesium stearate on tablet dissolution in acidic medium. Int J Pharm. 2016;511(2):757–64. https://doi.org/10.1016/j.ijpharm.2016.07.034.
7. Assifaoui A, Chambin O. Pectin as drug-release vehicle. In: Pectin: technological and physiological properties. Springer; 2020. p. 189–207. https://doi.org/10.1007/978-3-030-53421-9_10.
8. Awad A, Madla CM, Gavins FKH, Allahham N, Trenfield SJ, Basit AW. Liquid dosage forms. In: Remington. Elsevier; 2021. p. 359–79. https://doi.org/10.1016/B978-0-12-820007-0.00020-9.
9. Banerjee R, Kumar KJ, Kennedy JF. Structure and drug delivery relationship of acidic polysaccharides: a review. Int J Biol Macromol. 2023;243:125092. https://doi.org/10.1016/j.ijbiomac.2023.125092.
10. Barik A, Dhar A. Introduction to different types of dosage forms and commonly used excipients. In: Dosage forms, formulation developments and regulations. Elsevier; 2024. p. 67–82. https://doi.org/10.1016/B978-0-323-91817-6.00002-4.
11. Batra V, Bhowmick A, Ray AR, Behera BK. Sustained release of ferrous sulfate from polymer-coated gum arabica pellets. J Pharm Sci. 1994;83(5):632–5. https://doi.org/10.1002/jps.2600830507.
12. Bedada SK, Appani R, Boga PK. Capsaicin pretreatment enhanced the bioavailability of fexofenadine in rats by P-glycoprotein modulation: in vitro, in situ and in vivo evaluation. Drug Dev Ind Pharm. 2017;43(6):932–8. https://doi.org/10.1080/03639045.2017.1285310.
13. Boddu SH, Renukuntla J, Rega A, Alexander K. Excipients and non-medicinal agents as active pharmaceutical ingredients. In: Excipient applications in formulation design and drug delivery. Springer; 2015. p. 613–36. https://doi.org/10.1007/978-3-319-20206-8_21.
14. Bolhuis GK, Anthony Armstrong N. Excipients for direct compaction—an update. Pharm Dev Technol. 2006;11(1):111–24. https://doi.org/10.1080/10837450500464255.
15. Bommenahalli Veerabhadrappa NB, Fernandes S, Panda R. A review of green purchase with reference to individual consumers and organizational consumers: a TCCM approach. Clean Responsible Consum. 2023;8:100097. https://doi.org/10.1016/j.clrc.2022.100097.
16. Braithwaite MC, Tyagi C, Tomar LK, Kumar P, Choonara YE, Pillay V. Nutraceutical-based therapeutics and formulation strategies augmenting their efficiency to complement modern medicine: an overview. J Funct Foods. 2014;6:82–99. https://doi.org/10.1016/j.jff.2013.09.022.
17. Caldwell J, Gardner I, Swales N. An introduction to drug disposition: the basic principles of absorption, distribution, metabolism, and excretion. Toxicol Pathol. 1995;23(2):102–14. https://doi.org/10.1177/019262339502300202.
18. Campo VL, Kawano DF, da Silva DB, Carvalho I. Carrageenans: biological properties, chemical modifications and structural analysis—a review. Carbohydr Polym. 2009;77(2):167–80. https://doi.org/10.1016/j.carbpol.2009.01.020.
19. Cardoso M, Costa R, Mano J. Marine origin polysaccharides in drug delivery systems. Mar Drugs. 2016;14(2):34. https://doi.org/10.3390/md14020034.
20. Cares Pacheco M-G, Jiménez Garavito M-C, Ober A, Gerardin F, Silvente E, Falk V. Effects of humidity and glidants on the flowability of pharmaceutical excipients. An experimental energetical approach during granular compaction. Int J Pharm. 2021;604:120747. https://doi.org/10.1016/j.ijpharm.2021.120747.

21. Chatterjee M, Mukherjee A. Evaluation of bioactivity of green nanoparticles synthesized from traditionally used medicinal plants: a review. In: Evidence based validation of traditional medicines. Springer Singapore; 2021. p. 799–815. https://doi.org/10.1007/978-981-15-8127-4_38.
22. Chen L-C, Chen Y-C, Su C-Y, Wong W-P, Sheu M-T, Ho H-O. Development and characterization of lecithin-based self-assembling mixed polymeric micellar (saMPMs) drug delivery systems for curcumin. Sci Rep. 2016;6(1):37122. https://doi.org/10.1038/srep37122.
23. Chen M-L. Lipid excipients and delivery systems for pharmaceutical development: a regulatory perspective. Adv Drug Deliv Rev. 2008;60(6):768–77. https://doi.org/10.1016/j.addr.2007.09.010.
24. Chen M-L, Sadrieh N, Yu L. Impact of osmotically active excipients on bioavailability and bioequivalence of BCS class III drugs. AAPS J. 2013;15(4):1043–50. https://doi.org/10.1208/s12248-013-9509-z.
25. Choudhary N, Khajuria V, Gillani Z, Tandon V, Arora E. Effect of Carum carvi, a herbal bioenhancer on pharmacokinetics of antitubercular drugs: a study in healthy human volunteers. Perspect Clin Res. 2014;5(2):80. https://doi.org/10.4103/2229-3485.128027.
26. Constantinides PP, Chakraborty S, Shukla D. Considerations and recommendations on traditional and non-traditional uses of excipients in oral drug products. AAPS Open. 2016;2(1):3. https://doi.org/10.1186/s41120-016-0004-3.
27. Constantinides PP, Han J, Davis SS. Advances in the use of tocols as drug delivery vehicles. Pharm Res. 2006;23(2):243–55. https://doi.org/10.1007/s11095-005-9262-9.
28. Constantinides PP, Wasan KM. Lipid formulation strategies for enhancing intestinal transport and absorption of P-glycoprotein (P-gp) substrate drugs: in vitro/in vivo case studies. J Pharm Sci. 2007;96(2):235–48. https://doi.org/10.1002/jps.20780.
29. Constantinides PP. Excipients as inhibitors of the absorption of dietary cholesterol: principles and case studies, symposium on excipients as atypical actives in nutraceuticals and pharmaceuticals: applications and development considerations. AAPS Annual Meeting, San Diego, CA; 2014.
30. Corvis Y, Négrier P, Espeau P. Physicochemical stability of solid dispersions of enantiomeric or racemic ibuprofen in stearic acid. J Pharm Sci. 2011;100(12):5235–43. https://doi.org/10.1002/jps.22727.
31. Crowley PJ. Excipients as stabilizers. Pharm Sci Technol Today. 1999;2(6):237–43. https://doi.org/10.1016/S1461-5347(99)00158-3.
32. Dave VS, Saoji SD, Raut NA, Haware RV. Excipient variability and its impact on dosage form functionality. J Pharm Sci. 2015;104(3):906–15. https://doi.org/10.1002/jps.24299.
33. Del Valle EMM. Cyclodextrins and their uses: a review. Process Biochem. 2004;39(9):1033–46. https://doi.org/10.1016/S0032-9592(03)00258-9.
34. Dey A, Mitra AK, Alam T. Natural and synthetic polymers in medicine. In: Handbook of polymers in medicine. Elsevier; 2023. p. 1–31. https://doi.org/10.1016/B978-0-12-823797-7.00001-0.
35. Diab R, Canilho N, Pavel IA, Haffner FB, Girardon M, Pasc A. Silica-based systems for oral delivery of drugs, macromolecules and cells. Adv Colloid Interf Sci. 2017;249:346–62. https://doi.org/10.1016/j.cis.2017.04.005.
36. Doogue MP, Polasek TM. The ABCD of clinical pharmacokinetics. Ther Adv Drug Saf. 2013;4(1):5–7. https://doi.org/10.1177/2042098612469335.
37. Duttagupta D, Jadhav V, Kadam V. Chitosan: a propitious biopolymer for drug delivery. Curr Drug Deliv. 2015;12(4):369–81. https://doi.org/10.2174/1567201812666150310151657.
38. Elder DP, Kuentz M, Holm R. Pharmaceutical excipients—quality, regulatory and biopharmaceutical considerations. Eur J Pharm Sci. 2016;87:88–99. https://doi.org/10.1016/j.ejps.2015.12.018.
39. Farag Y, Leopold CS. Development of shellac-coated sustained release pellet formulations. Eur J Pharm Sci. 2011;42(4):400–5. https://doi.org/10.1016/j.ejps.2011.01.006.
40. Fiume MM, Boyer I, Bergfeld WF, Belsito DV, Hill RA, Klaassen CD, Liebler DC, Marks JG, Shank RC, Slaga TJ, Snyder PW, Andersen FA. Safety assessment of talc as used in cosmetics. Int J Toxicol. 2015;34(1_suppl):66S–129S. https://doi.org/10.1177/1091581815586797.

41. Gazi AS, Begum A, Begum S. An overview on pharmaceutical excipients—their roles and applications. Int J Pharm Sci Med. 2022;7(12):84–96. https://doi.org/10.47760/ijpsm.2022.v07i12.005.
42. George M, Abraham TE. Polyionic hydrocolloids for the intestinal delivery of protein drugs: alginate and chitosan—a review. J Control Release. 2006;114(1):1–14. https://doi.org/10.1016/j.jconrel.2006.04.017.
43. Gershkovich P, Darlington J, Sivak O, Constantinides PP, Wasan KM. Inhibition of intestinal absorption of cholesterol by surface-modified nanostructured aluminosilicate compounds. J Pharm Sci. 2009;98(7):2390–400. https://doi.org/10.1002/jps.21616.
44. Glicksman M, editor. Food hydrocolloids. CRC Press; 2019. https://doi.org/10.1201/9780429290374.
45. Goel R, Bhardwaj S, Bana S. Pharmaceutical excipients. In: Dosage forms, formulation developments and regulations. Elsevier; 2024. p. 311–48. https://doi.org/10.1016/B978-0-323-91817-6.00003-6.
46. Gomes CSF, Santos DFG, Amaral MHR. Minerals in pharmacy and cosmetics. In: Minerals latu sensu and human health. Springer; 2021. p. 405–41. https://doi.org/10.1007/978-3-030-65706-2_9.
47. Grube S, Wolfrum U, Langguth P. Characterization of the epithelial permeation enhancing effect of basic butylated methacrylate copolymer—in vitro studies. Biomacromolecules. 2008;9(5):1398–405. https://doi.org/10.1021/bm701337q.
48. Haydel SE, Remenih CM, Williams LB. Broad-spectrum in vitro antibacterial activities of clay minerals against antibiotic-susceptible and antibiotic-resistant bacterial pathogens. J Antimicrob Chemother. 2007;61(2):353–61. https://doi.org/10.1093/jac/dkm468.
49. Heaney RP. Calcium supplements: practical considerations. Osteoporos Int. 1991;1(2):65–71. https://doi.org/10.1007/BF01880445.
50. Illanes A, Guerrero C, Vera C, Wilson L, Conejeros R, Scott F. Lactose-derived prebiotics. Elsevier; 2016. https://doi.org/10.1016/C2014-0-03393-X.
51. Irie T, Uekama K. Pharmaceutical applications of cyclodextrins. III. Toxicological issues and safety evaluation. J Pharm Sci. 1997;86(2):147–62. https://doi.org/10.1021/js960213f.
52. Jose S, Fangueiro JF, Smitha J, Cinu TA, Chacko AJ, Premaletha K, Souto EB. Cross-linked chitosan microspheres for oral delivery of insulin: Taguchi design and in vivo testing. Colloids Surf B: Biointerfaces. 2012;92:175–9. https://doi.org/10.1016/j.colsurfb.2011.11.040.
53. Kalasz H, Antal I. Drug excipients. Curr Med Chem. 2006;13(21):2535–63. https://doi.org/10.2174/092986706778201648.
54. Kalepu S, Manthina M, Padavala V. Oral lipid-based drug delivery systems—an overview. Acta Pharm Sin B. 2013;3(6):361–72. https://doi.org/10.1016/j.apsb.2013.10.001.
55. Kar M, Chourasiya Y, Maheshwari R, Tekade RK. Current developments in excipient science: implication of quantitative selection of each excipient in product development. In: Basic fundamentals of drug delivery. Elsevier Inc; 2018. https://doi.org/10.1016/B978-0-12-817909-3.00002-9.
56. Khurana IS, Kaur S, Kaur H, Khurana RK. Multifaceted role of clay minerals in pharmaceuticals. Future Sci OA. 2015;1(3):FSO6. https://doi.org/10.4155/fso.15.6.
57. Kim S-K, editor. Chitin and chitosan derivatives. CRC Press; 2013. https://doi.org/10.1201/b15636.
58. Kumari L, Choudhari Y, Patel P, Gupta GD, Singh D, Rosenholm JM, Bansal KK, Kurmi BD. Advancement in solubilization approaches: a step towards bioavailability enhancement of poorly soluble drugs. Life. 2023;13(5):1099. https://doi.org/10.3390/life13051099.
59. Lee KY, Mooney DJ. Alginate: properties and biomedical applications. Prog Polym Sci. 2012;37(1):106–26. https://doi.org/10.1016/j.progpolymsci.2011.06.003.
60. Li J, Wu Y. Lubricants in pharmaceutical solid dosage forms. Lubricants. 2014;2(1):21–43. https://doi.org/10.3390/lubricants2010021.
61. Li L, Ni R, Shao Y, Mao S. Carrageenan and its applications in drug delivery. Carbohydr Polym. 2014;103:1–11. https://doi.org/10.1016/j.carbpol.2013.12.008.

62. Li S, Han Q, Qiao C, Song J, Lung Cheng C, Xu H. Chemical markers for the quality control of herbal medicines: an overview. Chin Med. 2008;3(1):7. https://doi.org/10.1186/1749-8546-3-7.
63. Liu L, Chen G, Fishman ML, Hicks KB. Pectin gel vehicles for controlled fragrance delivery. Drug Deliv. 2005;12(3):149–57. https://doi.org/10.1080/10717540590929966.
64. Luo Z, Dai Y, Gao H. Development and application of hyaluronic acid in tumor targeting drug delivery. Acta Pharm Sin B. 2019;9(6):1099–112. https://doi.org/10.1016/j.apsb.2019.06.004.
65. Maleki A, Kettiger H, Schoubben A, Rosenholm JM, Ambrogi V, Hamidi M. Mesoporous silica materials: from physico-chemical properties to enhanced dissolution of poorly water-soluble drugs. J Control Release. 2017;262:329–47. https://doi.org/10.1016/j.jconrel.2017.07.047.
66. Martinez AW, Caves JM, Ravi S, Li W, Chaikof EL. Effects of crosslinking on the mechanical properties, drug release and cytocompatibility of protein polymers. Acta Biomater. 2014;10(1):26–33. https://doi.org/10.1016/j.actbio.2013.08.029.
67. Mathew AA, Mohapathra S, Panonnummal R. Pectin-based drug delivery systems for biomedical applications. In: Natural biopolymers in drug delivery and tissue engineering. Elsevier; 2023. p. 301–46. https://doi.org/10.1016/B978-0-323-98827-8.00018-7.
68. Mittal DS, Pawar S. Herbal excipients: an overview. Int J Pharmacogn Pharm Res. 2019;1(1):05–9. https://doi.org/10.33545/27067009.2019.v1.i1a.8.
69. Moondra S, Raval N, Kuche K, Maheshwari R, Tekade M, Tekade RK. Sterilization of pharmaceuticals. In: Dosage form design parameters. Elsevier; 2018. p. 467–519. https://doi.org/10.1016/B978-0-12-814421-3.00014-2.
70. Mughal MA, Iqbal Z, Neau SH. Guar gum, xanthan gum, and HPMC can define release mechanisms and sustain release of propranolol hydrochloride. AAPS PharmSciTech. 2011;12(1):77–87. https://doi.org/10.1208/s12249-010-9570-1.
71. Munarin F, Tanzi MC, Petrini P. Advances in biomedical applications of pectin gels. Int J Biol Macromol. 2012;51(4):681–9. https://doi.org/10.1016/j.ijbiomac.2012.07.002.
72. Munday DL, Cox PJ. Compressed xanthan and karaya gum matrices: hydration, erosion and drug release mechanisms. Int J Pharm. 2000;203(1–2):179–92. https://doi.org/10.1016/S0378-5173(00)00444-0.
73. Muruganantham S, Krishnaswami V, Anitha Manikandan D, Aravindaraj N, Suresh J, Murugesan M, Kandasamy R. Gums as pharmaceutical excipients: an overview; 2022. p. 145–89. https://doi.org/10.1007/978-3-030-91378-6_7.
74. Nezamdoost-Sani N, Khaledabad MA, Amiri S, Mousavi Khaneghah A. Alginate and derivatives hydrogels in encapsulation of probiotic bacteria: an updated review. Food Biosci. 2023;52:102433. https://doi.org/10.1016/j.fbio.2023.102433.
75. Nong Y, Maloh J, Natarelli N, Gunt HB, Tristani E, Sivamani RK. A review of the use of beeswax in skincare. J Cosmet Dermatol. 2023;22(8):2166–73. https://doi.org/10.1111/jocd.15718.
76. Ojewole E, Mackraj I, Akhundov K, Hamman J, Viljoen A, Olivier E, Wesley-Smith J, Govender T. Investigating the effect of aloe vera gel on the buccal permeability of didanosine. Planta Med. 2012;78(04):354–61. https://doi.org/10.1055/s-0031-1280431.
77. Pal D, Raj K. Properties of plant polysaccharides used as pharmaceutical excipients. In: Plant polysaccharides as pharmaceutical excipients. Elsevier; 2023. p. 25–44. https://doi.org/10.1016/B978-0-323-90780-4.00003-6.
78. Pal RS, Pal Y, Wal A, Wal P. Current review on plant based pharmaceutical excipients. Open Med J. 2019;6(1):1–5. https://doi.org/10.2174/1874220301906010001.
79. Panuška P, Smejkal J, Štofik M, Žmudová Z, Španbauerová K, Havlica J, Harrandt V, Vinopal S, Aubrecht P, Malý J. Advanced microfluidic platform for tumor spheroid formation and cultivation fabricated from OSTE+ polymer. Biochip J. 2024;18(3):393–409. https://doi.org/10.1007/s13206-024-00167-x.
80. Peptu C, Humelnicu AC, Rotaru R, Fortuna ME, Patras X, Teodorescu M, Tamba BI, Harabagiu V. Chitosan-based drug delivery systems. In: Chitin and chitosan. Wiley; 2019. p. 259–89. https://doi.org/10.1002/9781119450467.ch11.

81. Peterson B, Weyers M, Steenekamp JH, Steyn JD, Gouws C, Hamman JH. Drug bioavailability enhancing agents of natural origin (bioenhancers) that modulate drug membrane permeation and pre-systemic metabolism. Pharmaceutics. 2019;11(1):33. https://doi.org/10.3390/pharmaceutics11010033.
82. Qi S, Gryczke A, Belton P, Craig DQM. Characterisation of solid dispersions of paracetamol and EUDRAGIT® E prepared by hot-melt extrusion using thermal, microthermal and spectroscopic analysis. Int J Pharm. 2008;354(1–2):158–67. https://doi.org/10.1016/j.ijpharm.2007.11.048.
83. Qiu C, Wang C, Gong C, McClements DJ, Jin Z, Wang J. Advances in research on preparation, characterization, interaction with proteins, digestion and delivery systems of starch-based nanoparticles. Int J Biol Macromol. 2020;152:117–25. https://doi.org/10.1016/j.ijbiomac.2020.02.156.
84. Rowe RC, Sheskey PJ, Quinn M. Handbook of pharmaceutical excipients. Pharmaceutical Press; 1994.
85. Sabalingam S, Jayasuriya WJABN. Pharmaceutical excipients of marine and animal origin: a review. Biol Chem Res. 2019;6(12):184–96. BCR2019090201
86. Saravanan D, Muthudoss P, Khullar P, Rose VA. Pregelatinized starch: variability in gelatinization and its influence on product performance. Int J Res Pharm Sci. 2021;12(1):141–9. https://doi.org/10.26452/ijrps.v12i1.3969.
87. Şenel S, Kremer M, Kaş S, Wertz P, Hıncal A, Squier C. Enhancing effect of chitosan on peptide drug delivery across buccal mucosa. Biomaterials. 2000;21(20):2067–71. https://doi.org/10.1016/S0142-9612(00)00134-4.
88. Sharma A, Anup N, Tekade RK. Achieving sterility in biomedical and pharmaceutical products (part-I): thermal, chemical, and filtration sterilization. In: The future of pharmaceutical product development and research. Elsevier; 2020. p. 695–788. https://doi.org/10.1016/B978-0-12-814455-8.00020-7.
89. Shirwaikar A, Shirwaikar A, Prabhu S, Kumar G. Herbal excipients in novel drug delivery systems. Indian J Pharm Sci. 2008;70(4):415. https://doi.org/10.4103/0250-474X.44587.
90. Shukla D, Chakraborty S, Singh S, Mishra B. Lipid-based oral multiparticulate formulations—advantages, technological advances and industrial applications. Expert Opin Drug Deliv. 2011;8(2):207–24. https://doi.org/10.1517/17425247.2011.547469.
91. Siahi MR, Barzegar-Jalali M, Monajjemzadeh F, Ghaffari F, Azarmi S. Design and evaluation of 1- and 3-layer matrices of verapamil hydrochloride for sustaining its release. AAPS PharmSciTech. 2005;6(4):E626–32. https://doi.org/10.1208/pt060477.
92. Siepmann J. Modeling of drug release from delivery systems based on hydroxypropyl methylcellulose (HPMC). Adv Drug Deliv Rev. 2001;48(2–3):139–57. https://doi.org/10.1016/S0169-409X(01)00112-0.
93. Silvério LAL, Coco JC, de Macedo LM, dos Santos ÉM, Sueiro AC, Ataide JA, Tavares GD, Paiva-Santos AC, Mazzola PG. Natural product-based excipients for topical green formulations. Sustain Chem Pharm. 2023;33:101111. https://doi.org/10.1016/j.scp.2023.101111.
94. Šimat V, Elabed N, Kulawik P, Ceylan Z, Jamroz E, Yazgan H, Čagalj M, Regenstein JM, Özogul F. Recent advances in marine-based nutraceuticals and their health benefits. Mar Drugs. 2020;18(12):627. https://doi.org/10.3390/md18120627.
95. Singh P, Mishra G, Dinda SC. Natural excipients in pharmaceutical formulations. In: Evidence based validation of traditional medicines. Springer Singapore; 2021. p. 829–69. https://doi.org/10.1007/978-981-15-8127-4_40.
96. Sivamaruthi BS, Nallasamy PK, Suganthy N, Kesika P, Chaiyasut C. Pharmaceutical and biomedical applications of starch-based drug delivery system: a review. J Drug Deliv Sci Technol. 2022;77:103890. https://doi.org/10.1016/j.jddst.2022.103890.
97. Six K, Verreck G, Peeters J, Brewster M, Van Mooter G, den. Increased physical stability and improved dissolution properties of itraconazole, a class II drug, by solid dispersions that combine fast- and slow-dissolving polymers. J Pharm Sci. 2004;93(1):124–31. https://doi.org/10.1002/jps.10522.

98. Steffan BN, Venkatesh N, Keller NP. Let's get physical: bacterial-fungal interactions and their consequences in agriculture and health. J Fungi. 2020;6(4):243. https://doi.org/10.3390/jof6040243.
99. Stielow M, Witczyńska A, Kubryń N, Fijałkowski Ł, Nowaczyk J, Nowaczyk A. The bioavailability of drugs—the current state of knowledge. Molecules. 2023;28(24):8038. https://doi.org/10.3390/molecules28248038.
100. Strickley RG. Solubilizing excipients in oral and injectable formulations. Pharm Res. 2004;21(2):201–30. https://doi.org/10.1023/B:PHAM.0000016235.32639.23.
101. Te Wierik GHP, Eissens AC, Bergsma J, Arends-Scholte AW, Bolhuis GK. A new generation starch product as excipient in pharmaceutical tablets. Int J Pharm. 1997;157(2):181–7. https://doi.org/10.1016/S0378-5173(97)00234-2.
102. Teschke R, Frenzel C, Glass X, Schulze J, Eickhoff A. Herbal hepatotoxicity: a critical review. Br J Clin Pharmacol. 2013;75(3):630–6. https://doi.org/10.1111/j.1365-2125.2012.04395.x.
103. Thakkar S, Sharma K, Khurana S, Bansal AK. Excipients and their functionality for enabling technologies in oral dosage forms. In: Pharmaceutical excipients. Wiley; 2016. p. 97–143. https://doi.org/10.1002/9781118992432.ch3.
104. Thoma K, Klimek R. Photostabilization of drugs in dosage forms without protection from packaging materials. Int J Pharm. 1991;67(2):169–75. https://doi.org/10.1016/0378-5173(91)90429-R.
105. Thoorens G, Krier F, Leclercq B, Carlin B, Evrard B. Microcrystalline cellulose, a direct compression binder in a quality by design environment—a review. Int J Pharm. 2014;473(1–2):64–72. https://doi.org/10.1016/j.ijpharm.2014.06.055.
106. Tomuta I, Porfire A, Iurian S, Casian T. Optimization techniques in pharmaceutical formulation and processing. In: Dosage forms, formulation developments and regulations. Elsevier; 2024. p. 257–84. https://doi.org/10.1016/B978-0-323-91817-6.00014-0.
107. Tønnesen HH, Karlsen J. Alginate in drug delivery systems. Drug Dev Ind Pharm. 2002;28(6):621–30. https://doi.org/10.1081/DDC-120003853.
108. van der Merwe J, Steenekamp J, Steyn D, Hamman J. The role of functional excipients in solid oral dosage forms to overcome poor drug dissolution and bioavailability. Pharmaceutics. 2020;12(5):393. https://doi.org/10.3390/pharmaceutics12050393.
109. Veronica N, Heng PWS, Liew CV. Ensuring product stability—choosing the right excipients. J Pharm Sci. 2022;111(8):2158–71. https://doi.org/10.1016/j.xphs.2022.05.001.
110. Wang H-J, Hsiong C-H, Ho S-T, Lin M-J, Shih T-Y, Huang P-W, Hu OY-P. Commonly used excipients modulate UDP-glucuronosyltransferase 2B7 activity to improve nalbuphine oral bioavailability in humans. Pharm Res. 2014;31(7):1676–88. https://doi.org/10.1007/s11095-013-1272-4.
111. Wang J, Wen H, Desai D. Lubrication in tablet formulations. Eur J Pharm Biopharm. 2010;75(1):1–15. https://doi.org/10.1016/j.ejpb.2010.01.007.
112. Wani MY, Hasan N, Malik MA. Chitosan and aloe vera: two gifts of nature. J Dispers Sci Technol. 2010;31(6):799–811. https://doi.org/10.1080/01932690903333606.
113. Wawoczny A, Gillner D. The most potent natural pharmaceuticals, cosmetics, and food ingredients isolated from plants with deep eutectic solvents. J Agric Food Chem. 2023;71(29):10877–900. https://doi.org/10.1021/acs.jafc.3c01656.
114. Williams AC, Barry BW. Penetration enhancers. Adv Drug Deliv Rev. 2004;56(5):603–18. https://doi.org/10.1016/j.addr.2003.10.025.
115. Yang Q-Q, Sui Z, Lu W, Corke H. Soybean lecithin-stabilized oil-in-water (O/W) emulsions increase the stability and in vitro bioaccessibility of bioactive nutrients. Food Chem. 2021;338:128071. https://doi.org/10.1016/j.foodchem.2020.128071.
116. Yeo Y, Kim B-K. Drug carriers: not an innocent delivery man. AAPS J. 2015;17(5):1096–104. https://doi.org/10.1208/s12248-015-9789-6.
117. Young S, Wong M, Tabata Y, Mikos AG. Gelatin as a delivery vehicle for the controlled release of bioactive molecules. J Control Release. 2005;109(1–3):256–74. https://doi.org/10.1016/j.jconrel.2005.09.023.

Regulatory Frameworks Governing Natural Excipients: Ensuring Safety and Compliance in Pharmaceutical Applications

14

Hardik Rana, Shachi Patel, Dhrusha Patel, Isha Tanna, and Kuldeep Vinchurkar

14.1 Introduction

The term "the substances employed as a medium for administering a medication" refers to excipients, which are essentially inert supports for the active component or ingredients. Substances known as excipients are widely used in the pharmaceutical sector. Excipients, which can sometimes make up almost the whole formulation, are the substances or compounds that affect the final product's quality in addition to the active pharmaceutical component and packaging materials. Excipients can be used in different ways for creating medicinal dosage formulation [1]. Currently, many dosage forms are complex systems that include a number of extra ingredients (excipients) in addition to the active pharmaceutical ingredient (API); these substances are usually added in conjunction with the active pharmaceutical ingredients to: Protect, aid, or enhance the stability of the formulation: The unadulterated active pharmaceutical ingredient (API) is often observed to have short-term stability, which causes it to denaturate or adhere to the container wall, rendering it unusable [2]. In order to improve the stability of API, excipients are added. This helps to maintain the stability of the product and guarantee that. The API stays stable for the longer period of time, which in turn extends the dosage formulations shelf life [3].

H. Rana (✉) · S. Patel · D. Patel · I. Tanna
Department of Pharmaceutics, Anand Pharmacy College, Anand, Gujarat, India

K. Vinchurkar
Department of Pharmaceutics, Sandip Institute of Pharmaceutical Science (SIPS), Affiliated to Savitribai Phule Pune University (SPPU, Pune), Nashik, Maharashtra, India

J. S et al. (eds.), *Innovative Pharmaceutical Excipients: Natural Sources*, https://doi.org/10.1007/978-981-96-7959-1_14

Using excipients, it helps to:

- Assist in the creation of an exact dosage form, increase the formulation quantity for powerful medications.
- Boost patient acceptance.
- Assist in improving the active drug's bioavailability.

14.1.1 Overview of Natural Excipients in Pharmaceuticals

Excipients generally enhance the bioavailability of the active pharmaceutical ingredient; for instance, in numerous situations, an active compound (like aspirin) is not readily absorbed by the human body; in these instances, the active component is either dissolved in or combined with an excipient that can function as a solvent or help with the drug's absorption in the human body. Excipients can enhance the formulation's general safety and effectiveness throughout application and storage [4]. These components are frequently referred to as excipients, which are defined by the International Pharmaceutical Excipient Council as "Any material besides the active drug or pro-drug that is part of the manufacturing process or is present in final pharmaceutical dosage forms." According to their functions in formulations—binders, disintegrants, etc.—excipients are categorized by the US Pharmacopoeia-National Formulary (USPNF). It is possible to classify excipients based on their source. Excipients can be categorized according to their place of origin [5].

Based on their source of origin excipients are classified as (Fig. 14.1):

- Mineral source: Talc, calamine, calcium phosphate, paraffin, asbestos, kaolin, etc.
- Vegetable source: Peppermint, starch, arginates, guar gum, turmeric, acacia, etc.
- Animal source: Gelatine, stearic acid, beeswax, honey, musk, lactose, etc.
- Synthetic: Saccharin, boric acid, lactic acid, polysorbates, polyethylene glycols, povidone, etc.

Natural excipients are materials used in a range of formulations that come from a number of natural sources, such as plants, animals, and minerals. Because of their decreased toxicity, fewer adverse effects, and affordability, natural excipients are frequently included in formulations [6]. A wide variety of plant-derived pharmaceutical excipients or natural excipients can be found today. Capable of generate a diverse array of substances according to their characteristics and molecular weight, natural polymers emerged as a focal point in most drug-related research. Muscle fibers, polysaccharides, enzymes, and gummy exudates are the natural polymers being utilized efficiently in creating various pharmaceutical formulations. For example, chitosan, carrageenan, acacia, agar, Ispaghula, and karaya are natural polymers commonly utilized in pharmaceuticals products development. Modern consumers seek out natural components in food, medications, and beauty products, as they think anything derived from nature will be safer and free from adverse effects [7]. Types of excipients are given in Table 14.1.

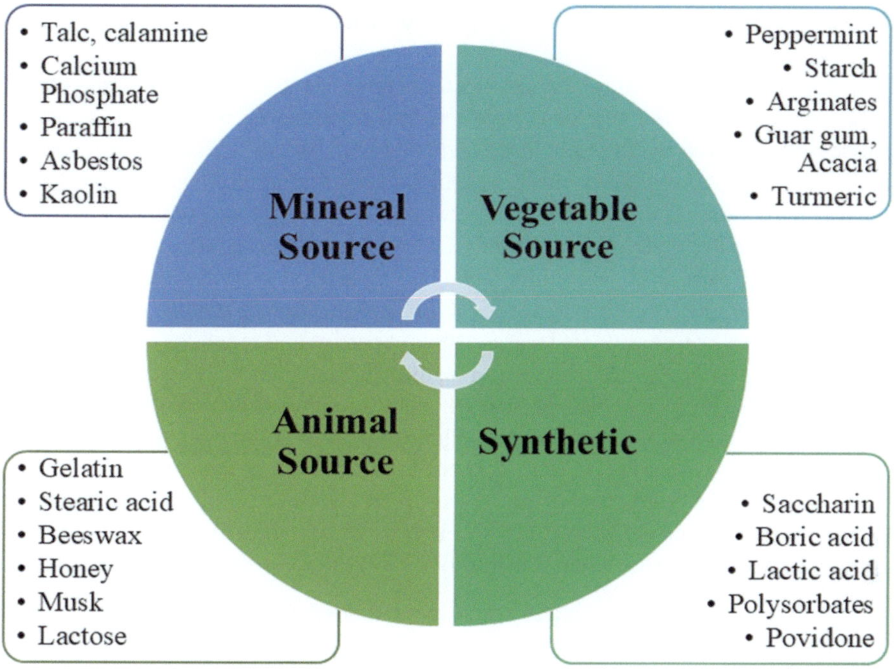

Fig. 14.1 Source of origin of excipients

Table 14.1 Types of excipients

Sr. no.	Type of excipients	Herbal excipients
1	Glidants	Talc, vitamin D
2	Fillers	Alginic acid, acacia, corn starch, alginate, polymers
3	Coagulating agents	Arabi, natural polymer, gelatine
4	Disintegrants	Agar, gellan gum, silicone
5	Lubricants	Castor oil, mineral oil, paraffin oil
6	Binders	Gelatin, lactose, plant cellulose, glucose
7	Preservatives	Clove oil, cumin seeds, neem oil, cayenne pepper
8	Antioxidants	Clove oil, cinnamon, turmeric, cocca
9	Sweetening agents	Glucose, lactose, honey
10	Coloring agents	Saffron, caramel, chlorophylls, red beetroot, turmeric
11	Flavoring agents	Raspberry, lemon, peppermint, orange
12	Chelating agents	Garlics, chlorella, onions, Brazil nuts
13	Solvents	Oils, purified water
14	Emulsify agents	Gum ghatti, acacia gum
15	Viscosity impairment agents	Gums, tragacanth, gelatin, aloe mucilage
16	Surface active agents	Tea saponins, ski waxes
17	Buffering agents	Acacia gum, gum ghatti

14.1.2 The Significance of Regulatory Frameworks

The field of regulatory affairs (RA) within the pharmaceutical industry is pivotal in obtaining approvals for new drug products and maintaining this authorization for the duration desired by the company. Regulatory affairs professionals are responsible for ensuring compliance with all applicable regulation and legal requirements. Their role includes acting as a liaison between the company and regulatory bodies. They ensure compliance with Current Good Manufacturing Practices (CGMP), International Council for Harmonization (ICH) guidelines Good Clinical Practices (GCP), and Good Laboratory Practices (GLP) by developing well-organized documentation. These professionals leverage their expertise and regulatory knowledge to translate regulatory requirements into feasible and practical strategies. Regulatory affairs are integral to the pharmaceutical sector, encompassing all phases of drug development, approval, and marketing. Pharmaceutical companies rely on the data collected during research and development to register and promote their products. To ensure the safety and efficacy of drug for patients, these companies must adhere to rigorous regulations and standards throughout the manufacturing process.

Regulatory affairs specialists are employed by numerous departments, including:

1. Preclinical studies: Preclinical research evaluated possible medication candidates using pharmacology and toxicology testing.
2. Research studies: Clinical research includes the collection of evidence, the interpretation of mathematical data, and the drafting of papers.
3. Production: There are numerous measures in place to guarantee that the products are effective and clean.
4. The process of examining every material for potency, safety, purity, and quality is known as quality control.
5. Assurance of quality entails managing tasks including record processing, auditing, fault audits, and complaints.

Significance of Regulatory Affairs

In the recent era, a company's and thus a products success relies on reducing the time it takes to connect with a consumer. Consequently, the firms RA takes need to be managed efficiently. Insufficient data reporting can hinder a prompt and effective assessment of a marketing application. Developing a new medication could potentially requires millions of Euros or pounds, and a mere 3 months postponement in its market launched with incorrect labeling. Any incident could lead to a loss of millions in revenue of dollars along with a decline in trust from investors, healthcare providers, and patients.

14.1.3 Challenges in Natural Excipient Usage

The use of natural excipients in drug products creates multiple challenges that need solutions to guarantee both patient safety and adherence to legal standards. Among the main issues raised in the literature are:

1. Data on Toxicity and Safety
   Pediatric drug development faces a major problem because natural excipients do not have complete safety data. Adults have acceptable dosage ranges but scientists often lack the needed evidence for children especially newborns. The missing data about safety poses a risk during drug development for young patients.
2. Compliance with Regulations
   Companies using natural excipients encounter official requirements that are hard to follow. The prolonged market entry for newly developed excipients mainly results from official approval regulations. When new excipients do not have their own approval process, the whole pharmaceutical product remains unable to get marketed approval. Different rules in each region make it harder to do multinational projects.
3. Manufacturing Difficulties
   Nature-based excipients could experience damage when environmental factors vary. Physical conditions where natural excipients are made along with climate and production standards affect their composition and performance outcomes. Natural excipient production moves more slowly than normal and has increased risk of microbial contamination during creation.
4. Adoption by the Industry
   The pharmaceutical industry accepts new natural excipients into use at a steady pace. Both producers and government agencies need many years to approve new excipients. New excipient development takes too long because of the extended work needed to create unique ingredient mixes.
5. Economic Factors
   Customers typically view natural excipients as more expensive due to their unique manufacturing needs. Their high production costs restrict their use as medicine in industry.

14.2 Global Regulatory Perspective

To match laws between countries regulatory agencies establish the global regulatory viewpoint as their operating framework. As trade boundaries fade, countries now need this perspective more than ever for running economic activities and health research across nations. Multiple national regulatory bodies, international groups, and private firms support proper governance through their joint work to build common guidance and standards for worldwide regions.

Regulatory Action - Important Elements in Worldwide Standards

- Alignment of Regulations: When nations work together to make their safety rules similar, they improve international trade connections and help trade flow smoothly. The International Council on Harmonization leads standard development efforts that unify scientific rules for pharmaceutical products globally to boost their safety and efficiency while enabling smoother industry access.

Table 14.2 Table showing what organizations control commerce in every market

Regulatory agencies	Countries
Ministry of Health and Family Welfare (MoHFW)	India
Food and Drug Administration (FDA)	United States
National Health Surveillance Agency (ANVISA)	Brazil
Therapeutics Goods Administration (TGA)	Australia
Health Products and Food Branch (HPFB)—Health Canada	Canada
National Medical Products Administration (NMPA)	China
European Medicines Agency (EMA)	European Union
National Agency for the Safety of Medicines and Health Products (ANSM)	France
Federal Institute for Drugs and Medical Devices	Germany
Pharmaceuticals and Medical Devices Agency (PMDA)	Japan
National Agency for Food and Drug Administration and Control (NAFDAC)	Nigeria
Health Sciences Authority (HSA)	Singapore
Medicines & Healthcare products Regulatory Agency (MHRA)	United Kingdom
South African Health Products Regulatory Authority (SAHPRA)	South Africa

Under the global regulatory system, enforcement relies on decentralized methods, unlike traditional international law which requires central authority control. Under this new approach, national organizations become important in applying worldwide standards across their regional areas.

- Cooperation Beyond Boundaries: Transnational companies and government agencies work together to create solutions for international financial rules environmental protection and disease control. When governmental institutions cooperate, they develop systems for exchanging industry updates and set international quality benchmarks.
- Impact of Globalization: Economic globalization has reshaped sovereignty practices because nations match their rules to international standards. The need to oversee cross-border activity has driven regulatory changes between nations according to this update.
- Stress Conformity Rather Than Punishment: Current legal systems prefer harsh punishment when companies break laws instead of basing actions on accepted standards. The partnership between industries and regulators builds cooperation that makes everyone benefit from following rules [8]. Regulatory agencies of different countries given are in Table 14.2.

14.2.1 United States: FDA Guidelines for Natural Excipients

FDA requires pharmaceutical excipients to meet rules that limit their use of natural ingredients such as plants minerals and animals. Without excipients drugs would not work effectively and would lack stability and safety traits. The FDA closely controls how natural excipients are used safely in medicines to protect patients.

- *FDA Regulatory Framework for Natural Excipients*
  FDA requires pharmaceutical excipients to meet rules that limit their use of natural ingredients such as plants minerals and animals. Without excipients, drugs would not work effectively and would lack stability and safety traits. The FDA closely controls how natural excipients are used safely in medicines to protect patients.
- *FDA Guidelines for Excipients in Drug Products*
  FDA requires pharmaceutical excipients to meet rules that limit their use of natural ingredients such as plants minerals and animals. Without excipients, drugs would not work effectively and would lack stability and safety traits. The FDA closely controls how natural excipients are used safely in medicines to protect patients.
  United States Pharmacopeia (USP) Monographs: Both the FDA and USP develop monographs to show what excipients need to meet regarding their identity content quantity and cleanliness standards. These monograph documents help make sure natural excipients remain safe and consistent for use in drug formulation. These publications ensure that pharmaceutical items contain the correct amounts of cellulose starch and magnesium stearate as inactive ingredients.
  Safety and Compatibility: FDA verifies that excipients in drugs do not decrease their benefits according to regulations. Natural ingredients need testing for side effects or drug reactions before combining them with pharmaceutical substances.
- *FDA's Role in GRAS and Non-GRAS Excipients*
  The FDA categorizes excipients into two groups: The FDA divides excipients into two groups called generally recognized as safe (GRAS) and non-GRAS subject to different safety regulations. GRAS excipients: Standardized scientific research demonstrates the security of these ingredients so they do not need FDA review before market introduction. They have to follow all required safety standards. The FDA creates and keeps a registry with GRAS substances which lists natural excipients confirmed as safe when used in their stated applications. Non-GRAS excipients: Only FDA-approved excipients without GRAS status can be used. The company must submit proof of excipient toxicology data before using it in their drug products. Plant-based natural excipients need to pass detailed tests for allergy and pollutant content assurance.
- *After Market Monitoring of Excipients*
  When pharmaceutical medicines reach the market, FDA watches excipient safety. Studying drug behavior in the public allows scientists to find side effects that happen when taking natural ingredients. People who take medicines containing excipients and their doctors can use MedWatch to report problems to the FDA. As per the FDA 2020, It uses MedWatch to gather information about excipient problems to take action quickly
- *Cooperation with Global Regulatory Authorities*
  The Federal Food, Drug, and Cosmetic Act (FDCA) ranks as the most significant regulatory instrument because it determines how drug components and excipients must obtain their licenses. Before making new drugs, the FDA requires complete safety testing and stability confirmation for all drug components including natural source excipients.

14.2.2 European Union Agency (EMA): Regulations and Monographs

Under European Union (EU) law and EMA oversight, the European Medicines Agency directs regulatory practices for natural excipients throughout the EU. The system exists to secure proper functioning of pharmaceutical products that use natural excipients.

- *Legal Framework and Guidelines According to EU*
- Under Directive 2001/83/EC pharmaceutical products need marketing authorization and it provides the regulatory base for excipients. Under Article 65 all product labels must show excipients even when they work in specific ways. Our directive exists to protect patients' safety through comprehensive pharmaceutical ingredient disclosure. The European Medicines Agency sets official requirements for excipient labeling in product materials. These guidelines' salient features include:

  All excipients are required to be enumerated in the package leaflet.

  Particular labeling stipulations are mandated for parenteral, ocular, and topical preparations.

  Excipients that possess a recognized pharmacological effect must also be incorporated into the labeling.

  Adhering to these requirements is crucial for ensuring transparency and safety in pharmaceutical products.
- *Monographs for Natural Excipients*

  Natural excipients undergo evaluation using guidelines written by the European Union and pharmacopeia's according to EU requirements. The European Medicines Agency encourages companies to follow these standards so all products containing excipients provide necessary safety and quality outcomes.

  Both the European Pharmacopoeia (EP) and national pharmacopoeias of EU countries demand that excipients pass defined quality control tests. Applicants can choose excipient standards from Japan's or America's Pharmacopeia when European Pharmacopeia's lack their preferred option provided, they justify their selection.
- *Safety and Quality Guidelines*

  EMA scientific guidelines cover tests for physical attributes plus methods to verify and check excipient purity against accepted impurity limits. Customers can trust their safety when using pharmaceutical goods that meet these natural excipient recommendations. Through EMA's scientific standards, companies can test excipients completely using established methods for measuring physical traits plus checking substance identities while detecting and setting allowed impurity levels. Medicines must follow these rules because they help make sure that natural ingredients are safe to eat. The scientific guidelines issued by the EMA specify a thorough set of requirements for the evaluation of excipient quality, which includes: They explain the needed tests for each excipient such as physical tests along with ways to confirm exact material identity and purity standards with defined impurity limits. These standards protect the safety of natural excipients for people who consume pharmaceutical products. The EMA has established scientific guidelines that articulate comprehensive standards for the assessment of excipient quality, which involve: The international guide lists exact standards on how to test physical quality and find out what a product contains while limiting harmful impurities. These rules help evaluate if natural ingredients in medicine products are safe to consume. According to EMA scientific guidelines, customers need to test excipient quality using defined standards that examine both physical properties plus chemical characteristics including limited levels of impurities. The recommended standards help protect product users from harm when natural compounds are used in medication formulas [9–11].

- *Role of EU Monographs*
  The European Union produces detailed monographs on herbal substances which summarize their purposes and safety data along with interactions with pharmaceutical products. Scientific consensus on safety depends on these documents which give references to use applications applying for commercial rights. These monographs are essential regarding safety and efficacy. Monographs are also important for selling a product as it is considered as reference standard. Supporting the harmonization of herbal medical goods throughout Member States is also supported by EU monographs. The European Union herbal monographs offer complete drug information about herbal substances used as excipients and detail their permitted medical use along with safety data and medication interactions. Our scientific studies depend on using these monographs for their research development. This system works as a reference guide when makers of herbal drugs prepare their marketing applications. Our organization helps all Member States apply standard rules when using herbal medicines. You can find all important details about herbal ingredients as excipients through EU herbal monographs which provide their medication listing and safety information along with drug interaction updates. Organizations need these reference materials as scientific proof to evaluate medicine safety and performance before giving marketing authorization. Actions, can be found in the European Union's herbal monographs. In order to provide a scientific foundation for safety and effectiveness, these monographs are essential. Serving as references for applications for marketing authorization. Promoting quality standards uniformity regarding to herbal medical goods among Member States is also the significant role of EU monographs [12].

14.2.3 Natural Excipients According to Indian Pharmacopeial Norms

As the primary guide for both pharmaceutical standards and excipients including natural materials, the Indian Pharmacopoeia sets essential rules for excipients management in India. The system ensures that natural excipients used in medicine manufacturing should meet essential quality standards of safety and effectiveness.

- *Indian Pharmacopoeia Overview*
  In 1940, the Drugs and Cosmetics Act made the Indian Pharmacopoeia an official document that defines rigorous quality standards for medical products and their added components. This document presents complete monographs on various substances including their definitions, features, identification, and quality criteria. Meeting pharmaceutical quality standards and regulatory rules in India depends fully on the rules defined in the Indian Pharmacopoeia.
- *Natural Excipient Roles*
  Pharmaceutical companies prefer natural excipients because they both help medication work effectively and are compatible with human biology. All formulations using natural excipients need to reach high safety and effectiveness standards before market release. To maintain safety and quality standards, natural excipients need to follow guidelines from the Indian Pharmacopeia. Natural excipients' superior biocompatibility and possible health advantages are the main reasons for their growing use in pharmaceutical formulations. These excipients must meet strict quality requirements in order to avoid compromising the final medication product's efficacy or safety. The Indian Pharmacopoeia serves as a guide for manufacturers, offering extensive standards to guarantee that natural excipients meet accepted safety and quality standards.

Adherence to Legal Framework
Main food safety rules come from the Food Safety and Standards Authority of India (FSSAI) and businesses must follow them. Every ingredient in food needs to match standards accepted by pharmacopeia's recognized by FSSAI. Food products containing natural excipients need to pass safety tests before they reach customer stores. FSSAI rules direct all food companies to operate by their standards. Under FSSAI rules all food product ingredients must follow guidelines published by recognized pharmacopeia's including the Indian Pharmacopoeia. Food products that contain natural excipients must meet safety requirements before becoming available to consumers. You must adhere to food safety guidelines particularly what FSSAI set. FSSAI states companies must only use substances from trusted pharmacopoeias for their food items including the Indian Pharmacopoeia. Natural excipients in food products need to meet safe product standards before they can be sold to consumers [13].

14.2.4 Regulatory Approaches in Other Regions (China and Japan)

To guarantee pharmaceutical safety and effectiveness, excipients in drugs need close observation. This chapter explores the particular rules and challenges that control natural excipients under China and Japan's comprehensive regulatory schemes.

- *Regulatory Approaches in China*
  The pharmaceutical regulations grew tougher when China became successful in international markets. The CFDA formerly NMPA exercises authority over drugs and manages all excipient regulations. Current regulatory updates are meant to

enhance oversight performance while matching global standards, yet producers experience significant difficulties because of these reforms.

Registration Is Required: China requires pharmaceutical supplies to register all their used excipients even if these materials came from imports. Businesses new to compliance issues in China face unique worldwide challenges when starting this process.

Pharmaceutical Administration Law: Under Article 11 of this law until all provinces work together to create a common rule, the medicinal use requirements drive separate regulations across different provinces.

New Changes: The 2019 update to the medication Administration Law better shields patients but creates fresh obstacles for companies making medication components. Despite NMPA's continuous work to align China's regulations with international standards challenges remain because China remains partly non-aligned with European Pharmacopoeia reference materials.

- *Regulatory Framework in Japan*

  Japan's PMDA agency watches over pharmaceutical excipients by using a detailed system centered on safety standards and quality assurance. The PMDA service solution works with domestic and international regulatory standards.

Japanese Regulation Features

- *Fundamental Elements of Japanese Regulatory Framework*

  Global Supply Chain Oversight: Japan's quality control standards protect its patients from contamination risks of drugs imported from worldwide pharmaceutical suppliers.

  Regulatory Guidance: Manufacturers need to meet both security evaluation criteria and product quality standards set by the Pharmaceuticals and Medical Devices Agency (PMDA) for detailed guidance on using excipients.

  International Harmonization Participation: To match global safety standards and better serve public health needs, Japan works actively within International Council for Harmonization (ICH) harmonization projects.

- *Essential Features of Japanese Regulatory Practices*

  Global Supply Chain Management: Under current quality standards, Japan places strict controls to protect the pharmaceutical supply chain from contamination issues.

  Documentation: The PMDA provides complete guidance about excipients that lists them which manufacturers need to follow when proving product quality and risk assessment.

  Harmonization Efforts:

  Japan joins international harmonization programs to create a global-standard and domestic-friendly regulatory system [10].

  Core Components of Japanese Regulatory Standards:

  Monitoring of Worldwide Supply Chain Management: Japan puts strong quality controls in place to prevent pharmaceutical problems after companies from different countries start working together.

Guidance Documentation: Monitoring of Worldwide Supply Chain Management:
Japan puts strong quality controls in place to prevent pharmaceutical problems after companies from different countries start working together.
Dedication to Harmonization:
Japan joins international standards setting efforts led by ICH to maintain both international and local standards for protecting public health [8].
Opportunities and Challenges:
Each country faces special rules when controlling ingredients in their pharmaceutical products. China demands extensive new market entrants undergo a challenging registration process because it includes many parts with no relief offered to experienced manufacturers. The regulatory systems present chances for companies who know how to effectively follow these requirements. Japanese manufacturers face strong regulation but must steadily update their activities to match international standards. The authorities that oversee excipient production in both China and Japan create specific difficulties for their respective industries. Businesses that enter the Chinese market must face a confusing registration process that seems too difficult for new entrants due to complex rules and no special rules for existing manufacturers. When companies understand and manage the challenging rules, they discover new chances to succeed. Japan updates its extensive rules to match international standards but this ongoing adjustment adds burden to producers working within Japan's market. Each country experiences unique difficulties when setting regulations for excipients. The official registration process in China tests new manufacturers severely through its challenging setup and lack of support options for established firms. Businesses that are specialist in understanding regulatory requirements find these standards beneficial for their growth. Japanese manufacturers must adapt their systems each time the government updates restrictions to meet international requirements [14].

14.3 Safety Assessment of Natural Excipients

Establishing safe and acceptable limits for these compounds is crucial, as is conducting risk-benefit analyses on suggested novel excipients in pharmaceutical products. It is necessary to assess the safety of natural excipients before using them in pharmaceutical formulations. Natural excipients are commonly believed to be safer since they originate from natural sources, such as plants, animals, or minerals, although this perception might be misleading. Elements like biological variability, potential contaminants, and impurities need to be carefully taken into account. Additionally important are toxicological studies, particularly when they involve specific populations with allergies or intolerances. Regulations and Good Manufacturing Practices (GMP) are crucial for lowering risks. A comprehensive analysis ensures that these excipients are both safe and effective for a variety of patient populations [15].

14.3.1 Toxicological Evaluation

Excipients are vital in pharmaceutical formulations, but their potential toxicity is a significant concern. Factors such as the route of administration, duration of exposure (acute or chronic), and other conditions can influence their toxic effects, which may include genotoxicity, carcinogenicity, reproductive toxicity, and irritation [16]. Certain excipients can cause skin toxicities when applied topically, such as hypersensitivity from lanolin and local anesthetics or phototoxicity from oils like bergamot and cinnamon, which are frequently used in fragrances. Recommendations for evaluating the safety of excipients were published in 1996 by the Safety Committee of the International Pharmaceutical Excipients Council (IPEC-Americas). These recommendations specified the toxicology tests necessary to detect any toxicities according to the planned route of administration. Excipient safety recommendations were also developed by the FDA/CDER's Inactive Ingredients Subcommittee of the Pharmacology/Toxicology Coordinating Committee. These guidelines were made public in 2002 and completed in 2005. This guidance, "Nonclinical Studies for the Safety Evaluation of Pharmaceutical Excipients," was created in conjunction with the safety testing guidelines of IPEC and the International Conference on Harmonization (ICH) M3 guidance.

Regulatory Toxicology Extensive safety testing may not be necessary for an excipient that is used in pharmaceutical products that have already received approval or that are generally regarded as safe (GRAS). According to FDA guidelines, the following studies should be used to gather toxicological data:

- Safety pharmacology with an emphasis on the central nervous, respiratory, and cardiovascular systems.
- ADME (absorption, distribution, metabolism, and excretion) and pharmacokinetic investigations.
- Battery for ICH S2B genetic toxicology.
- Researches in reproductive toxicology.
- Research using two species (rodents and non-rodent animals) with repeated doses for the proper length of time.
- Research on carcinogenicity, if required.

A "weight-of-evidence" (WOE) methodology is also taken into account in the FDA excipient guidance. The excipient may be exempt from carcinogenicity testing if it satisfies specific requirements, including no structural alerts, no pharmacological activity in safety pharmacology or other assays, no genotoxicity, no reproductive toxicity, no severe toxicities in 90- or 180-day repeat dose tests, a large safety margin, and potentially negative results from transgenic assays. It is crucial to remember that even if the Center for Drug Evaluation and Research review division grants a waiver, the excipient may still result in cancers in rats during long-term bioassays [17].

Evaluation Test of Gums Several plant-based mucilages and gums, such as carrageenan, thaumatin, lard, storax, agar, gum acacia, and tragacanth, meet numerous fundamental conditions for usage as pharmacological excipients [18].

Acute Toxicity Study In order to evaluate toxicity and behavioral effects, oral dosages of gum solutions ranging from 0 to 5000 mg/kg were administered to male and nonpregnant female mice. The animals were then watched for 6 h. The median lethal dosage (LD50) was calculated by monitoring mortality within 24 h and any delayed toxic effects over the next 14 days.

Sub-chronic Toxicity Study *Cussonia arborea* or *Cordia mellinii* gum solutions at subtherapeutic, therapeutic, and supratherapeutic dosages were administered to 64 male albino rats in groups, while water was provided to the control groups. Rats' weight fluctuations, food and water consumption, organ weights, and blood sample analysis were all tracked. After being harvested, the organs were inspected and stored for histological analysis.

Hematological Analysis Using an automated analyzer, around 2 mL of whole blood was extracted in Ethylene diamine tetraacetic acid tubes and examined for hematological parameters, including platelets, white blood cells, hemoglobin, hematocrit, and other indices.

Serum Biochemistry For biochemical examination, around 3 mL of blood was drawn, centrifuged to extract the serum, and then kept at −20 °C. An automated analyzer was used to assess a number of parameters, including albumin, total protein, urea, creatinine, bilirubin levels, liver enzymes like alkaline phosphatase (ALP), alanine transaminase (ALT), aspartate aminotransferase (AST), and fasting glucose [19].

14.3.2 Allergenicity and Hypersensitivity Reaction

Excipients in pharmaceutical formulations have varying risks of allergenicity and hypersensitivity depending on the excipient and patient characteristics. In sensitive people, excipients including parabens, saccharin, and menthol have been reported to induce hypersensitivity responses like urticaria, bronchospasm, or contact dermatitis. Patients with lactase deficiency may develop sensitivity to lactose, while those with phenylketonuria may be at danger from aspartame. These possible problems can be decreased by selecting different excipients and keeping an eye on patient-specific hazards. Sensitization is the production of allergic reactions, which can be long-lasting because of immunologic memory. Allergy is defined as any excessive immune response to a foreign antigen, independent of the mechanism of response, with the antigen being safe for most individuals [20].

Different Types of Hypersensitivity Reactions

Type I responses make up around 48% of all hypersensitivity reactions, followed by type IV reactions (18%), and types III and II reactions (10% and 6%, respectively).

Type I: Within minutes of being exposed to an allergen, type I hypersensitivity, commonly referred to as "allergy," is characterized by quick immunoglobulin E (IgE)-mediated responses. When allergens enter mucosal tissues, they activate antigen-presenting cells, which in turn cause Th2-cell polarization and the synthesis of IgE. This is the beginning of sensitization. In order to make mast cells and basophils more sensitive to the allergen in the future, these IgE antibodies attach to them. This causes degranulation and the release of inflammatory mediators. Food allergens are classified as either moderate cross-reactive allergens (Class II) or severe primary sensitizers (Class I), depending on the kind of allergen and the site of exposure.

Type IV: Type IV responses, also referred to as "delayed hypersensitivity," happen when sensitized people have their worst symptoms 48–72 h after being exposed to an allergen. These reactions can be divided into three or four subcategories according to the effector cells and mediators, the time of start, the clinical presentation, and the cells involved. Although Th2 and other T-cells may also be involved, antigen-specific Th1 cells and cytotoxic T-lymphocytes are the main mediators of delayed hypersensitivity responses. Smaller molecules can function as haptens, which only attach to proteins (prohaptens) to become allergens. Certain compounds are prehaptens, which means they require external activation to become haptens.

Linalool, a substance present in the essential oil of several plants, is an illustration of this; it undergoes autoxidation and turns into an allergy. Both activation routes are known for other compounds, such geraniol, which is likewise found in plant essential oils. These ingredients, which are frequently found in flavoring agents, might make the finished product more allergic. Photoallergy is a particular kind of type IV response in which allergic reactions are brought on by contact with photoallergens, primarily through the skin. Frequently, photoallergens are haptens that, when exposed to ultraviolet (UV) light, produce reactive species and attach covalently to proteins such as human serum albumin to become complete allergens.

Pseudoallergy/anaphylactoid intolerance: Irritators are defined by the US Occupational Safety and Health Administration as noncorrosive materials that, when they come into direct contact with the skin, eyes, nose, or respiratory system, temporarily produce irritation because of a chemical reaction at the point of contact. It is unclear if type IV hypersensitivity can develop as a result of contact allergens' irritating effects. A substantial dosage is necessary for phototoxicity to occur, and drugs can be delivered systemically or via direct contact. After absorbing photons (energy), these substances may produce stable phototoxic chemicals or free radicals, which can be harmful. Skin that has been exposed to the sun usually develops erythema [21].

14.3.3 GRAS (Generally Recognized as Safe) Status

GRAS (Generally Recognized as Safe) is an FDA designation indicating that a substance or ingredient added to food is considered safe by experts and is exempt from the usual food additive requirements outlined in the Federal Food, Drug, and Cosmetic Act (FFDCA). For a substance to be granted GRAS status, various factors such as testing, exposure, and manufacturing processes must be evaluated by a GRAS expert panel. Once approved by the panel, the substance can only be used in specific food products and concentrations for which approval was granted. Substances with GRAS status, officially recognized by the FDA, are listed in the regulations under 21C.F.R. 182, 184, and 186. The GRAS notice procedure was created by the FDA to allow manufacturers to designate substances as GRAS and inform the agency. Although the FDA examines the submission, the GRAS classification is not formally confirmed. Compared to official GRAS clearance, this method gives less confidence for pharmaceutical additions but helps food makers. The Federal Food, Drug, and Cosmetic Act requires premarket clearance for substances that do not fit GRAS standards, and the FDA retains enforcement jurisdiction over noncompliant products [22].

14.4 Compliance Requirements

14.4.1 Documentation and Dossier Preparation

To allow producers to designate chemicals as GRAS and notify the FDA, the FDA created the GRAS notification method. The GRAS designation is not fully verified until the FDA has reviewed the proposal. Compared to formal GRAS clearance, this approach offers pharmaceutical additions less confidence, but it is advantageous to food businesses. Pre-market approval is required under the Federal Food, Drug, and Cosmetic Act for chemicals that do not fit GRAS standards, and the FDA has enforcement jurisdiction over goods that do not comply. According to this guideline, applications for marketing authorizations or changes concerning an excipient in approved pharmaceutical goods must include information on excipients, such as antioxidants and antimicrobial preservatives [23].

The drug product's description and composition: Excipients should include the excipients' common names, amount, purpose, and a reference to an applicable standard. The brand name with commercial grade should be supplied when the common name is insufficient to convey functional features.

Pharmaceutical development: Excipients ought to be specified with their common name, amount, purpose, and a reference to an applicable standard. The brand name with commercial grade should be supplied when the common name is insufficient to convey functional features. EU Directives 78/25/EEC, 94/36/EC, and 95/45/EC require coloring agents used in pharmaceuticals to be pure and in accordance with food safety standards. Excipients in sterile goods must have bioburden and endotoxin limits stated, and residual solvent data must adhere to predetermined standards.

- Excipients listed in an EU Member State's Pharmacopoeia or the European Pharmacopoeia: The dossier for marketing authorization should include reference to the most recent version of the pharmacopoeia. When a set of related materials (polymers) is covered in a monograph, the precise excipient specification and the reasoning behind it should be presented.

- Excipients described in a third country pharmacopoeia: The monograph from a third-country pharmacopoeia (such as USP or Japanese Pharmacopoeia) may be accepted if an excipient is not listed in either the European Pharmacopoeia or a Member State pharmacopoeia. The applicant must offer specifications in accordance with the general monograph for Substances for Pharmaceutical Use of the European Pharmacopoeia and justify this reference.
- Excipients not described in any pharmacopoeia: The following kinds of testing ought to be used to determine a suitable excipient specification: Physical attributes, Tests for identification, Purity tests, which include individual and overall impurity limitations that need to be identified, for example, using a chromatographic relative retention time. Physical, chemical, biological, and, if necessary, immunological testing can all be used to determine purity.
- If required, assay or limit tests and the associated validation parameters.
- Additional pertinent testing, such as quantitative tests on factors that have been found to affect the dosage form's performance.

The rationale for the specifications: An excipient specification's justification is based on its particular usage and selection, citing pertinent criteria (CPMP/ICH/367/96). Justification is usually not needed for excipients listed in the European Pharmacopoeia or EU Member State pharmacopoeias unless certain acceptance requirements or functionality-related traits need to be proven.

Human or animal-derived excipients: Chapters 5.1.7 Viral Safety and 5.2.8. Reducing the Risk of Animal Spongiform Encephalopathy Agents Spreading via Human and Veterinary Medicinal Products in the General European Pharmacopoeia should be followed when documenting viral safety and TSE risk.

Novel excipients: For new excipients, the drug substance format should be followed, including complete manufacturing, characterization, and control information along with cross-references to supporting safety data.

(a) A thorough explanation of the excipient.

(b) The following should be taken into account for both new excipients and excipients that are offered as a combination of substances. Any bibliographic data pertaining to the chemistry, toxicology, and real-world application of the substance.

The community's rules for food additives: any standards derived from toxicological studies, with references to such data. The quality requirements specified in the directives are acceptable as long as the routine control tests are validated.

The Food Chemical Codex and other publications, as well as the international requirements (FAO/WHO/JECFA).

Information on the chemical used in cosmetic goods for cutaneously applied medications (according to Directive 76/768/EEC, as modified).

Information on the new excipient's toxicity based on the medication's dose form and, if relevant, its administration method in Module 4; safety.

CPMP Guideline on New Active Substance Chemistry. A broad description of the production and purifying processes;

- The excipient's place of origin, including the manufacturer's name and address.
- Tests for purity, identity, and physical and chemical characteristics.
- Analytical techniques were validated by the display of batch findings.
- Other information, such as microbiological testing.
- Pollution, foreign material presence, and leftover solvents.
- The quality of each component and the physico-chemical test results for the combination should be mentioned when an excipient is made from a combination of multiple components.
- For the active components indicated in the Notice for Guidance on Stability Testing of New Drug Substances and Products (CPMP/ICH/2736/99), stability data should be supplied when needed.

Regulation of drug products: Excipient identity testing and assay are often not necessary upon release, with the exception of antioxidants and antimicrobial preservatives, which need to adhere to strict regulations. Tests for these substances' identity and content determination should be included of the release requirements, together with explanations for the suggested limitations.

Consistency: A medication's stability is dependent on the characteristics of its excipients, and stability standards should make sure that antioxidants and antimicrobial preservatives are tracked during the course of the product's shelf life. The preservative effectiveness of multidose nonsolid solutions has to be evaluated in settings that mimic actual use.

Labeling: The following guidance documents must be considered for each excipient used in a medicinal product: Thiomersal, CPMP Position Paper on the Application of the Sensitization Warning Statement (CPMP/2612/99), and Excipients on the Label and Package Leaflet of Human Medicinal Products (Eudralex 3BC7A) [24, 25].

14.4.2 GMP for Natural Excipients

With an emphasis on Good manufacturing practices (GMP) and a suitable system for quality management, this paper aims to offer guidelines for the manufacture, control, storage, and distribution of excipients used in pharmaceutical products. During their manufacturing, packing, repackaging, labeling, quality control, release, storage, distribution, and other associated processes, GMP should be created, put into practice, and maintained. When manufacturing excipients for which scientific literature, publicly available information, or historical data suggest that they pose a higher risk due to the possibility of toxic impurity formation during the manufacturing process, or of contamination during storage and distribution, additional precautions should be taken [26]. IPEC-Americas states that it is crucial to confirm that the excipient's producer is following good manufacturing procedures (GMPs) suitable for excipients intended for use as pharmaceutical ingredients. "Lastly, it is critical for sponsors to comprehend the entire supply chain back to the manufacturer for excipients that are not bought directly from the manufacturer or a manufacturer-authorized distributor," adds IPEC-Americas [27].

An excipient is defined by the WHO as a material other than active ingredients that has undergone a suitable safety evaluation and/or is included into a medication delivery system to assist in the manufacturing process of a medication delivery device; preserve, strengthen, and improve bioavailability, stability, or patient acceptance; help in identifying the product [28]; and improve any other aspect of the medication's general efficacy and safety when it is being used or stored.

Following are the important points taken into consideration:

- Quality management
- Quality risk management
- Complaints
- Recalls
- Returns
- Self-inspection, quality audits and supplier's audits and approvals
- Personnel
- Sanitation and hygiene
- Documentation
- Premises
- Equipment and utilities
- Materials
- Production
- Qualification and validation
- Quality control
- Life cycle and continuous improvement principles
- Storage and distribution [26]

14.4.3 Analytical Method and Quality Assurance

A pharmaceutical formulation's excipients are substances other than the active pharmaceutical ingredient (API), according to the International Pharmaceutical Excipients Council (IPEC). The safety of these substances is thoroughly assessed. They can be used to help with manufacturing and processing, protect the product, improve stability, bioavailability, and patient acceptability, make it easier to identify the product, or improve any aspect of the drug delivery system's efficacy or safety during use or storage. The form of the dosage form, its functioning, and, most importantly, its atmosphere's compatibility with the active/inactive ingredient and preparation method have all been taken into consideration when choosing excipients. They serve as the fundamental and structural support of all dose forms, including liquid, semi-solid, and solid forms utilized for topical and traditional applications.

The following section discusses the analytical methods frequently used to analyze pharmaceutical excipients in both their raw and pharmaceutical dose forms.

Capillary Electrophoresis

Using capillary electrophoresis (CE), analysis of pharmaceutical excipients and raw materials can be performed with advantages including automation, quick separation, and cost effectiveness. CE is a high-resolution, adaptable method that is frequently used to analyze a variety of substances, including ions, medications, and biopolymers. Its flexibility and modes, including micellar electrokinetic capillary chromatography (MEKC), make it ideal for evaluating pharmaceutical excipients and raw materials with low UV absorbance. The reported applications of capillary electrophoresis in the analysis of various pharmaceutical excipients are described in the section that follows.

- *Carbohydrate analysis:* To assess carbohydrates for effective separation and detection, capillary electrophoresis using high pH electrolytes, borate complexes, and specific derivatization procedures can be employed.
- *Fatty acid analysis:* By employing surfactant solutions with a high concentration of organic solvents and indirect UV detection, capillary electrophoresis may be utilized to evaluate free fatty acids, especially those present in pharmaceutical excipients.
- *Evaluation of flavoring substances:* CE and MEKC may test ionic chemical substances like Bitrex and flavoring additives like aspartame using specific electrolytes, internal standards, and UV detection for precise separation and sensitivity.
- *Chitosan analysis:* The molecular weight and distribution of chitosan were determined using Size Exclusion Chromatography and Gel Permeation Chromatography with acidic preparation and sodium nitrate buffer, revealing performance-altering differences in the samples' Mw values.

Scanning Thermal Microscopy and Differential Scanning Calorimetry The pharmaceutical industry regularly utilizes differential scanning calorimetry (DSC) to ascertain the physiochemical properties of raw materials and APIs to aid in new chemical entity development. Scanning thermal microscopy (SThM), which uses a thermal probe and a modified AFM approach, looks at the surface and thermal properties of materials to provide information on topography and local conductivity.

Mannitol analysis: Mannitol's DSC examination showed a melting range of 163–168 °C, which is consistent across analytical methods and well-aligned with values seen in the literature.

Near Infrared (NIR) Reflectance Spectroscopy To identify pharmaceutical excipients, NIR uses either supervised techniques that need previous sample information for classification or unsupervised techniques based on spectral patterns to identify samples.

Lactose analysis: For APIs and excipients, particle size is analyzed using methods like FT-NIR spectroscopy since it has a substantial influence on flowability, compression, dissolution, and tablet integrity [29].

14.5 Standardization and Quality Control

The incorporation of natural excipients in pharmaceutical formulation development has both special potential and constraints. Pharmaceutical goods rely on the safety, effectiveness, and consistency of these excipients, even though they are valued for their biocompatibility, biodegradability, and low toxicity. Standardization and quality control are essential procedures to ensure that natural excipients meet legal criteria and preserve the integrity of the finished formulation [30]. This part discusses the methods and procedures used to standardize natural excipients, focusing on the critical steps that must be taken to ensure their proper quality assessment. The aim is to ensure that natural excipients meet the highest standards of safety, efficacy, and compliance by setting purity requirements and using advanced analytical techniques, which would allow for their smooth integration into pharmaceutical applications [1, 29].

14.5.1 Role of Pharmacopoeias in Setting Standards

Pharmaceutical excipients are a wide range of substances that are important in the formulation of drugs. They comprise a broad range of substances, including gases, liquids, and solids, and can be either natural or man-made. The intricate interactions between the chemical, physical, and particle structures of these excipients determine how well they work. The establishment of precise, standard quality requirements for excipients calls for pharmacopoeias. This is necessary to establish safe, effective, and consistent pharmaceuticals. National and international pharmacopoeias are regarded as key regulatory tools that enable pharmaceutical products to meet the desired requirements on safety and quality [31]. Comprehensive specifications regarding excipients and APIs are defined in the compendia that detail quality standards like content, purity, and possible contaminants in such grades. Internationally recognized pharmacopoeias, like EP, USP, and JP, establish standardized production, testing, and quality control procedures for pharmaceutical formulations. These standards are of extreme importance in the international pharmaceutical industry, where excipients and APIs are often sourced from many regions and uniformity in the supply chain is of utmost importance [32, 33].

The Pharmacopeial Discussion Group (PDG), which strives to harmonize excipient monographs and ensure that excipients meet the same high standards irrespective of their country of origin, is keen on harmonizing excipient standards across different pharmacopoeias. This partnership streamlines the development and manufacturing processes, thus increasing efficiency and ensuring the safety and quality of the final goods. As pharmaceutical markets grow and sourcing of excipients becomes more worldwide, the activities of the PDG are important in creating consistency in excipient laws. Excipients have in the past not been as stringently tested for safety as active components. This is no longer the case because of the increasing complexity of pharmaceutical formulations. Excipients may interact with other excipients or APIs, thus altering the drug's efficacy or, worse, cause adverse side effects. This is why pharmacopoeias evolved to include deeper safety assessments of excipients taking into account the potential toxicity, interactions, and impact on the final formulation [34].

Production, distribution, and storage procedures are stressed to be important to ensure the quality of excipients along with chemical safety. Regulations like Good Manufacturing Practices (GMP), Good Laboratory Practices (GLP), and Good Distribution Practices (GDP) guide the pharmacopoeias in formulating these guidelines. Right from manufacturing up to the final pharmaceutical product, these frameworks have provided an orderly approach for ensuring the quality of excipients. A growing focus in the pharmaceuticals industry on making less use of excipients in their drug products reflects the continued drive on effective and simplification of drug formulations. This tendency towards simplified formulation is sustained by the uncompromising set of rules embodied in pharmacopoeias along with all vital activities of an excipient. Ultimately, these pharmacopoeias sustain an excipient functionality, safety along with the general well-being of the patient that are achieved since natural excipients meet the stringiest quality compliance requirements [35].

14.5.2 Challenges in the Standardization of Natural Excipients

Excipients play an important role in pharmaceutical formulation by combining the active pharmaceutical ingredient with it so as to ensure the finished product attains the required bulk and other qualities. Their presence influences the efficiency of varied dose forms including solid, liquid, semisolid, and nanoformulations. However, the inherent complexity and variety of natural excipients pose a number of special problems for standardization compared to synthetic excipients, making their consistent use in pharmaceutical formulations more challenging. Another difficulty in establishing norms for natural excipients is the fact that they are a very diverse group. Bio-derived excipients are derived from naturally occurring products, including plant, animal, or mineral origin. It also necessary to establish that manufactured excipients are generally described by their chemical content. The physical and chemical properties of the excipient material may be significantly different in a number of ways based on a variety of factors, such as harvest conditions, soil type, environmental influences, and seasonal effects. The batch-to-batch and supplier-to-supplier variations inherent with natural excipients make it even more challenging to develop repeatable and reproducible specifications. This is a critical requirement for pharmaceutical applications, in which uniformity must be assured to ensure safety and efficacy of the product [36].

Lack of a well-recognized framework for quality control is a major obstacle when using natural excipients. Natural excipients frequently don't meet the same strict standards as synthetic excipients, which are regulated by standardized monographs from pharmacopoeias like the USP and the EP. The stability, potency, and purity of these excipients can vary depending on how they are sourced, extracted, and processed. Establishing uniform quality standards is challenging due to this heterogeneity, which also complicates the regulatory clearance process. Furthermore, a lot of natural excipients include different active ingredients, such phytochemicals derived from plants, which add to batch-to-batch variation. Ultimately, these discrepancies may impact the stability, bioavailability, and performance of the excipient as well as its interactions with other excipients or active pharmaceutical ingredients (APIs) [37].

Another area of significant issues relates to the safety evaluation of natural excipients. Most natural excipients have limited safety information available, whereas many synthetic excipients have been extensively studied to determine their toxicological profile. In addition, the stringent, modern safety testing required by current pharmaceutical standards does not always align with the ancient uses of these natural substances in herbal products. Therefore, to ensure the safe incorporation of natural excipients into pharmaceutical formulations, comprehensive toxicological studies, clinical studies, and adherence to current regulatory requirements are necessary. The regulatory environment around natural excipients is changing, too. However, development of standardized methods for efficacy, safety, and quality evaluation remains very challenging. Uniformity is being gained due to efforts of global groups, such as the Pharmacopoeial Discussion Group (PDG), that seek to standardize excipient requirements between different pharmacopoeias. However, the lack of generally applicable testing procedures and the need for tailor-made strategies for every excipient can hinder the regulatory acceptance of natural excipients. This complexity makes it harder to create international regulatory frameworks [5].

Despite these challenges, work in the discovery of natural, biocompatible, biodegradable, and environmentally friendly excipients as replacements for their synthetic counterparts continues to date. Thanks to advances in analytical techniques such as mass spectrometry, near-infrared spectroscopy, and high-performance liquid chromatography (HPLC), natural excipients can be characterized and quality controlled much more accurately. Such progress allows for a partial alleviation of the natural unpredictability by attempting to standardize and legitimize the usage of natural material in pharmaceutical formulations. Despite significant challenges in terms of standardizing natural excipients due to issues of unpredictability, safety issues, and constant changes in regulations, research around the world has been opening windows for more standardization. The creation of strong regulatory frameworks will be critical in ensuring the safe and effective integration of natural excipients into medication formulations as the pharmaceutical industry increasingly looks at them for their potential advantages [38].

14.5.3 Case Studies: Successful Standardization of Specific Natural Excipients

The ability of natural components to enhance the efficacy, safety, and consumer appeal of skincare and therapeutic products is reflected in the successful use of plant-based excipients in pharmaceutical formulations, especially in topical dosage forms. These case studies not only exemplify the diversity of applications for natural excipients, but they also make alert to the constant attempts to standardize the substances so that their performance as well as quality will be uniform across various formulations. The following lists a number of plant-based excipients that have been efficiently added to topical dosage forms, as well as the efforts to standardize their application [39].

Turmeric (*Curcuma longa*) Extract in Anti-inflammatory Creams A lotion was made that reduces inflammation by using turmeric oleoresin, which is high in curcuminoids. The mixture showed notable anti-inflammatory benefits in preclinical animals, indicating the potential of turmeric extract to reduce skin inflammation and support skin health. Standardizing turmeric extract for topical application involves establishing constant quantities of curcuminoids, guaranteeing the extract's purity, and managing variables like the extraction technique and solvent employed. The regulatory agencies, in order to ensure uniform therapeutic efficacy, realized the need for standardized monographs defining the quality attributes of turmeric extract and, in particular, its chemical composition and potency in topical preparations.

Green Tea (*Camellia sinensis*) Polyphenols in Antiaging Serums An antiaging serum was prepared using green tea polyphenols, which are well known for their antiaging properties. The serum reduced fine wrinkles, enhanced skin elasticity, and showed high antioxidant activity. This example illustrates how green tea polyphenols can enhance skin health and longevity. To determine the optimal concentration of polyphenols and ensure batch-to-batch reproducibility, the standardization of green tea polyphenols in cosmetic formulations requires extensive testing. The use of reliable extraction processes and the screening of polyphenol content and stability, plus potential impurities would be part of quality control, and standardized approaches would be followed to ensure final product consistency in terms of outcomes and efficacy as use of green tea-based skincare develops.

Aloe Vera (*Aloe barbadensis* Miller) Gel in Wound-Healing Ointments It has been investigated that the addition of aloe vera gel to ointments for wound healing improves tissue regeneration, reduces inflammation, and accelerates wound closure. Although the medicinal benefits of aloe vera have been known for ages, its use in pharmaceutical goods needs to be standardized due to its diversity based on the source, harvest period, and processing techniques. Standardization of aloe vera gel is characterized by setting accurate extraction methods, ensuring the consistent concentration of active ingredients such as polysaccharides, and defining the permissible levels of purity. All these standards ensure that aloe vera gel delivers steady therapeutic effects and works effectively in wound healing formulations.

Soy Lecithin in Topical Antifungal Creams Soy lecithin has been added to topical antifungal formulations so that the delivery of the drug and absorption in the skin are improved. Soy lecithin is used immensely as a natural emulsifier and penetration enhancer. Pharmaceutical applications of soy lecithin can be evidenced through enhanced effectiveness of antifungal medications. Guarantees in terms of purity, as well as the determination of the optimal concentration that can improve drug delivery while ensuring safety, must be done to standardize soy lecithin in topical applications. Among analytical techniques used in determining the consistency and quality of soy lecithin across batches is HPLC, a high-performance liquid chromatography method.

Tea Tree (*Melaleuca alternifolia*) Oil in Antimicrobial Creams A cream with tea tree oil was made that demonstrated high antibacterial activity against skin infections by bacteria and fungi. Tea tree oil has a complex composition, which is dependent on the conditions of growth of the plant and extraction method, so standardization is important for topical use. Pharmacopoeial monographs for tea tree oil are being developed to define its chemical composition, including the most significant active principles such as terpinen-4-ol, to ensure uniformity and safety. These guidelines ensure the safety and effectiveness of oil in antibacterial compositions.

Lavender (*Lavandula angustifolia*) Oil in Relaxing Massage Lotions The effect of lavender oil in massage lotions was assessed, and it was found that the lotion relieved emotional states and autonomic nerve activity. It is often added to wellness products because its properties have well-documented aromatherapeutic benefits; however, quality and consistency will have to be established in order to ensure the product's efficacy and safety. To standardize lavender oil, limits have been established for the important bioactive substances such as linalool and linalyl acetate, and quality control tests that are quite stringent have been used to ensure the purity of the oil.

Jojoba (*Simmondsia chinensis*) Oil in Moisturizing Creams It was found that jojoba oil improved the lotions' moisture and flexibility properties. Being an emollient, this component is pretty useful for skincare formulations, but in the context of its use as a cosmetic ingredient, purification, composition, and stability requirements are to be followed to ensure that the product is delivered to the market with consistent quality. Standardization activities focus on the identification and determination of the jojoba oil's esters, fatty acids, and other major constituents besides ensuring stability in diverse environmental conditions.

Chamomile (*Matricaria chamomilla*) Extract in Soothing Ointments Calming ointments were made using chamomile extract, which illustrated marked anti-inflammatory and calming benefits for skin that had been irritated. Standardization is required to ensure that the consistent therapeutic benefits of flavonoids and terpenoids present in chamomile extract are achieved if applied topically. The standardization procedure involves verifying the chemical profile of the extract, especially the amounts of important active ingredients such as apigenin and chamazulene, and ensuring that it is free from impurities that would compromise its efficacy or safety.

These case studies indicate that these excipients have growing interest due to their medicinal quality and their improving performance in the topical formulation, thus proving successful standardization is required for such natural excipients for pharmaceutical products to be constant, safe, and effective. Strong regulatory frameworks put in place with the most advanced analytical techniques will allow the pharmaceutical industry to continue to seize the promise of natural excipients while upholding the highest standards of quality and patient safety [40].

14.6 Regulatory Challenges and Solutions

Some of the advantages of using natural excipients, especially those from plants, in pharmaceutical formulations include reduced side effects, higher patient compliance, and sustainability. To ensure that these components are used safely and effectively in pharmaceutical products, a number of issues related to the regulatory environment surrounding these excipients need to be addressed. Herbal medicines and their preparations are increasingly gaining popularity worldwide, especially in developing countries. However, getting these products to market involves overcoming a number of financial, legal, and technical hurdles. The key regulatory barriers to the development and use of natural excipients are addressed in this chapter, as well as potential solutions and ongoing efforts to address them.

The development and application of natural excipients in pharmaceutical products are complex because there are no widely acknowledged regulations. The manufacturing of natural excipients, of course, is most often associated with the lack of clear laws and controls. In contrast, defined standards and regulatory frameworks have already proven helpful for synthetic excipients. With the quality and safety of natural excipients differing greatly depending on harvest conditions, extraction techniques, and place of origin, they are more difficult to regulate. Recognizing the particular challenges of managing natural products—especially those utilized in traditional medicine—is also essential. Usually, complicated plant parts are used to make these medications, and the active ingredients can vary based on the plant's origin, the methods used for extraction, and the surrounding environment. There are particular difficulties with traditional herbal products because their efficacy and safety are often based on historical use and cultural practices, rather than on current scientific evidence [41].

14.6.1 Variability in Natural Sources

In pharmaceutical formulations, natural excipients provide a number of advantages over synthetic ones, such as sustainability, biocompatibility, and a decreased risk of side effects. However, because natural excipients are inherently variable, it can be difficult to guarantee the uniformity, effectiveness, and safety of finished pharmaceutical products. Harvesting practices, the intricate makeup of natural materials, and environmental considerations are some of the causes of this variety.

Understanding and addressing these causes of variability are crucial for regulatory systems to ensure that natural excipients meet strict quality standards [11]. Variability in natural sources is given in Table 14.3.

14.6.2 Addressing Stability and Shelf Life Issues

Because of their good safety profile, biocompatibility, and biodegradability, natural excipients are used frequently in the formulation of pharmaceutical products. Stability and shelf life are major barriers to their use since these are essential factors in ensuring constant quality, efficacy, and safety for pharmaceutical products. Because of their chemical composition, sensitivity to the environment, and intrinsic variability, natural excipients may be more susceptible to several types of degradation compared to synthetic excipients. These stability issues have an important impact on the regulatory systems put in place to ensure the safe use of natural excipients in pharmaceutical applications [42].

Oxidative Sensitivity and Environmental Sensitivity
Most of the plant-based excipients are prone to oxidation, which is a serious stability issue. Many natural excipients, such as plant oils and extracts, contain unsaturated chemicals that are very susceptible to oxidation when exposed to heat, light, or air. Oxidation may degrade these bioactive substances, thus altering the stability of the finished pharmaceutical product and affecting the efficacy of the excipient. The reason plant-based excipients like essential oils can present a risk to drug safety is that they might lose their medicinal properties or create dangerous by products upon oxidation. In addition to oxidative degradation, natural excipients are often extremely sensitive to environmental factors such as light, humidity, and temperature. Incompatible environmental conditions may cause the excipients and APIs to degrade faster, leading to unstable products as shown in Fig. 14.2. Differences in lighting and temperature, mainly during storage and transport, exacerbate these problems. This is a critical element of stability testing and, more often than not, proving to be more demanding for natural excipients than synthetic excipients [43].

Microbial Contamination and Batch-to-Batch Variation
Natural excipients are more susceptible to microbial contamination because they are derived from organic sources. Some natural excipients, including plant-based oils and herbal extracts, can provide an ideal environment for microbial growth since they are known to contain moisture or nutrients. Bacterial, fungal, or mold contamination at any stage during production or storage can compromise the quality and safety of the excipient, thus severely threatening patient safety. Another problem associated with natural excipients is batch-to-batch variance. As the excipients are derived from natural sources, their cultivation and environmental conditions might affect the chemical and physical characteristics of each batch. This could result in changes to the drug's formulation, which could impact its stability and effectiveness. Variability in excipient quality can also lead to variation in drug

Table 14.3 Variability in natural sources

Challenge	Description	Impact on drug product	Regulatory considerations
Intrinsic variability in natural sources	Natural excipients exhibit inherent diversity due to factors like geographic location, soil composition, climate, and seasonal fluctuations. These factors lead to variations in excipient properties	Variability in the chemical structure, physical characteristics, and biological activity of natural excipients can affect the consistency and efficacy of pharmaceutical products	Regulatory systems need to understand and address this variability to ensure product safety, efficacy, and quality
Complex composition	Plant extracts and other natural excipients contain various bioactive compounds, making extraction and quantification challenging. The composition can vary significantly based on the plant source and processing methods	Variations in active ingredients (e.g., alkaloids, flavonoids) can impact excipient functionality, affecting stability, solubility, and bioavailability of the final product	Standardized extraction methods and testing protocols should be developed to control variability and ensure consistent product performance
Variability due to raw material sourcing	Excipients from renewable biological resources are influenced by seasonal and environmental conditions, leading to composition variation. Differences in climate and growing methods also play a role	Variability in raw materials (e.g., plant species, harvest time, environmental conditions) can affect excipient characteristics, such as bioavailability and dissolution rate, impacting drug efficacy	Proper documentation of raw material sourcing is necessary to minimize variability and ensure consistent quality
Manufacturing variability in excipients	Variability in excipient production arises from differences in extraction, purification, and processing methods and can vary between suppliers and within a single supplier's process	Discrepancies in production can affect excipient quality, impacting the consistency and performance of the final pharmaceutical product, leading to batch-to-batch irregularities	Detailed documentation of production processes and quality control measures should be mandated to minimize variability in manufacturing
Impact of excipient variability on drug product quality	Variability in excipient characteristics (e.g., solubility, particle size, moisture content) can significantly influence drug product performance	Changes in excipient properties can affect drug solubility, bioavailability, tablet hardness, and disintegration time, leading to inconsistencies in drug release and therapeutic outcomes	A quality-by-design approach should be implemented, focusing on excipient control to ensure consistent product quality and therapeutic efficacy

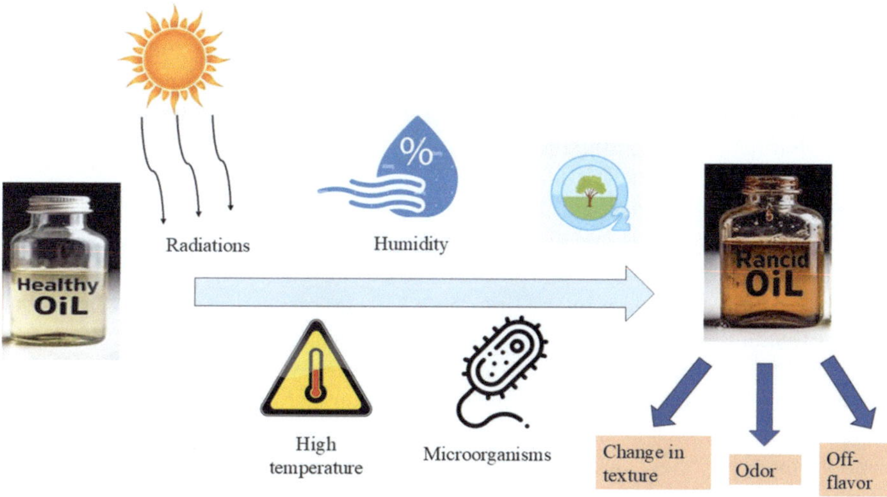

Fig. 14.2 Oxidation sensitivity

release profiles, which is especially troublesome when formulations have limited bioavailability.

Stability Testing and Regulatory Compliance

Stability testing is particularly crucial to the regulatory process due to the challenges that natural excipients present. In this regard, factors such as oxidative degradation, microbiological contamination, and environmental sensitivity are of utmost importance to ensure that the therapeutic product maintains its intended functionality, safety, and efficacy over time. However, natural excipients are far more difficult to stability test compared to synthetic excipients due to their chemical complexity and unpredictability of their basic components. Stability assessment of natural excipients is complicated by various circumstances:

- *Chemical Complexity:* Various chemicals, including extracts of plants or essential oils, constitute the natural excipient mix, which is complex in its chemical nature. This complexity presents a challenge in establishing reliable quality control processes, wherein stability testing markers are hardly available.
- *Biochemical Differentiation:* On the basis of plant species and production circumstances, harvesting might be significantly varying. These biochemically contrasting excipients vary in stability compatibility with APIs, depending upon each other.
- *Enzyme Activity:* The stability of many herbal-based excipients may be affected by the enzymes they contain. The efficacy or safety profile of the formulation may change due to enzyme-mediated degradation of the excipient or the API.

In regulatory systems, it is indispensable to ensure conformity to stability testing requirements. Stringent regulations were put in place by organizations, such as the

European Medicines Agency (EMA) and the US Food and Drug Administration (FDA), for stability testing of excipients, further emphasizing the necessity of proper examination under both accelerated and real-time conditions. Regulations compliance ensures stability and safety over the shelf life of natural excipients used within pharmaceutical formulations [44].

Addressing Safety Concerns

Because they are nontoxic and biodegradable, natural excipients are often considered to be safe, but safety issues can still arise. To ensure that batch-to-batch variation, microbial contamination, or interactions with APIs do not lead to adverse effects, the safety of natural excipients must be evaluated. For instance, the mass ratio of a formulation to API is usually high. Thermodynamic activity of the excipient may bring about interactions likely to compromise medication stability or unanticipated side effects; therefore, such natural excipients need comprehensive toxicological investigations regarding their safety and regulatory requirements like WHO/FAO Joint Expert Committee on Food Additives as well as International Toxicological Committees. These studies generally involve testing on laboratory animals for potential toxicity and ensuring the excipient does not have adverse effects, particularly on sensitive populations such as children or the elderly. Although natural excipients are believed to be relatively nontoxic, issues such as heavy metal contamination, variable hydration rates, and microbiological contamination must be closely controlled. Accordingly, regulatory rules need to require high-stringent safety evaluations, for example, on heavy metal screening, microbiological testing, and research on pharmacological effects from excipient-API interactions. Excipients of natural origin must, for the storage conditions to allow them to stay stable over the shelf life of the product, be kept properly stored. Particular light, humidity, and temperature conditions need to be preserved for some of the excipients to be stable. The regulatory agencies enforce strict criteria to ensure that pharmaceutical items, especially those requiring cold chain storage, remain within their quality range throughout storage and transit. This is because natural excipients are required to maintain their efficacy and safety over the lifecycle of the product; hence, compliance with these requirements is essential to prevent deterioration. It helps in preventing the deterioration of heat or light to the cold chain technique by precise temperature regulation both in storage and transportation. Loss of functionality by the excipients or the effect of stability on the active component is expected if those parameters are not met. The logistics companies and pharmaceutical manufacturers need to collaborate to ensure that the cold chain requirements are met and the natural excipients are kept within the required temperature range for ensuring the safety and efficacy of the product [45].

14.6.3 Overcoming Regulatory Hurdles with Innovative Approaches

The pharmaceutical business presents opportunities as well as challenges due to the increasing use of natural excipients in formulations. Plant-based or other natural excipients have many benefits such as sustainability, biocompatibility, and biodegradability, but their regulatory framework mostly fails to match synthetic ones. This disparity generates a lot of regulatory hurdles, making it difficult to seek approval of those products or ensure their safety and that they comply with industry standards. Novelty in formulation techniques, standardization, and ensuring regulatory compliance may help overcome some of these impediments and facilitate the successful marketing of natural excipients in drug products.

Regulatory Hurdles and Product Approval

One of the major challenges with plant-based excipients is that there is no global uniform and defined regulatory framework. Plant-based excipients often face different and less uniform regulatory standards depending on the location, whereas synthetic excipients are generally well-regulated with established safety profiles. The laws related to natural excipients can be cumbersome and not uniform in developed countries such as the US and the EU. For instance, many herbal products are marketed as dietary supplements in the US, which saves them from the rigorous safety and efficacy testing required for prescription drugs. Since these products may not be held to the same level of scrutiny, this approach may lead to safety issues. The Traditional Herbal Medicinal Products Directive 2004/24/EC of the European Union, however, demands extensive documentation from manufacturers, which would include preclinical safety data.

In order to overcome these regulatory obstacles, creative solutions should concentrate on creating uniform approval processes for natural excipients, guaranteeing that they satisfy essential safety and quality requirements without impeding the development of new products. This can entail working together to expedite the approval procedure and harmonize standards among industry players, regulatory agencies, and research teams worldwide. Furthermore, a number of formulation and compatibility issues arise when using plant-based excipients in contemporary drug delivery systems. To guarantee constant medication release and therapeutic efficacy, these excipients need to work with cutting-edge delivery technologies such lipid-based carriers, nanotechnology, and controlled-release systems. However, the inherent physical and chemical properties of excipients derived from plants, for example, changes in molecular weight, polarity, and solubility, may impact compatibility with synthetic excipients as well as with sophisticated delivery techniques [46].

Among the innovative methods for addressing formulation problems are:

Advanced Methods of Formulation: Natural excipients are more compatible with contemporary delivery systems when sophisticated formulation processes, such as liposomal formulations, nanoencapsulation, or co-processed excipients, are used.

Excipients Modification: Plant-based excipients can be made more soluble, bioavailable, or stable by altering their chemical makeup or particle size. This can enhance their compatibility with advanced delivery systems.

Synergistic Interactions: More stable and efficient products may be formulated if the natural excipient improves the performance of the API or other excipients.

This is known as a synergistic interaction. For example, some natural excipients have antioxidant qualities that improve stability or aid in drug absorption.
The pharmaceutical industry can overcome compatibility issues and formulate products that take advantage of the benefits of natural excipients while retaining product efficacy and quality by focusing on these innovative approaches.

Innovative Regulatory and Industry Approaches for Overcoming Challenges
A few creative strategies can be used to circumvent regulatory obstacles and ensure effective creation and marketing of natural excipients [47]:

Harmonization of Regulations on a Global Scale: The regulation of natural excipients across geographies can be harmonized for better streamlining of the approval procedure and access to markets. The International Council for Harmonization and the World Health Organization are excellent international organizations to facilitate the harmonization of standards.
Better Supply Chain Management: Traceable and certified supply chains for plant-based excipients can ensure quality consistency and reduce variability due to inappropriate sourcing or environmental factors. Cooperation between makers of excipients, suppliers of raw materials, and regulatory bodies can promote more efficient quality assurance procedures.
Education and Awareness: Improving the development of the segment depends on small- and medium-sized herbal companies understanding the importance of GMP, Good Agricultural and Collection Practices, and safety norms. Government and industries should invest on outreach activities and workshops as a means of making people adopt compliant regulatory measures.

14.7 Sustainability and Ethical Consideration

14.7.1 Eco-friendly Sourcing of Natural Excipients

There is growing pressure on the pharmaceutical sector to implement more environmentally friendly procedures, particularly with regard to lowering supply networks' carbon footprints. The use of environmentally friendly excipients, which are crucial components in medication formulations, is one important area that has to be improved [48]. Eco-friendly excipients are made to have as little of an impact as possible on the environment at every stage of their lifespan, from manufacturing to disposal. These excipients are biodegradable and frequently made from renewable resources. With its line of environmentally friendly lubricants, which includes rice bran-based magnesium stearate and stearic acid and palm-based magnesium stearate and stearic acid approved by RSPO, Sudeep Pharma Limited is meeting this need.

Magnesium Stearate and Stearic Acid Made from Rice Bran: In addition to items made from palm oil, rice bran, a byproduct of milling rice, is an environmentally benign and sustainable resource by using rice bran to produce stearic acid and magnesium stearate.

Surfactants Using *Daphnia magna* and *Dendrocephalus brasiliensis* as test species, this study evaluated the toxicity of biosurfactants (surfactin, iturin, and fengycin) and synthetic surfactants (sodium dodecyl sulfate and Triton X-100). The findings demonstrated that while both species were susceptible to SDS, *D. brasiliensis* exhibited greater sensitivity to all of the substances. Surfactin was the most harmful of the biosurfactants, whereas iturin and fengycin were the least poisonous. The study emphasizes how crucial *D. brasiliensis* is for keeping an eye on contaminants in freshwater ecosystems that are tropical and subtropical [49].

Gums The increasing interest in Pickering emulsions stabilized by biopolymeric particles derived from natural sources, especially plant gums, is highlighted in this study. These gums are said to be perfect for making nontoxic, biodegradable, and environmentally friendly stabilizers since they are made from plant exudates or seeds. The varieties and characteristics of plant gums, the processes used to produce gum-based particles, and their creative uses in low-fat food formulations, active packaging, medicine delivery, and plant-based goods are all covered in the article. It also looks at possible avenues for further study in the areas of environmental science, tissue engineering, and food applications [50].

14.7.2 Ethical Aspects in the Global Supply Chain

Excipient users must have access to excipients made with reduced embedded greenhouse gas (GHG) emissions, little to no waste, and little effect on water availability and quality if they are to meet their sustainability goals. As a result, excipient manufacturers must optimize their operations and transition to low-carbon or renewable energy sources in the near future. Excipient users and suppliers must work together and exchange data in order to find ways to offer net zero ingredients within the parameters of existing laws and to make step change improvements in order to meet the pharmaceutical industry's long-term goals. The guidelines and criteria for best practices in environmental sustainability and information sharing across the pharmaceutical supply chain are outlined in this blog post.

An international organization called the IPEC Federation works to improve the safety and quality of pharmaceutical excipients. The following topics are included in the draft IPEC Sustainability information package: Supplier ESG Program, Supplier supply chain and procurement, Safety and health and Continuity of business [51].

It talks about how rising understanding of the complexity of sourcing processes has caused the cosmetics and home care industry to move toward natural components. It draws attention to the demand on stakeholders, regulatory agencies, and consumers to address concerns about environmental preservation, social justice, and geopolitical considerations. Along with highlighting the dearth of efficient compliance systems for guaranteeing the ethical sourcing of natural ingredients, the essay looks at how new standards and methodologies may enhance supply chain procedures [52].

14.7.3 Regulatory Implications of Sustainability Initiatives

The difficulties and factors to be taken into account while using excipients in medication formulations and their packaging are covered in this paragraph. Excipients in smaller packs can increase packing material and transit weight, but they also save waste. Innovative packaging options that enable the testing or usage of excipients without sacrificing the remaining material may aid in achieving sustainability objectives. Since better excipients may lessen adverse interactions and enhance medication stability, resulting in advantages like lower pill sizes and less need for cold storage, sustainability should be assessed per medicine dose. This will eventually minimize carbon emissions. Supply networks may eventually need to measure these environmental advantages as "avoided GHG emissions." [51].

14.8 Future Trends in Regulatory Approaches

Future innovations in drug delivery, ethical sourcing, personalized treatment, AI integration, and international regulatory harmonization will shape the regulatory framework controlling plant-based excipients in pharmaceuticals. New developments such as nanotechnology and plant-derived nanoparticles provide improved medicine delivery methods, and there will be a growing emphasis on the ecological and ethical source of plant materials. Excipient combination optimization using AI will soon revolutionize drug development with customized formulations suited to the specific needs of every patient. For these developments, however, there must be an inter-agency approach by regulatory bodies in formulating uniform, worldwide standards that guarantee sustainability, safety, and efficacy. The investigation of various plant sources, backed by strict laws, will propel the development and worldwide availability of plant-based excipients in pharmaceuticals [53, 54].

14.8.1 Harmonization of Global Regulatory Frameworks

This includes the integration of international regulatory regimes to help in ensuring safety of patients as well as ensuring compliance with growing global demand for plant-based medicines. As medicinal plants are widely utilized in developing nations, and at the same time, plant-based therapy is widely adopted in advanced nations, integration in this manner becomes very vital. The issue of balancing access to a diverse set of alternatives and patient safety provides regulatory hurdles. These products are now regulated by a variety of national and regional regulatory regimes, which differ from one another. To harmonize the regulation of medicinal plants, the WHO is necessary in promoting international cooperation and providing direction. However, true global harmonization remains challenging to attain because regional laws, safety assessments, and product categories remain diverse. Efforts toward the standardization of regulations will be ongoing to ensure safety, effectiveness, and accessibility in the use of plant-based medicines globally, by developing transparent and widely recognized frameworks. This would help expand the plant-based pharmaceutical sector while meeting stringent safety and legal requirements [55].

14.8.2 Emerging Trends in Digital Traceability and Blockchain for Compliance

Blockchain technology and new developments in digital traceability are transforming compliance frameworks, particularly for natural excipients used in pharmaceuticals. Due to its intrinsic decentralization, transparency, and immutability, blockchain is a perfect way to improve supply chain traceability [56]. Blockchain enables accurate tracking of natural excipient items from point of origin to ultimate distribution by tokenizing goods and incorporating Smart Contracts (SCs), guaranteeing total transparency and accountability at every turn. This method lets for both traceability forward and traceability backward along with real-time tracking of a product's location and its transformation history, as well as provenance [57]. Blockchain traceability methods-based technologies, which may include use of RFID connected with the use of the Ethereum blockchain, make possible identification and authenticity verification for the fake [58]. Adoption of the GS1 V2.0 standard also provides a basis for linking digital and physical objects, offering complete traceability information through timestamps, GLNs, and GTINs. The solutions address the challenges of ensuring safety and quality compliance by automating critical processes such as risk management and compliance verification. Blockchain further enhances global regulatory harmonization by making cross-border cooperation between regulatory authorities easier. Therefore, there is a great potential for the improvement in safety, effectiveness, and compliance in the pharmaceutical industry, especially for natural excipients, through the integration of blockchain technology and digital traceability systems [55].

14.9 Conclusion

Regulatory frameworks are critical to ensuring the safety, quality, and compliance of natural excipients in pharmaceutical applications. While obstacles like as unpredictability in natural sources, stability concerns, and standardization continue, worldwide policies are constantly evolving to meet this complexity. Adhering to Good Manufacturing Practices (GMP), undertaking toxicological evaluations, and applying tight quality control procedures are still critical. In addition, ecological and ethical issues are increasingly influencing regulatory decisions. In the future, worldwide regulatory harmonization and digital technology integration will increase transparency and expedite excipient regulation.

References

1. Singh P, Mishra G, Dinda SC. Natural excipients in pharmaceutical formulations. In: Evidence based validation of traditional medicines. Springer; 2021. p. 829–69.
2. Patel R, Barker J, Elshaer A. Pharmaceutical excipients and drug metabolism: a mini-review. Int J Mol Sci. 2020;21(21):8224.
3. Mittal DS, Pawar S. Herbal excipients: an overview. Int J Pharm Pharm Sci. 2019;1(1):05–9.
4. Gazi AS, Begum A, Begum S. An overview on pharmaceutical excipients—their roles and applications. Int J Pharm Sci Med. 2022;7(12):84–96.
5. Ganguly D, Choudhury A, Chanda R, Das H, Mitra S, Ghosh R, et al. A comprehensive overview of edible natural excipients and their potential use in pharmaceutical formulation development. J Young Pharm. 2023;15(4):689–94.
6. Milić J, Čalija B, Dordević SM. Diversity and functionality of excipients for micro/nanosized drug carriers. In: Microsized and nanosized carriers for nonsteroidal anti-inflammatory drugs: formulation challenges and potential benefits; 2017. p. 95–132. https://doi.org/10.1016/B978-0-12-804017-1.00004-2.
7. Narang AS, Mantri RV, Raghavan KS. Excipient compatibility and functionality. In: Developing solid oral dosage forms: pharmaceutical theory and practice. 2nd ed; 2017. p. 151–79. https://doi.org/10.1016/B978-0-12-802447-8.00006-6.
8. Regulatory framework definition: 124 samples | Law Insider.
9. Bozza A, Campi C, Garelli S, Ugazio E, Battaglia L. Current regulatory and market frameworks in green cosmetics: the role of certification. Sustain Chem Pharm. 2022;30:100851.
10. Roshan K. Role of regulatory affairs in pharmaceutical company: an overview. Glob J Res Anal. 2018;7(1):2277–8160.
11. Yu YB, Taraban MB, Briggs KT, Brinson RG, Marino JP. Excipient innovation through precompetitive research. Pharm Res. 2021;38(12):2179.
12. Poonam V, Sagar G, Abhishek K, Yuvraj S. Remarkable contribution of natural excipients in finished pharmaceutical products (FPPs). Int J Pharm Sci Rev Res. 2018;52(1):7–14.
13. Singh KK, Rajeev SR, Mohammad A. The exploring role and responsibility of Indian pharmacopoeia commission: an introduction. J Educ Technol Health Sci. 2024;10(3):64–8.
14. Singh S, Bhardwaj S, Singhal S. Current trends and regulatory aspects of nutraceutical: a comprehensive review. Int J Green Pharm. 2023;17(2):108.
15. Torović L, Vojvodić S, Lukić D, Srđenović Čonić B, Bijelović S. Safety assessment of herbal food supplements: elemental profiling and associated risk. Food Secur. 2023;12(14):2746.
16. Pandey V, Polaka S, Nalla LV, Tekade M, Sharma MC, Tekade RK. Excipient toxicity and safety. Pharmacokinet Toxicokinet Considerations. 2022;II:487–511.
17. Goda Y. Regulatory science of natural products. J Nat Med. 2022;76(4):732–47.
18. Hulke Y, Bhange M. Naturally occurring excipients and their pharmaceutical and engineering applications. AIP Conf Proc. 2024;3188(1):100078.
19. Kipo SL, Ofori-Kwakye K, Kuntworbe N, Gbedema SY, Mensah KB, Ossei PPS. Preliminary health risk assessment of two exudate gums as pharmaceutical excipients. Sci Afr. 2023;20:e01648.
20. Bruusgaard-Mouritsen MA, Nasser S, Garvey LH, Krantz MS, Stone CA. Anaphylaxis to excipients in current clinical practice: evaluation and management. Immunol Allergy Clin N Am. 2022;42(2):239.
21. Wiesner J. Sensitization and allergies of herbal products. In: Toxicology of herbal products; 2017. p. 237–69. https://doi.org/10.1007/978-3-319-43806-1_11.
22. Manchanda S, Chandra A, Bandopadhyay S, Deb PK, Tekade RK. Formulation additives used in pharmaceutical products: emphasis on regulatory perspectives and GRAS. In: Dosage form design considerations, vol. I. Academic; 2018. p. 773–831.
23. Ml C, Quirce S. Immediate hypersensitivity reactions caused by drug excipients: a literature review allergy to drug excipients. J Investig Allergol Clin Immunol. 2020;30(2):86–100.

24. USP Guideline for submitting requests for revision (RFR) to USP-NF submission guideline for excipients. Submission guideline for excipients 2012.
25. Scientific guideline, "Excipients in the dossier for application for marketing authorisation of a medicinal product—scientific guideline | European Medicines Agency (EMA)." EMA, Europe.
26. WHO good manufacturing practices for pharmaceutical.
27. Stoothoff WH, Johnson GVW. In equal measures: the importance of excipient quality. Biochim Biophys Acta Mol basis Dis. 2023;47(2):12–6.
28. He TT, Ung COL, Hu H, Wang YT. Good manufacturing practice (GMP) regulation of herbal medicine in comparative research: China GMP, cGMP, WHO-GMP, PIC/S and EU-GMP. Eur J Integr Med. 2015;7(1):55–66.
29. Kumar M, Bhatia R, Rawal RK. Applications of various analytical techniques in quality control of pharmaceutical excipients. J Pharm Biomed Anal. 2018;157:122–36.
30. Noviana E, Indrayanto G, Rohman A. Advances in fingerprint analysis for standardization and quality control of herbal medicines. Front Pharmacol. 2022;13:853023.
31. Sarma N, Upton R, Rose U, Guo D-A, Marles R, Khan I, et al. Pharmacopeial standards for the quality control of botanical dietary supplements in the United States. J Diet Suppl. 2023;20(3):485–504.
32. Pharmacopoeial Harmonisation—European Directorate for the Quality of Medicines & HealthCare. Council of Europe Portal, 2025
33. WHO Expert Committee on specifications for pharmaceutical preparations fiftieth report.
34. Jadaun GPS, Rastogi S, Kumar A, Chauhan J, Sharma SK, Kumar M, et al. Ensuring the quality of medicines in India: an update on the development, modernization, and harmonization of drug standards in the Indian Pharmacopoeia. Saudi Pharm J SPJ. 2023;31(12):101825.
35. Mark Wiggins J, Albanese JA. A brief history of pharmacopoeias: a global perspective. PharmTechcom; 2019.
36. Barouki R, Kogevinas M, Audouze K, Belesova K, Bergman A, Birnbaum L, et al. The COVID-19 pandemic and global environmental change: emerging research needs. Environ Int. 2021;146:106272.
37. Wang H, Chen Y, Wang L, Liu Q, Yang S, Wang C. Advancing herbal medicine: enhancing product quality and safety through robust quality control practices. Front Pharmacol. 2023;14:1265178.
38. Yadav N, Singh Chandel S, Venkatachalam T, Fathima SN. Herbal medicine formulation, standardization, and commercialization challenges and sustainable strategies for improvement. Ref Ser Phytochem. 2024;Part F3164:1769–95.
39. Sandhaanam SD, Pandikumar P, Ignacimuthu S. Standardization of natural products and drugs. Acta Sci Pharm Sci. 2019;3(2):19–20.
40. Nammas M. Systematic review of plant-based excipients in topical drug delivery. J Med Biomed Sci. 2021;16:162–8.
41. Saito J, Agrawal A, Patravale V, Pandya A, Orubu S, Zhao M, et al. The current states, challenges, ongoing efforts, and future perspectives of pharmaceutical excipients in pediatric patients in each country and region. Child. 2022;9(4):453.
42. Selvadurai S, Divya S, Geetha A, Rohini K, Anbazhagan S. Herbal excipients-significance of substances of natural origin as excipients. Int J PharmTech Res. 2020;2020(3):242–52.
43. Dinda SC. A review & research on natural excipients in drug delivery. Scholar's Press; 2020.
44. Sharma SM, Sharma R, Chandrateya P, Sardana S. Advancing drug chemical stability: unravelling recent developments and regulatory considerations. Pharm Regul Aff (Open Access). 2024; https://doi.org/10.37421/2167-7689.2024.13.408.
45. Eltayeb Suliman AMA, Elhadi Ahmed MA. Natural excipients applications in conventional pharmaceutical formulations—part I. Med Aromat Plan. 2021;10(7):397.
46. Pinco RG, Sullivan TM. Regulation of pharmaceutical excipients. Excip Dev Pharm Biotechnol Drug Deliv Syst. 2006;4(September):37–50.
47. Comoglu T, Ozyilmaz ED. Pharmaceutical excipients in pediatric and geriatric drug formulations: safety, efficacy, and regulatory perspectives. Pharm Dev Technol. 2024;30(1):1–9.

48. Sumit Mehetre SPL. Sustainable solutions in pharma supply chains: reducing carbon footprint with sustainable excipients—articles. Pharma Ind (PIINDIA), Farma Forum; 2024.
49. Santos VSV, Silveira E, Pereira BB. Ecotoxicological assessment of synthetic and biogenic surfactants using freshwater cladoceran species. Chemosphere. 2019;221:519–25.
50. Sharkawy A, Rodrigues AE. Plant gums in Pickering emulsions: a review of sources, properties, applications, and future perspectives. Carbohydr Polym. 2024;332:121900.
51. Wood R. Excipient sustainability—a user and a supplier perspective | Croda Pharma. Croda Pharma, IPEC Fed; 2022.
52. accentmicrocell. The environmental responsibility of excipient manufacturers—Accent Microcell Ltd. Accent Bonding Precision; 2023.
53. Nammas M. Systematic review of plant-based excipients in topical drug delivery. Ibnosina J Med Biomed Sci. 2024;16(04):162–8.
54. Mali A, Kuvar V, Bharadwaj S. Bridging the gap: a comparative investigation of pharmaceutical excipient regulations. Ther Innov Regul Sci. 2024;58(2):258–72.
55. Picking D. The global regulatory framework for medicinal plants. In: Fundamentals, applications, and strategies. 2nd ed. Academic; 2024. p. 769–82.
56. Kamran R, Khan N, Sundarakani B. Blockchain technology development and implementation for global logistics operations: a reference model perspective. J Glob Oper Strateg Sourc. 2021;14(2):360–82.
57. Azevedo P, Gomes J, Romão M. Supply chain traceability using blockchain. Oper Manag Res. 2023;16(3):1359–81.
58. DeMerlis CC, Smith A, Schoneker DR. Regulatory information for excipients. In: Pharmaceutical excipients: properties, functionality, and applications in research and industry. Wiley; 2016. p. 241–67.

Innovative Pathways in Natural Polymer-Based Excipients: Shaping the Future of Drug Formulation

15

Sakshi Wani, Shounak Kulkarni, Anjali Pingale, Amarjitsing Rajput, and Prashant Pingale

15.1 Introduction

Natural or manmade polymers are large molecules with simple structures. To tailor them to specific needs, one can mix and match low- and high-molecular-weight (MW) chains with a variety of functional groups. Medication delivery systems and the medical field make use of both natural and synthetic polymers for the purposes of site-specific targeting and controlled medication release. Additionally, they hide flavor while maintaining and safeguarding oral drug delivery devices. Many food items and medicine dose forms contain polymers, either natural or synthetic [72].

Biopolymers (DNA, RNA, proteins, as well as polysaccharides) have been integral to metabolic activity from the inception of life. The utilization of naturally derived polymers for the production of goods as well as necessary items, including textiles,

The original version of the chapter has been revised. A correction to this chapter can be found at https://doi.org/10.1007/978-981-96-7959-1_16

S. Wani · S. Kulkarni
Department of Pharmaceutics, GES's Sir Dr. M. S. Gosavi College of Pharmaceutical Education and Research, Nashik, Maharashtra, India

A. Pingale
Department of Pharmacognosy, Adivasi Seva Samittee's Institute of Industrial and Pharmaceutical Technology, Nashik, Maharashtra, India

A. Rajput
Department of Pharmaceutics, Poona College of Pharmacy, Bharati Vidyapeeth Deemed to Be University, Pune, Maharashtra, India

P. Pingale (✉)
Department of Pharmaceutics, Gokhale Education Society's Sir Dr. M. S. Gosavi College of Pharmaceutical Education and Research, Nashik, Maharashtra, India

J. S et al. (eds.), *Innovative Pharmaceutical Excipients: Natural Sources*, https://doi.org/10.1007/978-981-96-7959-1_15

ornamental items, utensils, weaponry, and educational materials, has been recognized for centuries. The growth in the polymer sector goes back to the nineteenth century, when significant researches were performed for the modification of the structural properties as well as the functionalities of widely used polymers by alterations in their chemical composition. Thomas Hancock modified the characteristics of conventional elastic materials by incorporating specific additions. Subsequently, Charles Goodyear improved the properties of standard elastics through the reaction of vulcanization with sulfur. The first synthetic polymer, Bakelite, was created, followed by the engineered fiber, rayon, in 1911. In 1919, Herman Staudinger began methodical investigations in polymer sciences and proposed that high molecular weight polymers are composed of long chains of covalently linked small molecules [9, 21].

The polymer that is used to make drug delivery systems (DDS) determines how effective they are. Different mechanisms of drug release of polymers are illustrated in Table 15.1. Synthetic polymers are commonly utilized for the development of nanomedicines because of their specific shapes, sizes, and mechanical properties. However, they can be toxic and incompatible, which can be problematic. On the other hand, they are biocompatible with biomolecules and have fewer side effects, which is why biomedical analysts are interested in them [28].

Natural polymers are prevalent throughout nature and can be isolated. These polymers have been synthesized by addition polymerization or condensation polymerization. The human body comprises several natural polymers, such as proteins and

Table 15.1 Mechanism of drug release from polymers

Drug delivery system	Mechanism of drug release
Diffusion-controlled release	Drug molecules diffuse across the polymer matrix under the impact of variables like drug solubility in the polymer, matrix thickness, and concentration gradient, therefore affecting the release rate [3]
Erosion-controlled release	Certain polymers are engineered to progressively dissolve upon exposure to human fluids or external influences, therefore releasing the encapsulated medicine [79]
Stimuli-responsive release	Polymers can be designed to respond to external stimuli like light, ultrasound, or magnetic fields, causing structural changes that trigger drug release [5]
Swelling-controlled release	Specific polymers, especially hydrogels, can expand upon exposure to water or biological fluids. It forms pores and channels, facilitating the diffusion of the medication [33]
Biodegradation-triggered release	Biodegradable polymers, such as PLGA, progressively decompose within the body, releasing the encapsulated medication over time [27]
Chemically responsive release	Polymers can be engineered to deliver medications in reaction to variations in pH, temperature, or additional external factors [44]
Hydrolysis of bonds	The breakdown of the medicine is regulated by the hydrolysis of covalent bonds in the polymer backbone. As the polymer undergoes degradation via bond cleavage, it liberates the medication [78]
Surface erosion	Certain drug delivery systems use a polymer coating that gradually erodes, revealing the drug core. This method is frequently employed in oral controlled-release pills [66]
Complexation and ion exchange	Some drug delivery systems employ a polymer coating that progressively degrades, exposing the drug core. This technique is commonly utilized in orally administered controlled-release tablets [15]

nucleic acids. Primarily, natural polymers are derived from condensation polymers, with water produced as a by-product. Harmful, nonbiodegradable, and physiologically incompatible synthetic polymers are used in the present pharmaceutical system. High production costs are a result of the precise chemical synthesis required of these polymers [61]. Natural polymers are the way to go because of their eco-friendliness, biodegradability, biocompatibility, and affordability. They are quickly broken down, metabolized, and digested by enzymes, and every living thing makes them. Additionally, biopolymers are less harmful since they are made from monomers that are naturally occurring in living things. Since the necessary components are abundant in naturally occurring resources, they are also practical from a financial standpoint. Furthermore, there is less processing required for the by-products and wastes of natural polymer production, which are more biodegradable in environmental situations. Because of their widespread production, plant-based polymers are ubiquitous and easy to come by [60].

Plant-derived polymers have found extensive application in the development of innovative pharmaceuticals, including microcapsules, microspheres, nanoparticles (NPs), cubosomes, liposomes, transferosomes, and hydrogels [28, 35]. The polymers favored encompass natural polysaccharides, including xanthan gum, alginate, starch, chitosan, gelatin, pectin, dextran, carrageenans, guar gum, gum arabic, cellulose, among others [87]. A significant supply of organic (plant) polymers is sustained through the cultivation of plant resources, ensuring a continuous delivery of raw materials [43]. Biodegradable polymers are effective in the formulation of sustained release medications. The bioavailability of drug release in a regulated manner through a delivery system necessitates optimization researches, ensuring both chemical and physical stability. The administration of biopharmaceutical products via ocular, nasal, transdermal, and oral routes facilitate a straightforward and efficient delivery within the body [74]. Because of their biodegradable nature and the ease of administering them directly within the body, naturally occurring polysaccharides are preferred for the controlled delivery of drugs as well as various bioactive substances [11].

This chapter of the book describes many natural polymers, including those that are derived from plants and those that are derived from animals. They have a wide range of uses as excipients and pharmacological adjuvants. Furthermore, when they are included, they provide a number of benefits compared to conventional methods.

15.2 Pharmaceutical Formulations Based on Natural Polymers: A Promising Substitute

Natural polymers have been explored and applied in the cosmetic, pharmaceutical, and medical spheres [10, 11]. Among polymeric forms, polymeric NPs and hydrogels based on natural polymers have attracted increasing attention for topical medication distribution, cosmetic active ingredients, and skin regeneration [86]. For instance, polymeric nanoparticles have been extensively applied to reduce stratum corneum damage and transport medications to the several layers of the skin. Controlled release rates, high entrapment efficiency (EE), and lowered enzymatic degradation define the benefits of polymeric NPs over alternative nanocarriers for topical or transdermal drug administration [55].

Mostly composed of hydrophilic polymers, hydrogels are three-dimensional networks with considerable water absorption capacity without dissolving (in most situations). Hydrogels can have either permanent (also known as chemical) or reversible (also known as physical) intercrossing. Molecular entanglements and secondary bonds—such as hydrogen bonds, ionic bonds, and hydrophobic interactions—keep physical crosslinking gels intact. Conversely, covalent bonds hold chemical crosslinked gels together. For many biomedical uses, including wound healing, drug delivery, tissue engineering and ophthalmic applications, hydrogels—with their high-water content, softness identical to extracellular matrix (ECM), adjustable physiochemical properties, and high permeability to small compounds—are interesting materials [54]. Particularly for wound healing, the high water content of hydrogels not only preserves a moist environment, which is advantageous to lower the danger of scarring, but also offers a cooling feeling and lowers tissue adhesion, therefore lowering patient suffering [24]. Conversely, an important feature of hydrogels is their swelling capacity, which is quite helpful for absorbing wound exudates (high in nutrients, favorable for bacterial development) [31]. Their great porosity helps oxygen penetration, thereby enabling the tissue to "breathe" and oxygen-requiring reparative processes to occur. Furthermore, laden with antimicrobial and anti-inflammatory medicines, growth factors, and nanoparticles to increase bioactivity, the hydrogels themselves create a barrier against the entrance of microorganisms into the wound bed. Moreover, the physiological surroundings of a skin cut define by a higher pH. Consequently, demand for hydrogel dressings with pH-responsive characteristics—that is, "smart" release of the encapsulated active agent—has been rising. Apart from that, hydrogels could provide a physical barrier at the bleeding site to aid in hemostasis [68].

Considering current in vitro, ex vivo, and in vivo investigations, the next sections will investigate the technological significance of natural polymers in DDS given their enormous potential for pharmaceutical use.

15.3 Types of Natural Polymers

Natural polymers can be sourced from plants as well as animals. Figure 15.1 illustrates different sources and examples of polymers from those sources.

15.3.1 Plant-Based Natural Polymers

15.3.1.1 Guar Gum

A polysaccharide derived from the endosperm of guar beans, guar gum goes by several names: galactomannan, guaran, and guar. Its building blocks are galactose and mannose. Some examples of gum modifications are yttrium crosslinked guar gum-g-acrylamide gel frameworks, carboxymethyl, and guar. A hypertension medication with a pulsatile discharge formulation was created in 2009 using a controlled-release combination of guar gum and sodium alginate [73].

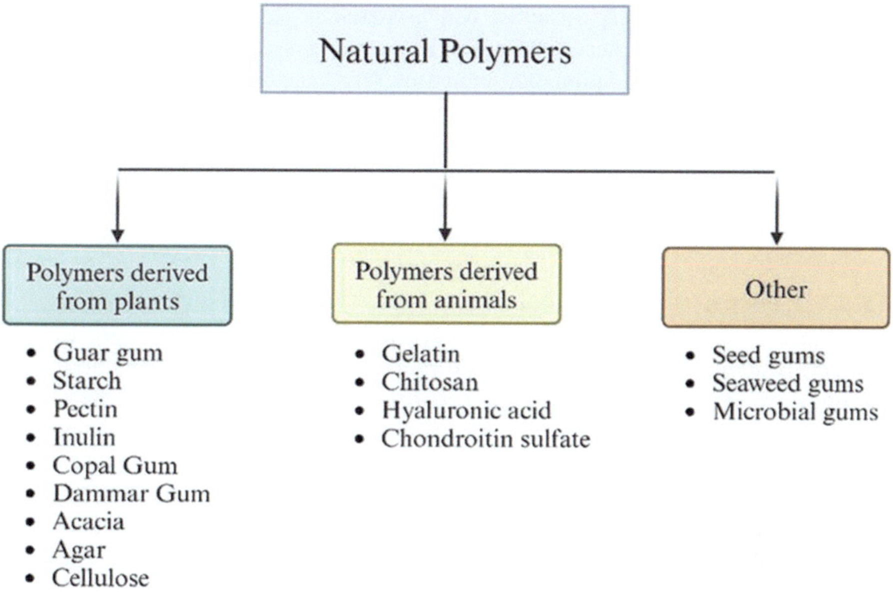

Fig. 15.1 Classification of naturally occurring polymers

To modulate calcium channel blockers such as nifedipine and verapamil hydrochloride, scientists have created microspheres made of poly-acrylamide-guar gum hydrogel. Sennoside is delivered to the colon specifically using matrix tablets that contain 50% w/w guar gum, and it is also delivered specifically to the colon by pellets coated with Eudragit FS 30D [47]. In humans, the absorption of matrix tablets containing guar gum is delayed for a longer period of time and the rate constant is reduced. Tablets encapsulating metronidazole and hydrocortisone hydrogels have both been made using various polysaccharides [30].

A synthetic biomimetic superabsorbent polymer is prevalent in sheets; Pugliese and Gelain studied how guar gum affected its composition and physicochemical characteristics. Their hydrogel scaffolds combine the benefits of guar gum and superabsorbent polymers to create structures that can mend themselves, respond to stimuli, and change shape in response to temperature changes. The results showed that guar gum-peptide assemblies with customizable properties might be formed autonomously [52].

15.3.1.2 Starch

Carbohydrates in plant storage organs mostly come from starch, a polysaccharide present in fully developed seeds of cereals such as wheat, oats, corn, and rice. Amylopectin and amylose are the two polymers that make it up. It comes from a variety of places. The major starch component, amylopectin, makes up 80% of starch, with amylose accounting for the remaining 20%. Its edema-inducing and enzyme-degrading characteristics, however, restrict its use in controlled-release biopharmaceutical delivery systems. Native starch is physically and chemically derivatized to overcome these constraints. Improved cross-linking properties, resistance to enzymatic degradation, and increased safety are all benefits of derivatization. Nanoparticles mixed with insulin, for example, have starch as a delivery mechanism because of their mucoadhesive characteristics, which improve the efficiency of trans-nasal insulin transfer. For effective transnasal distribution using mucoadhesive carriers, these nanoparticles feature an expanded surface area, a delayed release rate, and an improved controlled discharge rate [13].

15.3.1.3 Pectin

Plants that grow on trees, such as apples, quinces, crests, gooseberries, grapes, and oranges, contain pectin, a complex hydrophilic polymer. Its low production costs and high concentration make it a popular choice for drug delivery systems. Solubility and gelation are influenced by the order of methyl ester substituents, which vary depending on the plant source [12].

In contrast to low-methoxy pectin, which is more hydrophilic and soluble, high-methoxy pectins are insoluble and form gels at pH 7.4. Pectin can be made less soluble by chemical processes like saponification without losing its biodegradability. Protecting the tablet centers from early drug release is possible with a high pectin coat-to-core ratio. There is an improvement in protection even at lower ratios when using chitosan at concentrations of 3–5%. Regulated release of the poorly soluble prednisolone has been achieved by the use of pectin in microcapsular delivery techniques. The drug's solubility and diffusion coefficients are improved with high methoxylated apple pectin. The ideal formulation for oral sustained-release tablets was determined by Malviya et al. [39] to be a drug-to-polymer ratio of 1:1.5 [39].

Using pectin grafted with Polyethylene Glycol (PEG) and Methacrylic Acid (MAA), Abbasi and colleagues (2019) created a hydrogel-derived regulated medicament delivery mechanism for treating ulcerative colitis. Using metamizole, Lemos and researchers produced magnetic hydrogel microspheres with an encapsulating effectiveness of 85% by coating chitosan with pectin. Medication release might be controlled with both hydrogel systems [2, 38].

15.3.1.4 Inulin

Rose found inulin two centuries ago; it is a polysaccharide that is present in many plants, such as *Polymnia sonchifolia*, *Taraxacum officinale*, *Cynara scolymus*, *Helianthus tuberosus*, and *Cichorium intybus*. A terminal group of (1↔2)-d-glucose and (2 → 1)-linked d-fructosyl residues makes it up [22]. The degree of polymerization of the hydrolyzed product, oligo-fructose, ranges from 2 to 20. Inulin is a low-calorie sweetener, fat substitute, and texture modifier in the food industry. It can be used as a diagnostic tool for kidney function and as a protein stabilizer, among other medical uses. Its physicochemical qualities are greatly affected by the amount of polymerization and its processing history [40, 76].

Studies have demonstrated that inulin can suppress the growth of colon cancer cells when taken orally, especially in conjunction with short-chain fatty acids such as propionic acid and butyric acid. Cancer is reduced as a result of this. Inulin increases immunoglobulin A and interleukin-10 levels and reduces neutrophil oxidative burst action; it also has immunological effects on the gastrointestinal system. Interferon-gamma production by peripheral blood mononuclear cells is also enhanced [65].

Reducing dosage regimens, side symptoms, and costs are the goals of drug delivery formulations that attempt to enhance the stability of different vaccines, proteins, as well as peptides. Because of its resistance to the enzymatic activity of the gastrointestinal system, inulin—a biochemically inert, nontoxic, gelling material—is perfect for drug delivery. Inulin hydrogels are cross-linked with succinic and methacrylic anhydrides by exposing them to ultraviolet light. A diflunisal-delivering cross-linked inulin hydrogel exhibited pH-sensitive swelling, with decreased swelling in acidic settings as well as increased swelling in neutral to basic ones [3].

15.3.1.5 Copal Gum

As a binding agent in dental materials and a pain reliever, this is a substance which is fibrous and derived from *Bursera bipinnata*. The cream-colored substance possesses distinct physicochemical properties, such as an acid value, ester value, moisture content, refractive index, saponification value, softening point, and glass transition temperature. When it comes to creating delayed delivery systems, copal resin (CR) is quite promising, especially for the colon. Propylene glycol improves the mechanical properties of CR films, making them stiffer and stickier. To lessen stickiness, sorbitan mono-oleate, glyceryl monostearate, and sorbitan monolaurate are added. Research shows that CR films exhibit remarkable swelling characteristics over a range of humidity conditions. Coating agents made of film-forming natural materials are ideal for the colon-targeted delivery of drugs [17, 49, 75].

15.3.1.6 Dammar Gum

Dammar, first utilized as varnish in 1826, and thus referred to as Dammar Varnish. Dammar gum, which is sourced from trees in East Asia, is employed in a variety of culinary products and for drug delivery. It is present in the following types: pebble-shaped dameri (Batu) and crystalline resin type (Matakucking). Dammar gums can exhibit a variety of colors, including translucent, faint yellow, and gray-brown. Dammar gum has been examined for the purpose of microencapsulation and the delayed new delivery carrier systems research. The encapsulation of drugs like ibuprofen as well as diltiazem hydrochloride has been modeled through microparticles using techniques like evaporation of solvent and oil-in-oil emulsion. Different methodologies for the evaluation of the microparticles were conducted. The drug's solubility and its ratios with the gum are the determining factors in its entrapment in dammar gums. An increase in the gum-to-drug ratio leads to larger microparticles and improved encapsulation efficiency, with zero-order drug release kinetics [45, 64]. Figure 15.2 illustrates natural polymers and derivatives.

Different formulations based on some plant-derived polymers are listed in Table 15.2.

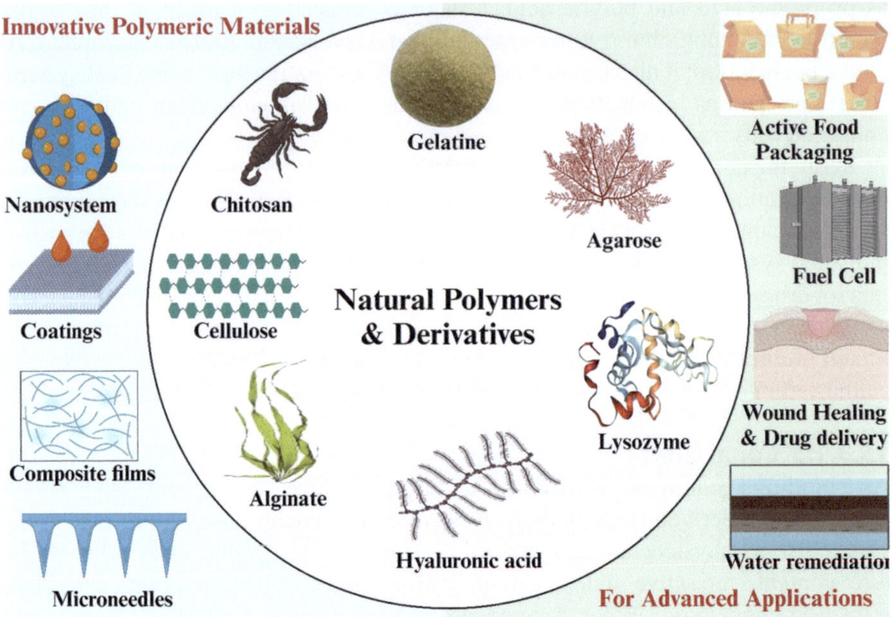

Fig. 15.2 Natural polymers and derivatives

Table 15.2 List of formulations produced from natural polymers sourced from various plants

Plant polymer	Origin of the polymer	Formulation	Reference
Xanthan gum	Microbial fermentation of glucose from the bacterial coat of *Xanthomonas campestris*	Diclofenac sodium Metoprolol tartrate	[69]
Pectin	Outer covering of grapes, apple, quince, plume, gooseberry	Matrix tablets of diclofenac sodium	[39]
Mimosa pudica seed mucilage	Seed of *Mimosa pudica*	Sustained release formulation diclofenac sodium	[70]
Tamarind seed polysaccharide	Seed kernel of *Tamarindus indica*	TSP matrix of paclitaxel	[53]
Guar gum	*Cyamopsis tetragonolobus*	Matrix tablets of zidovudine	[82]
Grewia gum	Inner bark of the tree *Grewia mollis*	Cimetidine tablet	[48]
Mucuna gum	Cotyledons of *Mucuna flagellipes*	Good suspending agent/stabilizer in dosage formulations such as suspensions and emulsion	[51]
Copal gum	Resin isolated from the plant *Bursera bipinnata*	Glibenclamide	[8]
Dammar gum	Hydrophobic gum obtained from plant *Shorea wiesneri*	Spherical microparticles of diltiazem hydrochloride	[45]
Shellac	Curcumin	String fibers using triaxial electrospinning	[84]
Subabul seed polysaccharide	Seed kernel of *Leucaena leucocephala*	Ibuprofen tablet	[77]

15.4 Natural Polymers Derived from Animals

15.4.1 Gelatin

It is synthesized by the regulated hydrolysis of collagen obtained from animal tissues, including skin, cattle, as well as porcine bone. The complex polypeptide mélange is the result of enzymatic hydrolysis, acid hydrolysis, and base hydrolysis [4]. A microbial expression method for human collagen fabrication was devised by researchers. This method is more consistent, less susceptible to diseases, and enables the production of desired molecular size and charge [26].

15.4.1.1 Gelatin-Derived Pharmaceutical Delivery Systems

Gelatin and derivates have also being examined for the purpose of pulmonary and sustained-release drug delivery. It has been demonstrated that mannosylation of gelatin nanoparticles is effective in transporting isoniazid to alveolar macrophages. The selective drug release to alveolar tissues, reduction in hepatotoxicity, and enhancement of antibacterial activity have all been demonstrated by these nanoparticles [41]. They can also serve as substitutes for substances such as potassium iodide, calcium chloride, ethylenediamine tetraacetic acid, and polymyxin on a peptide. The anticancer drug chloro-tetracycline hydrochloride was released without significant alteration by genipin crosslinking, which was achieved by developing a highly biocompatible gelatin-hydrogel composite [23]. Ramanathan et al. conducted a study that demonstrated the rapid and increased release of highly hydrophilic 5-fluorouracil (5-FU) and methotrexate using polyhydroxybutyric acid/gelatin nanofibers. The hydrophilic 5-FU was released at a higher rate than methotrexate [42].

15.4.1.2 Gelatin Tissue Scaffolds As Well As Hydrogels

To improve tissue targeting capacity, gelatin has been combined with other organic and chemically synthetic components. Greatest osteoconductive affinity for Sarcoma Osteogenic cells (Sao-2) cells has been shown by complex scaffolds composed of gelatin, chitosan, and sintered hydroxyapatite. Reduced breakdown rate and improved mechanical properties abound from gelatin/montmorillonite chitosan nanocomposite scaffolds. Three percent Iron Oxide (Fe_3O_4) magnetic nanoparticles in ferrogels exhibit controlled pulsatile drug delivery capacity. Using 5-FU, Anirudhan et al. created a pH-responsive hydrogel drug carrier that enhanced drug release in the colon and medicated encapsulation capability. Furthermore increasing with gelatin dosage was the rate of medication loading [14, 29].

15.4.2 Chitosan

A heteropolysaccharide made from chitin, chitosan is frequently utilized in drug delivery systems because of its bioactive qualities and suitability for human usage. It is a preferred option for pharmacological and medicinal applications due to its biodegradability. According to studies, chitosan taken orally can decrease inflammation and promote skin porosity. The polymer is a useful resource in many different industries because of its thorough study and proven usefulness [47].

The units of β-(1 → 4)-2-acetamido-D-glucose and β-(1 → 4) 2-amino-D-glucose make up chitosan, a polysaccharide. It has three crystalline varieties—α, β, as well as γ—based on its linear structure. The outer layers of shrimp and crab include the most well-known variation, α-sort. An amino/acetamido group, as well as primary and secondary hydroxyl groups, makes up its functional moieties. The structural factors that impact the sub-atomic weight and degree of deacetylation determine the varied uses of chitosan, a basic polysaccharide containing a primary amino group [19, 34].

Because of their non-toxicity, biodegradability, and compatibility with living organisms, chitin and its derivative chitosan are gaining prominence. Analgesic, anticancer, hemostatic, hypolipidemic, and antibacterial effects are among its documented medicinal, restorative, and pharmaceutical uses. Nevertheless, delineating, isolating, and purifying these characteristics from different compounds are necessary for comprehending their features. A chitosan's bioactive potential is greatly affected by its molecular weight (MW) and chain conformation [6].

The cationic characteristic of chitosan makes it an ideal component of drug delivery systems such as hydrogels, microspheres, nanoparticles, and tablets. The pharmaceutical industry makes use of it to deliver drugs to the body's organs through the skin. Because of its importance in the biomedical and cosmetic sectors, shorter chitosan chains enhance water solubility. Develop delayed drug delivery systems, extend pharmacological effects, improve therapeutic efficacy, and mitigate side effects by selecting the appropriate type of chitosan with specified features. Research shows that chitosan's molecular weight (MW) and degree of deacetylation (DD) have great potential as medicines. Drug consolidation is greatly affected by the gelation properties of microspheres that have been crosslinked with glutaraldehyde. Researchers evaluated the effectiveness of a polyelectrolyte complex for gene delivery and industrial uses. The complex consisted of trimethyl chitosan, hyaluronate/dextran/alginate nanoparticles [36].

Researchers have found that chitosan nanoparticles can effectively transport polytadine, which could be used to treat type 2 diabetes. Researchers developed electrospun nanofibers and a chitosan-grafted-poly(N-vinylcaprolactam) copolymer to regulate the release of cisplatin and deal with MCF-7 breast cancer cells [46].

15.4.3 Hyaluronic Acid

All living things contain hyaluronic acid, a carbohydrate with a "Jello"-like polysaccharide structure. Although it is more commonly found outside of cells, it can sometimes be found inside them [1]. Elastoviscosity maintenance in fluid connective tissues, regulation of water transport and tissue hydration, cell separation, metastasis, cancer advancement, and mitosis are just a few of its many biological activities [20]. Many different types of soft connective tissues contain hyaluronate, which is the sodium salt of hyaluronic acid. These tissues include skin, umbilical cord, kidneys, lungs, synovial fluid, and vitreous humor. Almost all warm-blooded organisms display their distinctive biological importance [67].

15.4.4 Chondroitin Sulfate

It is obtained from the cartilage of chicken, pig, cow, whale, etc. The majority of the supply comes from shark cartilage in the commercial sector. As the cost of raw materials rises and other biological factors come into play, the process of extracting chondroitin sulfate (CS) from shark ligaments becomes more risky [85]. CS has the ability to produce two cell receptors: syndecan, which interacts with adhesion polypeptides and cells, and spongy aggrecan, which is present in cartilage and acts as a stress absorber. Regeneration of the central nervous system is dependent on CS, which also improves ECM formation by human chondrocytes [58]. Majd et al. created copper oxide nanoparticles in 2023; these particles have the potential to encapsulate Imatinib Mesylate, an anticancer drug, for the purpose of targeted delivery of cancer therapy. An encouraging new method of medication delivery is pH-sensitive nanocarriers, which Khodayari et al. demonstrated with the development of a pH-sensitive magnetic nanocarrier for doxorubicin delivery [18].

15.4.5 Silk Fibroin

Extracted naturally from domestic silkworm, Bombyx mori (B. mori), silk fibroin is a naturally occurring protein with various beneficial effects that has drawn particular interest in the biomedical and pharmaceutical domains. Its remarkable biocompatibility, regulated biodegradation (producing non-inflammatory by-products), appropriate mechanical qualities, and drug stabilization capacity define such [83]. Thus, among other uses, silk fibroin has been generally employed to produce sustained release DDS, tissue regeneration, biosensors, among other uses. Particularly, numerous silk fiber-based wound dressings including hydrogels and membranes have been developed. Since silk fibroin hydrogels may be readily attached to the surface of the skin, they have drawn most attention. High water content, gelation behavior, and molecular structure of silk fibers help to maintain skin desirably hydrated, stop cracking, and encourage wound healing. Furthermore, certain silk fibroin-based hydrogels feature a three-dimensional network formed by their sponge-like structure that allows the loaded and stabilized nanocarriers to diffuse through and execute gradual drug release [25]. Furthermore carried by this structure are the drug molecules themselves (not included into nanosystems). Fascinatingly enough, these hydrogels can also be utilized without any drugs included. Different silk fibroin concentrations—1.5%, 3% and 6%—were used to make several silk fibroin-based hydrogels. Greater porosity in in vitro experiments revealed corresponding greater swelling for the 1.5% concentration. The 6% formulation showed problematic application to the skin surface and great hardness [50].

Comparatively with a negative control, in vivo tests in rabbits revealed that both tested silk fibroin-based hydrogel (1.5% and 3%) groups and the scar removal cream-treated group reduced the scar hyperplasia index. By whitening the scar's color and lowering its thickness, the 3% silk fibroin-based hydrogel group and the scar removal cream-treated group demonstrated a statistically significant therapeutic effect on the hypertrophic scar, by whitening the scar's color and reducing its The researchers discovered that the decrease in expression levels of α-smooth muscle actin might be the reason silk fibroin-based hydrogels had therapeutic effect. Furthermore, compared to the control group—better than in past studies—the scar hyperplasia index in the 3% silk fibroin-based hydrogel group dropped by 16.6% [37, 71].

15.5 Seaweed Polysaccharides

15.5.1 Alginates

Natural polysaccharide polymers derived from brown seaweed, known as alginates, have multiple uses in the delivery of drugs. In linear blocks, you can find D-mannuronic acid and L-guluronic acid residues. The intranasal administration of metoprolol tartrate in the form of bio-adhesive sodium alginate microspheres is one such example. The emulsification-crosslinking approach was used to synthesis these microspheres, and they had superior therapeutic efficacy in the management of hypertension and angina pectoris. When compared to medicine distribution by the nose or mouth, in vivo investigations demonstrated that the suppression of isoprenaline-induced tachycardia was more sustained [56, 81].

15.5.2 Carrageenan

Carrageenan (CR) is a polymer obtained from red algae in the Atlantic Ocean, utilized in food, medicines, and experimental medicine [16]. It serves as a viscosity enhancer, gelling agent, and emulsifier, frequently supplemented with gelatin to formulate capsules and obscure gritty flavors in antacid gels [62]. CR-chitosan fibers, embedded with active medicinal ingredients, are utilized for wound healing and hygiene maintenance. CRs serve as a coalescing element in lotions as well as shampoos, underscoring their significance in cosmetology. CRs are obtained from multiple species of Rhodophyta and are predominantly composed of D-galactose units, carbohydrate residues, and sulfates [7].

Figure 15.3 illustrates structures of different natural polymers.

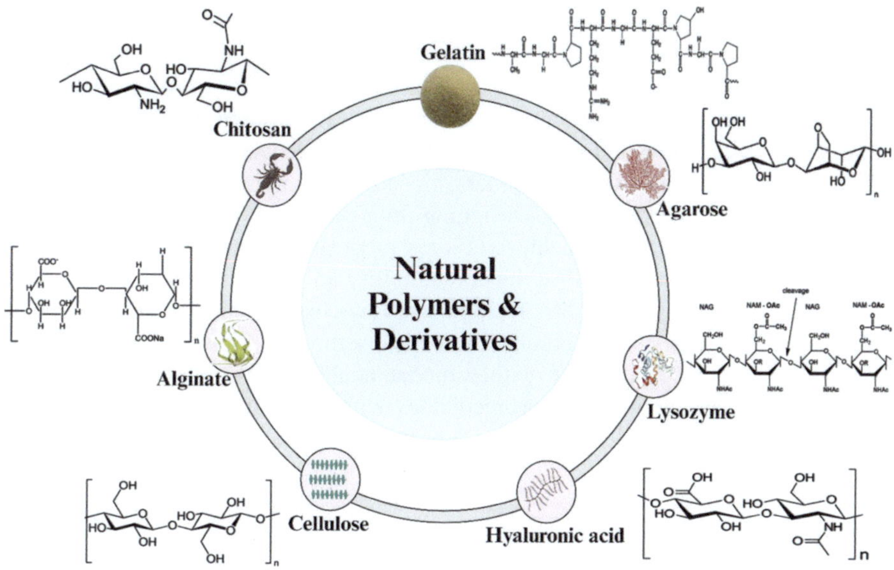

Fig. 15.3 Structures of different natural polymers

15.6 Future Directions and Prospects

Polymers including chitosan, alginate, starch, collagen, and gelatin are used in different sectors due to their safety, biocompatibility, and ease of enzymatic processing. They are abundant in nature and economically advantageous, having lesser toxicity compared to synthetic polymers. These polymers are particularly valuable in medication delivery systems due to their biocompatibility with humans [63].

The efficacy of natural polymers, such as chitosan, in drug delivery systems is restricted by their susceptibility to acid degradation. Nevertheless, the functionality and performance of polymers can be enhanced through the implementation of structural modifications [80]. The investigation of nanostructured biomaterials is essential for the advancement of nanomedicine, as the efficacy and durability of products are influenced by the interaction between polymers and drugs [59]. The complex structure, immunogenic characteristics, variability in purity levels, insufficient strength, and unpredictable disintegration are among the challenges. Comprehending their interactions with other biomolecules and their physical and biochemical properties can assist in the mitigation of immunogenic responses [32]. Some of the other concerns are water infiltration, swelling, leached substances, wear detritus, and sterilizing methods. Recent technological advancements have enhanced the practical and configurational qualities of polymers [57].

15.7 Conclusion

The practical and configurational properties of polymers have been enhanced by new technological developments. Natural monomers and polymers have emerged as a more eco-friendly substitute for conventional drug delivery systems, thanks to advancements in engineered biopolymers, bio-inspired polymer design, and synthetically inspired polymer research. Designing drug delivery systems that use natural polymers has the potential to achieve tailored medication distribution with delayed release. Because of their intrinsic biocompatibility and relative ease of production and processing, natural polymers are appealing low-cost sources for use in biomedicine and clinical settings. In a number of medical settings, natural polymeric materials have shown great promise due to their biodegradability and unique properties. We anticipate that new and innovative medicines with the ability to greatly improve patient outcomes will be developed as a consequence of further research in this field.

References

1. Abatangelo G, Vindigni V, Avruscio G, Pandis L, Brun P. Hyaluronic acid: redefining its role. Cells. 2020;9(7):1743.
2. Abolghasemi-Fakhri L, Ghanbarzadeh B, Dehghannya J, Abbasi F, Adun P. Styrene monomer migration from polystyrene based food packaging nanocomposite: Effect of clay and ZnO nanoparticles. Food and Chemical Toxicology. 2019;129:77–86.
3. Adepu S, Ramakrishna S. Controlled drug delivery systems: current status and future directions. Molecules. 2021;26(19):5905.
4. Alipal J, Pu'Ad NM, Lee TC, Nayan NHM, Sahari N, Basri H, et al. A review of gelatin: properties, sources, process, applications, and commercialisation. Mater Today Proc. 2021;42:240–50.
5. Alsehli M. Polymeric nanocarriers as stimuli-responsive systems for targeted tumor (cancer) therapy: recent advances in drug delivery. Saudi Pharm J. 2020;28(3):255–65.
6. Aranaz I, Alcántara AR, Civera MC, Arias C, Elorza B, Heras Caballero A, Acosta N. Chitosan: an overview of its properties and applications. Polymers. 2021;13(19):3256.
7. Asadzadeh N, Ghorbanpour M, Sayyah A. Effects of filler type and content on mechanical, thermal, and physical properties of carrageenan biocomposite films. Int J Biol Macromol. 2023;253:127551.
8. Attama A, Nwabunze O. Mucuna gum microspheres for oral delivery of glibenclamide: evaluation. Acta Pharma. 2007;57(2):161–71.
9. Baranwal J, Barse B, Fais A, Delogu GL, Kumar A. Biopolymer: a sustainable material for food and medical applications. Polymers. 2022;14(5):983.
10. Behrooznia Z, Nourmohammadi J. Polysaccharide-based materials as an eco-friendly alternative in biomedical, environmental, and food packaging. Giant. 2024;19:100301.
11. Benalaya I, Alves G, Lopes J, Silva LR. A review of natural polysaccharides: sources, characteristics, properties, food, and pharmaceutical applications. Int J Mol Sci. 2024;25(2):1322.
12. Beneke CE, Viljoen AM, Hamman JH. Polymeric plant-derived excipients in drug delivery. Molecules. 2009;14(7):2602–20.
13. Bertoft E. Understanding starch structure: recent progress. Agronomy. 2017;7(3):56.
14. Contessi Negrini N, Tarsini P, Tanzi MC, Farè S. Chemically crosslinked gelatin hydrogels as scaffolding materials for adipose tissue engineering. J Appl Polym Sci. 2019;136(8):47104.
15. Daihom BA, Bendas ER, Mohamed MI, Badawi AA. Domperidone resinate complex as new formulation for gastroretentive drug delivery. J Drug Deliv Sci Technol. 2020;58:101868.
16. Das IJ, Bal T. Exploring carrageenan: from seaweed to biomedicine—a comprehensive review. Int J Biol Macromol. 2024;268:131822.

17. Doshi P, Dickersin K, Healy D, Vedula SS, Jefferson T. Restoring invisible and abandoned trials: a call for people to publish the findings. Bmj. 2013;346.
18. dos Santos Batista JG, Vigilato Rodrigues MÁ, de Freitas LF, Moreira Fonseca AC, de Souza Rodrigues A, Cunha da Cruz CP, et al. Copper-based nanomaterials for biologically relevant compounds. In: Copper-based nanomaterials in organic transformations; 2024. p. 305–38. https://doi.org/10.1021/bk-2024-1466.ch012.
19. El Hadrami A, Adam LR, El Hadrami I, Daayf F. Chitosan in plant protection. Mar Drugs. 2010;8(4):968–87.
20. Fallacara A, Baldini E, Manfredini S, Vertuani S. Hyaluronic acid in the third millennium. Polymers. 2018;10(7):701.
21. Flórez M, Cazón P, Vázquez M. Selected biopolymers' processing and their applications: a review. Polymers. 2023;15(3):641.
22. Flückiger FA, Hanbury D. Pharmacographia: a history of the principal drugs of vegetable origin, met with in Great Britain and British India. Macmillan; 1874.
23. Foox M, Zilberman M. Drug delivery from gelatin-based systems. Expert Opin Drug Deliv. 2015;12(9):1547–63.
24. Gounden V, Singh M. Hydrogels and wound healing: current and future prospects. Gels (Basel, Switzerland). 2024;10(1):43.
25. Hassan MA, Basha AA, Eraky M, Abbas E, El-Samad LM. Advancements in silk fibroin and silk sericin-based biomaterial applications for cancer therapy and wound dressing formulation: a comprehensive review. Int J Pharm. 2024;662:124494.
26. Haug IJ, Draget KI. Gelatin. In: Handbook of hydrocolloids. Woodhead Publishing; 2009. p. 142–63.
27. Idrees H, Zaidi SZJ, Sabir A, Khan RU, Zhang X, Hassan SU. A review of biodegradable natural polymer-based nanoparticles for drug delivery applications. Nano. 2020;10(10):1970.
28. Jain S, Jain A, Jain R, Chauhan NS. Potential of natural polymeric materials in pharmaceutics. Pharmacol Res Nat Prod. 2024;2:100014.
29. Jaipan P, Nguyen A, Narayan RJ. Gelatin-based hydrogels for biomedical applications. Mrs Commun. 2017;7(3):416–26.
30. Kajjari PB, Manjeshwar LS, Aminabhavi TM. Novel interpenetrating polymer network hydrogel microspheres of chitosan and poly (acrylamide)-grafted-guar gum for controlled release of ciprofloxacin. Ind Eng Chem Res. 2011;50(23):13280–7.
31. Kamoun EA, Kenawy ERS, Chen X. A review on polymeric hydrogel membranes for wound dressing applications: PVA-based hydrogel dressings. J Adv Res. 2017;8(3):217–33.
32. Khan TK, Potnuri N, Sarvadnya AA, Bharti A. Polymers in controlled release drug delivery systems: an exploration of classification, properties, and applications. In: Novel drug delivery system from fundamental principles to advanced applications; 2024. https://doi.org/10.58532/nbennurndch2.
33. Korsmeyer RW, Gurny R, Doelker E, Buri P, Peppas NA. Mechanisms of solute release from porous hydrophilic polymers. Int J Pharm. 1983;15(1):25–35.
34. Kou SG, Peters LM, Mucalo MR. Chitosan: a review of sources and preparation methods. Int J Biol Macromol. 2021;169:85–94.
35. Kulkarni Vishakha S, Butte Kishor D, Rathod Sudha S. Natural polymers—a comprehensive review. Int J Res Pharmaceut Biomed Sci. 2012;3(4):1597–613.
36. Li Q, Dunn ET, Grandmaison EW, Goosen MF. Applications and properties of chitosan. In: Applications of chitan and chitosan. CRC Press; 2020. p. 3–29.
37. Ma L, Dong W, Lai E, Wang J. Silk fibroin-based scaffolds for tissue engineering. Front Bioeng Biotechnol. 2024;12:1381838.
38. Magallanes-Cruz PA, Flores-Silva PC, Bello-Perez LA. Starch structure influences its digestibility: a review. J Food Sci. 2017;82(9):2016–23.
39. Malviya R, Srivastava P, Bansal M, Sharma PK. Formulation and optimization of sustained release tablets of diclofenac sodium using guar gum as release modifier. Int J Pharm Sci Res. 2010;1(6):83.
40. Mensink MA, Frijlink HW, van der Voort Maarschalk K, Hinrichs WL. Inulin, a flexible oligosaccharide. II: Review of its pharmaceutical applications. Carbohydrate polymers. 2015;134:418–28.

41. Milano F, Masi A, Madaghiele M, Sannino A, Salvatore L, Gallo N. Current trends in gelatin-based drug delivery systems. Pharmaceutics. 2023;15(5):1499.
42. Mohammed AM, Osman SK, Saleh KI, Samy AM. In vitro release of 5-fluorouracil and methotrexate from different thermosensitive chitosan hydrogel systems. AAPS PharmSciTech. 2020;21:1–11.
43. Mohan P, Doshi S, Khinchi MP, Sharma N, Agrawal D. An review on natural polymers approaches to floating drug delivery system. Asian J Pharm Res Dev. 2013;1:145–59.
44. Mollazadeh S, Mackiewicz M, Yazdimamaghani M. Recent advances in the redox-responsive drug delivery nanoplatforms: a chemical structure and physical property perspective. Mater Sci Eng C. 2021;118:111536.
45. Morkhade D, Joshi S. Evaluation of gum damar as a novel microencapsulating material for ibuprofen and diltiazem hydrochloride. Indian J Pharm Sci. 2007;69(2):263–8.
46. Mourya VK, Inamdar NN. Chitosan-modifications and applications: opportunities galore. React Funct Polym. 2008;68(6):1013–51.
47. Nokhodchi A, Raja S, Patel P, Asare-Addo K. The role of oral controlled release matrix tablets in drug delivery systems. Bioimpacts. 2012;2(4):175–87.
48. Odeniyi MA, Jaiyeoba KT. Optimization of ascorbic acid tablet formulations containing hydrophilic polymers. Optimization. 2009;57:2.
49. Osete-Cortina L, Doménech-Carbó MT. Analytical characterization of diterpenoid resins present in pictorial varnishes using pyrolysis–gas chromatography–mass spectrometry with on line trimethylsilylation. J Chromatogr A. 2005;1065(2):265–78.
50. Pollini M, Paladini F. The emerging role of silk fibroin for the development of novel drug delivery systems. Biomimetics. 2024;9(5):295.
51. Prasanthi NL, Manikiran SS, Rama Rao N. Effect of hydrophilic polymers on the formulation and evaluation of Propranolol hydrochloride matrix tablets. Indian Pharm. 2010;8(12):61–6.
52. Pugliese R, Gelain F. Characterization of elastic, thermo-responsive, self-healable supramolecular hydrogel made of self-assembly peptides and guar gum. Mater Des. 2020;186:108370.
53. Rajesh KS, Venkataraju MP, Gowda DV. Effect of hydrophilic natural gums in formulation of oral-controlled release matrix tablets of propranolol hydrochloride. Pak J Pharm Sci. 2009;22(2):211–9.
54. Rao T, Chvs PHANINDRA, Yamini M, Prasad C. Hydrogels the three dimensional networks: a review. Int J Curr Pharm Res. 2021;13(1):12–7.
55. Raszewska-Famielec M, Flieger J. Nanoparticles for topical application in the treatment of skin dysfunctions—an overview of dermo-cosmetic and dermatological products. Int J Mol Sci. 2022;23(24):15980.
56. Raus RA, Nawawi WMFW, Nasaruddin RR. Alginate and alginate composites for biomedical applications. Asian J Pharm Sci. 2021;16(3):280–306.
57. Ren J, Lin H, Zhang Y, Li X, Zhang Z, You C, Wang F. The chitosan/cellulose nanocrystal cross-linked carriers effectively encapsulate ursolic acid to enhance the delivery of bioactive natural products. J Drug Deliv Sci Technol. 2024;97:105687.
58. Restaino OF, Schiraldi C. Chondroitin sulfate: are the purity and the structural features well assessed? A review on the analytical challenges. Carbohydr Polym. 2022;292:119690.
59. Saini L, Dubey A, Pal R, Pandey P, Mandal RK. Synthetic and natural polymers enhancing drug delivery and their treatment: a comprehensive review. J Drug Deliv Ther. 2024;14(10):153–65.
60. Samir A, Ashour FH, Hakim AAA, et al. Recent advances in biodegradable polymers for sustainable applications. npj Mater Degrad. 2022;6:68.
61. Satchanska G, Davidova S, Petrov PD. Natural and synthetic polymers for biomedical and environmental applications. Polymers. 2024;16(8):1159.
62. Sedayu BB, Cran MJ, Bigger SW. A review of property enhancement techniques for carrageenan-based films and coatings. Carbohydr Polym. 2019;216:287–302.
63. Sharaf M, Liu CG. Chitin and chitosan applications in medication delivery systems. In: Chitin and chitosan. Jenny Stanford Publishing; 2025. p. 283–319.
64. Sharma J, Kaur L, Kanuja N, Nagpal M, Bala R. Natural polymers-promising potential in drug delivery. Int J Pharmtech Res. 2013;5(2):684–99.
65. Sheng W, Ji G, Zhang L. Immunomodulatory effects of inulin and its intestinal metabolites. Front Immunol. 2023;14:1224092.

66. Stiepel RT, Pena ES, Ehrenzeller SA, Gallovic MD, Lifshits LM, Genito CJ, et al. A predictive mechanistic model of drug release from surface eroding polymeric nanoparticles. J Control Release. 2022;351:883–95.
67. Sudha PN, Rose MH. Beneficial effects of hyaluronic acid. Adv Food Nutr Res. 2014;72:137–76.
68. Suvarnapathaki S, Wu X, Lantigua D, et al. Breathing life into engineered tissues using oxygen-releasing biomaterials. NPG Asia Mater. 2019;11:65.
69. Syed IA, Mangamoori LN, Rao YM. Formulation and characterization of matrix and triple-layer matrix tablets for controlled delivery of metoprolol tartrate. Int J Pharm Sci Drug Res. 2011;3(1):23–8.
70. Tapia C, Costa E, Terraza C, Munita AM, Yazdani-Pedram M. Study of the prolonged release of theophylline from polymeric matrices based on grafted chitosan with acrylamide. Pharmazie. 2002;57(11):744–9.
71. Tariq Z, Zhu Z, Tariq MH, Wang Y, Liu J, Hai AM, et al. Incorporation of nanoparticles in silk fibroin solution: toward water-stable silk fibroin films possessing silk I structure. Polym Adv Technol. 2024;35(4):e6392.
72. Thang NH, Chien TB, Cuong DX. Polymer-based hydrogels applied in drug delivery: an overview. Gels (Basel, Switzerland). 2023;9(7):523.
73. Thombare N, Jha U, Mishra S, Siddiqui MZ. Guar gum as a promising starting material for diverse applications: a review. Int J Biol Macromol. 2016;88:361–72.
74. Tsung TH, Tsai YC, Lee HP, Chen YH, Lu DW. Biodegradable polymer-based drug-delivery systems for ocular diseases. Int J Mol Sci. 2023;24(16):12976.
75. Umekar MJ, Yeole PG. Characterization and evaluation of natural copal gum-resin as film forming material. Int J Green Pharm. 2008;2(1):37–42. https://doi.org/10.4103/0973-8258.39163.
76. Vergauwen R, Van Laere A, Van den Ende W. Properties of fructan: fructan 1-fructosyltransferases from chicory and globe thistle, two Asteracean plants storing greatly different types of inulin. Plant Physiol. 2003;133(1):391–401.
77. Verma PRP, Razdan B. Studies on disintegrant action of Leucaena leucocephala seed gum in ibuprofen tablet and its mechanism. J Sci Ind Res. 2007;66(07):550–7.
78. Visan AI, Popescu-Pelin G, Socol G. Degradation behavior of polymers used as coating materials for drug delivery—a basic review. Polymers. 2021;13(8):1272.
79. Walker J, Albert J, Liang D, Sun J, Schutzman R, Kumar R, et al. In vitro degradation and erosion behavior of commercial PLGAs used for controlled drug delivery. Drug Deliv Transl Res. 2023;13(1):237–51.
80. Wargantiwar S, Bhattacharya S. Biocompatible natural polymers and cutting-edge fabrication techniques in the development of next-generation oral thin films for enhanced drug delivery systems. Curr Pharm Des. 2024;31(5):331–43.
81. Wawszczak A, Kocki J, Kołodyńska D. Alginate as a sustainable and biodegradable material for medical and environmental applications—the case studies. J Biomed Mater Res B Appl Biomater. 2024;112(9):e35475.
82. Yadav AS, Kumar PA, Vinod R, Rao BS, Kulkarni SV. Design and evaluation of guar gum based controlled release matrix tablets of Zidovudine. J Pharm Sci Technol. 2010;2(3):156–62.
83. Yang H, Zhu Z, Ni S, Wang X, Nie Y, Tao C, et al. Silk fibroin-based bioelectronic devices for high-sensitivity, stable, and prolonged in vivo recording. Biosens Bioelectron. 2025;267:116853.
84. Yang Y, Chen W, Wang M, Shen J, Tang Z, Qin Y, Yu DG. Engineered shellac beads-on-the-string fibers using triaxial electrospinning for improved colon-targeted drug delivery. Polymers. 2023;15(10):2237.
85. Zhang W, Xu R, Chen J, Xiong H, Wang Y, Pang B, et al. Advances and challenges in biotechnological production of chondroitin sulfate and its oligosaccharides. Int J Biol Macromol. 2023;253:126551.
86. Zhao L, Zhou Y, Zhang J, Liang H, Chen X, Tan H. Natural polymer-based hydrogels: from polymer to biomedical applications. Pharmaceutics. 2023;15(10):2514.
87. Zhao M, Han P, Mu H, Sun S, Dong J, Sun J, et al. Food packaging films from natural polysaccharides and protein hydrogels: a comprehensive review. Food Chem. 2025;X:102174.

Correction to: Innovative Pathways in Natural Polymer-Based Excipients: Shaping the Future of Drug Formulation

Sakshi Wani, Shounak Kulkarni, Anjali Pingale, Amarjitsing Rajput, and Prashant Pingale

Correction to:
Chapter 15 in: J. S et al. (eds.), *Innovative Pharmaceutical Excipients: Natural Sources*, https://doi.org/10.1007/978-981-96-7959-1_15

The affiliation for the corresponding author Prashant Pingale was updated incorrectly. It has now been corrected. The correct affiliation is:

Department of Pharmaceutics, Gokhale Education Society's Sir Dr. M. S. Gosavi College of Pharmaceutical Education and Research, Nashik, Maharashtra, India

The updated version of this chapter can be found at
https://doi.org/10.1007/978-981-96-7959-1_15

J. S et al. (eds.), *Innovative Pharmaceutical Excipients: Natural Sources*,
https://doi.org/10.1007/978-981-96-7959-1_16

MIX
Papier aus verantwortungsvollen Quellen
Paper from responsible sources
FSC® C105338

If you have any concerns about our products,
you can contact us on
ProductSafety@springernature.com

In case Publisher is established outside the EU,
the EU authorized representative is:
Springer Nature Customer Service Center GmbH
Europaplatz 3, 69115 Heidelberg, Germany

Printed by Libri Plureos GmbH
in Hamburg, Germany